Lecture Notes in Mathematics 1486

L. Arnold H. Crauel J.-P. Eckmann (Eds.)

Lyapunov Exponents

Proceedings of a Conference held in Oberwolfach, May 28 - June 2, 1990

Springer-Verlag
Berlin Heidelberg New York
London Paris Tokyo
Hong Kong Barcelona
Budapest

Editors

Ludwig Arnold
Institut für Dynamische Systeme
Universität Bremen
Postfach 330 440
W-2800 Bremen 33, Germany

Hans Crauel
Fachbereich 9 Mathematik
Universität des Saarlandes
W-6600 Saarbrücken 11, Germany

Jean-Pierre Eckmann
Département de Physique Théorique
Université de Genève
CH-1211 Genève 4, Switzerland

Front cover:
V. I. Oseledets: A multiplicative ergodic theorem.
Trudy Moskov. Mat. Obsc. 19 (1968), 179-210

Mathematics Subject Classification (1980):
Primary: 58F
Secondary: 34F, 35R60, 58G32, 60F, 60G, 60H, 70L05, 70K, 73H, 73K, 93D, 93E

ISBN 3-540-54662-6 Springer-Verlag Berlin Heidelberg New York
ISBN 0-387-54662-6 Springer-Verlag New York Berlin Heidelberg

Printed in Germany

Typesetting: Camera ready by author
Printing and binding: Druckhaus Beltz, Hemsbach/Bergstr.
46/3140-543210 - Printed on acid-free paper

Preface

These are the Proceedings of a conference on Lyapunov Exponents held at Oberwolfach May 28 – June 2, 1990. The volume contains an introductory survey and 26 original research papers, some of which have, in addition, survey character.

This conference was the second one on the subject of Lyapunov Exponents. The first one took place in Bremen in November 1984 and lead to the Proceedings volume Lecture Notes in Mathematics # 1186 (1986). Comparing those two volumes, one can realize pronounced shifts, particularly towards nonlinear and infinite-dimensional systems and engineering applications.

We would like to thank the 'Mathematisches Forschungsinstitut Oberwolfach' for letting us have the conference at this unique venue.

March 1991

Ludwig Arnold (Bremen)   Hans Crauel (Saarbrücken)   Jean-Pierre Eckmann (Genève)

Table of Contents

Chapter 3: Infinite-dimensional Random Dynamical Systems

Chapter 4: Deterministic Dynamical Systems

Chapter 5: Engineering Applications and Control Theory

Random Dynamical Systems

Ludwig Arnold
Institut für Dynamische Systeme
Universität Bremen
2800 Bremen 33

Hans Crauel
Fachbereich 9 MATHEMATIK
Universität des Saarlandes
6600 Saarbrücken 11

1 Introduction

The main purpose of this survey is to present and popularize the notion of a *random dynamical system* (RDS) and to give an impression of its scope. The notion of RDS covers the most important families of dynamical systems with randomness which are currently of interest. For instance, products of random maps — in particular products of random matrices — are RDS as well as (the solution flows of) stochastic and random ordinary and partial differential equations.

One of the basic results for RDS is the Multiplicative Ergodic Theorem (MET) of Oseledec [38]. Originally formulated for products of random matrices, it has been reformulated and reproved several times during the past twenty years. Basically, there are two classes of proofs. One makes use of Kingman's Subadditive Ergodic Theorem together with the polar decomposition of square matrices. The other one starts by proving the assertions of the MET for triangular systems, and then enlarges the probability space by the compact group of special orthogonal matrices, so that every matrix cocycle becomes homologous to a triangular one.

Let us emphasize that the MET is a *linear* result. It is possible to introduce Lyapunov exponent-like quantities for nonlinear systems directly à la (9) below, or, much more sophisticated, as by Kifer [27]. However, the wealth of structure provided by the MET is available for linear systems only. Speaking of an "MET for nonlinear systems" always means the MET for the *linearization* of a nonlinear system. What is new for nonlinear systems is the fact that the linearization lives on the tangent bundle of a manifold (instead of the flat bundle $\mathbb{R}^d \times \Omega$ as for products of random matrices). The MET yields nontrivial consequences for deterministic systems already. This case has been dealt with by Ruelle [39]. Ruelle's argument proceeds by trivialization of the nonflat tangent bundle. It is exactly the same argument that works for nonlinear random systems: infer the MET for the linearization of the system from the ordinary MET together with a trivialization argument. We reproduce the argument below.

Stochastic flows have entered the scene a couple of years ago. They are related to RDS, but they are not the same. We describe their relations, and point out their differences.

The final Section briefly reviews all contributions to the present volume.

2 Random Dynamical Systems and Multiplicative Ergodic Theory

2.1 RDS

Consider a set T (time), $T = \mathbb{R}$, $\mathbb{Z}$; $\mathbb{R}^+$, or $\mathbb{N}$, and a family $\{\vartheta_t : \Omega \to \Omega \mid t \in T\}$ of measure preserving transformations of a probability space $(\Omega, \mathcal{F}, P)$ such that $(t, \omega) \mapsto \vartheta_t\omega$ is measurable, $\{\vartheta_t \mid t \in T\}$ is ergodic, and $\vartheta_{t+s} = \vartheta_t \circ \vartheta_s$ for all $t, s \in T$ with $\vartheta_0 = \text{id}$. Thus $(\vartheta_t)_{t\in T}$ is a flow if $T = \mathbb{R}$ or $\mathbb{Z}$, and a semi-flow if $T = \mathbb{R}^+$ or $\mathbb{N}$. The set-up $((\Omega, \mathcal{F}, P), (\vartheta_t)_{t\in T})$ is a (measurable) dynamical system.

Definition A *random dynamical system* on a measurable space $(X, \mathcal{B})$ over $(\vartheta_t)_{t\in T}$ on $(\Omega, \mathcal{F}, P)$ is a measurable map

$$\varphi : T \times X \times \Omega \to X$$

such that $\varphi(0, \omega) = \text{id}$ (identity on X) and

$$\varphi(t + s, \omega) = \varphi(t, \vartheta_s\omega) \circ \varphi(s, \omega) \tag{1}$$

for all $t, s \in T$ and for all ω outside a P-nullset, where $\varphi(t, \omega) : X \to X$ is the map which arises when $t \in T$ and $\omega \in \Omega$ are fixed, and $\circ$ means composition. A family of maps $\varphi(t, \omega)$ satisfying (1) is called a *cocycle*, and (1) is the cocycle property.

We often omit mentioning $((\Omega, \mathcal{F}, P), (\vartheta_t)_{t\in T})$ in the following, speaking of a random dynamical system (abbreviated RDS) φ.
We do not assume the maps $\varphi(t, \omega)$ to be invertible a priori. By the cocycle property, $\varphi(t, \omega)$ is automatically invertible (for all $t \in T$ and for P-almost all ω) if $T = \mathbb{R}$ or $\mathbb{Z}$, and $\varphi(t, \omega)^{-1} = \varphi(-t, \vartheta_t\omega)$.

The following examples are quite distinct in many respects. However, they all are RDS.

1. The simplest case of a random dynamical system is a non-random — viz., deterministic — dynamical system. An RDS is deterministic if φ does not depend on ω, i. e., $\varphi(t, x, \omega) = \varphi(t, x)$. Then the cocycle property (1) reads $\varphi(t+s) = \varphi(t) \circ \varphi(s)$, hence $(\varphi(t))_{t\in T}$ consists of the iterates of a measurable map on X if $T = \mathbb{Z}^{(+)}$,and $(\varphi(t))_{t\in T}$ is a measurable (semi-) flow if $T = \mathbb{R}^{(+)}$, respectively.

2. Let $\vartheta : \Omega \to \Omega$ be a measure preserving transformation, and let $\psi : X \times \Omega \to X$ be a measurable map. Put $\psi_n = \psi \circ \vartheta^{n-1}$. Then

$$\varphi(n, \omega) = \begin{cases} \psi_n(\omega) \circ \psi_{n-1}(\omega) \ldots \circ \psi_1(\omega) & \text{for } n > 0 \\ \text{id} & \text{for } n = 0 \\ \psi_{n+1}^{-1}(\omega) \circ \psi_{n+2}^{-1}(\omega) \circ \ldots \circ \psi_0^{-1}(\omega) & \text{for } n < 0, \end{cases}$$

defines an RDS (of course, defining $\varphi(n, \omega)$ for $n < 0$ needs ϑ and $\psi(\cdot, \omega)$ invertible P-a. s.). In particular, if $X = \mathbb{R}^d$ and $x \mapsto \psi(x, \omega)$ is linear, then φ is a product of random matrices.

3. Suppose $T = \mathbb{R}$, and M is a C^1 manifold. Denote by TM the total space of the tangent bundle of M, and let $Y : M \times \Omega \to TM$ be a measurable map such that for P-almost all ω the map $Y(\cdot, \omega)$ is a smooth vector field. Then the random differential equation

$$\dot{x}(t) = Y(x(t), \vartheta_t\omega), \qquad x(0) = x_0, \tag{2}$$

induces a map $\varphi(t,\omega) : M \to M$, such that $x(t,\omega) = \varphi(t,\omega)x$ solves (2) with $x(0) = x$ for $t \in (t^-(x,\omega), t^+(x,\omega))$, where $t^-(x,\cdot) < 0 < t^+(x,\cdot)$ (P-a.s.) describe the maximal intervals of definition of solutions. If $t^-(x,\cdot) = -\infty$ and $t^+(x,\cdot) = +\infty$ (for all $x \in M$ P-a.s.) then φ is an RDS. In addition, $x \mapsto \varphi(t,\omega)x$ is a diffeomorphism for all $t \in \mathbb{R}$ (P-a.s.) in this case. The maximal interval of definition is automatically all of $\mathbb{R}$ if M is compact. If $-\infty < t^-(x)$ or $t^+(x) < \infty$ for some x with positive probability we speak of a *local RDS* or *local random flow*.

4. Suppose M is a C^2 manifold, and Y_i, $0 \le i \le n$, are smooth vector fields on M. Then the stochastic differential equation

$$dx(t) = Y_0(x(t))\,dt + \sum_{i=1}^{n} Y_i(x(t)) \circ dW_i(t), \qquad x(0) = x_0, \tag{3}$$

induces a *(local) stochastic flow*. Usually (3) is understood for $t \ge 0$. We will describe below how to give (3) a meaning on the whole time axis. Once this is done, maximal intervals of solutions, containing $t = 0$ as an interior point, exist and have the same properties as for random flows described in the previous example.

We have introduced *local* random and stochastic flows because they play a role in stochastic bifurcation. For details see below.

As for deterministic systems, RDS may be classified according to their spatial properties.

If X is a topological space (with Borel σ-algebra), a random dynamical system is said to be *continuous* if $\varphi(t,\omega) : X \to X$ is continuous for all $t \in T$ and all $\omega \in \Omega$ outside a P-nullset.
If X is a C^r manifold, $r \ge 1$, an RDS φ on X is said to be *differentiable* or *smooth* if $\varphi(t,\omega) : X \to X$ is C^r differentiable for all $t \in T$ and all ω outside a P-nullset.
A random dynamical system on a topological vector space X is said to be linear if $\varphi(t,\omega) : X \to X$ is linear for all $t \in T$ and all ω outside a P-nullset.

If an RDS consists of non-invertible maps then T cannot contain negative times. An RDS φ consisting of invertible maps need not allow negative time, since ϑ_t need not be invertible. So we have to distinguish between two kinds of invertibility. An RDS is said to be *two sided* if $T = \mathbb{R}$ or $T = \mathbb{Z}$. It is said to be *invertible* if, for all $t \in T$ and P-almost all ω, $\varphi(t,\omega)$ is invertible in the corresponding class (measurable, continuous, smooth). Clearly 'two sided' is stronger than 'invertible'.

Any RDS induces a measurable *skew product (semi-) flow*

$$\begin{aligned} \Theta_t : X \times \Omega &\to X \times \Omega \\ (x,\omega) &\mapsto (\varphi(t,\omega)x, \vartheta_t\omega), \end{aligned} \tag{4}$$

$t \in T$, where $\varphi(t,\omega)x = \varphi(t,x,\omega)$. The flow property $\Theta_{t+s} = \Theta_t \circ \Theta_s$ follows from the cocycle property of φ (see (1); we use the term flow for both continuous and discrete time T).

From the point of view of abstract ergodic theory, an RDS is nothing but an ordinary dynamical system $(\Theta_t)_{t\in T}$ with a factor $(\vartheta_t)_{t\in T}$ together with the extra bit of structure provided by the fact that the ergodic invariant measure P for the factor is given a priori. (This observation might serve as an abstract definition of RDS.)

A probability measure μ on $X \times \Omega$ (on the product σ-algebra $\mathcal{B} \otimes \mathcal{F}$) is said to be an *invariant measure for* φ if μ is invariant under Θ_t, $t \in T$, and if it has marginal P on Ω. Invariant measures always exist for continuous RDS on a compact X (which is in complete analogy with deterministic dynamical systems).

Denote by $Pr(X)$ the space of probability measures on X, endowed with the smallest σ-algebra making the maps $Pr(X) \to \mathbb{R}$, $\nu \mapsto \int_X h\,d\nu$, measurable with h varying over the bounded measurable functions on X.
Given a measure $\mu \in Pr(X \times \Omega)$ with marginal P on Ω, a measurable map $\mu_. : \Omega \to Pr(X)$, $\omega \mapsto \mu_\omega$ will be called a *disintegration of* μ *(with respect to* P*)* if

$$\mu(B \times C) = \int_C \mu_\omega(B)\,dP(\omega)$$

for all $B \in \mathcal{B}$ and $C \in \mathcal{F}$.
Disintegrations exist and are unique (P-a. s.), e. g., if X is a Polish space. We will assume existence and uniqueness of a disintegration in the following.
A measure μ is invariant for the RDS φ if and only if

$$E(\varphi(t,\cdot)\mu_. \mid \vartheta_t^{-1}\mathcal{F})(\omega) = \mu_{\vartheta_t\omega} \qquad P\text{-a. s. for every } t \in T. \tag{5}$$

If T is two sided then $\vartheta_t^{-1}\mathcal{F} = \mathcal{F}$, hence for T two sided (5) reads

$$\varphi(t,\omega)\mu_\omega = \mu_{\vartheta_t\omega} \qquad P\text{-a. s. for every } t \in T.$$

2.2 Lyapunov exponents and the Multiplicative Ergodic Theorem

For a differentiable manifold M denote by TM the total space of its tangent bundle. The linearization of a differentiable map $\psi : M \to M$ is denoted by $T\psi : TM \to TM$ with $T_x\psi : T_xM \to T_{\psi(x)}M$, $x \in M$, denoting the action of $T\psi$ on individual fibers.

Suppose φ is a smooth RDS on a d-dimensional Riemannian manifold M. The chain rule yields

$$T_x\varphi(t+s,\omega) = T_{\varphi(s,\omega)x}\varphi(t,\vartheta_s\omega) \circ T_x\varphi(s,\omega) \tag{6}$$

for all $t, s \in T$ and $x \in M$ with P-measure 1. Consequently, the linearization $T\varphi : T \times TM \times \Omega \to TM$ is a cocycle over the skew product flow $\Theta_t(x,\omega) = (\varphi(t,\omega)x, \vartheta_t\omega)$ on $M \times \Omega$ (cf (4)).

Suppose μ is an invariant measure for φ such that

$$(x,\omega) \mapsto \sup_{0<t\leq t_0} \log^+(\|T_x\varphi(t,\omega)\|) \in L^1(\mu), \tag{7}$$

where $\log^+ = \max\{\log, 0\}$, and $\|\cdot\|$ denotes the norm induced by the Riemannian metric. Denote by $\lambda_1^\mu(x,\omega) \geq \lambda_2^\mu(x,\omega) \geq \ldots \geq \lambda_d^\mu(x,\omega)$ the *Lyapunov exponents of φ associated with μ*, where the $\Theta_\cdot$-invariant maps $(x,\omega) \mapsto \lambda_i^\mu(x,\omega)$ are defined via

$$\sum_{i=1}^{k} \lambda_i^\mu(x,\omega) = \lim_{t\to\infty} \frac{1}{t} \log \|\bigwedge^k T_x\varphi(t,\omega)\|, \tag{8}$$

$1 \leq k \leq d$. Here $\bigwedge^k$ denotes the k-fold exterior product of $T_x\varphi$. Existence of the limits in (8) follows from Kingman's subadditive ergodic theorem (Kingman [29]). Though (7) guarantees $\lambda_1 < \infty$, the last λ_i's may equal $-\infty$. If μ is ergodic, the Lyapunov exponents do not depend on (x,ω). If M is compact, the Lyapunov exponents do not depend on the choice of the Riemannian metric.

Sometimes it is more convenient to count only the distinct Lyapunov exponents, denoted here by $\Lambda_1 > \Lambda_2 > \ldots > \Lambda_r$, where r is the number of distinct exponents, $1 \leq r \leq d$ (we assume μ ergodic to ease notation). Denote by $d_i = \max\{p - q + 1 \mid \lambda_p = \lambda_q = \Lambda_i\}$ the multiplicity of Λ_i.

There is another classical way to introduce Lyapunov exponents (see for instance Arnold and Wihstutz [8]). Put

$$\lambda(v,\omega) = \limsup_{t\to\infty} \frac{1}{t} \log \|T_x\varphi(t,\omega)v\|. \tag{9}$$

The map $\lambda(\cdot,\omega) : TM \to \mathbb{R} \cup \{-\infty\}$ satisfies $\lambda(cv) = \lambda(v)$ for all $c \neq 0$, $v \in TM$, and $\lambda(c_1v_1 + c_2v_2) \leq \max\{\lambda(v_1), \lambda(v_2)\}$ for all $c_1, c_2 \in \mathbb{R}$ and $v_1, v_2 \in T_xM$, $x \in M$ (sometimes called a *characteristic exponent*); we dropped ω, which is fixed here. These two properties imply that λ takes only finitely many values $\tilde\Lambda_1 > \tilde\Lambda_2 > \ldots > \tilde\Lambda_{\tilde r}$ as v varies over T_xM, $v \neq 0$. The Lyapunov exponents in this approach are the $\tilde\Lambda_i$. By definition of $v \mapsto \lambda(v,\omega)$, the sets

$$V_\delta(x,\omega) = \{v \in T_xM \mid \lambda(v,\omega) \leq \delta\}$$

are linear subspaces of T_xM for $\delta \in \mathbb{R}$ arbitrary. Put $V_i = V_{\tilde\Lambda_i}$ and $\tilde d_i = \dim V_i - \dim V_{i+1}$.

The two definitions of Lyapunov exponents presented above are in general not equivalent. However, they are equivalent if (and only if) (x,ω) is a *forward regular point* for $T_x\varphi(t,\omega)$ (see Arnold and Wihstutz [8]) pp. 2–3). In terms of the present paper, (x,ω) is forward regular if $\sum_1^r d_i\Lambda_i = \sum_1^{\tilde r} \tilde d_i\tilde\Lambda_i$. It is clear from (6) that the bundle of linear subspaces

$$V_i(x,\omega) = \{v \in T_xM \mid \lambda(v,\omega) \leq \tilde\Lambda_i(x,\omega)\},$$

is invariant under $T\varphi$ in the sense that $T_x\varphi(t,\omega)V_i(x,\omega) \subset V_i(\Theta_t(x,\omega))$. We refer to the family

$$T_xM = V_1(x,\omega) \supset V_2(x,\omega) \supset \ldots \supset V_r(x,\omega) \supset \{0\}$$

as the *Oseledec flag* associated with φ.
In the following we will be concerned with regular systems only, so that we need not distinguish between Λ and $\tilde{\Lambda}$.

We now recall the Multiplicative Ergodic Theorem (MET) of Oseledec for nonlinear RDS and sketch Ruelle's trivialization argument which reduces the nonflat bundle case to the flat bundle one.

Theorem
(i) (Multiplicative Ergodic Theorem without invertibility)
Suppose φ is a smooth RDS on a d-dimensional Riemannian manifold M and let μ be an invariant measure for φ such that (7) is satisfied. Then the linearization $T_x\varphi(t,\omega)$ of φ is forward regular at μ-almost all points $(x,\omega) \in M \times \Omega$.
(ii) (Multiplicative Ergodic Theorem with invertibility)
Suppose φ is a smooth two sided RDS on a d-dimensional Riemannian manifold M and let μ be an invariant measure for φ such that

$$(x,\omega) \mapsto \sup_{0<t\leq t_0} \{\log^+(\|T_x\varphi(t,\omega)\|) + \log^+(\|(T_x\varphi(t,\omega))^{-1}\|)\} \in L^1(\mu). \tag{10}$$

Then the linearization $T_x\varphi(t,\omega)$ of φ is bi-regular[1] *at μ-almost all points $(x,\omega) \in M \times \Omega$.*

Note that in the invertible case regularity implies that for μ-almost all (x,ω) the spaces

$$E_i(x,\omega) = \{v \in T_xM \mid \lambda^+(v,\omega) = \lambda^-(v,\omega) = \Lambda_i(x,\omega)\}$$

form a splitting of T_xM (with $\lambda^+(v,\omega) = \lambda(v,\omega)$ as in (9), and $\lambda^-(v,\omega)$ defined as in (9) with $t \to -\infty$). $TM = \bigoplus E_i$ is referred to as the *Oseledec splitting*.

PROOF OF THE MET Denote the tangent bundle by (TM,π,M). Choose a countable covering of M by bundle charts (M_i,ψ_i) trivializing TM locally in an isometrical manner. That means, M_i is an open subset of M, and $\psi_i : \pi^{-1}(M_i) \to M_i \times \mathbb{R}^d$, where $\pi : TM \to M$ denotes the canonical bundle projection, such that ψ_i restricted to $\pi^{-1}\{x\}$ is linear for all $x \in M_i$. In addition, ψ_i may be chosen to be an isomorphism with respect to the scalar product on $\pi^{-1}\{x\}$ induced by the Riemannian structure on M and the standard scalar product on $\mathbb{R}^d$ for all $x \in M_i$, see Klingenberg [31] Theorem 1.8.20. (It is not really essential to choose isometric bundle charts, it is simply more convenient.)
Next put $B_0 = M_0$ and $B_n = M_n \setminus \bigcup_{j<n} B_j$ to obtain a countable covering $\{B_n \mid n \in \mathbb{N}\}$ of M by disjoint Borel sets. Putting

$$\begin{aligned} \Sigma : TM &\to M \times \mathbb{R}^d \\ u &\mapsto \psi_i(x) \qquad \text{for } x \in B_i \end{aligned}$$

yields a bimeasurable bundle map from TM to the flat bundle $M \times \mathbb{R}^d$ such that $\Sigma_x : T_xM \to \{x\} \times \mathbb{R}^d$ is an isomorphism for all $x \in M$. Finally, put

$$\Psi(t;x,\omega) = \Sigma_{\varphi(t,\omega)x} \circ T_x\varphi(t,\omega) \circ \Sigma_x^{-1}.$$

[1] 'regular' in the terminology of Arnold and Wihstutz [8] p. 4

Then Ψ is a linear RDS on $\mathbb{R}^d$ over the (enlarged) probability space $(M \times \Omega, \mu)$. Since Σ_x is an isomorphism for all $x \in M$,

$$\|\wedge^k \Psi(t; x, \omega)\| = \|\wedge^k T_x\varphi(t, \omega)\|$$

for all k, $1 \leq k \leq d$, and

$$\|\Psi(t; x, \omega)y\| = \|T_x\varphi(t, \omega)(\Sigma_x^{-1} y)\|$$

for all $y \in \mathbb{R}^d$, $t \in T$, $x \in M$, and P-almost all $\omega \in \Omega$. Thus regularity of Ψ implies regularity of $T\varphi$. But Ψ satisfies the integrability conditions of the 'ordinary' MET by (7) and (10), respectively, hence Ψ is forward or bi-regular, respectively, for μ-almost all (x, ω). □

Note that the Theorem does not require compactness of the manifold M. For a given smooth RDS with an invariant measure it thus only remains to check the integrability conditions (7) or (10), respectively, to infer the conclusions of the MET. This has been done directly for white noise systems on compact manifolds by Carverhill [19] (without imposing any further assumptions). Later Kifer [28] has shown that white noise systems on compact manifolds satisfy much stronger integrability conditions.

Multiplicative ergodic theory becomes much more difficult when considering *infinite dimensional* RDS. Recall that for a finite dimensional linear deterministic system the Lyapunov exponents are precisely the real parts of the eigenvalues of A (for continuous time, $\dot{x} = Ax$) or the logarithms of the eigenvalues of A (for discrete time, $x_{n+1} = Ax_n$), respectively. Thus, the Lyapunov exponents are determined by the spectrum. The definition (see (8)) suggests that it is essential to have a 'well behaved' top part of the spectrum: isolated eigenvalues of finite multiplicity. Since spectra of infinite dimensional operators in general have a considerably more complicated structure than finite dimensional ones, it is clear that much less is to be expected for infinite dimensional RDS. For a survey on infinite dimensional systems see 4.3 below.

3 Random Dynamical Systems and Markov Processes

3.1 Two cultures

In the theory of RDS two well-established mathematical cultures meet, overlap, and sometimes collide:

- *Dynamical Systems*: the *flow* point of view. Typically $T = \mathbb{R}$ or $\mathbb{Z}$ unless mappings are non-invertible which typically happens only for discrete time. Invariance of a measure is defined as invariance with respect to the mappings of the system.

- *Markov processes, stochastic analysis*: Here time is almost exclusively $\mathbb{Z}^+$ or $\mathbb{R}^+$ (or part of it). Markov processes are defined and studied through their transition semigroups forward in time. The necessity to really *construct* stochastic processes with prescribed transition semigroups (their existence follows from Kolmogorov's theorem) created the theory of stochastic differential equations (SDE's) (which are really ODE's with white noise input). Continuous time is $\mathbb{R}^+$, and a filtration $\mathcal{F}_t$ (i. e., an increasing family of sub σ-algebras of $\mathcal{F}$) collects the information available at time t. Everything has to be adapted, i. e., $\mathcal{F}_t$-measurable. 'Invariant measure' in the Markov context means invariance with respect to the transition semigroup.

The door from Markov processes to dynamical systems was really opened around 1980, when several people (Elworthy [24], Bismut [12], Ikeda and Watanabe [25], Kunita [33]) realized that writing down an SDE for a Markov process means much more than originally thought of by the pioneers K. Itô et al. It means the construction of a *stochastic flow* (or, as we will see, of an RDS with independent increments) whose one-point motions are Markov with the prescribed transition semigroup or its generator, respectively.

Probabilists sometimes criticize moving from an SDE to an RDS as 'forgetting' some of the probabilistic structure of the original, e. g., the fact that coming from an SDE implies in particular that all n-point motions are Markov. We think that the contrary is true, as many contributions to this volume show.

First, the concept of an RDS allows to address completely *new questions on SDE's*, as for instance the problem of finding *all* invariant measures (and not only those solving the Fokker-Planck equation), the problem of random invariant manifolds, random normal forms etc.

Second, the underlying Markov structure gives rise to problems which do not make sense for deterministic dynamical systems, such as the interplay of measurability and adaptedness properties with dynamics, see Crauel [20], [21], [22].

There seems to be some need for describing the connection between RDS and Markov processes in some detail.

3.2 RDS with independent increments, Brownian RDS

An RDS $\varphi(t,\omega)$ over $(\Omega, \mathcal{F}, P, (\vartheta_t)_{t\in T})$ is said to have *independent increments* if for all $t_0 \le t_1 \le \ldots \le t_n$ the random variables

$$\varphi(t_1 - t_0, \vartheta_{t_0}\omega), \varphi(t_2 - t_1, \vartheta_{t_1}\omega), \ldots, \varphi(t_n - t_{n-1}, \vartheta_{t_{n-1}}\omega) \tag{11}$$

are independent. If, in addition, for $T = \mathbb{R}^+$ or $\mathbb{R}$ the map $t \mapsto \varphi(t,\omega)x$ is continuous for all $x \in X$ P-a. s., then the RDS or cocycle is said to be a *Brownian* RDS or cocycle.

Remarks (i) An RDS with independent increments automatically has *stationary* (time homogeneous) increments, as, by the ϑ_t invariance of P, $\varphi(h, \vartheta_t\omega) \stackrel{\mathcal{D}}{=} \varphi(h,\omega)$ for all $t \in T$.
(ii) If $\varphi(t,\omega)$ consists of invertible mappings then, by the cocycle property,

$$\varphi_{s,t}(\omega) := \varphi(t,\omega) \circ \varphi(s,\omega)^{-1} = \varphi(t-s, \vartheta_s\omega)$$

for $s \le t$, so (11) means that for $t_0 \le t_1 \le \ldots \le t_n$

$$\varphi_{t_0,t_1}, \varphi_{t_1,t_2}, \ldots, \varphi_{t_{n-1},t_n} \tag{12}$$

are independent.

3.3 RDS and Markov chains, discrete time $T = \mathbb{Z}^+$ or $\mathbb{Z}$

Case $T = \mathbb{Z}^+$: Here $\varphi(n,\omega) = \varphi(1,\vartheta^{n-1}\omega) \circ \ldots \circ \varphi(1,\omega)$, so the cocycle has independent increments if and only if $\varphi(1,\omega), \varphi(1,\vartheta\omega), \ldots$ are iid. We thus have a product of iid random mappings, i. e., a classical 'iterated function system'. The mapping

$$x \mapsto \varphi(n,\omega)x$$

defines a homogeneous Markov chain with transition kernel

$$P(x,B) = P\{\omega \mid \varphi(1,\omega)x \in B\}. \tag{13}$$

Putting $x_n = \varphi(n,\omega)x_0$ we have

$$x_{n+1} = \varphi(1,\vartheta^n\omega)x_n, \tag{14}$$

i. e., a stochastic difference equation generating the Markov chain.

Conversely, given a transition kernel $P(x,B)$ on X, we want to construct an RDS with independent increments over a dynamical system $(\Omega, \mathcal{F}, P, \vartheta)$, i. e., a cocycle $\varphi(n,\omega)$ with $(\varphi(1,\vartheta^n\omega))_{n\in\mathbb{Z}^+}$ iid, such that (13) holds. This question has been dealt with by Kifer [26] Section 1.1. It always has a positive answer as soon as X is a Borel subset of a Polish space and if we are content with a measurable mapping $(x,\omega) \mapsto \varphi(1,\omega)x$. If we want $x \mapsto \varphi(1,\omega)x$ to be continuous or homeomorphisms or smooth etc., a general answer to this representation problem is not known up to now (compare Kifer [26] p. 12).

Case $T = \mathbb{Z}$: Now $\varphi(n,\omega)$ is invertible, and the cocycle has independent increments if and only if $(\varphi(1,\vartheta^n\omega))_{n\in\mathbb{Z}}$ is iid. The mapping $x \mapsto \varphi(n,\omega)x$ defines a homogeneous Markov chain on all of $\mathbb{Z}$ starting at $x_0 = x$, and (14) can be inverted to give

$$x_n = \varphi(1,\vartheta^n\omega)^{-1}x_{n+1} = \varphi(-1,\vartheta^{-n}\omega)x_{n+1}.$$

We can now look at the *forward transition kernel*

$$P^+(x,B) = P\{\omega \mid \varphi(1,\omega)x \in B\}$$

and the *backward transition kernel*

$$P^-(x,B) = P\{\omega \mid \varphi(-1,\omega)x \in B\} = P\{\omega \mid \varphi(1,\omega)^{-1}x \in B\}.$$

Note that in general P^+ and P^- do not have the same invariant measures: a forward invariant measure ν^+ has to satisfy

$$\nu^+ = \int P^+(x,\cdot)\, d\nu^+(x) = E\varphi(1,\omega)\nu^+,$$

whereas a backward invariant measure ν^- is characterized by

$$\nu^- = \int P^-(x,\cdot)\, d\nu^-(x) = E\varphi(-1,\omega)\nu^- = E\varphi(1,\omega)^{-1}\nu^-.$$

How are measures $\nu^\pm$ related to invariant measures of the RDS, i. e., measures μ on $X \times \Omega$ whose disintegration satisfies $\varphi(1,\omega)\mu_\omega = \mu_{\vartheta\omega}$? For one sided time Ohno [37] has proved that if $\nu = \nu^+$ is an invariant measure for the forward transition kernel P^+, then $\mu = \nu \times P$ is invariant for the RDS. Conversely, if a product measure $\nu \times P$ is invariant for the RDS, then ν is invariant for P^+.

For two sided time, a product measure is invariant for the RDS if and only if ν is fixed under φ, i. e., $\varphi(1,\omega)\nu = \nu$ P-almost surely. If a measure $\nu^\pm$ is $P^\pm$ invariant then the measures μ^+ and μ^-, given by

$$\mu_\omega^\pm = \lim_{n\to\mp\infty} \varphi(n,\omega)^{-1}\nu^\pm,$$

are invariant for the RDS (so-called *Markov* measures). They are the ones which 'remember' the Markov kernels $P^\pm$. This equally applies for continuous time $T = \mathbb{R}^+$ or $\mathbb{R}$. These questions have been studied systematically by Crauel [21].

However, typically an RDS has more invariant measures not coming from the Markov chain — and those measures are needed for a systematic study of the RDS.

3.4 RDS and Markov processes, continuous time $T = \mathbb{R}^+$ or $\mathbb{R}$

Here the situation is much nicer and its description more complete than in the discrete time case. Basic results are due to Baxendale [10] and Kunita [33], [34], [35]. We describe the situation conceptually, i. e., without stating all technical assumptions, and quote freely from the above sources. We mainly emphasize the RDS point of view.

Case $T = \mathbb{R}^+$: Assume $X = \mathbb{R}^d$ (similar things hold on manifolds). Let $\varphi(t,\omega)$ be a Brownian RDS of homeomorphisms (or diffeomorphisms of some smoothness class). Then $(\varphi(t,\omega))_{t\in\mathbb{R}^+}$ is a Brownian motion with values in the group $\mathrm{Hom}(\mathbb{R}^d)$ (or $\mathrm{Diff}^*(\mathbb{R}^d)$ with a suitable $*$) in the sense of Baxendale [10], or

$$\varphi_{s,t}(\omega) = \varphi(t,\omega) \circ \varphi(s,\omega)^{-1},$$

$s,t \in \mathbb{R}^+$, is a temporally homogeneous Brownian flow in the sense of Kunita [35] p. 116.

By studying the infinitesimal mean

$$\lim_{h\searrow 0} \frac{1}{h} E(\varphi_{t,t+h}(\omega)x - x) = b(x)$$

and the infinitesimal covariance

$$\lim_{h\searrow 0} \frac{1}{h} E(\varphi_{t,t+h}(\omega)x - x)(\varphi_{t,t+h}(\omega)y - y)' = a(x,y),$$

Kunita constructs a vector field valued Brownian motion $(F(x,t,\omega))_{x\in\mathbb{R}^d,t\in\mathbb{R}^+}$, i. e., a continuous (in t) Gaussian process $(F(\cdot,t,\omega))_{t\in\mathbb{R}^+}$ with values in the space of vector fields on $\mathbb{R}^d$ (so $x \mapsto F(x,t,\omega)$ is a vector field), which has additively stationary independent increments and satisfies $F(\cdot,0,\omega) = 0$ (P-a. s.). The Brownian motion F is related to the

Brownian flow φ by $EF(x,t,\omega) = tb(x)$ and $\operatorname{cov}(F(x,t,\omega), F(y,s,\omega)) = \min\{t,s\}\, a(x,y)$. This implies that for all $s \in \mathbb{R}^+$

$$\varphi_{s,t}(\omega)x = x + \int_s^t F(\varphi_{s,u}(\omega)x, du, \omega), \quad t \in [s,\infty), \tag{15}$$

which has to be understood in the sense that (15) has a solution which coincides in distribution with the original Brownian flow φ. In short: The (forward) flow satisfies an Itô SDE driven by vector field valued Gaussian white noise. F is called the random infinitesimal generator of φ.

All n-point motions $(\varphi(t,\omega)x_1, \ldots, \varphi(t,\omega)x_n)$ are homogeneous Feller-Markov processes. In particular, $(\varphi(t,\omega)x)_{t\in\mathbb{R}^+}$ is a Markov process whose transition semigroup has generator

$$L = \sum_{i=1}^d b^i(x)\frac{\partial}{\partial x^i} + \frac{1}{2}\sum_{i,j=1}^d a^{ij}(x,x)\frac{\partial^2}{\partial x^i \partial x^j}. \tag{16}$$

The backward flow $\varphi_{s,t}(\omega)^{-1} = \varphi(s,\omega)\circ\varphi(t,\omega)^{-1}$, $0 \le s \le t$, satisfies for each $t \in \mathbb{R}^+$ a backward Itô equation in $s \in [0,t]$,

$$\varphi_{s,t}(\omega)^{-1}x = x - \int_s^t \hat{F}(\varphi_{u,t}(\omega)^{-1}x, \hat{du}, \omega),$$

where

$$\hat{F}(x,t,\omega) = F(x,t,\omega) - t\,c(x), \qquad c_i(x) = \sum_{j=1}^d \frac{\partial a^{ij}}{\partial x^j}(x,y)\Big|_{y=x}, \tag{17}$$

and the backward integral $\int_s^t \hat{F}(\varphi_{u,t}(\omega)^{-1}x, \hat{du}, \omega)$ is formally defined by the same definition as the forward integral — the difference being the inverted measurability counting from t backward to s.

As usual, things get more symmetric if we use Stratonovich forward and backward integrals. Put

$$F^0(x,t,\omega) = F(x,t,\omega) - \frac{t}{2}c(x),$$

then F^0 is the forward as well as the backward Stratonovich infinitesimal generator of φ.

Conversely, given a temporally homogeneous $\mathcal{V}(\mathbb{R}^d)$ (= vector fields on $\mathbb{R}^d$) -valued Brownian motion F, we can write down the SDE (15) to generate a Brownian flow with generator F. We can easily construct a Brownian RDS describing the same object. Indeed, put

$$\begin{aligned}
\Omega &= \{\omega \mid \omega(0) = 0, \omega(\cdot) \in \mathcal{C}(\mathbb{R}^+, \mathcal{V}(\mathbb{R}^d))\}\\
\mathcal{F} &= \text{Borel field}\\
P &= \text{distribution of } F = \text{'Wiener measure'}\\
\vartheta_t\omega(\cdot) &= \omega(\cdot + t) - \omega(t), \quad t \in \mathbb{R}^+,
\end{aligned}$$

so ϑ_t leaves P invariant and is ergodic. Moreover, representing $F(\cdot, t, \omega) \equiv \omega(t)$, the uniqueness of the solution flow of (15) yields

$$\varphi_{s,t}(\omega) = \varphi(t - s, \vartheta_s\omega), \qquad 0 \leq s \leq t$$

(for a rigorous proof see, e. g., Wentzell [40] p. 192), so that the flow property $\varphi_{0,t+s} = \varphi_{s,t+s} \circ \varphi_{0,s}$ reads

$$\varphi(t + s, \omega) = \varphi(t, \vartheta_s\omega) \circ \varphi(s, \omega),$$

which is nothing but the cocycle property. As (12) is satisfied, and (12) is equivalent to (11), we have constructed a Brownian RDS from the given $\mathcal{V}(\mathbb{R}^d)$-valued temporally homogeneous Brownian motion F.

The *result* is (modulo technical assumptions):
There is a one to one relation between Brownian RDS and temporally homogeneous vector field valued Brownian motions (defined via the SDE (15)).

Since the law of $F(\cdot, t)$ (Gaussian!) is uniquely determined by the infinitesimal characteristics $b(x)$ and $a(x, y)$, it follows that (a, b) uniquely determine the law of the corresponding Brownian RDS. Consequently, as a and b appear in the generator of the two-point motion, the law of the two-point motion uniquely determines the law of the Brownian RDS. (The last statement is in general not true in case of discrete time.)

Note that, however, the law of the Brownian RDS is in general not determined by the law of the one-point motion. The point is that the generator (16) only contains $a(x, x)$ instead of $a(x, y)$. In general there are thus many distinct Brownian RDS whose one-point motions have the same law. For instance, if $a(x, x) \in C_b^2(\mathbb{R}^d; \mathbb{R}^{d\times d})$ is given, it can be factorized as $a(x, x) = \sigma(x)\sigma(x)'$ with $\sigma : \mathbb{R}^d \to \mathbb{R}^{d\times d}$ Lipschitz. Now put $a(x, y) = \sigma(x)\sigma(y)'$, and generate a Brownian RDS by solving

$$d\varphi_{s,t}(\omega)x = b(\varphi_{s,t}(\omega)x)\, dt + \sigma(\varphi_{s,t}(\omega)x)\, dB(t, \omega),$$

where B is a standard Brownian motion in $\mathbb{R}^d$. In this case

$$F(x, t, \omega) = t\, b(x) + \sigma(x)B(t, \omega).$$

In general countably many B_j's are necessary to represent F.

Remarks: (i) By replacing the driving Brownian motion F in (15) by a continuous vector field valued semimartingale, Kunita [35] constructs a more general class of flows. However, these flows do not correspond to RDS, unless the semimartingale has stationary increments. A stochastic flow φ generated by (15) is a Brownian RDS if and only if F is a Brownian motion.
(ii) There are more stochastic flows than covered by (i). See Example 3 in Section 2.1.

Case $T = \mathbb{R}$: As working on $T = \mathbb{R}$ is quite unusual in stochastic analysis, we develop the connection between Brownian RDS and SDE on $\mathbb{R}$ in detail.

Suppose we are given a Brownian RDS on $\mathbb{R}$. Then $\varphi_{s,t}(\omega) := \varphi(t, \omega) \circ \varphi(s, \omega)^{-1}$, $s, t \in \mathbb{R}$ is a temporally homogeneous Brownian flow in the sense of Kunita [35] on *all* of $\mathbb{R}$. Our aim is to write down an SDE for φ on $\mathbb{R}$. We split the cocycle $(\varphi(t,\omega))_{t\in\mathbb{R}}$ into two independent halves which are both temporally homogeneous Brownian cocycles on $\mathbb{R}^+$:

(i) $(\varphi^+(t,\omega))_{t\in\mathbb{R}^+}$, defined by $\varphi^+(t,\omega) = \varphi(t,\omega)$, is a cocycle on $\mathbb{R}^+$ over the dynamical system $(\Omega, \mathcal{F}, P, (\vartheta_t)_{t\in\mathbb{R}^+})$,
(ii) $(\varphi^-(t,\omega))_{t\in\mathbb{R}^+}$, defined by $\varphi^-(t,\omega) = \varphi(-t,\omega)$, is a cocycle on $\mathbb{R}^+$ over the dynamical system $(\Omega, \mathcal{F}, P, (\vartheta_{-t})_{t\in\mathbb{R}^+})$.

The two cocycles φ^+ and φ^- are tied together by

$$\varphi^+(t,\omega)^{-1} = \varphi^-(t,\vartheta_t\omega), \qquad t\in\mathbb{R}^+.$$

Both $\varphi^\pm$ generate homogeneous Brownian flows

$$\varphi^\pm_{s,t}(\omega) = \varphi^\pm(t,\omega)\circ\varphi^\pm(s,\omega)^{-1} \qquad s,t\in\mathbb{R}^+.$$

There are thus two independent homogeneous vector field valued Brownian motions $(F^\pm(\cdot,t,\omega))_{t\in\mathbb{R}^+}$ such that for each fixed $s\in\mathbb{R}^+$

$$\varphi^\pm_{s,t}(\omega)x = x + \int_s^t F^\pm(\varphi^\pm_{s,u}(\omega)x, du, \omega), \qquad t\in[s,\infty), \tag{18}$$

and for each fixed $t\in\mathbb{R}^+$

$$\varphi^\pm_{s,t}(\omega)^{-1}x = x - \int_s^t \hat{F}^\pm(\varphi^\pm_{u,t}(\omega)^{-1}x, \hat{d}u, \omega), \qquad s\in[0,t]. \tag{19}$$

We want to determine the covariances $\operatorname{cov}(F^\pm(x,t,\omega), F^\pm(y,t,\omega)) = \min\{t,s\}\, a^\pm(x,y)$ and the means $EF^\pm(x,t,\omega) = t\, b^\pm(x)$. We have for $h>0$

$$E\varphi^-_{t,t+h}(\omega)x = E\varphi^+_{t,t+h}(\omega)^{-1}x$$

and

$$E(\varphi^-_{t,t+h}(\omega)x)(\varphi^-_{t,t+h}(\omega)y)' = E(\varphi^+_{t,t+h}(\omega)^{-1}x)(\varphi^+_{t,t+h}(\omega)^{-1}y)'.$$

Using the representation (18) for $\varphi^-(h,\omega)x$ and (19) for $\varphi^+(h,\omega)^{-1}x$, we obtain

$$a^-(x,y) = \hat{a}^+(x,y) = a^+(x,y) =: a(x,y)$$

and

$$b^-(x) = -\hat{b}^+(x) = -b^+(x) + c(x)$$

with $c(x)$ defined in (17). For the Stratonovich generators $F^{0\pm}$ this means $b^{0-}(x) = -b^{0+}(x)$, $a^{0-} = a^{0+} = a$, where $b^{0+} = b^+ - \frac{1}{2}c$.

We now piece the two independent vector field valued Brownian motions $F^\pm$ on $\mathbb{R}^+$ together to produce one on all of $\mathbb{R}$:

$$F^0(x,t,\omega) = \begin{cases} F^{0+}(x,t,\omega), & t\geq 0,\\ F^{0-}(x,-t,\omega), & t\leq 0. \end{cases} \tag{20}$$

This is a homogeneous Brownian motion on $T = \mathbb{R}$ with $F^0(x,0,\omega) = 0$, $EF^0(x,t,\omega) = t\,b^{0+}(x)$, and covariance given by $a(x,y)$.

As an intermediate step we introduce the notation

$$\int_0^t F^0(\varphi(u,\omega)x, \circ du, \omega) := \begin{cases} \int_0^t F^{0+}(\varphi^+(u,\omega)x, \circ du, \omega), & t \geq 0 \\ \int_0^{-t} F^{0-}(\varphi^-(u,\omega)x, \circ du, \omega), & t \leq 0. \end{cases}$$

As a result, the original cocycle $(\varphi(t,\omega))_{t\in\mathbb{R}}$ satisfies

$$\varphi(t,\omega)x = x + \int_0^t F^0(\varphi(u,\omega)x, \circ du, \omega)$$

on all of $\mathbb{R}$.

Similarly we put

$$\int_0^t F^0(\varphi_{u,t}(\omega)^{-1}x, \circ d\hat{u}, \omega) := \begin{cases} \int_0^t F^{0+}(\varphi^+_{u,t}(\omega)^{-1}x, \circ d\hat{u}, \omega), & t \geq 0 \\ \int_0^{-t} F^{0-}(\varphi^-_{u,-t}(\omega)x, \circ d\hat{u}, \omega), & t \leq 0. \end{cases}$$

Again as a result, the inverse $(\varphi(t,\omega)^{-1})_{t\in\mathbb{R}}$ of the original cocycle satisfies

$$\varphi(t,\omega)^{-1}x = x - \int_0^t F^0(\varphi_{u,t}(\omega)^{-1}x, \circ d\hat{u}, \omega)$$

on all of $\mathbb{R}$.

Our last step is the convention

$$\int_t^0 F^0(\varphi_{t,u}(\omega)x, \circ du) = -\int_0^t F^0(\varphi_{u,t}(\omega)^{-1}x, \circ d\hat{u}), \qquad t \in \mathbb{R}.$$

Now we can write down a unified SDE-representation of the Brownian flow $\varphi_{s,t}(\omega) = \varphi(t,\omega) \circ \varphi(s,\omega)^{-1}$, $s,t \in \mathbb{R}$, as follows

$$\varphi_{s,t}(\omega)x = x + \int_s^t F^0(\varphi_{s,u}(\omega)x, \circ du, \omega), \qquad s,t \in \mathbb{R}, \tag{21}$$

or, symbolically,

$$d\varphi = F^0(\varphi, \circ dt).$$

If properly interpreted, (21) is satisfied for all $s, t \in \mathbb{R}$. For instance, if $s < 0 < t$, then

$$x + \int_s^t F^0(\varphi_{s,u}(\omega)x, \circ du) = x + \int_s^0 F^0(\varphi_{s,u}(\omega)x, \circ du) + \int_0^t F^0(\varphi_{s,u}(\omega)x, \circ du)$$

$$= x - \int_0^s F^0(\varphi_{u,s}(\omega)^{-1}x, \circ d\hat{u}) + \int_0^t F^0(\varphi(u,\omega)\varphi(s,\omega)^{-1}x, \circ du)$$

$$= \varphi(s,\omega)^{-1}x + \int_0^t F^0(\varphi(u,\omega)\varphi(s,\omega)^{-1}x, \circ du) = (\varphi(t,\omega) \circ \varphi(s,\omega)^{-1})x = \varphi_{s,t}(\omega)x.$$

Conversely, given a vector field valued homogeneous Brownian motion $(F(x,t,\omega))_{x\in\mathbb{R}^d, t\in\mathbb{R}}$, characterized by $EF(x,t,\omega) = tb(x)$ and

$$\operatorname{cov}(F(x,t,\omega), F(y,s,\omega)) = \begin{cases} \min\{|t|,|s|\}\, a(x,y), & ts \geq 0, \\ 0, & ts < 0, \end{cases}$$

we write down equation (21), which consistently defines a Brownian flow on $\mathbb{R}$. The corresponding RDS is one over

$$\begin{aligned} \Omega &= \{\omega \mid \omega \in C(\mathbb{R}, \mathcal{V}(\mathbb{R}^d)), \omega(0) = 0\} \\ P &= \text{`Wiener measure' on } \Omega \text{ generated by } F \\ \vartheta_t\omega(\cdot) &= \omega(\cdot + t) - \omega(t), \qquad t \in \mathbb{R}. \end{aligned}$$

The cocycle property for $\varphi(t,\omega) := \varphi_{0,t}(\omega)$, $t \in \mathbb{R}$, follows from the fact that

$$\varphi_{s,t}(\omega) = \varphi(t-s, \vartheta_s\omega), \qquad s,t \in \mathbb{R}.$$

As a *result*, we again have a one to one correspondence between Brownian RDS φ for $T = \mathbb{R}$ and homogeneous Brownian motions F on $T = \mathbb{R}$. Again, (a, b) uniquely determines the law of the Brownian RDS.

All n-point motions of a Brownian RDS are homogeneous Markov processes on $\mathbb{R}$. In particular, the one-point motions $(\varphi(t,\omega)x)_{t\in\mathbb{R}}$ have the forward semigroup

$$P_t^+(x,B) = P\{\omega \mid \varphi(t,\omega)x \in B\} = P\{\omega \mid \varphi^+(t,\omega)x \in B\}, \qquad t \geq 0,$$

with generator

$$L^+ = \sum_{i=1}^d b^{+i}(x)\frac{\partial}{\partial x^i} + \frac{1}{2}\sum_{i,j=1}^d a^{ij}(x,x)\frac{\partial^2}{\partial x^i \partial x^j},$$

while for $-t \leq 0$ they have the backward semigroup

$$\begin{aligned} P_t^-(x,B) &= P\{\omega \mid \varphi(-t,\omega)x \in B\} = P\{\omega \mid \varphi^-(t,\omega)x \in B\} \\ &= P\{\omega \mid \varphi(t,\omega)^{-1}x \in B\} = P\{\omega \mid \varphi^+(t,\omega)^{-1}x \in B\}, \qquad t \geq 0, \end{aligned}$$

with generator

$$L^- = \sum_{i=1}^{d} b^{-i}(x)\frac{\partial}{\partial x^i} + \frac{1}{2}\sum_{i,j=1}^{d} a^{ij}(x,x)\frac{\partial^2}{\partial x^i \partial x^j},$$

where $b^- = -b^+ + c$. Note that both L^+ and L^- describe the *same* Markov process, but L^+ forward and L^- backward in time. In general $L^+ \neq L^-$, so the Fokker-Planck equations $L^{+*}\nu = 0$ and $L^{-*}\nu = 0$ usually have different solutions. For the relation between $\nu^\pm$ and invariant measures for the RDS φ cf. the discrete time case $T = \mathbb{Z}$.

Example Suppose a homogeneous vector field valued Brownian motion is given by

$$F^0(x,t,\omega) = tX_0(x) + \sum_{i=1}^{m} B_i(t,\omega)X_i(x), \qquad t \in \mathbb{R},$$

with $X_0, \ldots, X_m$ vector fields and $B-1, \ldots, B_m$ independent standard Brownian motions on $\mathbb{R}$ (this is the lucky case: the general one needs $m = \infty$). Then

$$d\varphi = X_0(\varphi)\,dt + \sum_{i=1}^{d} X_i(\varphi) \circ dB_i(t), \qquad t \in \mathbb{R},$$

gives a Brownian flow (and RDS) on $T = \mathbb{R}$, and the one-point motions have generators

$$L^\pm = \pm X_0 + \frac{1}{2}\sum_{i=1}^{m} X_i^2.$$

Remark: We have not discussed a generalization of the above case which runs under 'real noise case' (Arnold, Kliemann, and Oeljeklaus [6], Arnold and Kliemann [5]) or 'Markovian multiplicative system' (Bougerol [14], [15]; Carmona and Lacroix [18] Chapter IV) or 'Markovian RDS' (Crauel [21]). Here we have a stationary Markov process $(\xi_t)_{t\in T}$, and, for discrete time, a difference equation $x_{n+1} = f(x_n, \xi_n)$, so that (x_n, ξ_n) is Markov for suitable x_0, e. g., for x_0 deterministic. In the continuous time case, e. g., $\dot{x} = f(x, \xi_t)$; again $(x(t), \xi_t)$ is Markov for suitable x_0. This reduces to the independent increments case if (ξ_n) is iid or if ξ_t is white, respectively.

4 Recent Developments

We will now briefly review the contributions to this volume and put them into the context of RDS laid out above.

4.1 Linear RDS

The MET is a statement about a *linear* cocycle (which may be a linearization) over a dynamical system. So the linear theory is the basis for any nonlinear extension.

An introduction into recent developments around the Multiplicative Ergodic Theorem is given by GOLDSHEID[2]. In particular, for products of iid matrices he discusses the Central Limit Theorem, and introduces into algebraic conditions for simplicity of the Lyapunov spectrum found recently. These yield one of the most beautiful formulations of a hyperbolicity result: If the support of the distribution of the matrices is Zariski dense in $Gl(d, \mathbb{R})$, then their products have d distinct Lyapunov exponents $\lambda_1 \ldots > \lambda_d$.

A linear invertible RDS on $\mathbb{R}^d$ induces nonlinear RDS on the Graßmann manifolds $Gr(k, d)$ of k-dimensional subspaces of $\mathbb{R}^d$. CRAUEL shows how to calculate the Lyapunov exponents of the induced Graßmann systems in terms of the exponents of the original linear one.

LEIZAROWITZ generalizes the theory of p-moment Lyapunov exponents (see Arnold, Kliemann, and Oeljeklaus [6], Arnold, Oeljeklaus, and Pardoux [7], and Baxendale [11]) to a certain class of Markov processes in $\mathbb{R}^d$, which includes linear random differential equations driven by a finite step Markov chain. He shows that under his more general conditions the p-moment Lyapunov exponents are approximate eigenvalues for an associated differential operator.

Products of iid matrices with finitely many values $\{A_i \mid 1 \leq i \leq N\}$ with $p_i = P\{A = A_i\}$ are being addressed by PERES. He proves several results on the dependence of the Lyapunov exponents on the probabilities p_i. For instance, the top Lyapunov exponent $p \mapsto \lambda_1(p)$ depends on $p = (p_1, \ldots, p_N)$ locally real-analytically, provided $\lambda_1(p)$ is simple (i. e., $\lambda_1(p) > \lambda_2(p)$ in the notation of (8)).

LE JAN proves a theorem on the asymptotic behaviour of linear second order quantities and suggests possible applications to curvature of stable foliations etc.

KNILL considers $Sl(2, \mathbb{R})$-valued cocycles (over an aperiodic automorphism). Previously he has proved that the set of cocycles with positive (top) Lyapunov exponent is dense in the set of essentially bounded ones (Knill [32]), which provides another of the few genericity results on existence of non-vanishing Lyapunov exponents. Here he shows that there always exist essentially bounded $Sl(2, \mathbb{R})$ cocycles where the Lyapunov exponent depends discontinuously on variation of the cocycle.

LATUSHKIN AND STEPIN are concerned with the spectral theory of Linear Skew Product Flows. Starting from a cocycle of operators of a Hilbert space over a flow on a compact metric space with a quasi-invariant measure, they investigate the relations between spectra of weighted composition operators, the Lyapunov spectrum of the cocycle, and the Sacker-Sell spectrum of the cocycle.

BOUGEROL proves exponential stability of the Kalman filter for a linear system with stationary (additive and multiplicative) coefficients using the MET together with a result of M. Wojtkowski (see WOJTKOWSKI).

4.2 Nonlinear RDS

The papers in this section touch upon several areas of the theory of differentiable nonlinear stochastic systems, in which important progress has been made recently. But let us first give a brief comment on the measurable case.

[2]Names in SMALL CAPITALS refer to contributions in the present volume

As mentioned before, a measurable RDS is not much more than a dynamical system with a factor. A considerable part of Kifer's book [26] is devoted to this setting. Kifer considers exclusively products of iid maps, restricting attention to one sided time invariant Markov measures μ. Hence $\mu = \rho \times P$, where P is a one sided infinite product measure and ρ is a probability measure on X satisfying

$$E(\varphi(1)\rho) = \rho.$$

For this setting Kifer [26] generalizes the notion of fiber entropy introduced by Abramov and Rohlin [1], and investigated further by Ledrappier and Walters [36]. Bogenschütz [13] extends this notion to general measurable RDS, and observes that certain properties of the factor entropy — in particular, a generalization of the Shannon-McMillan-Breiman theorem for relative entropy — need the two sided time point of view.

Although not made explicit by the author, we consider BAXENDALE's contribution as important for stochastic bifurcation theory. One of the basic pictures of stochastic bifurcation is the following: Suppose we have a family $\varphi_\alpha(\omega)$ of cocycles with a family of invariant reference measures μ_α and corresponding top Lyapunov exponents $\lambda(\mu_\alpha)$. Suppose further $\lambda(\mu_\alpha) < 0$ for $\alpha < \alpha_0$, $\lambda(\mu_{\alpha_0}) = 0$, and $\lambda(\mu_\alpha) > 0$ for $\alpha > \alpha_0$, so that the reference measure μ_α looses its stability at $\alpha = \alpha_0$. We then expect bifurcation of a new invariant measure $\nu_\alpha \neq \mu_\alpha$ for $\alpha > \alpha_0$ with $\nu_\alpha \longrightarrow \mu_{\alpha_0}$ for $\alpha \searrow \alpha_0$ and, hopefully, $\lambda(\nu_\alpha) < 0$.
There are instructive examples in dimension $d = 1$ supporting this picture (see Arnold and Boxler [2]). It also turns out that necessarily $\lambda(\mu_{\alpha_0}) = 0$ if there is a family ν_α bifurcating from μ_α at $\alpha = \alpha_0$ (see Boxler [17]).
BAXENDALE now gives more evidence to the above paradigma by proving that an SDE in $\mathbb{R}^d$ with invariant (reference) measure $\delta_0 \times P$ (i. e., with a fixed point at $x = 0$), and corresponding top exponent λ has the following property: $\lambda < 0$, or $\lambda = 0$, or $\lambda > 0$ according to the diffusion process on $\mathbb{R}^d \backslash \{0\}$ being transient, or null-recurrent, or recurrent. In particular, for $\lambda > 0$ there exists another invariant measure besides the unstable $\delta_0 \times P$.

Considerable progress has recently been made on the problem of invariant manifolds for RDS. Boxler [16] has developed a stochastic center manifold theory. The most general results on invariant manifolds (e. g., those tangent to individual spaces of the Oseledec splitting) were obtained by Dahlke [23].

The method of Boxler [16] works on the level of the cocycle. However, if an RDS is given by a random or stochastic differential equation (as in many applications), one would like to obtain an approximation of the stochastic center manifold by manipulating the corresponding vector fields. This is not (yet) possible for the white noise case due to measurability problems (which call for the use of a stochastic calculus which can handle non-adapted processes). It is, however, possible for the real noise case, see BOXLER.
We mention in passing that the problem of simplifying a random diffeomorphism by smooth coordinate transformations, i. e., a *stochastic normal form theory*, has been dealt with by Xu Kedai [41] and Arnold and Xu Kedai [9].

ARNOLD AND BOXLER give a stochastic analogue of the well-known fact that small perturbations turn a hyperbolic fixed point into a bounded solution. The proof uses the MET, random norms and a recent result of Arnold and Crauel [3] for the affine case.

XUERONG MAO proves exponential stability of $x = 0$ of a general stochastic flow in the sense of Kunita [35] (see Section 3.4), and provides applications to Brownian flows.

KIFER discusses large deviations for products of random expanding maps in Markovian dependence. Since here 'state dependent transition probabilities' are allowed, the processes under consideration in this contribution are more general than RDS.

4.3 Infinite dimensional RDS

Multiplicative Ergodic Theory in infinite dimensions, if possible at all, should simultaneously generalize the finite dimensional random case (i. e., the MET) and the infinite dimensional deterministic spectral theory of operators. So far, not much of such a theory has come to existence. An up-to-date account of what has come to existence is provided by the survey paper of SCHAUMLÖFFEL. Multiplicative Ergodic Theory for a class of stochastic parabolic PDE's is worked out by FLANDOLI.
The top Lyapunov exponent of a product of iid infinite matrices is investigated by DARLING. These matrices arise in percolation theory.

4.4 Deterministic dynamical systems

One dimensional systems, in particular interval maps, have been one of the most active areas of interest during the past decade. KELLER gives a brief introduction into recent developments on piecewise monotone maps (usually understood as maps which are monotone on each of finitely many open intervals whose union is the whole unit interval except for finitley many one point sets, the endpoints of the intervals). In particular, Keller discusses the 'Lyapunov exponent maximizes entropy' formula $h_\mu \leq \max\{\lambda_\mu, 0\}$. It turns out that in case the Lyapunov exponent vanishes more subtle notions are needed, and Keller introduces and discusses algorithmic complexity for this kind of maps.

Also HOFBAUER is concerned with piecewise monotone interval maps. He proves $h_\mu \leq \max\{\lambda_\mu, 0\}$ for these maps, where the assumptions on the maps are weakened insofar countably many instead of finitely many intervals of monotonicity are allowed.

THIEULLEN introduces α-entropy for a smooth dynamical system, which is a modification of Katok's local entropy. He then extends Pesin's formula for α-entropy.

WOJTKOWSKI discusses Hamiltonian systems. He shows existence of non-zero Lyapunov exponents for two systems: a gas of hard balls interacting by elastic collisions, and a system of falling balls on a vertical line interacting by elastic collisions, with the bottom ball bouncing back elastically from a hard floor.

HOLZFUSS AND PARLITZ deal with the 'inverse problem' of extracting Lyapunov exponents of a linearized (deterministic) flow from a 'time series' of observations of the nonlinear system. They propose a method to extract the whole spectrum.

4.5 Engineering applications and control theory

The present volume gathers four contributions of engineers who investigate stability and other qualitative features of systems with random loads, random impurities etc., by means of Multiplicative Ergodic Theory.

ARIARATNAM AND XIE's contribution can serve as a survey. They present very well worked-out case studies. SRI NAMACHCHIVAYA, PAI, AND DOYLE look at stochastic stability of an electric power system with harmonically and stochastically varying network conditions. The (nonlinear) random oscillator is dealt with by WEDIG. He proposes several techniques for calculating invariant measures and Lyapunov exponents. BUCHER presents a method for the approximate calculation of the top exponent of a linear system in $\mathbb{R}^d$ and shows that it yields good results for certain models.

The survey paper of COLONIUS AND KLIEMANN describes the use of the concept of Lyapunov exponents in nonlinear control theory. Originally, nonlinear control theory and Multiplicative Ergodic Theory came into touch via support theorems: the supports of solutions of the Fokker-Planck equation can be characterized as invariant control sets. This way it is possible to obtain uniqueness results for invariant (Markov) measures, see Kliemann [30] and Arnold and Kliemann [5]. Here COLONIUS AND KLIEMANN consider control systems on a smooth manifold M, given by differential equations whose right hand sides depend on a control parameter from a compact subset U of some $\mathbb{R}^k$. This system turns out to be a (topological) skew product flow over the set of Lebesgue measurable $\{t \mapsto u(t) \in U\}$ with time shift. Colonius and Kliemann present a spectral theory for the linearization (in $x \in M$) of the system, which they use for discussing stabilizability, stability radii, and robustness. This view opens the door to an area which is very rich in structure and sheds new light on the MET.

References

[1] L. M. Abramov and V. A. Rohlin, The entropy of a skew product of measure-preserving transformations, *Amer. Math. Soc. Transl. Ser. 2*, 48 (1966) 255–265

[2] L. Arnold and P. Boxler, Stochastic bifurcation: Instructive examples in dimension one, in *Stochastic Flows*, M. Pinsky and V. Wihstutz (eds.). Birkhäuser, Bosten 1991

[3] L. Arnold and H. Crauel, Iterated function systems and multiplicative ergodic theory, in *Stochastic Flows*, M. Pinsky and V. Wihstutz (eds.). Birkhäuser, Bosten 1991

[4] L. Arnold and W. Kliemann, Large deviations of linear stochastic differential equations, pp. 117–151 in *Stochastic Differential Systems*, H. J. Engelbert and W. Schmidt (eds.), Lecture Notes in Control and Information Sciences 96. Springer, Berlin 1987

[5] L. Arnold and W. Kliemann, On unique ergodicity for degenerate diffusions, *Stochastics* 21 (1987) 41–61

[6] L. Arnold, W. Kliemann, and E. Oeljeklaus, pp. 85–125 in *Lyapunov Exponents. Proceedings. Bremen 1984*, L. Arnold and V. Wihstutz (eds.), Lecture Notes in Mathematics 1186. Springer, Berlin 1986

[7] L. Arnold, E. Oeljeklaus, and E. Pardoux, Almost sure and moment stability for linear Itô equations, pp. 129–159 in *Lyapunov Exponents. Proceedings. Bremen 1984*, L. Arnold and V. Wihstutz (eds.), Lecture Notes in Mathematics 1186, Springer, Berlin 1986

[8] L. Arnold and V. Wihstutz, pp. 1–26 in *Lyapunov Exponents. Proceedings. Bremen 1984*, L. Arnold and V. Wihstutz (eds.), Lecture Notes in Mathematics 1186, Springer, Berlin 1986

[9] L. Arnold and Xu Kedai, Normal forms for random dynamical systems, Preprint, Bremen 1991

[10] P. Baxendale, Brownian motions in the diffeomorphism group I, *Compositio Mathematica* 53 (1984) 19–50

[11] P. Baxendale, Moment stability and large deviations for linear stochastic differential equations, pp. 31–54 in *Proceedings of the Taneguchi Symposium on Probabilistic Methods in Mathematics*, N. Ikeda (ed.), Katata & Kyoto 1985

[12] M. Bismut, A generalized formula of Itô and some other properties of stochastic flows, *Z. Wahrscheinlichkeitstheorie Verw. Geb.* 55 (1981) 331–350

[13] T. Bogenschütz, Entropy, pressure and a variational principle for random dynamical systems, Preprint, Bremen 1991

[14] P. Bougerol, Comparaison des exposants de Lyapounov des processus markoviens multiplicatifs, *Ann. Inst. Henri Poincaré* 24 (1988) 439–489

[15] P. Bougerol, Théorèmes limites pour les systèmes linéaires à coefficients markoviens, *Probab. Th. Rel. Fields* 78 (1988) 193–221

[16] P. Boxler, A stochastic version of center manifold theory, *Probab. Th. Rel. Fields 83* (1989) 509–545

[17] P. Boxler, A necessary condition for a stochastic bifurcation, Preprint, Bremen 1990

[18] R. Carmona and J. Lacroix, *Spectral Theory of Random Schrödinger Operators*, Birkhäuser, Boston 1990

[19] A. Carverhill, Flows of stochastic dynamical systems: ergodic theory, *Stochastics* 14 (1985) 273–317

[20] H. Crauel, Extremal exponents of random dynamical systems do not vanish, *J. Dynamics Differential Equations* 2 (1990) 245–291

[21] H. Crauel, Markov measures for random dynamical systems, Preprint, Bremen 1989

[22] H. Crauel, Non-Markovian invariant measures are hyperbolic, Preprint, Bremen 1990

[23] S. Dahlke, *Invariante Mannigfaltigkeiten für Produkte zufälliger Diffeomorphismen*, Dissertation, Universität Bremen 1989

[24] K. D. Elworthy, Stochastic dynamical systems and their flows, pp. 79–95 in *Stochastic Analysis*, A. Friedman and M. Pinsky (eds.). Academic Press 1978

[25] N. Ikeda and S. Watanabe, *Stochastic Differential Equations and Diffusion Processes.* North Holland – Kodansha, Tokyo 1981

[26] Yu. Kifer, *Ergodic Theory of Random Transformations*, Birkhäuser, Boston 1986

[27] Yu. Kifer, Characteristic exponents for random homeomorphisms of metric spaces, pp. 74–84 in *Lyapunov Exponents. Proceedings. Bremen 1984*, L. Arnold and V. Wihstutz (eds.), Lecture Notes in Mathematics 1186, Springer, Berlin 1986

[28] Yu. Kifer, A note on integrability of C^r-norms of stochastic flows and applications, pp. 125–131 in *Stochastic Mechanics and Stochastic Processes, Proceedings Swansea 1986*, Lecture Notes in Mathematics 1325, Springer, Berlin 1988

[29] J. F. C. Kingman, The ergodic theory of subadditive stochastic processes, *J. Roy. Statist. Soc. Ser. B* 30 (1968) 499–510

[30] W. Kliemann, Recurrence and invariant measures for degenerate diffusions, *Ann. Probab.* 15 (1987) 690–707

[31] W. Klingenberg, *Riemannian Geometry*, de Gruyter, Berlin 1982

[32] O. Knill, Positive Lyapunov exponents for a dense set of bounded measurable $\mathrm{Sl}(2,\mathbb{R})$ cocycles, Preprint, Zürich 1990

[33] H. Kunita, *Stochastic Differential Equations and Stochastic Flows of Diffeomorphisms*, pp. 143–303 in Ecole d'Eté de Probabilités de Saint-Flour 1982, Lecture Notes in Mathematics 1097. Springer, Berlin 1984

[34] H. Kunita, *Lectures on Stochastic Flows and Applications*, Tata Institute of Fundamental Research, Bombay. Springer, Berlin 1986

[35] H. Kunita, *Stochastic Flows and Stochastic Differential Equations*, Cambridge University Press 1990

[36] F. Ledrappier and P. Walters, A relativised variational principle for continuous transformations, *J. London Math. Soc. (2)*, 16 (1977) 568–576

[37] T. Ohno, Asymptotic behaviors of dynamical systems with random parameters, *Publ. RIMS, Kyoto Univ.* 19 (1983) 83–98

[38] V. I. Oseledec, A multiplicative ergodic theorem. Lyapunov characteristic numbers for dynamical systems. *Trans. Moscow Math. Soc.* 19 (1968) 197–231

[39] D. Ruelle, Ergodic theory of differentiable dynamical systems, *Publ. Math. I.H.E.S.* 50 (1979) 27–58

[40] A. D. Wentzell, *Theorie zufälliger Prozesse.* Akademie-Verlag, Berlin 1979

[41] Xu Kedai, *Normalformen für zufällige dynamische Systeme*, Dissertation, Universität Bremen 1990

Lyapunov exponents and asymptotic behaviour of the product of random matrices

I.Ya. Goldsheid*

Fakultät für Mathematik
Ruhr-Universtät-Bochum (FRG)
SFB-237 Bochum-Essen-Düsseldorf

The existence of different Lyapunov exponents is an important property of the product of random matrices. This fact is well known to specialists. We will explain it here once more and this will be done through discussing asymptotic properties of the products of random matrices. There are several reasons to do that. First of all we want to give a relatively complete picture concerning this subject. Secondly, we present here some new results. Finally, one of our goals is to justify the following conception: most of asymptotic properties of the products of random matrices basically rely on the existence of different Lyapunov exponents. All of that is the content of section 1.

In turn there arises a question about the description of Lyapunov exponents. The second part of the paper contains some algebraic results. We think that it is very important to realise that the algebraic language gives an adequate description of possible relations between Lyapunov exponents. Combination of algebraic and probabilistic techniques allows to obtain results which can be applied to some problems arising in mathematical physics. We do not discuss this subject here and refer to [GM2] and [KLS]. The description of Lyapunov exponents is given in section 3.

It is easy to see that formulations of main results, presented in sections 2 and 3 are different from the ones given in [GM2]. Still the main ideas – to "mix" and "how to mix" algebraic and probabilistic techniques – were explained already in [GM1], and the essence of our work is due to [GM2]. We hope that the form of explanation which is used here is simpler than in [GM2].

Many important ideas go back to works of Furstenberg, Tutubalin, Virtser, Guivarc'h, Raugi ([F1], [F2], [T], [V], [G], [GR1], [GR2]), but our list of references is far from being exhaustive. The interested reader should find further references in the reviews and books [BL], [GM2], [GR1], [L].

I am grateful to L.Arnold for his encouragement and patience. Kind hospitality of S.Albeverio and W.Kirsch made my stay at the Ruhr University, Bochum pleasant, interesting and useful. The financial support of SFB-237 program is acknowledged with thanks.

1.Asymptotic behaviour of the product of random matrices

1.1.Preliminaries

Let us introduce several notations and formulate some definitions and conditions which will be used in the sequel sometimes without reminding that they are explained in this section.

* On leave from Math. Institute, Academy of Sciences USSR, 450057 Ufa, USSR

First of all throughout the paper the notation $a := b$ or $b =: a$ means that a is set equal to b by definition.

Let $(\Omega, \mathcal{B}, P)$ be a probability space, where $\Omega = \{\omega\}$ is the space of elementary events, $\mathcal{B}$ is a σ-algebra of measurable subsets of Ω, and P is a probability measure on the measurable space $(\Omega, \mathcal{B})$.

Let T denote an ergodic transformation preserving measure P.

Let $A : \Omega \to GL(m, \mathbb{R})$ be a measurable map of the space Ω into the group of invertible m-dimensional matrices and

$$A_n \equiv A_n(\omega) := A(T^{n-1}\omega),$$

We will always suppose that

$$(C1) \qquad \int_\Omega \ln^+ \|A(\omega)\| \, dP(\omega) < \infty,$$

where $^+$ denotes the positive part of our function.

For some theorems we need stronger conditions:

$$(C2) \qquad \int_\Omega (|\ln \|A(\omega)\| \,| + |\ln \|A^{-1}(\omega)\| \,|) dP(\omega) < \infty,$$

$$(C3) \qquad \int_\Omega (\|A(\omega)\|^\delta + \|A^{-1}(\omega)\|^\delta) dP(\omega) < \infty, \text{for some } \delta > 0$$

The next condition describes the mixing properties of T. Let

$$\mathcal{B}_n^- := \sigma\text{-algebra generated by functions } \{A_i : i \leq n\}$$

$$\mathcal{B}_{n+r}^+ := \sigma\text{-algebra generated by functions } \{A_i : i \geq n + r\}$$

then uniformly in n, $B_1 \in \mathcal{B}_n^-$ and $B_2 \in \mathcal{B}_{n+r}^+$

$$(C4) \qquad |P(B_1 B_2) - P(B_1)P(B_2)| \to 0 \text{ if } r \to +\infty.$$

We set also

$$A(n) := A_n A_{n-1} ... A_1 \tag{1}$$

$$A(n, k) := A_n A_{n-1} ... A_{k+1}, \tag{2}$$

where $n > k$.

By $O(m)$ we denote the group of real orthogonal matrices of order m. We will use the following decomposition of $A(n)$:

$$A(n) = U(n)D(n)V(n), \tag{3}$$

where $U(n), V(n) \in O(m)$ and $D(n) := diag(d_1(n), \cdots, d_m(n))$ is a diagonal matrix with diagonal elements $0 < d_1(n) \leq d_2(n) \leq \cdots \leq d_m(n)$.

It is evident that the numbers $d_i(n)$ are the square roots of the eigenvalues of the matrix $A(n)A^*(n)$ and the columns of $U(n)$ coincide with orthonormalized system of eigenvectors of the same matrix: if u_i is the i-th column of $U(n)$ then

$$A(n)A^*(n)u_i(n) = d_i^2(n)u_i(n). \tag{4}$$

Similarly columns of $V^*(n)$ coincide with the eigenvectors of $A^*(n)A(n)$.

If $U(n)$ is chosen then $V(n)$ is defined in a unique way. But for the choice of $U(n)$ we still have some degrees of freedom, especially in the case when some of the numbers $d_i(n)$ coincide. The investigation of the asymptotic behaviour of $A(n)$ can be achieved by describing the behaviour of the components of the decomposition of (3). To avoid the arising non-uniqueness it is better to deal with the mentioned eigensubspaces (or vectors). Moreover it turns out that it is more convenient to use the language of the so called space of flags.

Definition. Let $\tau = (\tau_1, \ldots, \tau_s)$ be an s-dimensional vector with positive integer components and suppose that $1 \le \tau_1 < \cdots < \tau_s = m$. *A flag of type* τ in $\mathbb{R}^m$ is a sequence of subspaces $\mathcal{W} := \{\mathcal{W}_1, \mathcal{W}_2, \cdots, \mathcal{W}_s\}$ such that $\mathcal{W}_1 \subset \mathcal{W}_2 \subset \ldots \subset \mathcal{W}_s = \mathbb{R}^m$ and $dim\mathcal{W}_i = \tau_i$ for all i. The set of all flags of type τ constitutes the *space of flags* $\mathcal{F}_\tau$.

There are different possibilities to equip $\mathcal{F}_\tau$ with a metric. Some of these metrics are very useful (see the proof of the multiplicative ergodic theorem in [GM2]). Here we will bear in mind a very simple metric. There is a natural one-to-one correspondence between every $\mathcal{W}$ and a sequence of orthogonal projectors $\pi(\mathcal{W}) := \{\pi_1, \pi_2, \cdots, \pi_s\}$, where π_i is a projector onto $\mathcal{W}_i$. Now, if $\mathcal{W}^{(i)} = \{\mathcal{W}_1^{(i)}, \mathcal{W}_2^{(i)}, \cdots, \mathcal{W}_s^{(i)}\} \in \mathcal{F}_\tau$, $i = 1, 2,,$ then

$$\rho(\mathcal{W}^{(1)}, \mathcal{W}^{(2)}) := \sum_{i=1}^{s-1} \left\| \pi_i^{(1)} - \pi_i^{(2)} \right\| \tag{5}$$

Finally we want to explain some algebraic condition but at first it is necessary to give several definitions, connected with the notion of Zariski closure.

Definition. If $\mathcal{H}$ is a set in $\mathbb{R}^N$ then the smallest algebraic manifold $Z(\mathcal{H})$ containing $\mathcal{H}$ is called the *Zariski closure* of $\mathcal{H}$.

Definition. We say that $\mathcal{H}$ is *algebraically (or Zariski) dense* in $\mathcal{R}$ if $Z(\mathcal{H}) = \mathcal{R}$.

Usually we will deal with the case when $\mathcal{H}$ will be a group or a semi-group generated by matrices, belonging to some set $\mathcal{T}$. This set $\mathcal{T}$ in turn will be the support of a measure ν, defined by the formula

$$\nu(\Gamma) = P\{A(\omega) \in \Gamma\} \tag{6}$$

where Γ is a Borel set, $\Gamma \subset GL(m, \mathbb{R})$.

Let $G_\mathcal{T}$ denote the group generated by matrices belonging to some set $\mathcal{T}$, and let $G := Z(G_\mathcal{T})$ be the Zariski closure of $G_\mathcal{T}$. If $g \in G$ then we consider again the polar decomposition $g = U(g)D(g)V(g)$, where $U(g), V(g) \in G \cap O(m)$ and $D(g)$ is a diagonal matrix with diagonal elements $d_1(g), \ldots, d_m(g)$. Now our algebraic condition is the following one:

(C5)
For every $g \in G$ we can find a polar decomposition such that $U(g), V(g) \in G \cap O(m)$.

It is trivial that if G satisfies (C5) then we have also: $D(g) \in G$ for every $g \in G$.

Remark. If G is a semi-simple Lie group, then (C5) is satisfied up to a change of the scalar product in $\mathbb{R}^m$. Our considerations do not depend on the choice of the norm in $\mathbb{R}^m$, and this means that they are applicable to the case of a semi-simple group.

1.2. Multiplicative ergodic theorem (MET)

The MET was proved by Oseledets [O]. More recent proofs can be found, in [R1], [R2], [GM2]. Here we want to point out some moments which are characteristic for this theorem and which will be useful for us.

Our formulation of the MET will be slightly different from the traditional one (but of course equivalent to it). Namely the theorem will be divided in two parts: random and deterministic.

Theorem 1. *If the condition (C1) is satisfied then with probability 1 there exists*

$$\lim_{n\to\infty} n^{-1} \ln d_i(n) =: \lambda_i \tag{7}$$

Evidently

$$\lambda_1 \le \lambda_2 \le ... \le \lambda_m \tag{8}$$

and it may happen that some of these *Lyapunov exponents* are equal to $-\infty$.

Theorem 1 has different proofs. The simplest one relies on the subadditive ergodic theorem (c.f.[R1], [GM]).

For the second part of the MET we need one more property of the sequence of matrices A_n. Namely it is easy to show that under condition (C1) with probability 1

$$\lim_{n\to\infty} n^{-1} \ln \|A_n\| = 0 \tag{9}$$

Now the second part of MET can be formulated as a purely deterministic result.

Theorem 2. *Suppose that a sequence of matrices A_n is such that relations (7) and (9) are satisfied. Let us denote by $\alpha_1 < ... < \alpha_s$ all the different values of λ_i in (8), and let k_i be the multiplicity of α_i in the sequence (8). Then there exists a flag*

$$\mathcal{V} = (\mathcal{V}_1 \subset \mathcal{V}_2 \subset ... \subset \mathcal{V}_s)$$

of type $\tau = (\tau_1, ..., \tau_s)$ with $\tau_i = k_1 + ... + k_i$ such that if $x \in \mathcal{V}_i \setminus \mathcal{V}_{i-1}$ then

$$\lim_{n\to\infty} n^{-1} \ln \|A(n)x\| = \alpha_i \tag{10}$$

(we suppose here that $\mathcal{V}_0 := \{0\}$).

We want now to explain an estimate which is actually equivalent to Theorem 2. This estimate is a consequence of the fact that we really have different Lyapunov exponents.

Let $v_1(n), v_2(n), ..., v_m(n)$ be an orthonormal system of eigenvectors of the matrix $A^*(n)A(n)$ with corresponding eigenvalues $d_1^2(n) \le d_2^2(n) \le ... \le d_m^2(n)$. Let us consider a sequence of flags $\mathcal{V}(n) = (\mathcal{V}_1(n), ..., \mathcal{V}_s(n))$ where

$$\mathcal{V}_i(n) := \text{subspace generated by vectors } v_1(n), ..., v_{\tau_i}(n) \tag{11}$$

Let $\pi_i(n)$ be an orthogonal projector onto $\mathcal{V}_i(n)$. Then for any fixed $\varepsilon > 0$ there exists $N(\varepsilon)$ such that for all $n > N(\varepsilon)$, arbitrary positive integer k and $x \in \mathcal{V}_i(n)$ we have:

$$\|(\pi_{i+r+1}(n+k) - \pi_{i+r}(n+k))x\| \leq e^{-(\alpha_{i+r+1}-\alpha_i-\varepsilon)n} \|x\| \tag{12}$$

This estimate shows that the flags $\mathcal{V}(n)$ converge to the flag $\mathcal{V}$ as $n \to \infty$ with an exponential speed: for sufficiently big n

$$\|\pi_i - \pi_i(n)\| \leq e^{-(\alpha_{i+1}-\alpha_i-\varepsilon)n} \|x\| \tag{13}$$

where π_i is an orthogonal projector onto the component $\mathcal{V}_i$ of the flag $\mathcal{V}$.

We finish this section by pointing out the following essential moment: as we already said the estimates (12) and (13) rely on the fact that among Lyapunov exponents there are *different* ones. Theorem 2 is an easy *consequence* of (9). In the case when all the numbers λ_i are equal Theorem 2 is *trivial.*

Sometimes it is claimed that the matrices $V(n)$ in (3) can be chosen in such a way that they converge to some matrix V. That is true if all λ_i are different. Otherwise it is not the case and this is one more reason why the usage of the space of flags is more adequate to the situation and in fact the behaviour of $V(n)$ is described by the behaviour of $\mathcal{V}(n)$.

1.3. The central limit theorem (CLT) and the asymptotic behaviour of the product of random matrices

The CLT describes more precisely the asymptotic behaviour of the diagonal part of decomposition (3). Namely, the following result can be proved.

Theorem 3. *Suppose that condition (C3) is satisfied, all the λ_i are different and matrices A_n are independent. Then the sequence of matrices $D(n)$ has asymptotically log-normal distribution:*

$$\lim_{n\to\infty} P\{\xi_n \in \Gamma\} = \int_\Gamma f(x)dx$$

where

$$\xi_n := \frac{1}{\sqrt{n}}(\ln d_1(n) - n\lambda_1, \ldots, \ln d_m(n) - n\lambda_m), \tag{14}$$

Γ is an open set, $\Gamma \subset \mathbb{R}^m$ and $f(x)$ is the density of a Gaussian distribution.

It may happen (and it happens very often) that $f(x)$ is degenerate in the sense that it is concentrated on a subspace of $\mathbb{R}_m$. Now the natural question is: what is the dimension of this subspace. We are going to explain the answer to this question. To do that we have to use the notion of Zariski closure (see section 1.1).

We denote by T the support of the common distribution of matrices A_n, $A_n \in GL(m, \mathbb{R})$. Let G_T be the group generated by matrices belonging to T and $G := Z(G_T)$ be its Zariski closure.

Theorem 4. *Suppose that A_n are independent identically distributed random matrices satisfying condition (C3), and the group G satisfies condition (C5). Then (14) is true and the dimension of the distribution f coincides with the dimension of the Lie subgroup of diagonal matrices belonging to G.*

Remark that in Theorem 4 we do not suppose that all the Lyapunov exponents are different. The proof of this theorem is now being prepared for publication [GG].

In the case when $G = GL(m, \mathbb{R})$, the distribution of matrices A_n is absolutely continuous with respect to Haar measure and has a compact support, this result was proved by Tutubalin [T]. If G is a semi-simple Lie group and again the distribution of matrices A_n is absolutely continuous with respect to Haar measure and has a compact support this is a result of Virtser [V]. There is a strong difference between the technique which is used to prove Theorem 4 and the one used in [T], [V].

It is worth to mention that Theorem 4 relies on the fact that the Lyapunov exponents of the considered products of matrices are also described by the diagonal subgroup of $Z(G_T)$.

To formulate the next result we need one more notation. Let $A(n)$ be defined by (1), and $u_i(n)$, $1 \le i \le m$, be an orthonormal system of eigenvectors of the matrix $A(n)A^*(n)$ with corresponding eigenvalues $d_1^2(n) \le d_2^2(n) \le ... \le d_m^2(n)$. Denote by $\mathcal{U}(n)$ the sequence of flags $\mathcal{U}(n) := (\mathcal{U}_1(n), ..., \mathcal{U}_s(n))$ where

$$\mathcal{U}_i(n) := \text{subspace generated by vectors } u_1(n), ..., u_{\tau_i}(n)$$

with τ_i defined in Theorem 2.

Theorem 5. *Let $A_n(\omega) := A(T^{n-1}\omega)$ be a sequence of matrices from $GL(m, \mathbb{R})$. Suppose that:*

1) T is invertible and satisfies condition (C4).

2) Among the Lyapunov exponents of the sequence of matrices $A(n)$ there exist different ones: $\alpha_1 < ... < \alpha_s$. The multiplicity of α_i is k_i and $\tau_i = k_1 + ... + k_i$.

3) The diagonal part of decomposition (3) is asymptotically log-normal distributed. Then:

I. *The sequence of flags $\mathcal{V}(n) \in \mathcal{F}_\tau$ converges with probability 1. We denote by $P_1(\cdot)$ its limiting distribution.*

II. *The sequence of flags $\mathcal{U}(n) \in \mathcal{F}_\tau$ converges in law to some distribution $P_2(\cdot)$.*

III. *If the distributions $P_1(\cdot)$ and $P_2(\cdot)$ are continuous (i.e. do not have atoms) then the three random variables $\mathcal{U}(n), \xi_n, \mathcal{V}(n)$ are asymptotically independent:*

$$\lim_{n\to\infty} P\{\mathcal{U}(n) \in \Gamma_1, \xi_n \in \Gamma, \mathcal{V}(n) \in \Gamma_2\} = P_1(\Gamma_1)P_\xi(\Gamma)P_2(\Gamma_2)$$

where $\Gamma_1, \Gamma, \Gamma_2$ are open sets, $\Gamma_1, \Gamma_2 \in \mathcal{F}_\tau$, $\Gamma \subset \mathbb{R}^m$, $P_\xi(\cdot)$ is a Gaussian distribution in $\mathbb{R}^m$.

IV. *If only conditions 1) and 2) are satisfied and $P_1(\cdot)$ and $P_2(\cdot)$ are continuous then $\mathcal{U}(n)$ and $\mathcal{V}(n)$ are asymptotically independent.*

Proof. I is a trivial consequence of estimate (13).

To prove II,III and IV we define at first one more sequence of flags. Namely, let us consider a the sequence of matrices $A(n, n-k)$, defined by formula (2). Now a flag $\mathcal{U}(n, n-k)$ is defined in the same way as $\mathcal{U}(n)$ with the only difference that instead of $A(n)$ we have to use $A(n, n-k)$. We need the following

Lemma 6. *For any fixed $\varepsilon > 0$ there exists $N(\varepsilon)$ such that for all $n > k > N(\varepsilon)$,*

$$P\{\rho(\mathcal{V}(n), \mathcal{V}(k)) < \varepsilon\} > 1 - \varepsilon, \qquad (15)$$

$$P\{\rho(\mathcal{U}(n), \mathcal{U}(n, n-k)) < \varepsilon\} > 1 - \varepsilon. \qquad (16)$$

Proof. (15) is an immediate consequence from (13).

To prove (16) let us remark that $A(n, n-k)$ and $A(n)$ have the same joint distribution as $A(0,-k) = A_0 \cdot \ldots \cdot A_{-k+1}$ and $A(0,-n) = A_0 \cdot \ldots \cdot A_{-n+1}$, because T is a measure preserving transformation (remember that $A_{-s}(\omega) := A(T^{-s-1}\omega)$). Thus

$$P\{\rho(\mathcal{U}(n), \mathcal{U}(n, n-k)) < \varepsilon\} = P\{\rho(\mathcal{U}(0,n), \mathcal{U}(0,k)) < \varepsilon\} \tag{17}$$

Now the same reason which makes the sequence $\mathcal{V}(n)$ tend to the limit provides the proof that the sequence $\mathcal{U}(0,n)$ has a limit with probability 1. This together with (17) proves (16).

Let $\xi_{n-2k,2k}$ and $d_i(n-2k, 2k)$ be defined as a function of $A(n-2k, 2k)$ in the same way as $\xi_n, d_i(n)$ are defined by the matrix $A(n)$ (see section 1.1).

Note now that if $k \to \infty$ and $n \to \infty$ in such a way that $k/\sqrt{n} \to 0$ then

$$\xi_n - \xi_{n-2k,2k} \to 0 \text{ as } n, k \to \infty \tag{18}$$

with probability 1. This follows immediately from an evident estimate:

$$\left\|A^{-1}(n, n-2k)\right\|^{-1} d_i(n-2k, 2k) \le d_i(n) \le d_i(n-2k, 2k)\, \|A(n, n-2k)\| \,.$$

(We use here the fact that $\frac{1}{\sqrt{n}} \ln \|A(n, n-2k)\| \to 0$ as $n \to \infty$.)

II is now a simple consequence of the relation

$$P\{\mathcal{U}(n) \in \Gamma_1\} = P\{\mathcal{U}(0,-n) \in \Gamma_1\}.$$

and the remark about the convergence of the sequence $\mathcal{U}(0,-n)$ with probability 1 (which was made in the proof of Lemma 6).

We are now prepared to prove III. It follows from (18) and Lemma 6 that

$$P\{\mathcal{U}(n) \in \Gamma_1, \xi_n \in \Gamma, \mathcal{V}(n) \in \Gamma_2\} -$$

$$P\{\mathcal{U}(n, n-k) \in \Gamma_1, \xi_{n-2k,2k} \in \Gamma, \mathcal{V}(k) \in \Gamma_2\} \to 0 \text{ as } n \to \infty, \tag{19}$$

(we use here the continuity properties of $P_1(\cdot)$ and $P_2(\cdot)$ and we suppose of course that $k/\sqrt{n} \to 0$).

But from the definition of $\mathcal{U}(n, n-k), \xi_{n+2k,2k}$ and $\mathcal{V}(k)$ it follows that the "distance" between σ-algebras on which these functions depend tends to ∞, because $k \to \infty$. Thus we can use property (C4) which actually says in our case that these three quantities are asymptotically independent. The last together with (19) finishes the proof of III.

IV is proved in the same but simpler way.

Theorem 5 is proved.

2. Algebraic results

In this section we explain the algebraic ideas which are used in further considerations. The results presented here enable us to consider the Zariski closure of a group generated by the support of a measure instead of a semi-group generated by the same set. This, of course, is very important because usually it is very hard to understand how a semi-group "looks". At the same time it is relatively easy to calculate the Zariski closure of a group. The algebraic approach gives us also a possibility to describe in many cases all natural relations between the Lyapunov exponents.

2.1 Quasi-projective transformations (QPT). List of results

The notion of QPT was introduced by H.Furstenberg [F1] for the case of projective space. The proofs of the results we list here can be found in [GM2].

Let $\mathcal{P}(\mathcal{L})$ be the projective space of the linear space $\mathcal{L}$ and $\mathcal{P} := \mathcal{P}(\mathbb{R}^m)$. By PGL($\mathcal{L}$) we denote a group of projective transformations acting on $\mathcal{P}(\mathcal{L})$.

If B is a linear operator with a kernel $kerB$, then $B(x)$ is defined in a natural way for every $x \in \mathcal{P}(\mathcal{L})\backslash\mathcal{P}(kerB)$.

Definition. A transformation b of a projective space $\mathcal{P}$ is called a QPT if there exists a sequence of non-degenerate matrices B_n such that for any $x \in \mathcal{P}$

$$\lim_{n\to\infty} B_n(x) = b(x).$$

Remark. Evidently $b(x)$ is a QPT but not a projective transformation iff

$$\lim_{n\to\infty} \|B_n\|\,|detB_n|^{-1/m} = \infty$$

We start with a study of the structure of QPTs. This also will give us the possibility to investigate the connection between the algebraic properties of groups and the properties of growth of random products of their elements.

1) From any sequence of projective transformations it is possible to extract a subsequence which converges to some QPT.

2) Any QPT b can be described in the following way: there exists a sequence of linear subspaces L_i and linear operators b_i with domains L_i, $b_i : L_i \to \mathbb{R}^m$ such that

$$\mathbb{R}^m = L_0 \supset L_1 \supset \ldots \supset L_k,\ dimL_i > dimL_{i+1} \tag{20}$$

$$kerb_i = L_{i+1}, \text{for } i \leq k-1, \text{and } kerb_k = 0 \tag{21}$$

And finally for $\bar{x} \in L_i\backslash L_{i+1}$ we have

$$b(x) = b_i(x) \tag{22}$$

where x is a line containing $\bar{x}$.

Vice versa if the operators b_i and spaces L_i are given and conditions (20), (21) are satisfied, then we can find a QPT with b defined by (22). Probably it is useful to bear in mind that the QPT b itself does not define the subspaces L_i and the operators b_i in a unique way.

3) The set of all QPT is a semi-group.

A stronger assertion is given by the following property.

4) Let H be a semi-group, $H \subset PGL(\mathbb{R}^m)$. By $\bar{H}$ we will denote the set of QPT which can be represented in the form

$$b = \lim_{n\to\infty} B_n,$$

where $B_n \in H$. Then $\bar{H}$ is a semi-group.

5) We denote by $M_0(b)$ the image of the points of continuity of QPT b and by $M_1(b)$, the closure of the set of discontinuity of b. If $M_0(b)$ consists of more than one point then $M_1(b) = \mathcal{P}(L_1(b))$ and $M_0(b) = \mathcal{P}(Imb_0)$. If $M_0(b)$ consists of one point then either $M_1(b) = \mathcal{P}(L_i(b))$ for some $i \geq 1$ or $M_1(b)$ is an empty set.

6) If $M_0(b) \cap M_1(b) = \emptyset$ then the restriction of b to $M_0(b)$ is the linear diffeomorphism defined by the restriction of b_0 to $M_0(b)$. The linear subspace in $\mathbb{R}^m$ which is formed by the straight lines contained in $M_0(b)$ is an invariant subspace for b_0 and the restriction of b_0 to this subspace is an invertible operator.

7) Suppose that $M_0(b) \cap M_1(b) \neq \emptyset$ and $M_0(b) \not\subset M_1(b)$ then $dimM_0(b^2) < dimM_0(b)$.

2.2. QPT on the spaces Gr_k and $\mathcal{F}$

The group G acts by linear transformation on the Grassman manifold of k-dimensional subspaces of $\mathbb{R}^m$. Let us denote these manifolds by Gr_k and let $\mathcal{P}_k := \mathcal{P}(\wedge^k \mathbb{R}^m)$, then Gr_k is embedded in $\mathcal{P}_k$. This allows immediately not only to introduce the notion of QPT for the space Gr_k but also to prove the analogues of properties 1) – 6) for these transformations. The situation with property 7) is more complicated if we want to consider the QPT generated by actions of our group G. It is worth to mention that if we consider the group of all linear transformations of $\wedge^k \mathbb{R}^m$ then nothing special can be said about the corresponding QPT (e.g. properties 1) – 7) hold for QPT of $\mathcal{P}_k$). But we are restricted to our group G and this leads to some special and important features of QPT acting on $\mathcal{P}_k$. We omit the further consideration of QPT acting on $\mathcal{P}_k$ because formally saying we will not use them here. More details can be found in [GM2]. The very special case which is of the most importance for us is the one when we deal with the space of flags. We need the properties of $M_0(b)$ and $M_1(b)$ for this case. For mere analogues of 1) – 6) one can again use the fact, that in a natural sense

$$\mathcal{F} \subset \oplus_{k=1}^m Gr_k.$$

But the following property is of a special character:

8) If $M_0(b) \cap M_1(b) \neq \emptyset$ then $M_0(b) \subset M_1(b)$.

We want to emphasize here that if the group G does not coincide with $GL(m, \mathbb{R})$, but is a nontrivial algebraic subgroup satisfying condition ($C5$) (e.g. is semi-simple), then to obtain the analogue of 8) we have to consider an algebraic submanifold of $\mathcal{F}$. This manifold can be described in different ways. The way we do it here is not the universal one, but is applicable to many important cases, and we hope that it is the shortest and the simplest one for readers which are not specialists in group theory. Namely let $g \in G$ and $g = UDV$ with U, D, V described in ($C5$), then for every orthogonal matrix $V(g) = (v_1, v_2, \ldots, v_m)$ with rows v_i we construct the flag

$$\mathcal{V}(g) = (\mathcal{V}_1, \mathcal{V}_2, \ldots, \mathcal{V}_m)$$

where $\mathcal{V}_k$ is the k-dimensional plane in $\mathbb{R}^m$ generated by vectors $v_1, v_2, \ldots, v_k$. All these flags constitute the mentioned manifold, which we will denote by $\mathcal{F}_G$. (Usually $\mathcal{F}_G$ is equivalent to an algebraic submanifold of some $\mathcal{F}_r$.) After this remark the proof of 8) which is given in [GM2], can be applied also to the general case.

We will finish this list of properties with the following one:

9) From any sequence of QPT b_n we can choose a subsequence that converges to some QPT c at all points $x \in \mathcal{P}$. Let H be a subset of $PGL(\mathbb{R}^m)$, and $\bar{H}$ the set of QPT that can be represented as $b = \lim_{n\to\infty} h_n, h_n \in H$. If $b_n \in \bar{H}$ and $c = \lim_{n\to\infty} b_n$, then $c \in \bar{H}$.

2.3. Main algebraic result

Definition. Let (X, ρ) be a compact metric space and H be a semi-group of continuous mappings of X into itself. The action of H onto X is called *proximal* if for any $x, y \in X$ there exists a sequence $h_n \in H$ such that

$$\lim_{n\to\infty} \rho(h_n x, h_n y) = 0.$$

Theorem 7. *Let G be a group satisfying condition (C5) and $H \subset G$ is a semi-group which is algebraically dense in G. If the action of G on the space of flags $\mathcal{F}_G$ is proximal then the action of H on $\mathcal{F}_G$ is also proximal.*

The proof of this theorem will not be given here for the general case. We restrict ourself to the case $G = GL(m, \mathbb{R})$. In this case $\mathcal{F}_G$ coincides with $\mathcal{F}$.

2.4. Proof of Theorem 7

The main ideas of the proof of this result are contained in [GM1] and explained in details in [GM2]. We follow the scheme which can be found in these papers. The next three lemmas enable us to use some simple elements of the theory of algebraic groups in our construction.

Lemma 8. [VO]. *The Zariski closure of a semi-group is a group.*

Lemma 9. [VO]. *A compact subgroup of a real algebraic group is an algebraic group.*

Lemma 10. [GM2]. *A compact sub-semigroup of a topological group is a group.*

First of all we have to mention that as far as we consider the case when $G = GL(m, \mathbb{R})$, condition (C5) is automatically satisfied.

Suppose now that the statement of Theorem 7 is not true. This, evidently, means that there exist two points $x, y \in \mathcal{F}$ such that whatever sequence $h_n \in H$ we take we will have:

$$\liminf_{n\to\infty} \rho(h_n x, h_n y) \geq c > 0. \tag{23}$$

for some constant c, which depends of course on x and y.

Let $\bar{H}$ be a set of QPT described in 4). Then (23) contradicts the following

Statement 11. *$\bar{H}$ contains a QPT b such that $M_0(b)$ consists of one point.*

Suppose first that this statement is true and denote $M_0(b)$ by x_b. By definition

$$b = \lim_{n\to\infty} b_n, \text{where } b_n \in H.$$

If $x, y \notin M_1(b)$ then $b(x) = b(y)$, or equivalently

$$\lim_{n\to\infty} b_n(x) = \lim_{n\to\infty} b_n(y) = x_b. \tag{24}$$

which is impossible if (23) is true.

If $b(x) \neq b(y)$ then at least one of these two points belongs to an algebraic manifold $M_1(b)$.

Lemma 12. *If M is a nontrivial algebraic submanifold in $\mathcal{F}$ and $x, y \in \mathcal{T}$, then there exists an element $h \in H$ such that $h(x) \notin M$ and $h(y) \notin M$.*

Proof. The condition "$h(x)$ or $h(y) \in M$ for every $h \in H$" is an algebraic one. If the statement of our lemma is not true then H satisfies this condition. But then $Z(H)$ also satisfies this condition. The last contradicts the condition of Theorem 7 saying that $Z(H) = G$, because it is evident that G contains an element g such that $g(x) \notin M$ and $g(y) \notin M$. Lemma 12 is proved.

As far as $M_1(b)$ is an algebraic manifold, we can find $h \in H$ such that $h(x)$ and $h(y) \notin M_1(b)$, but then

$$(bh)(x) = (bh)(y)$$

which also contradicts (23).

Now to finish the proof of our theorem we have now to prove Statement 11. Let us suppose that it is not true. We select from $\bar{H}$ an element b with the smallest value of $dim M_0(b)$, $dim M_0(b) \geq 1$.

We can suppose that b is chosen so that

$$M_0(b) \cap M_1(b) = \emptyset. \tag{25}$$

If this is not the case, then we fix $x \in M_0(b)$ and consider $h \in H$ such that $h(x) \notin M_1(b)$ (we use here Lemma 12). Then by property 8) we have (25). Thus replacing b by hb if necessary we suppose from now on that (8) holds. (We note that $M_1(hb) = M_1(b)$ and $M_0(bh) = M_0(b)$ for all $h \in G$).

Let $\mathcal{L}$ be the smallest linear subspace such that $M_0(b)$ is embedded into $\mathcal{P}(\mathcal{L})$. Let $G(b)$ denote the set of those $h \in G$ for which $M_0(bh) \cap M_1(bh) = \emptyset$. (It is evident that $G(b)$ is an open and algebraically dense subset of G). We denote by Φ the semi-group generated by the element of the form bh, $h \in G(b)$. Property 6) allows us to assign to each $\varphi \in \Phi$ a uniquely defined element $\beta(\varphi)$ of PGL($\mathcal{L}$) such that the action of $\beta(\varphi)$ on $M_0(b)$ coincides with the action of φ on $M_0(b)$. It is clear that β is a homomorphism of Φ into the group PGL($\mathcal{L}$).

Let Φ_0 denote the semi-group generated by the elements of the form bh, $h \in H \cap G(b)$. Let us show that the semi-group $\beta(\Phi_0)$ is relatively compact in PGL($\mathcal{L}$). Let φ_n be a sequence of elements from Φ_0. By property 9) we can suppose that $\lim_{n\to\infty} \varphi_n(x) = \varphi(x)$ for all $x \in \mathcal{F}$, where $\varphi \in \bar{H}$. We can also suppose that $M_0(\varphi) \cap M_1(\varphi) = \emptyset$, because if this is not the case, then we can replace φ by φh and φ_n by $\varphi_n h$ so that $M_0(\varphi h) \cap M_1(\varphi h) = \emptyset$. (Here we use again property 8), but this time we shift $M_1(\varphi)$ while $M_0(\varphi) = M_0(\varphi h)$). Now $M_0(\varphi) \subset M_0(b)$ because $M_0(\varphi_n) = M_0(b)$. However by the assumption that $dim M_0(b)$ is minimal we have: $dim M_0(\varphi) \geq dim M_0(b)$ and thus $M_0(\varphi) = M_0(b)$. Hence, going over to the limit with respect to the sequence $\beta(\varphi_n)$ preserves the dimension of $M_0(b)$, which in turn is equivalent to relative compactness of this sequence in PGL($\mathcal{L}$).

According to Lemmas 9 and 10 the elements $\beta(bh)$ are contained in a compact algebraic subgroup of PGL($\mathcal{L}$) if $h \in H \cap G(b)$. Since the map $h \to \beta(bh)$, $h \in G(b)$, is a rational map of $G(b)$ into PGL($\mathcal{L}$), and the semi-group H is Zariski dense in G, the elements $\beta(bh)$ belong to a compact subgroup of PGL($\mathcal{L}$) for $h \in G(b)$. However, it is easy to select from $G(b)$ a sequence of matrices h_n such that if $b' := \lim_{n\to\infty} h_n$, then

$dimM_0(bb') < dimM_0(b)$ (here we use the assumption that $dimM_0(b) \geq 1$). The latter fact contradicts the relative compactness of the sequence $\beta(bh_n)$ in PGL($\mathcal{L}$).

This finishes the proof of Statement 11 and also of Theorem 7.

3. Relations between Lyapunov exponents. Simplicity conditions

We discuss in this section the fundamental theorem on Lyapunov exponents. In section 3.1 we give formulation of the results. In section 3.2 we present a "mixture" of algebraic and probabilistic ideas which lead to the proof of the main theorem on Lyapunov exponents.

Many proofs are omitted here. We aim to expose the ideas. Details and some other results can be found in [GM2], [GR1], [GR2].

3.1 Formulation of results

To explain the theorem about the relations between Lyapunov exponents we need one simple but very useful example of a Zariski closure. Namely let D be a diagonal matrix with positive diagonal elements. We want to describe the Zariski closure of the group $G_D := \{D^n, n = 0, \pm 1, \pm 2, \cdots\}$.

It turns out that $Z(G_D)$ is characterized by a sequence of k integer numbers $I := (i_1, \ldots, i_k), 1 \leq i_1 < \ldots < i_k \leq m$ and a sequence of $m - k$ vectors $(a_{j1}, \ldots, a_{jk})$ with rational coordinates, where $j \in (1, 2, \ldots, m)\backslash I$. Now $Z(G_D)$ consists of all diagonal matrices g such that the elements $d_i(g), \ldots, d_{i_k}(g)$ can take arbitrary positive values and

$$d_{i_1}(g) = d_{i_1}^{a_{j1}}(g) d_{i_2}^{a_{j2}}(g) \cdot \ldots \cdot d_{i_k}^{a_{jk}}(g). \tag{26}$$

It is easy to see that every Zariski closed diagonal subgroup DC, consisting of matrices with positive diagonal elements, is generated by some element D, i.e. $\mathcal{D} = Z(G_D)$.

Let $A_n(\omega), n = 1, 2, \ldots$, be a stationary sequence of matrices with distribution ν. As usually $\mathcal{T}$ is the support of the measure ν in $GL(m, \mathbb{R})$ and $G = Z(G_\mathcal{T})$.

Suppose that conditions (C2) and (C5) are satisfied. Finally let k be the dimension of the submanifold of diagonal matrices $\mathcal{D} \subset G$. Then we can choose exactly k Lyapunov exponents such that all the other Lyapunov exponents are linear combinations of the chosen ones with rational coefficients:

$$\lambda_j = \sum_{r=1}^{k} a_{jr} \lambda_{i_r}. \tag{27}$$

Relations (27) are elementary consequences of the MET and (26) (in contrast with relation (8) we do not suppose that λ_i form a growing sequence).

The following question arises immediately: are the relations (27) the only possible ones? We do not know the answer to this question if the sequence $A_n(\omega)$ is just stationary. But if the matrices $A_n(\cdot)$ are in addition independent much more can be said. Namely, the corresponding result can be formulated in the following way (see for [GM2]):

Theorem 13. *If conditions (C2) and (C5) are satisfied and A_n are independent identically distributed matrices, then the vector $(\lambda_1, \lambda_2, \ldots, \lambda_m)$ of Lyapunov exponents is an internal point of the Weyl chamber of the group G.*

Moreover, the following result is true:

Theorem 14. *Relations (27) are really the only possible ones in the sense that given a vector $\lambda = (\lambda_1, \lambda_2, \ldots, \lambda_m)$ which is an internal point of the Weyl chamber of the group G (described above) we can find a distribution ν_ε such that:*

I. *The group generated by the support of the distribution ν_ε is Zariski dense in G:*

II. $\|\lambda - \lambda_\varepsilon\| < \varepsilon$, *where λ_ε is the vector of Lyapunov exponents of the product of independent matrices with distribution ν_ε.*

From now on we consider the case when $G = GL(m, \mathbb{R})$. Then Theorem 13 can be read as follows:

Theorem 15. *If A_n are independent and distributed according to ν, condition (C2) is satisfied and $G = GL(m, \mathbb{R})$, then all the Lyapunov exponents of the sequence A_n are different:*

$$\lambda_1 < \lambda_2 < \ldots < \lambda_m.$$

(λ_i do not depend on ω with probability 1).

3.2. Main ideas

Throughout this section we suppose that the conditions of Theorem 15 are satisfied.

Our aim is to explain very briefly the main ideas of the proof of this theorem. We start with the probabilistic part.

By $\mathcal{F}$ we denote the space of flags of type $\tau = (1, 2, \ldots, m)$.

Definition. A measure μ on the space of flags $\mathcal{F}$ is called *stationary* (or *invariant*) with respect to $A_n(\cdot)$ if for any Borel set $\Gamma, \Gamma \subset \mathcal{F}$,

$$\int_{\mathcal{F}} d\mu(x) P\{A_n(\omega)(x) \in \Gamma\} = \mu(\Gamma).$$

It is evident that μ is an invariant measure of a Markov chain on the direct product $\mathcal{F} \times G$. The corresponding random walk is given by

$$(W_n, A_n) \to (A_n(W_n), A_{n+1}),$$

where $W_n \in \mathcal{F}, A_n$ is our sequence of matrices. The measure μ has the following properties.

1) μ exists and is unique.

2) If M is a nontrivial algebraic submanifold in $\mathcal{F}$, then $\mu(M) = 0$.

3) Let $\eta := \sum_{k=1}^{\infty} 2^{-k} \nu^k$, where ν^k is the distribution of the product $A_k A_{k-1} \ldots A_1$. Then there exists a measure μ_ω such that with probability 1

$$\mu_\omega = \lim_{n \to \infty} A_1(\omega) A_2(\omega) \ldots A_n(\omega) g \mu,$$

where the convergence is understood in the weak sense and holds for almost all $g \in G$ with respect to the measure η. Moreover for any continuous function $f : \mathcal{F} \to \mathbb{R}$

$$\int_{\mathcal{F}} f(x) d\mu(x) = E \int_{\mathcal{F}} f(x) d\mu_\omega(x).$$

(We emphasize that μ_ω does not depend on g.)

Now our main algebraic result, Theorem 7, enters the game. Together with 2) and 3) it leads to

Lemma 16. *The sequence of measures $A(n,\omega)\mu$ converges weakly with probability 1 to a Dirac measure, that is, a measure concentrated at one point.*

It is easy to show using Lemma 16 together with property 2) that with probability 1

$$d_{m-i+1}^{-1}(n,\omega)d_{m-i}(n,\omega) \to \infty \text{ as } n \to \infty, \tag{28}$$

where $d_i(n,\omega)$ is defined by (3).

After that we have to turn again to our Markov chain and show that

$$\ln d_{m-i+1}(n,\omega)d_{m-i}^{-1}(n,\omega) = c(n,\omega) + \sum_{s=1}^{n} F_i(\mathcal{W}_s, A_s). \tag{29}$$

Here $c(n,\omega)$ is some function, which with probability 1 has a finite limit as $n \to \infty$. $F_i(\cdot,\cdot)$ is a functional, responsible for the behaviour of the left hand side of (28), and $F_i(\mathcal{W}, A)$ is defined as follows. Let $\mathcal{W} = (\mathcal{W}_1, \dots, \mathcal{W}_m) \in \mathcal{F}$. Consider the components $\mathcal{W}_{i-1}$, $\mathcal{W}_i$ and $\mathcal{W}_{i+1}$ of $\mathcal{W}$ and let $f_1, f_2, \dots, f_{i-1}$ be an orthonormal basis in $\mathcal{W}_{i-1}$, $f_1, f_2, \dots, f_i$ be an orthonormal basis in $\mathcal{W}_i$, $f_1, f_2, \dots, f_{i+1}$ be an orthonormal basis in $\mathcal{W}_{i+1}$. Then

$$F_i(\mathcal{W}, A) = \ln(det((Af_s Af_j))_{s,j=1}^{i} det^{-1/2}((Af_s Af_j))_{s,j=1}^{i+1} det^{-1/2}((Af_s Af_j))_{s,j=1}^{i-1})$$

(we put $det((Af_s Af_j))_{s,j=1}^{m+1} = 1$ and $det((Af_s Af_j))_{s,j=1}^{0} = 1$). If we consider now $\mathcal{W}_1$ as a random variable distributed according to the invariant measure of our chain then the right hand side of (29) can be considered like a sum of stationary random variables. This sum tends to $+\infty$, because of (28).Finally the result follows from

Lemma 17. *Let T be a measure P-preserving transformation of the space Ω, let $F : \Omega \to I\!R$ be a function on Ω such that $F^+ \in L_1(\Omega)$, and suppose that $\sum_{k=0}^{n} F(T^k\omega) \to +\infty$ as $n \to \infty$ with probability 1. Then*

$$\int_\Omega F(\omega)dP(\omega) > 0.$$

Really, from this lemma it follows that

$$\lim_{n\to\infty} n^{-1} \ln d_{m-i+1}(n,\omega)d_{m-i}^{-1}(n,\omega) > 0.$$

or, equivalently,

$$\lambda_{m-i+1} - \lambda_{m-i} > 0.$$

Clearly, what was explained in this section is just a scheme of a proof. Detailed explanations can be found in [GM2], and altogether they are rather long.

References

[BL]. Bougerol P., Lacroix J. Products of Random Matrices with Applications to Schrödinger Operators. - Boston-Basel-Stuttgart, 1985.

[F1]. Furstenberg H. Boundary theory and stochastic processes on homogenetic spaces. -Proc. Symp. Pure Math. -1972.- P.193-229.

[F2]. Furstenberg H. Non commuting random products. T.A.M.S., vol.108. -P.377-428.

[GG]. Goldsheid I.Ya., Guivarc'h Y. in preparation.

[GM1]. Goldsheid I.Ya., G.A. Margulis G.A. Conditions of simplicity of the spectrum of Lyapunov indices, Dokl. Akad. Nauk SSSR 293 (1987), 297-301. = Soviet Math. Dokl. 35 (1987), 309-313.

[GM2]. Goldsheid I.Ya., G.A. Margulis G.A. Lyapunov indices of products of random matrices. Uspekhi Mat. Nauk 44:5 (1989), 13-60 = Russian Math. Surveys 44:5 (1989), 11-71.

[G]. Guivarc'h Y. Quelques proprietes asymptotiques des produits de matrices alea - toires. Springer Lecture Notes in Math. - 1980. -774. -P.176-250.

[GR1]. Guivarc'h Y., Raugi A. Products of random matrices: convergence theorems. Contemporary Math. - 1986.- Vol.50.- P.31-54.

[GR2]. Guivarc'h Y., Raugi A. Propriete de contraction d'un semigroupe de matrices inversibles. Coefficients de Liapunoff d'un produit de matrices aleatoires independantes. Preprint. - 1987.

[KLS]. Klein A., Lacroix J., Speis A. Localization for the Anderson model on a strip with singular potentials. K. Functional Anal. (to appear) 1989.

[L]. Ledrapier F. Quelques proprétés des exposants caracté ristiques. Ecole d'Eté de Probabilites de Saint Flour. Lect. Notes in. Math. 1097.

[O]. Oseledets V.I. A multiplicative ergodic theorem, Characteristic Lyapunov exponents of dynamical systems, Trudy Moskov. Mat. Obshch. 19 (1968), 179-210. = Trans. Moscow Math. soc. 19 (1968), 197-231.

[R1]. Ruelle D. Ergodic theory of differentiable dynamical systems. Publ. Math. Paris. - 1979.- Vol.50.- P.27-58.

[R2]. Ruelle D. Characteristic exponents and invariant manifolds in Hilbert space. Ann. Math. - 1982.- Ser.2, v.115:2. - P.243-290.

[T]. Tutubalin V.N. Limit theorems for a product of random matrices, Teor. Veroyatnost. i Primenen. 10 (1965), 19-32. = Theory Probab. Appl. 10 (1965), 15-27.

[VO]. Vinberg E.B., Onishchik A.L. Seminar on algebraic groups and Lie groups (Russian), Moscow State University, Moscow 1969.

[V]. Virtser A.D. Central limiting theorem for semisimple Lie groups. Theory Probab. Appl. 15 (1970), 667-687.

Lyapunov Exponents of Random Dynamical Systems on Grassmannians

Hans Crauel*
Fachbereich 9 MATHEMATIK
Universität des Saarlandes
6600 Saarbrücken 11
Federal Republic of Germany

Abstract

We investigate the relations between the Lyapunov exponents of a linear random dynamical system and the Lyapunov exponents of the (nonlinear) systems generated by the linear one on the projective space as well as on the higher Graßmann manifolds.

1 Introduction

Let Φ be a linear random dynamical system on $\mathbb{R}^d$ — for instance, Φ could be a product of random matrices $\{A_n\}$ in stationary dependence,

$$\Phi(n,\omega) = A_n(\omega) \circ A_{n-1}(\omega) \circ \ldots \circ A_1(\omega)$$

for $n \geq 1$, and $\Phi(0) = \mathrm{id}$ almost surely. (This is the discrete time case.)

In case of continuous time, Φ could be the fundamental matrix of a random differential equation

$$\dot{x} = A(\xi_t)x, \qquad t \in \mathbb{R},$$

where $\{\xi_t\}$ is a stationary ergodic process, and $A(\cdot)$ is a matrix valued function. Also, linear stochastic differential equations

$$dx = A_0 x\, dt + \sum_{i=1}^{n} A_i x\, dW_t^i,$$

where $W_t = (W_t^1, \ldots, W_t^n)$ is a Wiener process in $\mathbb{R}^n$, and $(A_i)_{0\leq i\leq n}$ are $(d \times d)$-matrices, induce an important class of linear random dynamical systems.

*Partially supported by Volkswagen-Stiftung. Part of this work was done while the author was with the Institut für Dynamische Systeme, Universität Bremen.

Associated with a linear random dynamical system (abbreviated RDS) there are Lyapunov exponents $\lambda_1(\Phi) \geq \lambda_2(\Phi) \geq \ldots \geq \lambda_d(\Phi)$, which may be characterized via the exponential growth rates of k-dimensional volumes, $1 \leq k \leq d$.

If Φ takes values only in the general linear group $\mathrm{Gl}(d, \mathbb{R})$, it induces nonlinear RDS φ_k on the *Graßmann manifolds* $\mathrm{Gr}(k, d)$, the manifolds of k-dimensional subspaces of $\mathbb{R}^d$, by assigning to a k-dimensional subspace $K \subset \mathbb{R}^d$ the subspace $\Phi(K)$. In particular, for $k = 1$, this renders the projective space $\mathbb{P}^{d-1} = \mathrm{Gr}(1, d)$.

Suppose φ is an arbitrary (smooth) RDS on a d-dimensional Riemannian manifold, and let μ be an invariant measure for φ. Then there are Lyapunov exponents $\lambda_1(\varphi, \mu) \geq \lambda_2(\varphi, \mu) \geq \ldots \geq \lambda_d(\varphi, \mu)$ associated with φ and μ.

In the present paper we investigate the relations between the Lyapunov exponents of a linear RDS Φ and those of the induced nonlinear systems φ_k on the Graßmann manifolds $\mathrm{Gr}(k, d)$, $1 \leq k \leq d$.
First we note that associated with each ergodic invariant measure for φ_k on $\mathrm{Gr}(k, d)$ there is a uniquely determined k-tuple $\lambda_{j_1}(\Phi) \geq \ldots \geq \lambda_{j_k}(\Phi)$ of exponents of Φ. It should be emphasized that the numbers λ_{j_i}, $1 \leq i \leq k$, are determined uniquely by μ — however, the indices j_i, $1 \leq i \leq k$, are not. It is possible to enforce uniqueness, e. g., by putting $\tilde{\jmath}_1 = \min\{i \mid \lambda_i = \lambda_{j_1}\}$ and $\tilde{\jmath}_n = \min\{i > \tilde{\jmath}_{n-1} \mid \lambda_i = \lambda_{j_n}\}$, starting with any suitable $(j_1, \ldots, j_k)$. For the purposes of this paper it is not necessary to do so.

Since φ_k is an RDS on the $k(d-k)$-dimensional manifold $\mathrm{Gr}(k, d)$, there must be $k(d-k)$ Lyapunov exponents associated with each invariant measure for φ_k. Our main result is that these exponents are

$$(\lambda_i(\Phi) - \lambda_j(\Phi) \mid i \notin \{j_1, \ldots, j_k\}, j \in \{j_1, \ldots, j_k\}), \tag{1}$$

ordered decreasingly.

Lyapunov exponents of linear systems have been investigated for long time. For some history and for references up to 1985 see Arnold and Wihstutz [1] and Bougerol [3].

Properties of the projective system induced by a linear one have been used by most of the authors concerned with linear systems. However, it seems that Lyapunov exponents of the projective system itself have not been investigated.

The induced systems on higher Graßmann manifolds have not found much interest up to now. Baxendale [2] has dealt with the action of stochastic flows on Graßmannians. However, he was not interested in the exponents of these nonlinear systems themselves, but used them to derive formulae for sums of Lyapunov exponents.

The paper is organized as follows. In Section 2, some basic definitions and notions on random dynamical systems, multiplicative ergodic theory, and on exterior products are introduced. In Section 3, we prove (1) for projective systems. The method is simple: we embed the unit sphere S^{d-1} into $\mathbb{R}^d \backslash \{0\}$ as a regular submanifold in the obvious way. This allows to express the linearization $T\varphi$ of $\varphi = \varphi_1$ in terms of Φ and the projection map $\mathbb{R}^d \backslash \{0\} \to S^{d-1}$. Straightforward arguments then give the desired result. We conclude Section 3 by describing how to construct a normal basis for φ from a normal basis for Φ, which proceeds essentially by projection onto the tangent space $T(S^{d-1})$.

In Section 4, we interprete the Graßmann manifold $\mathrm{Gr}(k,d)$ as a submanifold of $S(\bigwedge^k \mathbb{R}^d)$, the unit sphere of the k-fold exterior product of $\mathbb{R}^d$ (up to identification of antipodal points). Clearly, on $\mathrm{Gr}(k,d)$ the system φ_k and the projective system of the k-fold exterior product $\bigwedge^k \Phi$ coincide. Next, we identify the tangent bundle of $\mathrm{Gr}(k,d)$ more or less explicitely as a subbundle of the tangent bundle of $S(\bigwedge^k \mathbb{R}^d)$. Finally, given an invariant measure for φ_k, we use Section 3 to construct a normal basis on the whole tangent space of $S(\bigwedge^k \mathbb{R}^d)$ in such a way that it contains a subbasis of $T\mathrm{Gr}(k,d)$. We thus obtain a normal basis for φ_k, and it remains only to read off the Lyapunov exponents of φ_k from this normal basis to prove (1).

2 Preliminaries

Let M be a d-dimensional Riemannian C^r manifold, $r \geq 1$, and denote by $\mathcal{B}$ its Borel σ-algebra. Let T be either $\mathbb{R}$, $\mathbb{R}^+$, $\mathbb{Z}$, or $\mathbb{N}$. Let $(\Omega, \mathcal{F}, P)$ be a probability space and let $\{\vartheta_t \mid t \in T\}$ be an ergodic family of measure preserving transformations of $(\Omega, \mathcal{F}, P)$ such that $\vartheta_{t+s} = \vartheta_t \circ \vartheta_s$ for all $t, s \in T$.

2.1 Definition A *random dynamical system* is a measurable map

$$\varphi : T \times M \times \Omega \to M$$

such that $\varphi(0,\omega) = \mathrm{id}$ and

$$\varphi(t+s,\omega) = \varphi(t,\vartheta_s\omega) \circ \varphi(s,\omega)$$

for all $s, t \in T$ outside some P-nullset in Ω. Here $\varphi(t,\omega) : M \to M$ is the map which arises when t and ω are kept fixed.

A random dynamical system φ is said to be *smooth* if $\varphi(t,\omega)$ is differentiable for all $t \in T$ and all $\omega \in \Omega$ outside a P-nullset. A smooth system is said to be *invertible* if $\varphi(t,\omega)$ is a diffeomorphism for all $t \in T$ and all $\omega \in \Omega$ outside a P-nullset. A random dynamical system $\varphi(t,\omega)$ is said to be *linear* if $\varphi(t,\omega)$ is a linear transformation of $M = \mathbb{R}^d$ for all $t \in T$ and all $\omega \in \Omega$ outside a P-nullset.

We shall abbreviate 'random dynamical system' by RDS in what follows. An RDS induces a measurable skew product flow

$$\begin{aligned} \Theta_t : M \times \Omega &\to M \times \Omega \\ (x,\omega) &\mapsto (\varphi(t,\omega)x, \vartheta_t\omega), \end{aligned} \tag{2}$$

where $\varphi(t,\omega)x = \varphi(t,x,\omega)$. In fact, $\Theta_{t+s} = \Theta_t \circ \Theta_s$; we use the term 'flow' for both continuous and discrete time T.

The most common examples of RDS are provided by products of random maps and the solution flows of stochastic or random differential equations. In this paper we are dealing with smooth RDS only, so we drop the 'smooth' henceforth.

A probability measure μ on $M \times \Omega$ (on the product σ-algebra $\mathcal{B} \otimes \mathcal{F}$) is said to be an *invariant measure for* φ if it is invariant under Θ_t, $t \in T$, and if it has marginal P on Ω. Invariant measures always exist if M is compact.
For linear RDS the measure $\delta_0 \times P$ is always invariant, where δ_0 denotes the Dirac measure in $0 \in \mathbb{R}^d$. Other invariant measures for linear systems exist only in very particular cases.

Exterior products

We take some exterior products terminology from Loomis and Sternberg [6]), Sections 7.7 and 7.8, specialized for euclidean space $(\mathbb{R}^d, \langle\cdot,\cdot\rangle)$. For an integer k, $1 \leq k \leq d$, denote by $\bigwedge^k \mathbb{R}^d$ the vector space of alternating k-linear forms. If $u_1, \ldots, u_k \in \mathbb{R}^d$ then $u_1 \wedge \ldots \wedge u_k \in \bigwedge^k \mathbb{R}^d$, where

$$(u_1 \wedge \ldots \wedge u_k)(v_1, \ldots, v_k) = \det[(\langle u_i, v_j\rangle)_{i,j}],$$

and $u_1 \wedge \ldots \wedge u_k = 0$ if and only if $\{u_i \mid 1 \leq i \leq k\}$ are linearly dependent. If $\{u_i \mid 1 \leq i \leq k\}$ and $\{v_i \mid 1 \leq i \leq k\}$ are two k-tuples of independent vectors, then $u_1 \wedge \ldots \wedge u_k = \lambda(v_1 \wedge \ldots \wedge v_k)$ if and only if $\operatorname{span}\{u_i \mid 1 \leq i \leq k\} = \operatorname{span}\{v_i \mid 1 \leq i \leq k\}$. A vector $u \in \bigwedge^k \mathbb{R}^d$ is said to be *decomposable* if $u = u_1 \wedge \ldots \wedge u_k$ for some $u_i \in \mathbb{R}^d$;

$$\textstyle\bigwedge_0^k \mathbb{R}^d = \{u_1 \wedge \ldots \wedge u_k \mid u_i \in \mathbb{R}^d\}$$

denotes the set of decomposable vectors. If $\{e_i \mid 1 \leq i \leq d\}$ is a basis of $\mathbb{R}^d$, then $\{e_{i_1} \wedge \ldots \wedge e_{i_k} \mid 1 \leq i_1 < i_2 < \ldots < i_k \leq d\}$ is a basis of $\bigwedge^k \mathbb{R}^d$, hence $\dim \bigwedge^k \mathbb{R}^d = \binom{d}{k}$. $\bigwedge^k \mathbb{R}^d$ inherits a canonical euclidean structure by putting

$$\langle u, v\rangle = \det[(\langle u_i, v_j\rangle)_{i,j}]$$

for $u = u_1 \wedge \ldots \wedge u_k$ and $v = v_1 \wedge \ldots \wedge v_k$, and extending to all of $\bigwedge^k \mathbb{R}^d$ by bilinearity.

We need the following lemma. The proof is straightforward.

2.2 Lemma *For $u_1, \ldots, u_k \in \mathbb{R}^d$ and $s \in \mathbb{R}^d$ with $|s| = 1$, denote by w_i the orthogonal projection of u_i to the hyperplane orthogonal to s, i. e., $w_i = u_i - \langle u_i, s\rangle s$, $1 \leq i \leq k$. Then*

$$|w_1 \wedge \ldots \wedge w_k|_k = |u_1 \wedge \ldots \wedge u_k \wedge s|_{k+1},$$

where $|\cdot|_k$ denotes the norm on $\bigwedge^k \mathbb{R}^d$.

A linear transformation Φ on $\mathbb{R}^d$ generates a linear transformation $\bigwedge^k \Phi$ on $\bigwedge^k \mathbb{R}^d$ by putting

$$\textstyle\bigwedge^k \Phi(u_1 \wedge \ldots \wedge u_k) = \Phi u_1 \wedge \ldots \wedge \Phi u_k$$

on $\bigwedge_0^k \mathbb{R}^d$, and extending to $\bigwedge^k \mathbb{R}^d$ by linearity. Clearly, $\bigwedge_0^k \mathbb{R}^d$ is invariant under $\bigwedge^k \Phi$, and the operator norm of $\bigwedge^k \Phi$ is realized on decomposable vectors already, i. e.

$$\|\textstyle\bigwedge^k \Phi\| = \sup\{|\bigwedge^k \Phi(u)| \mid u \in \bigwedge_0^k \mathbb{R}^d, |u| = 1\}. \qquad (3)$$

The Graßmann manifold $\operatorname{Gr}(k, d)$ is the manifold of all k-dimensional subspaces of $\mathbb{R}^d$. We identify $\operatorname{Gr}(k, d)$ with the image of $\bigwedge_0^k \mathbb{R}^d$ under the projection of $\bigwedge^k \mathbb{R}^d$ onto its projective space. Hence, $\mathrm{Gr}(k,d)$ is considered a $k(d-k)$-dimensional closed submanifold of the $(\binom{d}{k}-1)$-dimensional projective space $\mathbb{P}(\bigwedge^k \mathbb{R}^d)$.

Lyapunov exponents

For a differentiable manifold M denote by TM the total space of its tangent bundle. The linearization of a differentiable map $\psi : M \to M$ is denoted by $T\psi : TM \to TM$ with $T_x\psi : T_xM \to T_{\psi(x)}M$, $x \in M$, denoting the action of $T\psi$ on individual fibres.

Suppose φ is a smooth invertible RDS on a d-dimensional Riemannian manifold and let μ be an invariant measure for φ such that

$$(x,\omega) \mapsto \sup_{0<t\leq t_0} \{\log^+(\|T_x\varphi(t,\omega)\|) + \log^+(\|(T_x\varphi(t,\omega))^{-1}\|)\}$$

is integrable with respect to μ. Denote by $\lambda_1^\mu(x,\omega) \geq \lambda_2^\mu(x,\omega) \geq \ldots \geq \lambda_d^\mu(x,\omega)$ the Lyapunov exponents of φ associated with μ; the $\Theta_\cdot$-invariant maps $(x,\omega) \mapsto \lambda_i^\mu(x,\omega)$ are defined via

$$\sum_{i=1}^{k} \lambda_i^\mu(x,\omega) = \lim_{t\to\infty} \frac{1}{t} \log \|\textstyle\bigwedge^k T_x\varphi(t,\omega)\|, \tag{4}$$

$1 \leq k \leq d$. Here $\bigwedge^k$ denotes the k-fold exterior product of $T_x\varphi$. Existence of the limits in (4) follows from Kingman's subadditive ergodic theorem (Kingman [5]). If μ is ergodic, Lyapunov exponents do not depend on (x,ω). If M is compact, Lyapunov exponents do not depend on the choice of the Riemannian metric.

By Oseledec' theorem, for μ-almost all (x,ω) the limits

$$\lambda(v,\omega) = \lim_{t\to\infty} \frac{1}{t} \log \|T_x\varphi(t,\omega)v\|$$

exist for all $v \in T_xM$, and $\lambda(v,\omega)$ is one of the Lyapunov exponents.

A family of measurable maps $v_i : M \times \Omega \to TM$ is said to be a *normal basis* if $\{v_i(x,\omega)\}_{1\leq i\leq d}$ is a basis of T_xM for μ-almost all (x,ω), and if

$$\lambda(v_i,\omega) = \lambda_i(x,\omega) \tag{5}$$

for $1 \leq i \leq d$. If v_i, $1 \leq i \leq d$, are linearly independent μ-a. s., then (5) is equivalent to

$$\lambda\Big(\sum_{i=1}^{d} c_i v_i\Big) = \max_{c_i\neq 0} \lambda(v_i) \qquad \mu\text{-a. s.} \tag{6}$$

for all $c_1, \ldots, c_d \in \mathbb{R}$. It is also equivalent to

$$\lambda(v_{i_1} \wedge \ldots \wedge v_{i_k}) = \sum_{j=1}^{k} \lambda(v_{i_j}) = \sum_{j=1}^{k} \lambda_{i_j} \qquad \mu\text{-a. s.} \tag{7}$$

for any $k \leq d$ and $1 \leq i_1 < \ldots < i_k \leq d$, where the exponent in the first expression of (7) is to be understood with respect to $\bigwedge^k T\varphi$.

Sometimes it is more convenient to count only the distinct Lyapunov exponents, denoted here by $\Lambda_1 > \Lambda_2 > \ldots > \Lambda_r$, where r is the number of distinct exponents, $1 \leq r \leq d$ (we assume μ ergodic to ease notation). Denote by $d_i = \max\{p - q + 1 \mid \lambda_p = \lambda_q = \Lambda_i\}$ the multiplicity of Λ_i. Put

$$V_i(x, \omega) = \{v \in T_x M \mid \lambda(v, \omega) \leq \Lambda_i\},$$

then $V_i(x,\omega)$ is a linear subspace of T_xM for μ-almost all (x,ω), invariant under $T\varphi$ (i. e., $T_x\varphi(t,\omega)V_i(x,\omega) = V_i(\Theta_t(x,\omega))$ for all $t \in T$ μ-a. s.), and $d_i = \dim V_i(x,\omega) - \dim V_{i+1}(x,\omega)$ for $1 \leq i \leq r$. We refer to the family

$$T_xM = V_1(x,\omega) \supset V_2(x,\omega) \supset \ldots \supset V_r(x,\omega) \supset \{0\}$$

as the *Oseledec flag* associated with (φ, μ).
A family of measurable maps $v_j : M \times \Omega \to TM$, $1 \leq j \leq d$, is a normal basis if the last $(\dim V_i)$ of the v_j's span V_i μ-almost surely, $1 \leq i \leq r$.

Suppose $v_1, \ldots, v_k$ are measurable maps from $M \times \Omega$ to TM such that μ-a. s. $v_i(x,\omega) \in T_xM$ and $\{v_i\}_{1\leq i\leq k}$ are linearly independent. Then there exist maps $\bar{v}_1, \ldots, \bar{v}_{d-k} : M \times \Omega \to TM$ such that $(v_1, \ldots, v_k, \bar{v}_1, \ldots, \bar{v}_{d-k})$ can be renumbered to become a normal basis if and only if μ-a. s. for any of the Lyapunov exponents the number of v_i's, $1 \leq i \leq k$, realizing this exponent does not exceed its multiplicity. This is immediate from the characterization of a normal basis in terms of the Oseledec flag.

In case of a linear invertible RDS, Lyapunov exponents are understood with respect to the invariant measure $\delta_0 \times P$ (this is only for definiteness: for linear systems, (4) does not depend on an invariant measure). The integrability condition for a linear invertible RDS Φ reads

$$\omega \mapsto \sup_{0<t\leq t_0} \{\log^+(\|\Phi(t,\omega)\|) + \log^+(\|\Phi^{-1}(t,\omega)\|)\} \in L^1(P). \tag{I}$$

Since we have assumed ergodicity of $(\vartheta_t)_{t\in T}$, Lyapunov exponents of linear systems do not depend on ω. Clearly, the Lyapunov exponents of a linear system are independent of the norm on $\mathbb{R}^d$. For the remainder of the paper we will assume the integrability condition (I) for linear systems.

Linearly induced RDS

The general linear group $\mathrm{Gl}(d,\mathbb{R})$ acts in a canonical way on Graßmann manifolds: if $G \in \mathrm{Gl}(d,\mathbb{R})$ and $K \subset \mathbb{R}^d$ is a linear subspace, then (G,K) is mapped to the subspace $G(K)$; due to invertibility of G there is no loss of dimension.
Thus, a linear invertible RDS $\Phi(t,\omega)$ induces an RDS on each Graßmann manifold $\mathrm{Gr}(k,d)$, $1 \leq k \leq d$. Fix k, and denote the induced system by φ. Since $\mathrm{Gr}(k,d)$ is compact, invariant measures for φ exist. Each ergodic invariant measure for φ on $\mathrm{Gr}(k,d)$ uniquely determines a k-tuple $\lambda_{j_1}(\Phi) \geq \lambda_{j_2}(\Phi) \geq \ldots \geq \lambda_{j_k}(\Phi)$ of Lyapunov exponents of Φ as follows. Consider the linear RDS Φ over the enlarged probability space $(\mathrm{Gr}(k,d)\times\Omega, \mathcal{B}\otimes\mathcal{F}, \mu)$, where Φ is extended trivially, $\mathcal{B}$ denoting the Borel sets of $\mathrm{Gr}(k,d)$. This does not change the Lyapunov exponents of Φ. However, the (measurable) vector bundle

$$U(u,\omega) = \{v \in \mathbb{R}^d \mid v \in u\} \tag{8}$$

is μ-almost surely invariant under Φ (i.e., $\Phi(t,\omega)U(u,\omega) = U(\Theta_t(u,\omega))$ for all $t \in T$ and μ-almost all (u,ω), Θ_t denotes the skew product flow induced by φ, see (2)). The Lyapunov exponents $\lambda_{j_1}(\Phi) \geq \lambda_{j_2}(\Phi) \geq \ldots \geq \lambda_{j_k}(\Phi)$ determined by μ are the exponents of this invariant subbundle. In particular,

$$\int \log \frac{|(\bigwedge^k \Phi(1,\omega))(u_1 \wedge \ldots \wedge u_k)|}{|u_1 \wedge \ldots \wedge u_k|} \, d\mu(u,\omega) = \lambda_{j_1}(\Phi) + \ldots + \lambda_{j_k}(\Phi), \tag{9}$$

where $u \mapsto u_1(u), \ldots, u_k(u)$ is a basis of $u \in \mathrm{Gr}(k,d)$, u_i measurable in u. We summarize these considerations in the first part of the following lemma.

2.3 Lemma *Let Φ be a linear invertible d-dimensional RDS with Lyapunov exponents $\lambda_1(\Phi) \geq \ldots \geq \lambda_d(\Phi)$.*
(i) *Every ergodic invariant measure μ for the k-fold Graßmannian RDS induced by Φ uniquely determines a k-tuple $\lambda_{j_1}(\Phi) \geq \lambda_{j_2}(\Phi) \geq \ldots \geq \lambda_{j_k}(\Phi)$ of Lyapunov exponents of Φ.*
(ii) *For any k-tuple of Lyapunov exponents $\lambda_{j_1}(\Phi) \geq \ldots \geq \lambda_{j_k}(\Phi)$ of Φ with $j_1 < j_2 < \ldots < j_k$ there exists an ergodic invariant measure μ for the k-fold Graßmannian RDS induced by Φ such that $\lambda_{j_1}(\Phi) \geq \ldots \geq \lambda_{j_k}(\Phi)$ are the exponents of Φ on the invariant bundle* (8).

The proof of (ii) proceeds by a Krylov-Bogolyubov type argument, and is omitted here. See Proposition 2.2.1 of Crauel [4].
For $k = 1$, the exponent $\lambda_j(\Phi)$ determined by an ergodic invariant measure μ for the projective RDS is determined by (9) already.

3 Projective systems

Let S^{d-1} be the unit sphere of $\mathbb{R}^d$, consisting of all vectors of length one. Denote by $\pi : \mathbb{R}^d_0 \to S^{d-1}$, $y \mapsto y/|y|$, the canonical projection, where $\mathbb{R}^d_0 = \mathbb{R}^d \setminus \{0\}$. We consider the unit sphere as an embedded compact $(d-1)$-dimensional submanifold of $\mathbb{R}^d_0$. We identify the tangent bundle TS^{d-1} with the subbundle of $T\mathbb{R}^d_0$ defined by

$$T_s S^{d-1} = \{v \in \mathbb{R}^d \mid \langle v, s \rangle = 0\}, \tag{10}$$

where the tangent bundle of $\mathbb{R}^d_0$ is globally trivial and is identified with $\mathbb{R}^d_0 \times \mathbb{R}^d$. Fix the Riemannian metric on S^{d-1} induced by this identification.

Since π is a submersion, $T\pi : T\mathbb{R}^d_0 \to TS^{d-1}$ acts as a projection along $\ker(T\pi)$. By (10) and with the particular choice of Riemannian metric, $T\pi$ becomes an orthogonal projection, hence for $y \in \mathbb{R}^d_0$ and $v \in T_y\mathbb{R}^d_0$

$$(T_y\pi)v = |y|^{-1}(v - |y|^{-2}\langle v, y\rangle y) \in T_{\pi y}(S^{d-1}). \tag{11}$$

An invertible linear transformation $\Phi \in \mathrm{Gl}(d,\mathbb{R})$ induces a diffeomorphism φ of S^{d-1} by $\varphi(s) = \frac{\Phi s}{|\Phi s|}$, $s \in S^{d-1}$, and

$$\varphi \circ \pi = \pi \circ \Phi. \tag{12}$$

3.1 Lemma *For any $s \in S^{d-1}$ and $1 \le k \le d$*

$$\|\bigwedge^k T_s\varphi\| = \sup\Big\{\frac{|(\bigwedge^{k+1}\Phi)(v \wedge s)|}{|\Phi(s)|^{k+1}} \mid v \in \bigwedge_0^k \mathbb{R}^d, |v| = 1\Big\}.$$

Proof From (12) we get $T(\varphi \circ \pi) = T(\pi \circ \Phi)$ (mapping $T\mathbb{R}_0^d$ to TS^{d-1}), hence, for $1 \le k \le d-1$, also

$$\bigwedge^k T(\varphi \circ \pi) = \bigwedge^k T(\pi \circ \Phi), \tag{13}$$

mapping $\bigwedge^k T\mathbb{R}_0^d$ to $\bigwedge^k TS^{d-1}$.

Since $T_s\pi$ acts as a projection with image T_sS^{d-1} for $s \in S^{d-1}$, we have

$$\|\bigwedge^k T_s\varphi\| = \|\bigwedge^k T_s(\varphi \circ \pi)\|,$$

and consequently $\|\bigwedge^k T_s\varphi\| = \|\bigwedge^k T_s(\pi \circ \Phi)\|$.

To proceed, fix $y \in \mathbb{R}_0^d$ and a decomposable vector $(u_1 \wedge \ldots \wedge u_k) \in \bigwedge^k \mathbb{R}^d$. By (11) we get

$$\begin{aligned}\Big(\bigwedge^k T_y(\pi \circ \Phi)\Big)(u_1 \wedge \ldots \wedge u_k) &= \Big(\bigwedge^k T_{\Phi y}\pi\Big)\Big((\bigwedge^k T_y\Phi)(u_1 \wedge \ldots \wedge u_k)\Big) \\ &= \bigwedge_{i=1}^k \big(|\Phi y|^{-1}(\Phi u_i - |\Phi y|^{-2}\langle \Phi u_i, \Phi y\rangle \Phi y)\big).\end{aligned}$$

Lemma 2.2 yields

$$|\bigwedge^k T_y(\pi \circ \Phi)(u_1 \wedge \ldots \wedge u_k)|_k = \frac{|\Phi u_1 \wedge \ldots \wedge \Phi u_k \wedge \Phi y|_{k+1}}{|\Phi y|^{k+1}}.$$

In view of (3) the proof is complete. □

Now consider the projective space $\mathbb{P}^{d-1}$. The invertible linear transformation $\Phi(t,\omega)$ generates a diffeomorphism on $\mathbb{P}^{d-1}$ by mapping the line element through a vector $u \in \mathbb{R}^d$, $u \neq 0$, to the line element through $\Phi(t,\omega)u$. Denote this diffeomorphism by $\bar\varphi$ for a moment. The unit sphere is a two-fold covering of the projective space. Choosing a Riemannian metric on $\mathbb{P}^{d-1}$ such that the covering projection becomes locally isometric, Lemma 3.1 can be verified to carry over to $\bar\varphi$. Hence

$$\|\bigwedge^k T_u\bar\varphi\| = \sup\Big\{\frac{|(\bigwedge^{k+1}\Phi)(v \wedge s)|}{|\Phi(s)|^{k+1}} \mid v \in \bigwedge_0^k \mathbb{R}^d, |v| = 1\Big\}$$

for any $u \in \mathbb{P}^{d-1}$, where s on the right hand side is any of the two elements of the unit sphere representing u.

3.2 Theorem *Let Φ be a linear invertible RDS on $\mathbb{R}^d$, satisfying the integrability condition (I), and denote by $\lambda_1(\Phi) \ge \ldots \ge \lambda_d(\Phi)$ its Lyapunov exponents. Let φ be the projective RDS induced by Φ on $\mathbb{P}^{d-1}$, and let μ be an ergodic invariant measure for φ. Then the Lyapunov exponents $\lambda_1(\varphi,\mu) \ge \ldots \ge \lambda_{d-1}(\varphi,\mu)$ associated with φ and μ are given by*

$$\lambda_i(\varphi) = \begin{cases} \lambda_i(\Phi) - \lambda(\mu) & \text{if } \lambda_i(\Phi) > \lambda(\mu) \\ \lambda_{i+1}(\Phi) - \lambda(\mu) & \text{if } \lambda_i(\Phi) < \lambda(\mu), \end{cases}$$

where

$$\lambda(\mu) = \lambda_j(\Phi) = \int \log \frac{|\Phi(1,\omega)u|}{|u|}\, d\mu(u,\omega) \tag{14}$$

is the Lyapunov exponent of Φ determined (uniquely) by μ by (14).

PROOF Let $\omega \mapsto v_1(\omega), \ldots, v_d(\omega)$ be a normal basis for Φ. Redefine Φ and v_i, $1 \le i \le d$, on the enlarged probability space $(\mathbb{P}^{d-1} \times \Omega, \mu)$ by trivial extension. Obviously, this procedure does not change the exponents of Φ, and the v_i remain a normal basis.
Pick a measurable map from $\mathbb{P}^{d-1}$ to S^{d-1}, $u \mapsto s(u)$, such that $s(u)$ is a unit vector representing u. Replace $(u,\omega) \mapsto v_{i(\mu)}(\omega)$ by $(u,\omega) \mapsto s(u)$, where $i(\mu)$ is such that $\lambda_{i(\mu)} = \lambda(\mu)$. The resulting d vectors are a normal basis again, possibly after modifying the v_i's from the Oseledec flag space corresponding to $\lambda(\mu)$ to avoid linear dependence. Finally, remove $(u,\omega) \mapsto s(u)$ and renumber the remaining v_i from 1 to $d-1$ according to their original order.

The Lyapunov exponents of φ are given by (4). We use Lemma 3.1 to calculate them. Fix k, $1 \le k \le d-1$. By construction of the maps $(u,\omega) \mapsto v_i(u,\omega)$ and $(u,\omega) \mapsto s(u)$,

$$\lim_{t\to\infty} \frac{1}{t} \log \left| (\textstyle\bigwedge^{k+1}\Phi(t,\omega))(v_1 \wedge \ldots \wedge v_k \wedge s) \right| = \begin{cases} \sum_{i=1}^{k+1} \lambda_i(\Phi) & \text{if } \lambda_k(\Phi) > \lambda(\mu) \\ \sum_{i=1}^{k} \lambda_i(\Phi) + \lambda(\mu) & \text{if } \lambda_k(\Phi) \le \lambda(\mu) \end{cases}$$

for μ-almost all (u,ω). Lemma 3.1 yields

$$\lim_{t\to\infty} \frac{1}{t} \log \|\textstyle\bigwedge^k T_u\varphi(t,\omega)\| \ge \begin{cases} \sum_{i=1}^{k+1} (\lambda_i(\Phi) - \lambda(\mu)) & \text{if } \lambda_k(\Phi) > \lambda(\mu) \\ \sum_{i=1}^{k} (\lambda_i(\Phi) - \lambda(\mu)) & \text{if } \lambda_k(\Phi) \le \lambda(\mu) \end{cases} \tag{15}$$

μ-almost surely. Since

$$\|\textstyle\bigwedge^k T_u\varphi(t,\omega)\| \le \begin{cases} \|\bigwedge^{k+1}\Phi(t,\omega)\| |\Phi(t,\omega)s|^{-(k+1)} & \text{if } \lambda_k(\Phi) > \lambda(\mu) \\ \|\bigwedge^{k}\Phi(t,\omega)\| |\Phi(t,\omega)s|^{-k} & \text{if } \lambda_k(\Phi) \le \lambda(\mu), \end{cases}$$

we get μ-a. s.

$$\lim_{t\to\infty} \frac{1}{t} \log \|\textstyle\bigwedge^k T_u\varphi(t,\omega)\| \le \begin{cases} \sum_{i=1}^{k+1} \lambda_i(\Phi) - (k+1)\lambda(\mu) & \text{if } \lambda_k(\Phi) > \lambda(\mu) \\ \sum_{i=1}^{k} \lambda_i(\Phi) - k\lambda(\mu) & \text{if } \lambda_k(\Phi) \le \lambda(\mu). \end{cases} \tag{16}$$

Combining (15) and (16) gives the desired result. □

Theorem 3.2 says that the Lyapunov exponents of φ (associated with a particular ergodic invariant measure μ) are obtained by removing $\lambda(\mu)$ from the d exponents $\lambda_1(\Phi) \geq \ldots \geq \lambda_d(\Phi)$ and subtracting it from the remaining $(d-1)$ exponents.

If all exponents of Φ are distinct, then any invariant measure μ for φ is 'hyperbolic' in the sense that no vanishing Lyapunov exponents occur. Existence of d distinct exponents of Φ implies that there are precisely d ergodic invariant measures for φ which are random point measures supported by the images of the d (one dimensional) Oseledec spaces on the projective space.

3.3 Remark Under the conditions of Theorem 3.2, suppose $\{v_i : \Omega \to \mathbb{R}^d \mid 1 \leq i \leq d\}$ is a normal basis for Φ. Given an ergodic invariant measure for φ, we want to construct a normal basis. Recall that a normal basis for φ is, by definition, a $(k-1)$-tuple of maps from $\mathbb{P}^{d-1} \times \Omega$ to $T\mathbb{P}^{d-1}$ which are μ-almost everywhere linearly independent and satisfy (5). In fact, it suffices to construct a normal basis for the sphere system. First, modify the basis as in the proof of Theorem 3.2 to obtain the maps $(u,\omega) \mapsto s(u)$ and $(u,\omega) \mapsto v_i(u,\omega)$, $1 \leq i \leq d-1$. Then

$$w_i(u,\omega) := v_i(u,\omega) - \langle v_i(u,\omega), s(u)\rangle s(u), \tag{17}$$

$1 \leq i \leq d-1$, is a normal basis for φ. To prove (17), it suffices to note that $T_u(\varphi \circ \pi)(w) = T_u\varphi(w)$ if $w \in T_u S^{d-1}$, hence

$$\begin{aligned} |T_u\varphi(w)| &= |T_u(\pi \circ \Phi)(w)| \qquad \text{by (12)} \\ &= \left| \frac{1}{|\Phi s(u)|}\left(\Phi w - \frac{1}{|\Phi s(u)|^2}\langle \Phi w, \Phi s(u)\rangle \Phi s(u)\right)\right| \\ &= \frac{|\Phi w \wedge \Phi s(u)|}{|\Phi s(u)|^2} \end{aligned}$$

by Lemma 2.2. Evaluating this expression for the w_i defined in (17) yields

$$|T_u\varphi(w_i)| = \frac{|\Phi w_i \wedge \Phi s(u)|}{|\Phi s(u)|^2} = \frac{|\Phi v_i \wedge \Phi s(u)|}{|\Phi s(u)|^2}.$$

Since the Lyapunov exponent realized by w_i is

$$\lambda^{\varphi}(w_i) = \lambda^{\Phi}(v_i) - \lambda(\mu), \tag{18}$$

the assertion follows (where we used superscripts in (18) to denote the exponents associated with φ and Φ, respectively.)

3.4 Remark The k-fold exterior product of Φ is a bona fide linear RDS. Its Lyapunov exponents are all sums of k of the exponents of Φ:

$$\{\lambda_{i_1}(\Phi) + \ldots + \lambda_{i_k}(\Phi) \mid 1 \leq i_1 < \ldots < i_k \leq d\},$$

which are $\binom{d}{k} = \dim(\bigwedge^k \mathbb{R}^d)$ many.

Suppose μ is an ergodic invariant measure for the projective system induced by $\bigwedge^k \Phi$ on

$\mathbb{P}(\bigwedge^k \mathbb{R}^d)$. According to Lemma 2.3, there are $\lambda_{j_1}(\Phi) \geq \lambda_{j_2}(\Phi) \geq \ldots \geq \lambda_{j_k}(\Phi)$ associated with μ, and (9) holds. By Theorem 3.2, φ has $\binom{d}{k} - 1$ Lyapunov exponents associated with μ, given by

$$\{\lambda_{i_1}(\Phi) + \ldots + \lambda_{i_k}(\Phi) \; - \; (\lambda_{j_1}(\Phi) + \ldots + \lambda_{j_k}(\Phi)) \mid 1 \leq i_1 < \ldots < i_k \leq d, \text{ and } i_p \neq j_p \text{ for some } p\}. \tag{19}$$

4 Graßmann systems

The Graßmann manifold $\mathrm{Gr}(k,d)$ is a $k(d-k)$-dimensional closed submanifold of the projective space $\mathbb{P}(\bigwedge^k \mathbb{R}^d)$, invariant under the projective system induced by $\bigwedge^k \Phi$, the k-fold exterior product of a linear RDS Φ on $\mathbb{R}^d$ as described in Section 2. On $\mathrm{Gr}(k,d)$ the projective RDS induced by $\bigwedge^k \Phi$ and the k-fold Graßmannian RDS induced by Φ coincide, and we will denote both of them by φ. Suppose μ is an ergodic invariant measure for φ on $\mathrm{Gr}(k,d)$. Denote by $\{\lambda_{j_1}(\Phi),\ldots,\lambda_{j_k}(\Phi)\}$ the Lyapunov exponents determined by μ according to Lemma 2.3. Remark 3.4 gives the Lyapunov exponents of φ associated with μ when considered as an RDS on the entire projective space. In this section, we consider φ as an RDS on the lower-dimensional invariant submanifold $\mathrm{Gr}(k,d)$. Clearly, the Lyapunov exponents of the Graßmann system form a subset of the exponents of the projective system of the k-fold exterior product of Φ.

4.1 Theorem *Let Φ be a linear invertible RDS on $\mathbb{R}^d$, satisfying the integrability condition* (I), *and denote by $\lambda_1(\Phi) \geq \ldots \geq \lambda_d(\Phi)$ its Lyapunov exponents. Let φ be the k-fold Graßmann RDS induced by Φ on* $\mathrm{Gr}(k,d)$, *and let μ be an ergodic invariant measure for φ. Then the Lyapunov exponents associated with φ and μ are given by the $k(d-k)$-tuple*

$$(\lambda_i(\Phi) - \lambda_j(\Phi) \mid i \notin \{j_1,\ldots,j_k\}, j \in \{j_1,\ldots,j_k\}), \tag{20}$$

put into decreasing order.

Note that (20) is nothing but

$$(\lambda_{i_1}(\Phi) + \ldots + \lambda_{i_k}(\Phi) \; - \; (\lambda_{j_1}(\Phi) + \ldots + \lambda_{j_k}(\Phi)) \mid 1 \leq i_1 < \ldots < i_k \leq d, \text{ and } i_p \neq j_p \text{ for exactly one } p),$$

describing a subset of the set given by (19).

Proof Consider the set of decomposable k-vectors of unit length. They form a closed submanifold of the unit sphere in $\bigwedge^k \mathbb{R}^d$, which two-fold covers $\mathrm{Gr}(k,d)$, the covering map being the restriction of the one between the unit sphere and the projective space of $\bigwedge^k \mathbb{R}^d$. We make the same identifications of the tangent spaces as in the previous section. Hence, the tangent space of the unit sphere $S(\bigwedge^k \mathbb{R}^d)$ is identified as

$$T_s S(\textstyle\bigwedge^k \mathbb{R}^d) = \{v \in \bigwedge^k \mathbb{R}^d \mid \langle v, s\rangle = 0\},$$

s a unit vector in $\bigwedge^k \mathbb{R}^d$. If s is decomposable, a straightforward argument shows that $T_s S(\bigwedge^k \mathbb{R}^d)$ is spanned by the decomposable vectors orthogonal to s,

$$T_s S(\textstyle\bigwedge^k \mathbb{R}^d) = \operatorname{span}\{v \in \bigwedge_0^k \mathbb{R}^d \mid \langle v, s\rangle = 0\}.$$

By minor misuse of notation, the tangent space of the pre-image of $\mathrm{Gr}(k,d)$ under the covering map is denoted by $T\mathrm{Gr}(k,d)$ again. It is not difficult to verify, using the implicit function theorem, that for $s = s_1 \wedge \ldots \wedge s_k$

$$\begin{aligned} T_s\mathrm{Gr}(k,d) &= \operatorname{span}\{(s_1 \wedge \ldots \wedge s_{i-1} \wedge y \wedge s_{i+1} \wedge \ldots \wedge s_k) \in T_s S(\textstyle\bigwedge^k \mathbb{R}^d) \mid \\ &\qquad\qquad y \in \mathbb{R}^d, 1 \le i \le k\} \\ &= \operatorname{span}\{(s_1 \wedge \ldots \wedge s_{i-1} \wedge y \wedge s_{i+1} \wedge \ldots \wedge s_k) \in T_s S(\textstyle\bigwedge^k \mathbb{R}^d) \mid \\ &\qquad\qquad y \notin \operatorname{span}\{s_1, \ldots, s_k\}\} \\ &= \operatorname{span}\{(s_1 \wedge \ldots \wedge s_{i-1} \wedge y \wedge s_{i+1} \wedge \ldots \wedge s_k) \in T_s S(\textstyle\bigwedge^k \mathbb{R}^d) \mid \\ &\qquad\qquad y \perp \operatorname{span}\{s_1, \ldots, s_k\}\}. \end{aligned} \tag{21}$$

We are going to construct a normal basis for $T\varphi$ on $T\mathrm{Gr}(k,d)$. First, redefine Φ on the probability space $(\mathrm{Gr}(k,d) \times \Omega, \mu)$ by trivial extension. Thus, Φ induces the skew product flow $(v; u, \omega) \mapsto (\Phi(t,\omega)v; \varphi(t,\omega)u, \vartheta_t\omega)$ on the (trivial) vector bundle $\mathbb{R}^d \times (\mathrm{Gr}(k,d) \times \Omega)$. By construction, the subbundle

$$\{(v; u, \omega) \mid v \in u\} \tag{22}$$

is μ-a. s. invariant under the skew product flow. Consequently, there exists a normal basis $s_i : \mathrm{Gr}(k,d) \times \Omega \to \mathbb{R}^d$, $1 \le i \le k$, for Φ on the invariant subbundle (22). In particular, $\operatorname{span}\{s_i(u,\omega) \mid 1 \le i \le k\} = u$ for μ-almost all (u, ω).
Since $\{s_i\}_{1 \le i \le k}$ is a normal basis for an invariant subbundle, it can be supplemented to a full normal basis for Φ on the whole bundle by vectors, say,

$$v_1, \ldots, v_{d-k} : \mathrm{Gr}(k,d) \times \Omega \to \mathbb{R}^d.$$

A normal basis for $\bigwedge^k \Phi$ is obtained from $\{s_1, \ldots, s_k, v_1, \ldots, v_{d-k}\}$ by taking a maximal number of linearly independent elements of

$$\begin{aligned} \{s_{i_1} \wedge \ldots \wedge s_{i_p} \wedge v_{j_1} \wedge \ldots \wedge v_{j_{k-p}} \mid \\ 1 \le p \le k,\ 1 \le i_1 < \ldots < i_p \le k,\ 1 \le j_1 < \ldots < j_{k-p} \le d-k\} \end{aligned}$$

and ordering them appropriately.
Denote by V the set of all elements of this normal basis except for $s = s_1 \wedge \ldots \wedge s_k$, so V has exactly $\binom{d}{k} - 1$ elements. Remark 3.3 implies that

$$W = \{w : \mathrm{Gr}(k,d) \times \Omega \to T(S(\textstyle\bigwedge^k \mathbb{R}^d)) \mid w(u,\omega) = v(u,\omega) - \langle v(u,\omega), s(u)\rangle s(u), v \in V\}$$

is a normal basis for $T\varphi$ on $T(S(\bigwedge^k \mathbb{R}^d))$. Having identified $T\mathrm{Gr}(k,d) \subset T(S(\bigwedge^k \mathbb{R}^d))$ in (21), it is immediate that for $v_* \in \{v_1 \ldots, v_{d-k}\}$ and $1 \le i \le k$

$$w = (s_1 \wedge \ldots \wedge s_{i-1} \wedge v_* \wedge s_{i+1} \wedge \ldots \wedge s_k) - \langle v, s\rangle (s_1 \wedge \ldots \wedge s_k) \tag{23}$$

satisfies $w(u,\omega) \in T\mathrm{Gr}(k,d)$ (μ-a. s.). Since there are exactly $k(d-k) = \dim \mathrm{Gr}(k,d)$ elements of W which are of the form (23), the set $W_0 = \{w \in W \mid (23) \text{ is satisfied}\}$ is a normal basis for $T\varphi$ on $T\mathrm{Gr}(k,d)$.

It remains to check which Lyapunov exponents are realized by elements of W_0. From (18) we obtain for w satisfying (23)

$$\begin{aligned}
\lambda^{\varphi}(w) &= \lambda^{\wedge^k \Phi}(s_1 \wedge \ldots \wedge s_{i-1} \wedge v_* \wedge s_{i+1} \wedge \ldots \wedge s_k) - (\lambda_{j_1} + \lambda_{j_2} + \ldots + \lambda_{j_k}) \\
&= \sum_{\substack{p=1 \\ p \neq i}}^{k} \lambda_{j_p} + \lambda(v_*) - \sum_{p=1}^{k} \lambda_{j_p} \\
&= \lambda(v_*) - \lambda_{j_i},
\end{aligned}$$

where the λ's denote exponents of Φ.
Since $\lambda(v_*)$ takes all values $\{\lambda_p \mid p \notin \{j_1,\ldots,j_k\}\}$, and λ_{j_i} takes all values $\{\lambda_p \mid p \in \{j_1,\ldots,j_k\}\}$, the proof is complete. □

References

[1] L. Arnold and V. Wihstutz, pp. 1–26 in *Lyapunov Exponents. Proceedings. Bremen 1984*, L. Arnold and V. Wihstutz (eds.), Lecture Notes in Mathematics 1186, Springer, Berlin 1986

[2] P. H. Baxendale, The Lyapunov spectrum of a stochastic flow of diffeomorphisms, p. 322–337 in *Lyapunov Exponents. Proceedings. Bremen 1984*, L. Arnold and V. Wihstutz (eds.), Lecture Notes in Mathematics 1186, Springer, Berlin 1986

[3] P. Bougerol and J. Lacroix, *Products of Random Matrices with Applications to Schrödinger Operators*, Birkhäuser, Boston 1985

[4] H. Crauel, *Random Dynamical Systems: Positivity of Lyapunov Exponents, and Markov Systems*, Dissertation, Universität Bremen 1987

[5] J. F. C. Kingman, The ergodic theory of subadditive stochastic processes, *J. Roy. Statist. Soc. Ser. B* 30 (1968) 499–510

[6] L. H. Loomis and S. Sternberg, *Advanced Calculus*, Addison-Wesley, Reading (Massachusetts) 1968

Eigenvalue representation for the Lyapunov exponents of certain Markov processes

Arie Leizarowitz
Department of Mathematics
Technion-Israel Institute of Technology
Haifa 32000

1 Introduction

We aim at obtaining explicit expressions for the Lyapunov exponents of certain Markov processes. As mentioned by Wihstutz in his survey paper [10] there are not many results in this direction. In fact all the explicit expressions refer to two-dimensional systems, and in particular to the random harmonic oscillator. Moreover, the framework of these results is that the system depends on some small parameter ϵ and one computes the behaviour of the Lyapunov exponent in the limit where $\epsilon \to 0$.

We consider space-homogeneous processes and study the relationship between their Lyapunov and p-moment Lyapunov exponents $g(p)$. Part of the discussion holds in quite general context. We will, however, obtain more explicit results for two-dimensional random evolution systems.

For a stochastic process $t \to x(t, x_0)$ in R^n, $x_0 \in R^n$ deterministic, λ is the *Lyapunov exponent* if

$$\lambda = \lim_{t\to\infty} \frac{1}{t} \log |x(t, x_0)| \text{ a.s., for a.e. } x_0,$$

and the p-moment Lyapunov exponents are

$$g(p, x_0) = \lim_{t\to\infty} \frac{1}{t} \log E|x(t, x_0)|^p. \tag{1.1}$$

(We will introduce below another definition for $g(p, x_0)$ which will not require the existence of a limit in (1.1).) The relationship between the Lyapunov and the p-moment Lyapunov exponents was observed and studied by Arnold (see Arnold [1] and Arnold, Oeljeklaus & Pardoux [3]). For the problem which he studied Arnold has shown that the limit in (1.1) does not depend on x_0, it exists for every p, $g(\cdot)$ is convex with $g(0) = 0$ and $\lambda = g'(0)$. We will show that most of these properties may be established for a large class of Markov processes.

A large part of the recent work on Lyapunov exponents was concerned with linear differential equations of various types. We mention just a few studies out of the vast literature on the subject: Has'minskii [8], [9], Arnold, Oeljeklaus & Pardoux [3] and Baxendale [5] for white driving noise; Arnold [1], Arnold, Kliemann & Oeljeklaus [2] and

Crauel [7] for real driving noise; Arnold, Papanicolaou & Wihstutz [4] and Blankenship & Li [6] for jump-processes driving noise.

Eigenvalues representations for Lyapunov exponents were derived, e.g., in the following studies. In Arnold [1] and Arnold, Kliemann & Oeljeklaus [2] linear differential equations driven by real noise are considered, where the driving noise is a diffusion process on a C^∞ or analytic compact manifold. Under hypoellipticity assumptions the Peron-Frobenius theory is employed to represent $g(p)$ as the isolated principal eigenvalue of a certain operator, which corresponds to a nonnegative eigenfunction (Theorem 1 in Arnold [1], Propsition 5.1 (ii) in Arnold, Kliemann & Oeljeklaus [2]).

In Arnold, Oeljeklaus & Pardoux [3] stochastic differential equations in R^d driven by white noise are considered. Under a certain non-degeneracy assumption the p-moment Lyapunov exponent is represented as an isolated principal eigenvalue of a certain operator which generates a semigroup of positive operators on the continuous functions on the unit sphere. Similar result appears in Baxendale [5] while considering the same framework. In these studies the semigroup under consideration is compact and irreducible, which enables the application of a generalized Peron-Frobenius theorem to deduce the existence of a nonnegative eigenfunction corresponding to the principal eigenvalue $g(p)$.

In section 2 we display the notations, introduce our definitions and present some general results. In section 3 we consider random evolution homogeneous differential equations and derive the main result of representing $g(p)$, the p-moment Lyapunov exponent, as an *approximate eigenvalue* of a certain operator. Section 4 is concerned with two-dimensional systems, in which case $g(p)$ are *genuine eigenvalues* (and not merely approximate eigenvalues) of certain differential operators on the unit circle. An application to the random-evolution harmonic oscillator is described in section 5.

2 Framework, notations and general results

We consider a Markov process $\{x_t\}_{t\geq 0}$ with a probability space $(\Omega, \mathcal{F}, P)$ in the state space X which is a cone endowed with a metric ρ. We denote $|x| = \rho(x, 0)$ for every $x \in X$ and assume that $|x| = 0$ if and only if $x = 0$, and $|\alpha x| = \alpha|x|$ for every $x \in X$ and $\alpha > 0$. We denote $C = \{x \in X : |x| = 1\}$ and consider C as a metric space with the metric ρ restricted to C. Let $\mu(d\theta)$ be a fixed Borel measure on C. A typical situation with which we will deal below is $X = R^n \times \{1, \cdots, m\}$ for some positive integers n and m, where we identify the m points $(0, i)$, $1 \leq i \leq m$, and denote all these points by 0. In this situation $|(x, i)| = ||x||$ for some norm $||\cdot||$ on R^n, and $\mu(d\theta)$ is the product measure of m Lebesgue measures on S^{n-1}.

We consider Markov processes which are *space homogeneous* in the sense that

$$P_{x_0}(x(t) \in A) = P_{\alpha x_0}(x(t) \in \alpha A) \tag{2.1}$$

for every $x_0 \in X$, $\alpha > 0$, $t > 0$ and every Borel set $A \subseteq X$. We define the random variables

$$Y_p(s, x_0) = \int_0^\infty e^{-st}|x(t, x_0)|^p dt \tag{2.2}$$

for a deterministic starting point $x_0 \neq 0$, which are well-defined (possibly $+\infty$) for every p and s.

Assumption A. *For every starting point $x_0 \neq 0$ there exist some $s_0 < s_1$ such that*

$$Y_p(s_0, x_0) \notin L^1(\Omega) \text{ and } Y_p(s_1, x_0) \in L^1(\Omega).$$

Definition 2.1. *For every p and x_0 we define $g(p, x_0)$ as the infimal value of the numbers s for which $Y_p(s, x_0)$ in (2.2) belongs to $L^1(\Omega)$.*

It is clear that whenever the limit in (1.1) exists then, since it expresses the *exact* exponential growth of $t \to E|x(t, x_0)|^p$, it is equal to $g(p, x_0)$.

It follows from Assumption A that $g(p, x_0)$ is finite. It may, however, depend on the initial value x_0, and we want to exclude this possibility. We consider the projected process $\theta(t) = x(t)/|x(t)|$ which is a stochastic process in the state space $C = \{x \in X : |x| = 1\}$. It follows that $t \to \theta(t)$ is a Markov process on C with the transition probability function

$$Q_{\theta_0}(t, B) = P_{x_0}(t, A). \tag{2.3}$$

The function $P_{x_0}(t, A)$ in (2.3) is the transition function of the process $t \to x(t, x_0)$, while $A \subseteq X$ and $x_0 \in X$ are given in terms of $B \subseteq C$ and $\theta_0 \in C$ by $A = \cup_{\alpha>0}\alpha B$ and $x_0 = \beta\theta_0$ for some $\beta > 0$. (By (2.1) the definition in (2.3) is independent of the choice of $\beta > 0$.)

Assumption B.

(i) *The process $t \to x(t)$ is such that $x(t) \neq 0$ a.s. for all $t \geq 0$ whenever $x(0) \neq 0$ a.s.*

(ii) *The process $t \to \theta(t)$ is such that for every $\theta_0 \in C$*

$$Q_{\theta_0}(\theta(t_0) \in B) > 0 \text{ for some } t_0 > 0$$

whenever $\mu(B) > 0$.

Definition 2.2. *We call $t \to x(t)$ an* homogeneous irreducible process *if it is a space homogeneous Markov process which satisfies Assumption B.*

The following result asserts that under Assumptions A and B the value of $g(p, x_0)$ is essentially independent of x_0. Since $g(p, x_0) = g(p, \alpha x_0)$ for every $\alpha > 0$, we restrict ourself to starting points $x_0 \in C$.

Proposition 2.3. *Let assumptions A and B hold. Then for every p there exists a number $g(p)$ such that for μ-almost every $x_0 \in C$*

$$s > g(p) \Rightarrow Y_p(s, x_0) \in L^1(\Omega), \; s < g(p) \Rightarrow Y_p(s, x_0) \notin L^1(\Omega).$$

Remark 2.4. Let $X = R^n$ and $x(t, x_0)$ depend linearly on x_0. Let $\tilde{m}(dx)$ be the translationally invariant Lebesgue measure on X, and $\mu(d\theta)$ the Lebesgue measure on S^{n-1}. We claim that if $p > 0$, then $g(p)$ is the threshold value for *every* x_0. To see this let A be the set of initial values x_0 for which, according to Proposition 2.3, $g(p, x_0) = g(p)$ holds.

Then the complement of A has null $\tilde{m}$-measure, hence $A+A=X$ (since $A\cap(x-A)$ has infinite measure, in particular is nonempty, for every $x\in X$). Then every $x_0\in X\setminus\{0\}$ can be written as $x_0=x_1+x_2$ where $x_1,x_2\in A$, and by linearity $x(t,x_0)=x(t,x_1)+x(t,x_2)$. For $0<p<1$ we obtain

$$|x(t;x_0)|^p\leq|x(t;x_1)|^p+|x(t;x_2)|^p$$

which implies that $g(p,x_0)\leq g(p)$. (The case $p\geq 1$ is similar.) The converse inequality always holds under Assumption B.

Proof of Proposition 2.3: The mapping $x_0\to g(p,x_0)$ is measurable and bounded and we have to show that it is essentially constant on C (in the $\mu(d\theta)$ measure). If B is such that $\mu(B)>0$ then by Assumption B

$$Q_{x_0}(\theta(t_0)\in B)>0 \tag{2.4}$$

for some $t_0>0$. We estimate $EY_p(s,x_0)$ by conditioning on the values of $x(t_0,x_0)$ as follows:

$$E\int_0^\infty e^{-st}|x(t,x_0)|^p dt\geq e^{-st_0}E[E\int_{t_0}^\infty e^{-s(t-t_0)}|x(t,x_0)|^p dt|x(t_0,x_0)]=$$

$$=e^{-st_0}\int_X E[\int_0^\infty e^{-st}|x(t,y)|^p dt]P_{x_0}(x(t_0)\in dy)\geq$$

$$\geq e^{-st_0}\int_A E[\int_0^\infty e^{-st}|x(t,y)|^p dt]P_{x_0}(x(t_0)\in dy)$$

where $A=\cup_{\alpha>0}\alpha B$. If s is such that $s<g(p,y)$ for every $y\in B$ then $E\int_0^\infty e^{-st}|x(t,y)|^p dt$ $=\infty$ for every $y\in A$. Since (2.4) implies that $P_{x_0}(x(t_0)\in A)>0$ it follows that $E\int_0^\infty e^{-st}|x(t,x_0)|^p dt=\infty$. We thus conclude that $s\leq g(p,x_0)$ whenver $s<g(p,y)$ for every $y\in B$. Since x_0 is an arbitrary point in C and B is any set with $\mu(B)>0$, there is a null set $N\subseteq C$ and a number $g(p)$ such that $g(p,y)=g(p)$ if $y\in C\setminus N$, and $g(p,y)\geq g(p)$ if $y\in N$, which is the assertion of the proposition.

□

It is immediate from the definition, using Hölder inequality, that $g(\cdot)$ is convex, which enables us to define

$$k_+=\lim_{p\to 0+}\frac{g(p)}{p},\quad k_-=\lim_{p\to 0-}\frac{g(p)}{p}.$$

Theorem 2.5. *Let Assumptions A and B hold for the homogeneous process* $t\to x(t,x_0)$. *Let* $\lambda_\pm(x_0)$ *be defined by*

$$\lambda_-(x_0)=\liminf_{t\to\infty}\frac{1}{t}\log|x(t,x_0)|,\quad \lambda_+(x_0)=\limsup_{t\to\infty}\frac{1}{t}\log|x(t,x_0)|. \tag{2.5}$$

Then

$$\lambda_-(x_0)\leq k_+ \text{ and } k_-\leq\lambda_+(x_0) \text{ a.s. for almost every } x_0\in C. \tag{2.6}$$

In particular if $\lambda=\lim_{t\to\infty}\frac{1}{t}\log|x(t;x_0)|$ *exists a.s. then*

$$k_-\leq\lambda\leq k_+ \text{ for a.e. } x_0\in C,$$

and if, furthermore, $g(\cdot)$ *is differentiable at* $p = 0$ *then*

$$\lambda(x_0) = g'(0) \text{ a.s. for a.e. } x \in C.$$

Proof: A real number k satisfies

$$g(p) \geq kp \text{ for every } p > 0 \ (p < 0) \tag{2.7}$$

if and only if it satisfies

$$k \leq k_+ \ (k \geq k_-). \tag{2.8}$$

For a given x_0 let $\kappa(x_0)$ be a random variable which satisfies

$$\kappa(x_0) < \lambda_-(x_0) \text{ a.s.} \tag{2.9}$$

Then by the definition in (2.5), there is a positive random variable $\omega \to c(\omega)$, such that *a.s.* $|x(t, x_0)| \geq c(\omega)e^{\kappa(x_0)t}$ for all $t \geq 0$. For a positive p and an $s > g(p)$ we thus have

$$e^{-st}|x(t, x_0)|^p \geq c(\omega)^p e^{(p\kappa(x_0)-s)t} \tag{2.10}$$

for every $t \geq 0$. From the fact that $E\int_0^\infty e^{-st}|x(t, x_0)|^p dt < \infty$ for μ-almost every x_0 (for this value of p) it follows that *a.s.* $\int_0^\infty e^{-st}|x(t, x_0)|^p dt < \infty$ for μ-almost every x_0. We consider only rational positive values for p and conclude that *a.s.* $\int_0^\infty e^{-st}|x(t, x_0)|^p dt < \infty$ for μ-almost every x_0 and every rational $p > 0$ which, in view of (2.10), yields that *a.s.* $p\kappa(x_0) - s < 0$ for μ-almost every x_0 and every rational $p > 0$. Since the last inequality holds for every $s > g(p)$ it follows that *a.s.* $g(p) \geq p\kappa(x_0)$ for every rational p, for μ-almost every x_0. But then it holds for every $p > 0$ for μ-almost every x_0. This being true for an arbitrary random variable $\kappa(x_0)$ which satisfies (2.9) we conclude that *a.s.* $g(p) \geq \lambda_-(x_0)p$ for every $p > 0$, for μ-almost every x_0. A similar argument for negative values of p implies that *a.s.* $g(p) \geq \lambda_+(x_0)p$ for every $p < 0$, for μ-almost every x_0. By the equivalence of (2.7) and (2.8) this implies (2.6).

□

3 Random evolution homogeneous differential equations

We consider differential equations of the form

$$\frac{dy}{dt} = f_{j_t}(y(t)), \ y(0) = y_0 \in R^n \tag{3.1}$$

in the state space $R^n \times \{1, \cdots, m\}$. The functions $f_i : R^n \to R^n$, $1 \leq i \leq m$ in (3.1) are homogeneous of order 1 so that

$$f_i(\alpha y) = \alpha f_i(y) \text{ for every } \alpha > 0 \text{ and } y \in R^n \setminus \{0\},\ 1 \leq i \leq m$$

and $\{j_t\}_{t\geq 0}$ is an irreducible Markov chain in $\{1, \cdots, m\}$. The Markov process under consideration is $x(t) = (y(t), j_t)$ with the initial value $x_0 = (y_0, j_0)$. The metric $\rho(x_1, x_2)$

for $x_1 = (y_1, i_1)$ and $x_2 = (y_2, i_2)$ is defined as $||y_1 - y_2||$ if $i_1 = i_2$, and as $||y_1|| + ||y_2||$ if $i_1 \neq i_2$, where $||\cdot||$ is some norm on R^n. The measure $\mu(dx)$ is the product measure of m Lebesgue measures on S^{n-1}. Assuming that the functions $f_i(\cdot)$ are continuous on $R^n \setminus \{0\}$ it follows that

$$||y_0||e^{-ct} \leq |x(t)| \leq ||y_0||e^{ct}$$

for some constant $c > 0$, every $t > 0$ and $\omega \in \Omega$, so that Assumption A holds.

We denote

$$\Phi_i(t, y_0) = E_{x_0}[|x(t, x_0)|^p \,|x_0 = (y_0, i)], 1 \leq i \leq m \tag{3.2}$$

and consider the Laplace transforms

$$\Psi_i(s, y_0) = \int_0^\infty e^{-st}\Phi_i(t, y_0)dt. \tag{3.3}$$

We observe that $\Psi_i(s, y_0)$ is finite (infinite) when $s > g(p)$ $(s < g(p))$. The functions $\{\Phi_i\}_{i=1}^m$ satisfy the backward equation

$$\frac{\partial \Phi_i}{\partial t} = \frac{\partial \Phi_i}{\partial y} \cdot f_i(y) + \sum_{j=1}^m g_{ij}\Phi_j(t, y),\ \Phi(0, y) = |y|^p,\ 1 \leq i \leq m,$$

where $(g_{ij})_{i=1}^m$ is the generator of $\{j_t\}_{t\geq 0}$. Then the Laplace transforms $\{\Psi_i\}_{i=1}^m$ satisfy

$$s\Psi_i(s, y) - |y|^p = \frac{\partial \Psi_i}{\partial y} \cdot f_i(y) + \sum_{j=1}^m g_{ij}\Psi_j(s, y),\ 1 \leq i \leq m. \tag{3.4}$$

The homogeneity implies that there exist functions $u_i(s, \theta)$ such that

$$\Psi_i(s, y) = u_i(s, \theta)|y|^p \text{ for } y = |y|\theta, \tag{3.5}$$

where $\theta \in S^{n-1}$. Let $(\theta_1, \cdots, \theta_{n-1})$ be a coordinate system for θ in S^{n-1}. Then every point $y \in R^n \setminus \{0\}$ is represented by the n-tupple $(r, \theta_1, \cdots, \theta_{n-1})$. There are then n function $\gamma_i : S^{n-1} \to R^1$ and first order differential operators $\{\tilde{L}_i(\theta)\}_{i=1}^n$ in the variables $\theta_1, \cdots, \theta_{n-1}$ such that

$$\frac{\partial \psi}{\partial y_j} = \gamma_j(\theta)\frac{\partial \psi}{\partial r} + \frac{1}{r}\tilde{L}_j(\theta)\psi$$

for C^1 real-valued functions ψ on R^n. For Ψ_i as in (3.5) we thus have

$$\frac{\partial \Psi_i}{\partial y_j} = p\gamma_j(\theta)u_i(s, \theta)r^{p-1} + [\tilde{L}_j(\theta)u_i(s, \theta)]r^{p-1}.$$

Using the last expression and denoting $f_i(y) = (f_{i1}(y), \cdots, f_{in}(y))^T$ where each $f_{ij} : R^n \setminus \{0\} \to R^1$ is an homogeneous function, we obtain

$$\frac{\partial \Psi_i}{\partial y} \cdot f_i(y) = [pG_i(\theta)u_i(s, \theta) + L_i(\theta)u(s, \theta)]r^p$$

denoting

$$G_i(\theta) = \sum_{j=1}^n \gamma_i(\theta)f_{ij}(\theta),\ \ L_i(\theta) = \sum_{j=1}^n f_{ij}(\theta)\tilde{L}_j(\theta).$$

Substituting the last expressions in (3.4) yields

$$L_i(\theta)u_i(s,\theta) + [pG_i(\theta) - s]u_i + \sum_{j=1}^{m} g_{ij}u_j + 1 = 0,\ 1 \le i \le m. \tag{3.6}$$

We thus consider for every real p the operator

$$(T_p u)_i(\theta) = L_i(\theta)u_i(\theta) + pG_i(\theta)u_i(\theta) + \sum_{j=1}^{m} g_{ij}u_j(\theta),\ 1 \le i \le m$$

whose domain and range are in $C(S^{n-1}, R^m)$. The following is our main result:

Theorem 3.1. *For every p the number $g(p)$ is an* approximate eigenvalue *of T_p in the sense that for every $\epsilon > 0$ there exists a $u_\epsilon(\cdot)$ satisfying*

$$||T_p u_\epsilon - g(p)u_\epsilon|| < \epsilon \text{ and } ||u_\epsilon|| = 1.$$

($||\cdot||$ is the sup-norm.)

Proof: We will prove that the number $g(p)$ is such that for every $s > g(p)$ there exists a function v^s which satisfies

$$T_p v^s = sv^s + \rho^s,\ ||v^s|| = 1 \tag{3.7}$$

where $||\rho^s|| \to 0$ as $s \downarrow g(p)$. Clearly this will imply the assertion of the theorem.

The function $s \to u(s,\theta)$ *is analytic* for $s > g(p)$ and *fails to be analytic at* $s = g(p)$. For every fixed i and θ the mapping $s \to u_i(s,\theta)$ is monotone decreasing for $s > g(p)$, hence $\lim ||u_i(s,\cdot)||$ exists as $s \downarrow g(p)$, where here $||\cdot||$ denotes the L^∞-norm for continuous functions on the unit circle. We consider the following two possibilities separately.

Case 1: $\lim_{s\downarrow g(p)} ||u(s,\cdot)|| = \infty$. In this case we define

$v^s(\theta) = u(s,\theta)/||u(s,\cdot)||$, and since (by (3.6)) $T_p u + \begin{pmatrix} 1 \\ \vdots \\ 1 \end{pmatrix} = su(s,\theta)$, it follows that (3.7) holds with $\rho^s = -\frac{1}{||u(s,\cdot)||}\begin{pmatrix} 1 \\ \vdots \\ 1 \end{pmatrix}$.

Case 2: In this case $\lim_{s\downarrow g(p)} ||u(s,\cdot)||$ is finite. We consider the function $u^{(1)}(s,\theta) = \frac{\partial u}{\partial s}(s,\theta)$ defined for $s > g(p)$ which satisfies

$$L_i u_i^{(1)} + [pG_i - s]u_i^{(1)} + \sum_{j=1}^{m} g_{ij}u_j^{(1)} - u_i = 0,\ 1 \le i \le m. \tag{3.8}$$

Also $|u_i^{(1)}(s,\theta)|$ is monotone decreasing for $s > g(p)$, for every fixed θ. If $\lim_{s\downarrow g(p)} ||u^{(1)}(s,\cdot)|| = \infty$ then the boundedness of $||u(s,\cdot)||$ for $s > g(p)$ implies, after dividing (3.8) by $||u^{(1)}(s,\cdot)||$ and using an argument similar to the one in Case 1, the validity of (3.7) for some $v^s(\cdot)$. More generally, we consider the functions

$$u^{(k)}(s,\theta) = \frac{\partial^{(k)} u}{\partial s^k}(s,\theta),\ k = 1,2,3,\cdots$$

defined for $s > g(p)$, and observe that $|u_i^{(k)}(s,\theta)|$ is monotone decreasing in s for every fixed i and θ. These functions satisfy the equations

$$L_i u_i^{(k)} + [pG_i - s]u_i^{(k)} + \sum_{j=1}^{m} g_{ij}u_j^{(k)} - ku_i^{(k-1)} = 0,\ 1 \le i \le m. \tag{3.9}$$

If for some $k \ge 1$ we have $\lim_{s \downarrow g(p)} ||u^{(k)}(s,\cdot)|| = \infty$ then we are done. If, however, this never occurs then for every k let $M_k = \sup_{s>g(p)} ||u(s,\cdot)||$, and denote $v^{(k)} = u^{(k)}/M_k$. Dividing (3.9) by M_k we obtain

$$L_i v_i^{(k)} + [pG_i - s]v_i^{(k)} + \sum_{j=1}^{m} g_{ij}v_j^{(k)} - v_i^{(k-1)}(\frac{kM_{k-1}}{M_k}) = 0,\ 1 \le i \le m. \tag{3.10}$$

If $\liminf_{k\to\infty} \frac{kM_{k-1}}{M_k} = 0$, then the assertion in (3.7) will follow from (3.10), since $||v^{(k)}|| \le 1$ for every $k \ge 1$ and every $s > g(p)$. If $\liminf_{k\to\infty} \frac{kM_{k-1}}{M_k} > 0$ then for some $\delta > 0$ we have $\frac{kM_{k-1}}{M_k} \ge \delta$ for all $k \ge 1$, which implies that $M_k \le Ck!(\frac{1}{\delta})^k$, for some $C > 0$ and every $k \ge 1$. Since M_k is a bound on $||\frac{\partial^k u(s,\theta)}{\partial s^k}||$ for all $s > g(p)$, it follows that $s \to u(s,\theta)$ can be extended to an analytic function on $s > g(p) - \delta/2$. This contradicts the fact that $s \to u(s,\theta)$ fails to be analytic at $s = g(p)$, which concludes the proof of the theorem.

□

4 Eigenvalue representation for the Lyapunov exponents of two-dimensional systems

In this situation the random evolution equation takes the form

$$\begin{cases} \dot{r}(t)/r(t) = G_{j_t}(\theta(t)) \\ \dot{\theta}(t) = F_{j_t}(\theta(t)) \end{cases} \tag{4.1}$$

in polar coordinates, for some known functions $G_i(\cdot)$, $F_i(\cdot)$, $1 \le i \le m$ defined on S^1, the unit circle in R^2. We assume the following:

A STANDING HYPOTHESIS

(i) The functions $F_i(\cdot)$, $1 \le i \le m$ are C^1 and the functions $G_i(\cdot)$, $1 \le i \le m$ are C^2.

(ii) Each function $F_i(\cdot)$ has a finite number of zeros in $[0, 2\pi]$.

(iii) Every $F_i(\cdot)$ changes signs at each of his zero points.

(iv) The functions $F_i(\cdot)$ and $F_j(\cdot)$ have distinct zeros whenever $i \ne j$.

The following is easily verified and we omit the proof.

Proposition 4.1. *A necessary and sufficient condition for the validity of Assumption B is that there do not exist two points θ_+ and θ_- such that*

$$F_i(\theta_+) > 0 \text{ and } F_i(\theta_-) < 0 \text{ for all } 1 \le i \le m. \tag{4.2}$$

For the solution $\theta(\cdot)$ of (4.1) the Markov process $t \to (\theta(t), j_t)$ has a unique equilibrium measure $\sigma(d\theta) = \{\sigma_i(d\theta)\}_{i=1}^m$, and by (4.1)

$$\frac{1}{t}\log r(t) = \frac{1}{t}\int_0^t G_{j_\tau}(\theta(\tau))d\tau. \tag{4.3}$$

The existence of a deterministic limit in the right-hand-side of (4.3) for $\sigma(d\theta)$-almost every θ_0 follows from the Ergodic Theorem, assuming (4.2). The one-dimensionality of S^1 actually implies that this limit exists for *every* θ_0:

Theorem 4.2. *Consider a two-dimensional random evolution system which satisfies Assumption B. Then* $\lim_{t\to\infty}\frac{1}{t}\int_0^t \log|x(\tau, x_0)|d\tau$ *exists for every* $x_0 \in R^2 \setminus \{0\}$ *and has a nonrandom value* λ, *the (top) Lyapunov exponent.*

We will next show that in the present situation there exist genuine eigefunctions for T_p corresponding to $g(p)$, not merely approximate eigenfunctions. The eigenvalue problem for T_p is

$$\begin{cases} F_i(\theta)u_i'(\theta) + pG_i(\theta)u_i(\theta) + \sum_{j=1}^m g_{ij}u_j(\theta) = g(p)u_i(\theta) \\ i = 1, \cdots, m, \end{cases} \tag{4.4}$$

for a continuous function $u(\cdot)$ with $||u|| = 1$. We denote by Z the set of all zeros of the functions $\{F_i(\cdot) : 1 \le i \le m\}$.

Theorem 4.3. *There exists* $p_0 > 0$ *such that for every* p *with* $|p| < p_0$ *there exists a nontrivial continuous function* v *on* S^1 *which satisfies*

$$T_p v = g(p)v \text{ on } S^1 \setminus Z.$$

Proof: By Theorem 3.1 there exists, for every $\epsilon > 0$, a function u_ϵ which satisfies

$$||T_p u_\epsilon - g(p)u_\epsilon|| < \epsilon \text{ and } ||u_\epsilon|| = 1.$$

For a positive integer k and $\epsilon = \frac{1}{k}$ denote this function by $u^k(\cdot)$. Let J be a compact subset of $S^1 \setminus Z$. Then both the functions $\{u^k(\cdot)\}_{k=1}^\infty$ and their derivatives $\{\frac{du^k}{d\theta}(\cdot)\}_{k=1}^\infty$ are uniformly bounded on J (recalling $||u^k|| = 1$). Thus a subsequence of $\{u^k\}_{k=1}^\infty$ converges uniformly to a continuous function on J. By choosing an increasing sequence $\{J_i\}_{i=1}^\infty$ of compact subsets of $S^1 \setminus Z$ whose union is $S^1 \setminus Z$, we obtain via Cantor diagonal process a function $v(\cdot)$ which satisfies (4.4) on $S^1 \setminus Z$. We prove next that $v(\cdot)$ is *continuous* on S^1.

Let $\theta_i \in Z$ be a zero of $F_i(\cdot)$ and we consider a closed interval I which contains θ_i in its interior but contains no other member of Z. It follows from $||u^k(\cdot)|| = 1$ that for a fixed $j \ne i$ the sequence $\{u_j^k(\theta)\}$ converges to $v_j(\theta)$ uniformly on I, implying the continuity of $v_j(\cdot)$ at θ_i for $j \ne i$.

The ith equation in (4.4) is

$$F_i(\theta)u_i'(\theta) + [pG_i(\theta) + g_{ii} - g(p)]u_i(\theta) = -\sum_{j\ne i} g_{ij}u_j(\theta). \tag{4.5}$$

We recall that $g_{ii} < 0$, $g(p) \to 0$ as $p \to 0$, and that $F_i(\theta) = c(\theta - \theta_i)[1 + o(1)]$ for some constant $c \neq 0$ (where $o(1)$ represents a function which tends to zero as $\theta \to \theta_i$). Then for sufficiently small $|p|$ equation (4.5) has the form

$$c(\theta - \theta_i)u_i'(\theta) - [g + \alpha(\theta)]u_i(\theta) = h(\theta) \tag{4.6}$$

where $g > 0$ and c are constants, $\alpha(\cdot)$ is continuously differentiable and such that $\alpha(\theta_i) = 0$, and $h(\cdot)$ is continuously differentiable. This equation can be solved by quadratures, which enables to establish the continuity of $u_i(\cdot)$ as follows. Suppose, for example, that $\alpha(\cdot)$ and $h(\cdot)$ are constant on I so that $\alpha(\theta) \equiv 0$ and $h(\theta) \equiv h(\theta_i)$. Then the solution of (4.6) is

$$u_i(\theta) = a|\theta - \theta_i|^{g/c} - h(\theta_i)/g$$

for some constant a. Depending on the sign of c this is bounded (if $c > 0$) or unbounded (if $c < 0$ and $a \neq 0$) in a neighborhood of θ_i. In the case $c < 0$ there is just one solution (the one corresponding to $a = 0$) which is bounded.

Generally, when $\alpha(\cdot)$ and $h(\cdot)$ in (4.6) are not necessarily constant, then we obtain a solution of the form

$$u_i(\theta) = K(\theta) + \rho a(\theta)|\theta - \theta_i|^{g/c} \tag{4.7}$$

for some continuous functions $K(\cdot)$ and $a(\cdot)$, and a constant ρ. (This can be established, e.g. for $\theta > \theta_i$, by considering the differential equation satisfied by the function $w(\theta) = [u_i(\theta) + \frac{h(\theta_i)}{g}](\theta - \theta_i)^{-g/c}$.) We consider the following two cases:

Case 1: $c > 0$. In this situation the limit function $v_i(\cdot)$ which is defined on $I \setminus \{\theta_i\}$ is a solution of (4.5), which is of type (4.6), hence must be of the form (4.7) with a positive exponent g/c, so that the limit of $v_i(\theta)$ as $\theta \to \theta_i$ exists and is equal to $K(\theta_i)$. This proves the continuity of $v_i(\cdot)$ in this case.

Case 2: $c < 0$. In this situation the limit function $v_i(\cdot)$ is a bounded solution of (4.5) on $I \setminus \{\theta_i\}$. But then it must be the only solution of (4.5) which is continuous and bounded on I, which proves the continuity of $v_i(\cdot)$ in this case as well.

In order to conclude that $v_i(\cdot)$ is nontrivial it is enough to know that the convergence of $\{u^k(\cdot)\}$ to $v(\cdot)$ is uniform on I. For this we consider the explicit expression (4.7) for solutions. In case 2 ($c < 0$) the second term in this expression (which is a solution to the homogeneous equation), does not appear. The first term, however, tends to zero in the L^∞-norm as the inhomogeneous term $h(\cdot)$ in (4.6) tends to zero in the L^∞-norm. Moreover, the norm of the difference between the inhomogeneous terms in the equations for v_i and u_i^k on I indeed tends to zero as k grows to infinity, which implies the uniform convergence on I in case 2.

In case 1 there appears an additional term $\rho a(\theta)|\theta - \theta_i|^{g/c}$, where the function $a(\cdot)$ does not depend on the inhomogeneous term in (4.6). There are terms $\rho_0 a(\theta)|\theta - \theta_i|^{g/c}$ and $\rho_k a(\theta)|\theta - \theta_i|^{g/c}$ in the expressions for v_i and u_i^k respectively, and the uniform convergence of $u_i^k(\cdot)$ to $v_i(\cdot)$ on I follows from the boundedness of the scalar sequence $\{\rho_k\}$ (which follows from the boundedness of $\{u_i^k(\cdot)\}$ on I). This concludes the proof of the theorem.

□

Remark 4.4. Let $\mathcal{L}$ be the following operator on continuous functions $u(\cdot)$

$$(\mathcal{L}u)_i(\theta) = F_i(\theta)u_i'(\theta) + \sum_{j=1}^{m} g_{ij}u_j(\theta).$$

Then $\mathcal{L}$ is the generator of the induced process on S^1, and in our notations $\mathcal{L} = T_0$. Considering the eigenvalue problem (4.4) for T_p and differentiating (4.4) formally with respect to p at $p = 0$ we obtain the equation

$$(\mathcal{L}u)_i = \lambda - G_i,\ 1 \le i \le m. \tag{4.8}$$

This is an *additive eigenvalue* problem for the unknown scalar λ and the function $u(\cdot)$ which may be considered in the spirit of Fredholm alternatives. Under Assumptions A and B there exists a unique equilibrium measure $\sigma(d\theta) = (\sigma_1(d\theta), \cdots, \sigma_m(d\theta))$, which satisfies $\mathcal{L}^*\sigma = 0$. Integrating (4.8) against $\sigma(d\theta)$ yields

$$\lambda = \sum_{i=1}^{m} \int_0^{2\pi} G_i(\theta)\sigma_i(d\theta)$$

where we recognize on the right-hand-side the Lyapunov exponent of the process. We thus conclude that if λ and $u(\cdot)$ satisfy (4.8) then indeed λ is the top Lyapunov exponent.

Remark 4.5. It can be shown that under Assumptions A and B Theorem 4.3 provides a *characterization* of the p-moments Lyapunov exponents in the sense that there exists a neighborhood $\mathcal{V}$ of the origin in R^2 such that if $(p, s) \in \mathcal{V}$ and s is an eigenvalue of T_p then $s = g(p)$. A similar result is stated in Corollary 2.3 in Baxendale [5], the proof of which can be adapted to our situation.

5 The nonsingular case

If none of the functions $F_i(\cdot)$ vanishes on $[0, 2\pi]$ then we say that the system is *nonsingular*. The eigenvalue problem (4.8) for the Lyapunov exponent λ discussed in Remark 4.4 becomes

$$\begin{cases} z_i'(\theta) + \frac{1}{F_i(\theta)} \sum_{j=1}^m g_{ij} z_j(\theta) + \frac{G_i(\theta)}{F_i(\theta)} - \frac{\lambda}{F_i(\theta)} = 0 \\ i = 1, \cdots, m. \end{cases} \tag{5.1}$$

Let $Z(\theta, \theta')$ be the fundamental solution of the linear equation which is associated with (5.1). Then the solution of (5.1) which corresponds to the initial condition $z(0) = z_0$ is

$$z(\theta) = Z(\theta, 0)z_0 + \int_0^\theta Z(\theta, \theta')a(\theta')d\theta' - \lambda \int_0^\theta Z(\theta, \theta')b(\theta')d\theta' \tag{5.2}$$

where the functions $a : S^1 \to R^m$ and $b : S^1 \to R^m$ are given by

$$a_i(\theta) = -\frac{G_i(\theta)}{F_i(\theta)},\ b_i(\theta) = -\frac{1}{F_i(\theta)},\ 1 \le i \le m. \tag{5.3}$$

We look for a λ for which $z(\cdot)$ in (5.2) is periodic of period 2π. Thus we look for an initial value z_0 which satisfies

$$[I - Z(2\pi, 0)]z_0 = \alpha - \lambda\beta \tag{5.4}$$

where I is the unit $m \times m$ matrix,

$$\alpha = \int_0^{2\pi} Z(2\pi, \theta)a(\theta)d\theta,\ \beta = \int_0^{2\pi} Z(2\pi, \theta)b(\theta)d\theta$$

and $a(\cdot)$, $b(\cdot)$ are as in (5.3). It is easy to see that under Assumption B the constant functions are the only periodic solutions of the homogeneous equation associated with (5.1). Therefore, under Assumption B, unity is a simple eigenvalue of $Z(2\pi,0)$, which implies that

$$Y = \mathrm{Im}(I - Z(2\pi,0))$$

is an $(m-1)$-dimensional subspace of R^m. If $\beta \notin Y$ then for every α equation (5.4) can be solved *uniquely* for λ. Let η be an eigenvector corresponding to the unit eigenvalue of $Z^T(2\pi,0)$:

$$Z^T(2\pi,0)\eta = \eta. \tag{5.5}$$

The next result follows from the fact that $Y = \{y \in R^m : \eta^T \cdot y = 0\}$.

Theorem 5.1. *Let Assumptions A and B hold. Then the Lyapunov exponent λ is determined by* (5.4) *as the unique real number for which $\alpha - \lambda\beta$ belongs to Y if and only if $\eta^T \cdot \beta \neq 0$, where η satisfies* (5.5). *In this case λ is given explicitly by*

$$\lambda = \frac{\eta^T \cdot \alpha}{\eta^T \cdot \beta}. \tag{5.6}$$

Remark 5.2. Let $\mathcal{U}$ be the set of all pairs $(F(\cdot), G(\cdot))$ of continuous functions from S^1 into R^m such that the components $F_i(\cdot)$ never vanish on S^1 for every $1 \le i \le m$. Let $\mathcal{U}_0$ consist of those pairs in $\mathcal{U}$ for which the condition $\beta \notin Y$ in Theorem 5.1 holds. Then it is easy to see that $\mathcal{U}_0$ is an open and dense subset of $\mathcal{U}$ in the topology induced by $C(S^1, R^m) \times C(S^1, R^m)$. In this sense the condition $\eta^T \cdot \beta \neq 0$ holds generically.

We consider a random harmonic oscillator which is modelled by the following equation

$$\ddot{\xi}(t) + (1 + \epsilon_{j_t})\xi(t) = 0, \ \xi(t) \in R^1$$

where $\epsilon_i > -1$ for every $1 \le i \le m$. Then the two-dimensional vector $x(t) = \begin{pmatrix} \xi(t) \\ \dot{\xi}(t) \end{pmatrix}$ satisfies the random evolution equation

$$\frac{dx}{dt} = A_{j_t} x(t), \ A_i = \begin{pmatrix} 0 & 1 \\ -(1+\epsilon_i) & 0 \end{pmatrix} \text{ for } 1 \le i \le m.$$

It turns out that

$$F_i(\theta) = -1 - \epsilon_i \cos^2(\theta), \ G_i(\theta) = -\epsilon_i \sin(\theta)\cos(\theta), \ 1 \le i \le m$$

so that $F_i(\cdot)$ never vanishes, the system is nonsingular and Theorem 5.1 can be applied to obtain a formula for λ. E.g. if we consider the situation where

$$\epsilon_i = \epsilon c_i \text{ for some constants } c_i, \ 1 \le i \le m$$

and $\epsilon \to 0$, then it follows from (5.6) that

$$\lambda(\epsilon) = \lambda_0 \epsilon^2 + o(\epsilon^2)$$

where λ_0 can be computed explicitly. This expression, however, does not admit a nice simple form.

Acknowledgement

I would like to thank an anonymous referee for many valuable comments which considerably improved the quality of this paper.

References

[1] L. Arnold, (1984). A formula connecting sample and moment stability of linear stochastic systems, *SIAM Journal on Applied Mathematics*, 44, 793-802.

[2] L. Arnold, W. Kliemann & E. Oeljeklaus, (1985). Lyapunov exponents of linear stochastic systems. In *Lecture Notes in Mathematics, 1186, Lyapunov Exponents*, Ed. L. Arnold & V. Wihstutz, 85-126. Springer-Verlag, New York.

[3] L. Arnold , E. Oeljeklaus & E. Pardoux, (1985). Almost-sure and moment stability for linear Ito systems. In *Lecture Notes in Mathematics, 1186, Lyapunov Exponents*, Ed. L. Arnold & V. Wihstutz, 129-159. Springer-Verlag, New York.

[4] L. Arnold, G. Papanicolaou & V. Wihstutz, (1986). Asymptotic analysis of the Lyapunov exponent and rotation number of the random oscillator and applications. *SIAM Journal on Applied Mathematics*, 46, 427-450.

[5] P.H. Baxendale, (1987). Moment stability and large deviations for linear stochastic differential equations. In *Proceedings of the Taniguchi Symposium on Probabilistic Methods in Mathematics.* Katata & Kyoto 1985, Ed. N. Ikeda, 31-54.

[6] G.L. Blankenship & C.W. Li, (1986). Almost-sure stability of linear stochastic systems with Poisson coefficients. *SIAM Jornal on Applied Mathematics*, 46, 875-911.

[7] H. Crauel, (1984). Lyapunov numbers of Markov solutions of linear stochastic systems, *Stochastics*, 14, 1-18.

[8] R.Z. Has'minskii, (1967). Necessary and sufficient conditions for the asymptotic stability of linear stochastic systems, *Theory of Probability & Applications*, 12, 144-147.

[9] R.Z. Has'minskii, (1980). *Stochastic Stability of Differential Equations.* Sijthoff & Noordhoof, Alphan aan den Rijn, the Netherland.

[10] V. Wihstutz, (1985). Parameter dependence of the Lyapunov exponent for linear stochastic systems. A survey. In *Lecture Notes in Mathematics, Lyapunov Exponents.* Ed. L. Arnold & V. Wihstutz, 200-215. Springer-Verlag, New York.

Analytic Dependence of Lyapunov Exponents on Transition Probabilities

Yuval Peres*†

The Hebrew University, Jerusalem

1 Introduction

Let $\{X_n\}_{n\geq 1}$ be independent identically distributed random variables, taking finitely many values $\{A_1, \ldots, A_b\}$ in the space of $d \times d$ real matrices. In this paper we study the dependence of the top Lyapunov exponent

(1) $$\gamma_1 = \lim_{n\to\infty} \frac{1}{n} E \log \|X_n X_{n-1} \ldots X_1\|$$

on the probability parameters

(2) $$p_j = \Pr\{X_1 = A_j\}, \qquad 1 \leq j \leq b.$$

Here E denotes the expectation and $\|\cdot\|$ is a norm on the space of $d\times d$ matrices. In Section 2 we show that this dependence is real–analytic whenever the top Lyapunov exponent is *simple*, i.e., when γ_1 differs from the other Lyapunov exponents $\gamma_2, \ldots, \gamma_d$. See [Kif] or [BL, Section III.5] for the definition of all Lyapunov exponents. More generally, we have

Theorem 1. *Assume the matrices $A_1, \ldots, A_b$ are invertible. If the kth Lyapunov exponent γ_k for the i.i.d. random product of the matrices X_n above is* simple *(i.e., $\gamma_{k-1} > \gamma_k > \gamma_{k-1}$) and the probabilities in (2) satisfy*

$$p_j > 0, \qquad 1 \leq j \leq b,$$

---

*Partially supported by a grant from the Landau Centre for Mathematical Analysis.

†Current Address: Mathematics Department, Stanford University, Stanford CA 94305

then γ_k is (locally) a real-analytic function of $(p_1,\ldots,p_b)$. (Note that there are no irreducibility assumptions.)

The recent results of Goldsheid and Margulis [GM] which are based on work of Furstenberg [F] and Guivarch and Raugi [GR1] make Theorem 1 quite constructive in the sense that the simplicity of the Lyapunov spectrum is usually easy to verify. However, the results in [GR1], [GM] are not *effective* as they do not provide explicit lower bounds for the differences between consecutive Lyapunov exponents. This is reflected in our inability to exhibit explicitly domains of analytic continuation for γ_1 as a function of $(p_1,\ldots,p_b)$ in Theorem 1. The importance of explicit estimates for such domains in perturbation theory is discussed in [Ka, Section II.3].

The situation is brighter for nonnegative matrices.

Definition: Let A be a square matrix with strictly positive entries a_{ij}. *Birkhoff's contraction coefficient* $\tau(A)$ is defined by

$$\tau(A)=\frac{1-\varphi(A)^{1/2}}{1+\varphi(A)^{1/2}},$$

where

$$\varphi(A)=\min_{i,j,k,\ell}\frac{a_{ik}a_{j\ell}}{a_{jk}a_{i\ell}}$$

(note that $0\le\tau(A)<1$ in this case). If A is not strictly positive, set $\tau(A)=1$. See [S, Chapter 3] for properties of τ and for references.

In [P] we establish

Theorem 2. *Let $F=\{A_1,\ldots,A_b\}$ be a set of nonnegative matrices such that no A_i has an all zero row and at least one A_i is strictly positive. Let $\{X_n\}$ be i.i.d. F-valued random variables. Then there is a relatively open, connected subset Ω of the complex hyperplane*

$$H=\left\{(z_1,\ldots,z_b)\in\mathbf{C}^b\;\middle|\;\sum_{j=1}^{b}z_j=1\right\}$$

defined by

$$\Omega=\left\{z\in H\;\middle|\;\sum_{j=1}^{b}\tau(A_j)\,|z_j|<1\right\}$$

such that

(i) Ω contains the open simplex of positive probability vectors:

$$\Delta^0_{b-1}=\left\{(p_1,\ldots,p_b)\;\middle|\;p_j>0,\ \sum_{j=1}^{b}p_j=1\right\}. \tag{3}$$

(ii) The top Lyapunov exponent γ_1 defined in (1), as a function of the probability vector $(p_1, \ldots, p_b)$ defined in (2), may be extended to a complex analytic function on Ω.

(Note that we do not assume that the matrices A_j are invertible.)

We do not prove Theorem 2 here but refer to [P] where an extension to random matrix products with Markovian dependence is established. Real-analytic dependence of the top Lyapunov exponent on the *matrix entries* for *positive* matrices was established by Ruelle [R].

For more general matrices, Le Page [LP2] established Hölder continuity of the top Lyapunov exponent as a function of the *matrix entries* (under the assumptions of "strong irreducibility" and the contraction property, see Section 2 for the definition). An example which appears in [ST] shows that this is the best regularity which can hold in this generality. In this paper we are concerned with the dependence on probabilities rather than on matrix entries, so the results are quite different.

The rest of the paper is organized as follows: In Section 2 we start by recalling some fundamental stability results due to Furstenberg and Kifer [FK]. In conjunction with the analysis of Le Page [LP] they yield the proof of Theorem 1. It is then shown that the "simplicity" assumption in Theorem 1 may be replaced by an irreducibility assumption but one cannot remove both assumptions. An extension of Theorem 1 to Markovian random matrix products is presented in Section 3.

2 Continuity and analyticity properties for i.i.d. products

To read this section it is helpful to have a copy of [BL] at hand. Let μ be a compactly supported probability measure on the group $GL(d, \mathbf{R})$ of $d \times d$ invertible matrices. Denote

$$\gamma_1(\mu) = \lim_{n\to\infty} \frac{1}{n} E[\log \|X_n \ldots X_1\|], \tag{4}$$

where X_j are i.i.d random matrices with distribution μ.

A subspace $V \subset \mathbf{R}^d$ is called *μ-invariant* if $MV = V$ for μ-almost every $M \in GL(d, \mathbf{R})$. A μ-invariant subspace V determines naturally measures μ_V and $\mu_{\mathbf{R}^d/V}$ on $GL(V)$ and $GL(\mathbf{R}^d/V)$ respectively. With this notation we have

Lemma A. ([FK, Lemma 3.6])

$$\gamma_1(\mu) = \max\{\gamma_1(\mu_V), \gamma_1(\mu_{\mathbf{R}^d/V})\}.$$

Corollary B. ([FK]) *If $\mu, \mu_1, \mu_2, \ldots$ are compactly supported probability measures on $GL(d,\mathbf{R})$ with $\mathrm{supp}(\mu_n) \subset \mathrm{supp}(\mu)$ for all n and $\mu_n \to \mu$ weakly as $n \to \infty$, then*

$$\gamma_1(\mu_n) \to \gamma_1(\mu).$$

In particular, when μ is supported on finitely many matrices, $\gamma_1(\mu)$ depends continuously on the probabilities $(p_1, \ldots, p_b)$ defined in (2), in the open simplex Δ^0_{b-1}.

Proof: If μ is irreducible, i.e., any μ–invariant subspace is trivial, the corollary is contained in Theorem B of [FK]. The general case follows by induction on the dimension d. Indeed, if V is a nontrivial μ–invariant subspace of $\mathbf{R}^d$ then V is also μ_n–invariant, and the induction hypothesis implies that

$$\gamma_1((\mu_n)_V) \to \gamma_1(\mu_V) \qquad \text{and} \quad \gamma_1((\mu_n)_{\mathbf{R}^d/V}) \to \gamma_1(\mu_{\mathbf{R}^d/V})$$

as $n \to \infty$. Using Lemma A for μ_n and for μ gives the desired conclusion. □

Remark: The assumptions of Corollary B may be weakened but we shall not need this.

We collect now some definitions required in the sequel.

Definitions: Let F be a subset of $GL(d,\mathbf{R})$

(i) F is *irreducible* if there is no proper nontrivial linear subspace V of $\mathbf{R}^d$ satisfying $MV = V$ for all matrices $M \in F$.

(ii) F is *strongly irreducible* if there is no finite union of proper nontrivial linear subspaces of $\mathbf{R}^d$ which is left invariant by each matrix in F.

(iv) The *index* of F is the minimal integer k such that there exists a sequence $\{M_n\}_{n\geq 1}$ in the semigroup generated by F for which $\|M_n\|^{-1}M_n$ converges to a rank k matrix.

(v) We say that F is *contracting* when its index is 1.

(vi) Given any stationary sequence of random $d \times d$ matrices $\{X_n\}$, the Lyapunov exponents $\gamma_1, \ldots, \gamma_d$ may be defined by the formula

$$\gamma_1 + \gamma_2 + \ldots + \gamma_k = \lim_{n\to\infty} \frac{1}{n} E[\log \| \wedge^k X_n \ldots \wedge^k X_1 \|] \tag{5}$$

(we assume that the expectations in (5) are finite). For any $d \times d$ matrix M, $\Lambda^k M$ denotes the $\binom{d}{k} \times \binom{d}{k}$ matrix representing the action of M on the kth exterior power of $\mathbf{R}^d$, $\Lambda^k(\mathbf{R}^d)$ (see [BL, Section III.5] for details).

(vii) For any nonzero vector $x \in \mathbf{R}^d$, we denote by $\bar{x}$ the direction determined by x in the projective space $\mathbf{P}(\mathbf{R}^d)$.

(viii) If x, y are *unit vectors* in $\mathbf{R}^d$ with directions $\bar{x}, \bar{y}$ we set

(6) $$\delta(\bar{x}, \bar{y}) = (1 - \langle x, y \rangle^2)^{1/2},$$

where $\langle \cdot, \cdot \rangle$ is the standard scalar product. $\delta(\bar{x}, \bar{y})$ is the sine of the angle between $\bar{x}$ and $\bar{y}$. See [BL, Section III.4] for the verification that $\delta(\cdot, \cdot)$ is a metric on $\mathbf{P}(\mathbf{R}^d)$.

After these preliminaries we can state two fundamental results due to Guivarch and Raugi [GR1] and Le Page [LP] respectively.

Theorem C. ([GR1], see also [BL, Theorem III.6.1].) *let μ be a probability measure on $GL(d, \mathbf{R})$ with Lyapunov exponents $\gamma_1 \geq \gamma_2 \geq \ldots \geq \gamma_d$, such that $\int \log(\|M\|)\, d\mu(M)$ is finite and* $\operatorname{supp}(\mu)$ *is irreducible. Then $\gamma_1 > \gamma_2$ if and only if* $\operatorname{supp}(\mu)$ *is strongly irreducible and contracting.*

Theorem D. ([LP], see also [BL, Proposition III.2.3].) *let μ be a probability measure on $GL(d, \mathbf{R})$ such that* $\operatorname{supp}(\mu)$ *is compact, strongly irreducible, and contracting. Denote by μ^n the nth convolution power of μ (i.e., μ^n is the distribution of an independent random product $X_n X_{n-1} \ldots X_1$ in which each factor has distribution μ).*

Then for every sufficiently small $\alpha > 0$

(7) $$\lim_{n \to \infty} \left[\sup_{\bar{x}, \bar{y}} \int \left(\frac{\delta(M \cdot \bar{x}, M \cdot \bar{y})}{\delta(\bar{x}, \bar{y})} \right)^{\alpha} d\mu^n(M) \right]^{1/n} < 1,$$

where the supremum is over pairs $\bar{x}, \bar{y} \in \mathbf{P}(\mathbf{R}^d)$ with $\bar{x} \neq \bar{y}$.

The notation $M \cdot \bar{x}$ in (7) refers to the natural action of $GL(d, \mathbf{R})$ on $\mathbf{P}(\mathbf{R}^d)$: $M \cdot \bar{x} = \overline{Mx}$.

We also need the following observation

Lemma E. ([F]) *Let μ be a probability measure on $GL(d, \mathbf{R})$ with* $\operatorname{supp}(\mu)$ *compact and irreducible. Define for $\bar{x} \in \mathbf{P}(\mathbf{R}^d)$,*

$$h(\bar{x}) = \int \log \frac{\|\tilde{M} x\|}{\|x\|}\, d\mu(\tilde{M}).$$

If for some $\bar{x} \in \mathbf{P}(\mathbf{R}^d)$ the sequence

$$\Gamma_n = \int h(M \cdot \bar{x})\, d\mu^n(M) \tag{8}$$

converges, then the limit must be the top Lyapunov exponent $\gamma_1(\mu)$. (Actually, when $\mathrm{supp}(\mu)$ is strongly irreducible the convergence of Γ_n is guaranteed, see [BL, Theorem III.4.3].)

Proof: Choose a unit vector x in the direction $\bar{x}$. Consider the Cesáro averages

$$\frac{1}{n}(\Gamma_1 + \ldots + \Gamma_n) = \frac{1}{n}\int \log \|Mx\|\, d\mu^n(M).$$

The right hand side converges to $\gamma_1(\mu)$ (see [FK, Proposition 3.8] or [BL, Proposition III.7.2]) which concludes the proof. □

Proof of Theorem 1: Any probability vector $p = (p_1, \ldots, p_b)$ defines a probability measure μ_p, supported on $F = \{A_1, \ldots, A_b\}$ by

$$\mu_p(A_j) = p_j\,, \qquad 1 \le j \le b. \tag{9}$$

We first study the dependence of the *top* exponent $\gamma_1(\mu_p)$ on p assuming that $F = \mathrm{supp}(\mu_p)$ is *irreducible*, and then derive the general case. Fix a probability vector p with positive components. Since F is irreducible and $\gamma_1 = \gamma_1(\mu_p)$ is simple, we know by Theorem C that F is *strongly* irreducible and contracting. Thus we may apply Le Page's theorem (Theorem D). For $0 < \alpha < 1$

$$\left(\frac{\delta(M \cdot \bar{x}, M \cdot \bar{y})}{\delta(x, y)}\right)^\alpha \le \delta(M \cdot \bar{x}, M \cdot \bar{y})$$

so we may infer from (7) that there exist $\theta > 1$, $C > 0$ with

$$\int \delta(M \cdot \bar{x}, M \cdot \bar{y})\, d\mu_p^n(M) \le C\,\theta^{-n} \tag{10}$$

for all $n \ge 1$ and $\bar{x}, \bar{y} \in \mathbf{P}(\mathbf{R}^d)$.

To every complex vector $z = (z_1, \ldots, z_b) \in \mathbf{C}^b$ we attach the operator T_z acting on the space of complex continuous functions on $\mathbf{P}(\mathbf{R}^d)$ by

$$(T_z f)(\bar{x}) = \sum_{j=1}^{b} z_j\, f(A_j \cdot \bar{x}). \tag{11}$$

Note that for fixed $\bar{x}$, $(T_z^n f)(\bar{x})$ is a homogeneous polynomial of degree n in $(z_1, \ldots, z_b)$.

Claim:

(i) If $f : \mathbf{P}(\mathbf{R}^d) \to \mathbf{C}$ satisfies a Lipschitz condition

$$|f(\bar{x}) - f(\bar{y})| \leq L\,\delta(\bar{x}, \bar{y}) \quad \textit{for all } \bar{x}, \bar{y} \tag{12}$$

(a Hölder condition would suffice) then for any fixed $\bar{x} \in \mathbf{P}(\mathbf{R}^d)$ the sequence $(T_z^n f)(\bar{x})$ converges uniformly in $z \in \Omega_1$, where

$$\Omega_1 = \left\{ (z_1, \ldots, z_b) \in \mathbf{C}^b \;\middle|\; \sum_{j=1}^{b} z_j = 1,\ \frac{|z_j|}{p_j} < \theta_1 \text{ for } 1 \leq j \leq b \right\} \tag{13}$$

and

$$1 < \theta_1 < \theta \tag{14}$$

(recall that θ was defined in (10)).

(ii) In particular, letting

$$f_j(\bar{x}) = \log \frac{\|A_j x\|}{\|x\|}, \qquad x \in \mathbf{R}^d,\ 1 \leq j \leq b,$$

the sequence of polynomials

$$\Gamma_n(z) = \sum_{j=1}^{b} z_j (T_z^n f_j)(\bar{x}) \tag{15}$$

converges uniformly in Ω_1 to an analytic function $\Gamma_\infty(z)$. If $q \in \Omega_1$ is a probability vector $(q_1, \ldots, q_b)$ then $\Gamma_\infty(q)$ is the top Lyapunov exponent associated with the probability measure μ_q.

Proof of the Claim:

(i) First observe that for $\bar{x}, \bar{y} \in \mathbf{P}(\mathbf{R}^d)$ and $z \in \Omega_1$, expanding $T_z^n f$ as a sum of b^n products of length $n+1$ each, and using $|z_j/p_j| < \theta_1$, gives

$$|(T_z^n f)(\bar{x}) - (T_z^n f)(\bar{y})| \leq \theta_1^n \int |f(M \cdot \bar{x}) - f(M \cdot \bar{y})|\, d\mu_p^n(M). \tag{16}$$

In conjunction with (10) and (12), this implies

$$|(T_z^n f)(\bar{x}) - (T_z^n f)(\bar{y})| \leq C\, L \theta_1^n \theta^{-n}. \tag{17}$$

From $\sum_{j=1}^{b} z_j = 1$ we get

$$|(T_z^{n+1} f)(\bar{x}) - (T_z^n f)(\bar{x})| = \left| \sum_{j=1}^{b} z_j \left[(T_z^n f)(A_j \cdot \bar{x}) - T_z^n f(\bar{x}) \right] \right| .$$

Applying (17) with $\bar{y} = A_j \cdot \bar{x}$ and remembering that $|z_j| \leq \theta_1$, we conclude that

(18) $$|(T_z^{n+1} f)(\bar{x}) - (T_z^n f)(\bar{x})| \leq b\theta_1 C L(\theta_1/\theta)^n ,$$

proving the desired uniform convergence. The projection to the first $b-1$ coordinates is a homeomorphism of Ω_1 onto an open subset of $\mathbf{C}^{b-1}$, which defines the analytic structure on Ω_1. The uniform convergence implies that $\lim_{n\to\infty}(T_z^n f)(\bar{x})$ is an analytic function of $z \in \Omega_1$. By (17), this limit does not depend on $\bar{x}$.

(ii) It is easy to verify directly that $f_1, \ldots, f_b$ are Lipschitz functions on $\mathbf{P}(\mathbf{R}^d)$ (see [BL, Lemma V.4.2]). Analyticity of Γ_∞ in Ω_1 follows. For any probability vector $q = (q_1, \ldots, q_b) \in \Omega_1$ we have

(19) $$(T_q^n f)(\bar{x}) = \int f_j(M \cdot \bar{x})\, d\mu_q^n(M) ,$$

and therefore

$$\Gamma_n(q) = \sum_{j=1}^{b} q_j (T_q^n f_j)(\bar{x}) = \int \left(\sum_{j=1}^{b} q_j f_j(M \cdot \bar{x}) \right) d\mu^n(M) .$$

Using the definition of $f_1, \ldots, f_b$, one may identify $\Gamma_n(q)$ with the Γ_n appearing in (8) which correspond to $\mu = \mu_q$. Thus Lemma E implies that

$$\Gamma_\infty(q) = \gamma_1(\mu_q)$$

as required. □

Returning to the proof of Theorem 1, we now drop the assumption that F is irreducible. Let V be a nontrivial F–invariant subspace of $\mathbf{R}^d$. The measure μ_p defines measures $\mu_{p,V}$ and $\mu_{p,\mathbf{R}^d/V}$ on $GL(V)$ and $GL(\mathbf{R}^d/V)$ respectively. From Lemma A

(20) $$\gamma_1(\mu_p) = \max\{\gamma_1(\mu_{p,V}), \gamma_1(\mu_{p,\mathbf{R}^d/V})\} .$$

Now the simplicity hypothesis $\gamma_1(\mu_p) > \gamma_2(\mu_p)$ implies that

(21) $$\gamma_1(\mu_{p,V}) \neq \gamma_1(\mu_{p,\mathbf{R}^d/V})$$

(see [FK, Section 5]).

Assume, for instance, that the left hand side in (21) is greater (the opposite case is similar). Then necessarily $\gamma_1(\mu_{p,V}) > \gamma_2(\mu_{p,V})$. Using induction on the dimension, we may conclude that $\gamma_1(\mu_{q,V})$ depends real-analytically on the probability vector q in a neighborhood of p. For q sufficiently close to p, the continuity result (Corollary B), or the induction hypothesis implies that $\gamma_1(\mu_q) = \gamma_1(\mu_{q,V}) > \gamma_1(\mu_{q,\mathbf{R}^d/V})$ finishing the proof of Theorem 1 for γ_1.

We now consider the kth exponent $\gamma_k = \gamma_k(\mu_p)$. Since $\gamma_k > \gamma_{k+1}$, the sum $\gamma_1 + \ldots + \gamma_k$ is the *simple* top Lyapunov exponent for the i.i.d. random product of $\{\wedge^k A_1, \ldots, \wedge^k A_b\}$ taken with probabilities $p_1, \ldots, p_b$. Therefore $\gamma_1 + \ldots + \gamma_k$ depends real-analytically on $(p_1, \ldots, p_b)$. Using $\gamma_{k-1} > \gamma_k$ we find similarly that $\gamma_1 + \ldots + \gamma_{k-1}$ is real analytic in the same parameters. Subtracting, we are done. □

The results of [GM] now imply

Corollary 3. *Let $F = \{A_1, \ldots, A_b\}$ be a finite set of invertible $d \times d$ matrices. For every positive probability vector $p = (p_1, \ldots, p_b)$, define the measure μ_p by (9). If the semigroup generated by the matrices $\{\det(A_i)^{-1/d} A_i\}_{i=1}^b$ is Zariski dense in the special linear group $SL(d, \mathbf{R})$, or alternatively if one of $A_1, \ldots, A_b$ has d eigenvalues of different modulus, then for each $1 \le k \le d$, the kth exponent $\gamma_k(\mu_p)$ depends real analytically on $(p_1, \ldots, p_b)$.*

Remark: The condition on the Zariski closure may be weakened, see [GM, Theorem 6.4].

We now alter the assumptions in Theorem 1.

Proposition 4. *Let $F = \{A_1, \ldots, A_b\}$ be an irreducible set in $GL(d, \mathbf{R})$. Then $\gamma_1(\mu_p)$ depends real-analytically on the positive probability vector $p = (p_1, \ldots, p_b)$.* (The measure μ_p was defined in (9).)

Proof: Here we only give the proof assuming that F is *strongly* irreducible. The general case uses Markovian random products and is deferred to Section 3. Denote the *index* of F (see Definition (iv)) by k. By a theorem of Guivarch and Raugi [GR1, Proposition III.6.2] we have

(22) $$\gamma_1 = \gamma_2 = \ldots = \gamma_k > \gamma_{k+1}\,.$$

If $k = d$, the rightmost inequality is meaningless, but the rest hold and so in this case

$$\gamma_1(\mu_p) = \frac{1}{d} \sum_{j=1}^{b} p_j \log|\det(A_j)|\,.$$

We may therefore assume that $k < d$, in which case $k\gamma_1(\mu_p) = \gamma_1(\mu_p) + \ldots + \gamma_k(\mu_p)$ is the *simple* top Lyapunov exponent for an i.i.d. random product of the matrices $\{\wedge^k A_j \mid 1 \le j \le b\}$ taken with probabilities $p_1, \ldots, p_b$. Invoking Theorem 1 concludes the proof. □

Example: When F is reducible and $\gamma_1(\mu_p) = \gamma_2(\mu_p)$ for a specific p, there is no reason to expect analyticity at p. Indeed, taking $F = \{A_1, A_2\}$ with

$$A_j = \begin{pmatrix} a_j & 0 \\ c_j & 1 \end{pmatrix},$$

Lemma A implies that

$$\gamma_1(\mu_{(p_1,p_2)}) = \max\{p_1 \log a_1 + p_2 \log a_2, 0\}$$

which has a point of nondifferentiability as a function of $p_1 \in (0,1)$ if $a_1 > 1 > a_2$.

3 Markovian random products

The goal of this section is to establish the analogue of Theorem 1 when the factors in the random matrix product are no longer i.i.d., but form a matrix valued, mixing, homogeneous finite state Markov chain.

The usual device of "blocking" allows one to view any finite memory Markov chain as a chain with memory 1 (and possibly a larger state space). The allowed transitions are determined by a fixed zero–one matrix $\mathcal{U} = (u_{ij} \mid 1 \le i, j \le b)$ and mixing means that some power of $\mathcal{U}$ has strictly positive entries (i.e., $\mathcal{U}$ is a *primitive* matrix). Denote by $S(\mathcal{U})$ the set of stochastic matrices with the same support as $\mathcal{U}$, i.e.,

$$S(\mathcal{U}) = \left\{ P = (p_{ij})_{i,j=1}^{b} \;\middle|\; p_{ij} \ge 0,\ \sum_{j=1}^{b} p_{ij} = 1,\ p_{ij} > 0 \Leftrightarrow u_{ij} = 1 \right\}. \tag{23}$$

Fix a finite set of matrices $F = \{A_1, \ldots, A_b\} \subset GL(d, \mathbf{R})$. For any $P \in S(\mathcal{U})$ and $1 \le k \le d$, denote by $\gamma_k(P)$ the kth Lyapunov exponent for the random product of matrices $X_1, X_2, X_3, \ldots$, which form a stationary F–valued Markov process with

$$\Pr\{X_{n+1} = A_j \mid X_n = A_i\} = p_{ij}, \qquad 1 \le i, j \le b, \tag{24}$$

$$\Pr\{X_1 = A_i\} = \pi_P(i), \qquad 1 \le i \le b, \tag{25}$$

where π_P is the unique probability vector satisfying

$$\pi_P P = \pi_P. \tag{26}$$

Theorem 5. *With the above notation, assuming $\mathcal{U}$ is primitive:*

(i) For any $P \in S(\mathcal{U})$ such that

$$\gamma_1(P) > \gamma_2(P).$$

the function $Q \to \gamma_1(Q)$ is real-analytic on a neighborhood of P in $S(\mathcal{U})$.

(ii) More generally, any simple Lyapunov exponent $\gamma_k(P)$ depends (locally) real-analytically on $P \in S(\mathcal{U})$.

(iii) If F is strongly irreducible with respect to $\mathcal{U}$ (see definition below) then $P \to \gamma_1(P)$ is real-analytic throughout $S(\mathcal{U})$.

Definition: A set of matrices $F = \{A_1, \ldots, A_b\}$ is called irreducible with respect to the zero–one matrix $\mathcal{U}$ if there do not exist subspaces $V_1, \ldots, V_b$ of $\mathbf{R}^d$ satisfying

$$u_{ij} = 1 \Rightarrow A_j V_i = V_j. \tag{27}$$

F is called *strongly irreducible with respect to* $\mathcal{U}$ if there are no $V_1, \ldots, V_b$ which are finite unions of nontrivial subspaces in $\mathbf{R}^d$ satisfying (27).

Remark: Markovian random products were studied by Virtser, Royer, Guivarch [G], and recently by Bougerol [Bo1] [Bo2]. See [Bo1] for detailed references. A very general condition for positivity of the top exponent (for random products in $SL(d, \mathbf{R})$) was given by Ledrappier [Led].

Extending work of Guivarch, Bougerol has succeeded in [Bo1] and [Bo2] to establish for Markovian random products results analogous to those obtained in the i.i.d. case by Guivarch, Raugi, and Le Page. This enables one to prove Theorem 5 along the lines of Theorem 1.

Proof of Theorem 5:

(i) The result of Furstenberg and Kifer, Lemma A above, was established in [FK] for i.i.d. products but the proof may be modified to cover the *mixing* Markovian case as well. Thus, as in the proof of Theorem 1, we may assume that F is irreducible with respect to $\mathcal{U}$. Condition (A) of [Bo2] is satisfied, so by [Bo2, Proposition 3.3], for $\alpha > 0$ sufficiently small, there exist $C > 0$ and $\theta > 1$ so that

$$E[\delta(X_n X_{n-1} \ldots X_1 \cdot \bar{u}, X_n X_{n-1} \ldots X_1 \cdot \bar{v})^\alpha] \leq C\,\theta^{-n} \tag{28}$$

for all $n \geq 1$ and $\bar{u}, \bar{v} \in \mathbf{P}(\mathbf{R}^d)$. Here $\{X_j\}$ is the Markov process defined by (24),(25), and the numbers C, θ in (28) may depend on the transition matrix P.

The rest of the proof proceeds as in Theorem 1. For the operators T_z used there, one substitutes operators T_Z, where $Z = (z_{ij})_{i,j=1}^b$ is a complex matrix with the same support as P. These operators act on the space of continuous functions on $\{1, \ldots, b\} \times \mathbf{P}(\mathbf{R}^d)$ by

$$(T_Z f)(i, \bar{x}) = \sum_{j=1}^{b} z_{ij}\, f(j, A_j \cdot \bar{x}).$$

In place of Lemma E, we use the expression for γ_1 in [Bo2, Proposition 3.8] together with [Bo2, Theorem 3.8].

(ii) Follows from (i) via exterior algebra.

(iii) Define the index of F with respect to the stochastic matrix P as the least integer k such that there exists a sequence of matrices M_n that arise with positive probability in the random product defined by (24) and (25), with $M_n/\|M_n\|$ converging to a rank k matrix. This index is identical for all $P \in S(U)$. Denoting this index by k, Theorem 1.7 in [Bol] implies that

$$\gamma_1(P) = \gamma_2(P) = \ldots = \gamma_k(P) > \gamma_{k+1}(P)$$

for all $P \in S(\mathcal{U})$, so

$$k\gamma_1 = \gamma_1 + \gamma_2 + \ldots + \gamma_k$$

is a simple top Lyapunov exponent for an appropriate Markovian random product. Invoking (i) concludes the proof. □

Remark: Relaxing the mixing condition.

The condition that $\mathcal{U}$ is a primitive matrix, i.e., that the Markov chain $\{X_n\}$ is mixing, was needed in the above proof in order to apply the results of [Bol]. The usual classification of states for Markov chains allows us to (almost) discard it. We use the language of [S, Section 1.2]. We may assume all indices $1, \ldots, b$ are *essential*, as any stationary vector assigns probability 0 to inessential indices. The essential indices divide into *communicating classes*, and with each class a *period* is associated. Suppose that $\mathcal{U}$ is a zero–one $b \times b$ matrix with a *unique* communicating class of period r. For $P \in S(\mathcal{U})$ the stationary vector π_P is still determined uniquely. If $r = 1$ then $\mathcal{U}$ is primitive. If $r > 1$ then by *blocking* (replacing F by $\mathcal{U}$–allowed products of length r thereof, and considering the random product in blocks of length r) we obtain a new Markovian random product with each Lyapunov exponent multiplied by r. The "blocked" Markov chain has r acyclic communicating classes with identical (corresponding) Lyapunov exponents (by ergodicity of the original process). Thus in the original random product, each exponent has multiplicity at least r, and those with multiplicity precisely r depend real-analytically on the transition matrix $P \in S(\mathcal{U})$. (Note that the transition probabilities for the blocked chain are homogeneous polynomials of degree r evaluated at the original transition probabilities.) If F is strongly irreducible with respect to $\mathcal{U}$, then each of the r components of the blocked process is strongly irreducible in the same sense, so by Theorem 5(iii) we get real analyticity of γ_1 in $P \in S(\mathcal{U})$.

The last case to consider is when $\mathcal{U}$ has more than one communicating class. The stationary vector π_P for $P \in S(\mathcal{U})$ is no longer unique, but fixing "weights" for the different communicating classes determines π_P, and the previous cases imply real analyticity of $\gamma_1(P)$ for $P \in S(\mathcal{U})$, provided the top Lyapunov exponent corresponding to *each communicating class* is simple.

Proof of Proposition 4 completed: We may assume that $F = \{A_1, \ldots, A_b\}$ is an irreducible set of matrices which is not strongly irreducible. Let r be the smallest integer such that there exists a finite union $W = \cup_{\ell=1}^{L} W_\ell$ of r-dimensional subspaces of $\mathbf{R}^d$ which is invariant under each matrix in F. Fixing r, we also take L to be minimal. Fix linear isomorphisms Φ_ℓ mapping W_ℓ onto $\mathbf{R}^r$ for $1 \leq \ell \leq L$. We may assume each Φ_ℓ is an isometry for the Euclidean metric. We now define a set $\tilde{F} = \{\tilde{A}_{j\ell} \mid 1 \leq j \leq b,\ 1 \leq \ell \leq L\}$ of $r \times r$ matrices and a $bL \times bL$ zero–one matrix $\mathcal{U}$ as follows: For each $1 \leq j \leq b$ and $1 \leq \ell \leq L$, there is a unique $1 \leq \ell' \leq L$, $\ell' = \ell'(j, \ell)$ such that

$$A_j(W_{\ell'}) = W_\ell \,. \tag{29}$$

Let

$$\tilde{A}_{ij} = \Phi_\ell A_j \Phi_{\ell'}^{-1} \,, \tag{30}$$

and set $\mathcal{U}(i, \ell' \,;\, j, \ell) = 1$ for *this* value of ℓ' and all $1 \leq i \leq b$. All other entries of $\mathcal{U}$ are taken to be zero. To any positive probability vector $p = (p_1, \ldots, p_b)$ we attach a $bL \times bL$ stochastic matrix $P \in S(\mathcal{U})$ with

$$P(i, \ell' \,;\, j, \ell) = p_j \,, \qquad 1 \leq i, j \leq b,\ 1 \leq \ell \leq L \,, \tag{31}$$

where ℓ' is defined from j, ℓ as above (all other entries of P are zero).

Claim:

(i) $\tilde{F}$ *is a strongly irreducible subset of* $GL(r, \mathbf{R})$ *with respect to the matrix* $\mathcal{U}$.

(ii) The transition matrix P *determines a unique communicating class.*

(iii) The top Lyapunov exponent $\gamma_1(P)$ *associated with* $\tilde{F}$ *and* P *(by (24) and (25)) equals the top Lyapunov exponent* $\gamma_1(\mu_P)$ *for the i.i.d. random product determined by* F *and* p. *(Here* P *and* p *are related by (31).)*

Let us note that establishing the claim will also complete the proof of Proposition 4, by virtue of Theorem 5 and the remark following its proof.

Proof of the claim:

(i) This is patterned after [BL, Lemma III.7.1]. Suppose that for each $1 \leq \ell \leq L$ there were a finite union $V(i,\ell)$ of nontrivial proper subspaces of $\mathbf{R}^r$ such that

$$\tilde{A}_{j\ell} V(i,\ell') = V(j,\ell)$$

for all $1 \leq i \leq b$, $1 \leq j \leq b$, $1 \leq \ell \leq L$ with $\ell' = \ell'(j,\ell)$ defined by (29). From (30) it would follow that

$$\bigcup_{i=1}^{b} \bigcup_{\ell=1}^{L} \Phi_\ell^{-1}(V(i,\ell))$$

is a finite union of subspaces of $\mathbf{R}^d$ with dimension less than r, which is invariant under $F = \{A_1, \ldots, A_b\}$, contradicting the minimality of r.

(ii) This follows easily from the postulated minimality of L.

(iii) Fix any nonzero vector $x_0 \in \mathbf{R}^d$. Map any trajectory

(32) $$x_n = X_n X_{n-1} \ldots X_1 x_0$$

of the Markovian random product, to a trajectory

(33) $$y_n = Y_n Y_{n-1} \ldots Y_1 y_0$$

of the i.i.d. random product as follows: If $X_1 = \tilde{A}_{j\ell} = \Phi_\ell A_j \Phi_{\ell'}^{-1}$ then let $y_0 = \Phi_{\ell'}^{-1}(x_0)$ and in general if

(34) $$X_n = \Phi_\ell A_j \Phi_{\ell'}^{-1}$$

(where here j, ℓ depend on n and $\ell' = \ell'(j,\ell)$) then simply take $Y_n = A_j$. Since we consider in (32) a sequence $X_1, X_2, \ldots, X_n$ which has positive probability under P, all the Φ's appearing implicitly in (32) cancel and one finds by induction that $y_n = \Phi_\ell^{-1}(x_n)$ where ℓ, which depends on n, is determined by (34). In particular, since each Φ_ℓ is an isometry

$$\|y_n\| = \|x_n\|. \tag{35}$$

The definition of the transition matrix P in (31) guarantees that this correspondence between trajectories is measure preserving as long as we take y_0 equidistributed among $\{\Phi_1^{-1}(x_0), \ldots, \Phi_L^{-1}(x_0)\}$ and X_1 is distributed according to

$$\Pr\{X_1 = \tilde{A}_{j\ell}\} = \pi_P(j,\ell) = p_j/L, \qquad 1 \le j \le b,\ 1 \le \ell \le L.$$

Note that this is the stationary vector π_P defined by (26). Finally, by a classical result of Furstenberg, irreducibility of F implies that

$$\lim_{n\to\infty} \frac{1}{n} \log \|y_n\| = \gamma_1(\mu_P) \quad \text{a.s.} \tag{36}$$

(see [FK, Theorem 3.5]).

The corresponding result for Markovian random products was established under a strong irreducibility condition by Guivarch [G] and assuming irreducibility only by Kifer [Kif, Theorem III.1.2] and Bougerol [Bo2, Lemma 2.6]. Thus

$$\lim_{n\to\infty} \frac{1}{n} \log \|x_n\| = \gamma_1(P), \quad \text{a.s.} \tag{37}$$

Combining (35), (36), and (37) concludes the proof. □

Acknowledgements

I am grateful to Shizuo Kakutani, François Ledrappier, Benjamin Weiss, and Ze'ev Rudnick for helpful conversations, and to Philippe Bougerol for his inspiring exposition in [BL].

References

[Bo1] P. Bougerol, *Comparaison des exposants de Lyapunov des processus Markoviens multiplicatifs*, Ann. Inst. Henri Poincaré **24** (1988), 439–489.

[Bo2] P. Bougerol, *Théorèmes limite pour les systèmes linéaires à coefficients Markoviens*, Prob. Th. Rel. Fields **78** (1988), 193–221.

[BL] P. Bougerol and J. Lacroix, *Products of Random Matrices with Applications to Schrödinger operators,* Birkhauser, Basel 1985.

[F] H. Furstenberg, *Non-commuting random products,* Trans. Amer. Math. Soc. **108** (1963), 377–428.

[FK] H. Furstenberg and Y. Kifer, *Random matrix products and measures on projective spaces,* Israel J. Math. **10** (1983), 12–32.

[GM] I.Ya Goldsheid and G.A. Margulis, *Lyapunov indices of a product of random matrices,* Russian Math. Surveys **44:5** (1989), 11–71.

[G] Y. Guivarch, *Exposant caractéristiques des produits de matrices aléatoires en dépendence Markovienne,* In "Probability measures on groups 7", ed. H. Heyer, Lecture Notes in Mathematics 1064, Springer–Verlag 1984, 161–181.

[GR1] Y. Guivarch and A. Raugi, *Frontière de Furstenberg, propriétés de contraction et théorèmes de convergence,* Zeit. für Wahr. und verw. Gebeite **69** (1985), 187–242.

[GR2] Y. Guivarch and A. Raugi, *Products of random matrices: convergence theorems,* In "Random Matrices and their Applications," J. Cohen, H. Kesten, and C. Newman editors, Contemporary Math. vol. 50, Amer. Math. Soc., Providence 1985.

[Ka] T. Kato, *Perturbation Theory for Linear Operators,* Second Edition, Springer–Verlag 1976.

[Kif] Y. Kifer, *Ergodic Theory of Random Transformations,* Birkhauser, Basel 1986.

[Led] F. Ledrappier, *Positivity of the exponent for stationary sequences of matrices,* In "Lyapunov Exponents Proceedings," L. Arnold and V. Whistutz editors, Lecture Notes in Mathematics 1186, Springer–Verlag 1986, 56–73.

[LP] E. Le Page, *Théorèmes limites pour les produits de matrices aléatoires,* In "Probability Measures on Groups," ed. H. Heyer, Lecture Notes in Math. 928, Springer–Verlag 1982, 258–303.

[LP2] E. Le Page, *Régularité du plus grand exposant caractéristique des produits de matrices aléatoires indépendantes et applications*, Ann. Inst. Henri Poincaré **25** (1989), 109–142.

[P] Y. Peres, *Domains of analytic continuation for the top Lyapunov exponent*, Preprint 1990.

[R] D. Ruelle, *Analyticity properties of the characteristic exponents of random matrix products*, Adv. Math. **32** (1979), 68–80.

[S] E. Seneta, *Non-negative matrices and Markov chains*, Second Edition, Springer-Verlag 1981.

[ST] B. Simon and M. Taylor, *Harmonic analysis on $SL(2, \mathbf{R})$ and smoothness of the density of states in the one-dimensional Anderson model*, Commun. Math. Phys. **101** (1985), 1–19.

A second order extension of Oseledets theorem

by *Y. LE JAN*

Laboratoire de Probabilités
UNIVERSITE PIERRE & MARIE CURIE
Tour 46-56,
4, Place Jussieu
75253 PARIS Cédex 05

Let $(\Omega,\mathcal{F},P)$ be a probability space and θ an invertible ergodic transformation.

Let E be a real vector space of dimension d.

Let A be a measurable map from Ω into the group of isomorphisms of E. We assume that $E\left(\mathrm{Log}^+ \|A\| + \mathrm{Log}^+ \|A^{-1}\|\right)$ is finite.

Set $A_n(\omega) = A(\omega)A(\theta\omega)\dots A(\theta^{n-1}\omega)$.

For $p \in \{1,2,\dots n\}$, define $s_p = \lim_{n\to\infty} \downarrow \frac{1}{n} \mathrm{Log} \|A_n^{\wedge p}\|$. Set $\lambda_1 = s_1$ and $s_p - s_{p-1} = \lambda_p$ for $2 \le p \le d$.

We assume that λ_1 is positive and that the Lyapounov exponents λ_p are distinct.

By Oseledets theorem (cf. [1]) we can find a random set of directions $\pm\,\tau_p(\omega)$, $1 \le p \le d$, such that $-\lim_{n\to+\infty} \frac{1}{n} \mathrm{Log} \|A_n^{-1}(\omega)\,\tau_p(\omega)\| = \lambda_p$ and

$$\frac{A_n^{-1}\,\tau_p}{\|A_n^{-1}\,\tau_n\|} = \tau_p \circ \theta_n .$$

Equivalently, we have

$$\lim_{n\to\infty} \frac{1}{n} \mathrm{Log} \|A_n(\omega)\,\tau_p(\theta_n\omega)\| = \lambda_p \quad\text{and}\quad \frac{A_n(\tau_p \circ \theta_n)}{\|A_n(\tau_p \circ \theta_n)\|} = \tau_p .$$

Let r, $1 \le r < d$ be such that $s_r > 0$.

Set $\sigma_r = \tau_1 \wedge \tau_2 \dots \wedge \tau_r$.

Let B be a random map from Ω into $\mathrm{Hom}\left(E^{\wedge r} \otimes E^{\wedge r},\ E^{\wedge r}\right)$, such that $E(\mathrm{Log}^+ \|B\|)$ is finite.

Proposition.

a) The series $R(\omega) = \sum_{n=0}^{\infty} \frac{A_n^{\wedge r}(\omega)\, B(\theta_n\omega)\, \sigma_r(\theta_n\omega)^{\otimes 2}}{\|A_n^{\wedge r}(\omega)\, \sigma_r(\theta_n\omega)\|^2}$

converges absolutely P *almost surely.*

R *is determined by the identity :*

$$R = \frac{B\,\sigma_r^{\otimes 2}}{\|\sigma_r\|^2} + \frac{A^{\wedge r}\, R \circ \theta}{\|A^{\wedge r}\,\sigma_r\|^2} .$$

b) For any r-*direction* v, *such that* $v \neq \sigma_r$ a.s,

$$R_N^v(\omega) = \sum_0^N \frac{A_n^{\wedge r}(\omega)\, B(\theta_n\omega)(A_{N-n}^{\wedge r}(\theta_n\omega)v)^{\otimes 2}}{\|A_N^{\wedge r}(\omega)v\|^2}$$

converges in probability towards $R(\omega)$.

c) This convergence holds a.s. for Lebesgue almost all r-*direction* v.

Proof : We take $r = 1$ for notational simplicity.

Fix $0 < \varepsilon < \frac{\lambda_1}{100} \wedge \frac{\lambda_1 - \lambda_2}{100}$

Set $H(\omega) = \sup_n \|B(\theta_n\omega)\|\, e^{-\varepsilon n}$ (finite by Borel-Cantelli's lemma) (1).

$$D_p(\omega) = \sup_n \left(\|A_n^{\wedge p}(\omega)\|\, e^{-(s_p+\varepsilon)n} \right) \tag{2}$$

$$K_p(\omega) = \inf_n \left(\|A_n\, \tau_p(\theta_n\omega)\|\, e^{-(\lambda_p-\varepsilon)n} \right) \tag{3}$$

$$Q_p(\omega) = \sup_n \|A_n\, \tau_p(\theta_n\omega)\|\, e^{-(\lambda_p+\varepsilon)n} \Big) \tag{4}$$

a) follows from the majoration

$$|R(\omega)| \le \sum \frac{\|A_n\|\,\|B_n\|}{\|A_n\, \tau(\theta_n\omega)\|^2} \le K_1^{-2}\, D_1 H(\omega) \sum e^{(4\varepsilon - \lambda_1)n}$$

To prove b) (c)) we show that $\|R_N - R_N^v\|$ converges in probability (a.s.) towards 0 where R_N is the sum of the (N+1) first terms of R.

For any vector $v \in E$, let $\sum_{i=1}^{d} \Pi_i^{\omega}(v)$ be its decomposition along the directions $\tau_i(\omega)$.

We have $\Pi_1^{\omega}(A_n v) = A_n(\omega)\, \Pi_1^{\theta_n \omega}(v)$ for all n.

Note that $\|\Pi_1^{\omega} v\| = \dfrac{\|v \wedge \tau_1 \ldots \wedge \hat{\tau}_1 \ldots \wedge \tau_d\|}{\|\sigma_d\|}$

We have
$$\frac{\tau_1(\theta_n \omega)}{\|A_n(\omega)\, \tau_1(\theta_n \omega)\|} = \frac{A_{N-n}(\theta_n \omega)\, \Pi_1^{\theta_N \omega} v}{\|A_N(\omega)\, \Pi_1^{\theta_N \omega} v\|}.$$

We use the identity
$$R_N(v) - R_N = \Sigma\, A_n\, B_n \left(\left(\frac{\alpha_n}{\|\beta_n\|}\right)^{\otimes 2} - \left(\frac{\alpha'_n}{\|\beta'_n\|}\right)^{\otimes 2} \right)$$

where $\alpha_n = A_{N-n}(\theta_n \omega) v$, $\alpha'_n = A_{N-n}(\theta_n \omega)\, \Pi_1^{\theta_N \omega} v$

$$\beta_n = A_N(\omega) v, \qquad \beta'_n = A_N(\omega)\, \Pi_1^{\theta_N \omega} v = \Pi_1^{\omega}\, \beta_n$$

which implies the majoration
$$\|R_N(v) - R_N\| \le \Sigma\, \|A_n\| \|B_n\| \left\| \frac{\alpha_n}{\|\beta_n\|} - \frac{\alpha'_n}{\|\beta'_n\|} \right\| \left(\frac{\|\alpha_n\|}{\|\beta_n\|} + \frac{\|\alpha'_n\|}{\|\beta'_n\|} \right) \qquad (*)$$

By (1) and (2) we have
$$\|B_n\| < H(\omega)\, e^{\varepsilon n} \quad (6) \qquad \text{and} \qquad \|A_n\| \le D_1(\omega)\, e^{(\lambda_1 + \varepsilon) n} \quad (7)$$

Note that
$$\left\| \frac{\alpha_n}{\|\beta_n\|} - \frac{\alpha'_n}{\|\beta'_n\|} \right\| \le \frac{\|\alpha_n - \alpha'_n\|}{\|\beta_n\|} + \frac{\|\alpha'_n\| \|\beta_n - \beta'_n\|}{\|\beta_n\| \|\beta'_n\|} \qquad (8).$$

It follows from a corollary to the subadditive ergodic theorem (cf [4] p.288) that $F_p(\omega) = \sup_n D_p(\theta_n \omega)\, e^{-n\varepsilon}$ is finite.

Hence
$$\|\alpha_n\| \le F_1(\omega)\, e^{n\varepsilon}\, e^{(\lambda_1 + \varepsilon)(N-n)} \|v\| \qquad (9)$$

and
$$\|\alpha'_n\| \le F_1(\omega)\, e^{n\varepsilon}\, e^{(\lambda_1 + \varepsilon)(N-n)} \|\Pi_1^{\theta_N \omega} v\| \qquad (10)$$

From (3) we get
$$\|\beta'_n\| \ge K_1(\omega)\, e^{(\lambda_1 - \varepsilon) N} \|\Pi_1^{\theta_N \omega} v\| \qquad (11).$$

Set $q_i(\omega) = \dfrac{|\tau_1(\omega) \wedge \ldots \wedge \hat{\tau}_i(\omega) \wedge \ldots \wedge \tau_d(\omega)|}{\|\sigma_d(\omega)\|}$. Then $\|\beta_n\| \ge q_1^{-1}(\omega)\ \|\beta'_n\|$ (12)

$\alpha_n - \alpha'_n = \sum_{i\ge 2} A_{N-n}(\theta_n\omega)\ \Pi_i^{\theta_N\omega} v$. Hence

$$\|\alpha_n - \alpha'_n\| \le \sum_{i\ge 2} \|\Pi_i^{\theta_N\omega} v\| \frac{\|A_N\ \tau_i(\theta_N\omega)\|}{\|A_n\ \tau_i(\theta_n\omega)\|} \quad \text{and by (3) and (4)}$$

$$\le \sum_{i>2} K_i Q_i(\omega)\ e^{(\lambda_i+\varepsilon)(N-n)}\ e^{2\varepsilon n}\ \|\Pi_i^{\theta_N\omega} v\| \tag{13}$$

It follows that $\|\beta_n - \beta'_n\| \le \sum_{i>2} K_i Q_i(\omega)\ e^{(\lambda_i+\varepsilon)N} \|\Pi_i^{\theta_N\omega} v\|$ (14).

Introducing (9) (11) (12) (13) (14) in (8) we get

$\exists\ C_1(\omega) > 0$ a.s. such that

$$\Big\| \frac{\alpha_n}{\|\beta_n\|} - \frac{\alpha'_n}{\|\beta'_n\|} \Big\| \le C_1(\omega) \Bigg(e^{(\lambda_2-\lambda_1+2\varepsilon)(N-n)}\ e^{(-\lambda_1+3\varepsilon)n} \left(\frac{\Sigma\|\Pi_i^{\theta_N\omega} v\|}{\|\Pi_1^{\theta_N\omega} v\|} \right)$$

$$+\ e^{(\lambda_2-\lambda_1+\varepsilon)(N-n)}\ e^{(\lambda_2-2\lambda_1+3\varepsilon)n} \left(\frac{\Sigma\|\Pi_i^{\theta_N\omega} v\|}{\|\Pi_1^{\theta_N\omega} v\|} \right)$$

$$< C_1(\omega) \left(\sum_2^d \frac{\|\Pi^{\theta_N\omega} v\|}{\|\Pi_1^{\theta_N\omega} v\|} \right) \Bigg)\ e^{-50\,\varepsilon N} \tag{15}$$

Similarly,

$$\frac{\|\alpha_n\|}{\|\beta_n\|} + \frac{\|\alpha'_n\|}{\|\beta'_n\|} \le C_2(\omega)\left(e^{2\varepsilon(N-n)} e^{(-\lambda_1+2\varepsilon)n}\right)\left(1 + \left(\frac{\|v\|}{\|\Pi_1^{\theta_N\omega} v\|}\right)\right) \tag{16}$$

Introducing (1), (2), (15), (16) in (*) we get

$$\|R_N(v) - R_N\| \le C_3(\omega)_N\, e^{-10\,\varepsilon N}\left(\sum_1^N \frac{\|\Pi_1^{\theta_N\omega}{}_v\|^2}{\|\Pi_1^{\theta_N\omega}{}_v\|^2}\right),$$ which ends the proof ot b).

To prove c) by a Borel Cantelli argument, it is enough to show that

$$E\left(\log^+\left(\frac{\|\Pi_1^\omega\theta\|^2}{\|\Pi_1^\omega\theta\|^2}\right)\right)$$

is finite for almost all direction θ.

By Fubini's theorem, this follows from the fact that $\int \log^+(\|\Pi_1^\omega\theta\|^{-2})d\theta$ is uniformly bounded by a constant independent fo ω. (After reduction to the two dimensional case, this follows from the integrability of $\log(\sin^2\varphi)$ on $[o\Pi[$).

Remarks :

This result can be applied to second derivative tensors of iterated maps.

$A(\omega) = D(\theta^{-1})(\theta\omega) = D\theta^{-1}(\omega)$. $B(\omega) = D^2(\theta^{-1})(\theta\omega)$

$= A\, D^2\theta[A^{\otimes 2}](\omega)$.

Random maps etc... can also be considered. $R(\omega)$ should then be related to local parameters of the stable foliation at ω (curvature, divergence...).

Actually, such geometric series appear in Pesin's theory.

It should be interesting to investigate the law of R. A first study has been done by the author in ([2], [3]), for the isotropic case.

REFERENCES

[1] **F. LEDRAPPIER** : Ecole d'été de Probabilités 1982, L.N.M. *1097*, Springer.

[2] **Y. LE JAN** : Asymptotic properties of isotropic Brownian flows. To appear in the Harris Festshrift.

[3] **Y. LE JAN** : Propriétés asymptotiques des flots browniens isotropes. C.R.A.S. Paris, t.309, Série I, *p. 63-65* (1989).

[4] **D. RUELLE** : Characteristic exponents and invariant manifolds in Hilbert space. Annals of Maths. t.115, *p. 243-289* (1982).

The upper Lyapunov exponent of Sl(2,R) cocycles: Discontinuity and the problem of positivity

Oliver Knill
Mathematikdepartement
ETH Zentrum
CH-8092 Zürich

Abstract

Let T be an aperiodic automorphism of a standard probability space (X,m). Let $\mathcal{P}$ be the subset of $\mathcal{A} = L^\infty(X, Sl(2,R))$ where the upper Lyapunov exponent is positive almost everywhere.
We prove that the set $\mathcal{P} \setminus int(\mathcal{P})$ is not empty. So, there are always points in $\mathcal{A}$ where the Lyapunov exponents are discontinuous.
We show further that the decision whether a given cocycle is in $\mathcal{P}$ is at least as hard as the following cohomology problem: Can a given measurable set $Z \subset X$ be represented as $Y \Delta T(Y)$ for a measurable set $Y \subset X$?

1 Introduction

We want to investigate the Banach manifold

$$\mathcal{A} = L^\infty(X, Sl(2,R))$$

of all measurable bounded $Sl(2,R)$-cocycles over a given aperiodic dynamical system (X,T,m). We are interested in the subset $\mathcal{P}$ of cocycles where the upper Lyapunov exponent

$$\lambda^+(A,x) = \lim_{n\to\infty} n^{-1} \log ||A^n(x)||$$

is positive almost everywhere.

We have shown [Kni 90] that $\mathcal{P}$ is dense in $\mathcal{A}$. This could give some explanation why one encounters so often positive Lyapunov exponents when making numerical simulations. Numerical experiments suggest also that the Lyapunov exponents behave irregular in dependence of parameters. We prove in this note that the set $\mathcal{P} \setminus int(\mathcal{P})$ is not empty. On this set the Lyapunov exponent is discontinuous. Discontinuity of Lyapunov exponents has been mentioned at different places (see [You 86]). The only published result we found is in [Joh 84] in the case of $sl(2,R)-$ cocycles over almost periodic flows. Johnson proved there that discontinuities can already occur when changing a real parameter of the cocycle. His situation is different from ours in that he has a special flow and special cocycles occurring in the theory of Schroedinger operators, where we have an aperiodic but else arbitrary discrete dynamical system.

The idea for producing examples where λ^+ is discontinuous, is to exchange the expanding and contracting directions of the cocycle. This idea is not new and has been used in [Kif 82] to give examples where the Lyapunov exponent of identically distributed independent random matrices depend discontinuously from the common distribution. Our situation is different, because Kifer changes the dynamical system and not the cocycle. We will see that the exchanging of expanding and contracting directions must be done carefully. It can happen that the exchanging is making stable a part of the unstable directions and unstable a part of the stable directions. This, we don't want. A cohomology condition for measurable sets will assure that the stable and unstable directions become indistinguishable. This will give zero Lyapunov exponents. For certain cocycles, which we call weak, we can make such an exchanging by small perturbations.

We mention now some results which concern the regularity of the Lyapunov exponent: Hölder continuity (and in some cases even C^∞ smoothness) of the Lyapunov exponent with respect to a real parameter has been shown by le Page [Pag 89] in the case of independent identically distributed random matrices.
Ruelle's ([Rue 79a]) results show that there is an open set in $\mathcal{A}$ where the Lyapunov exponent is real analytic. It is the set

$$\mathcal{S} = \{A \in \mathcal{A} | \; \exists C \in \mathcal{A} \;\; \exists \epsilon > 0, \; [C(T)AC^{-1}(x)]_{ij} \geq \epsilon\}$$

which is contained in $int(\mathcal{P})$. One could call $\mathcal{S}$ the uniform hyperbolic part of $\mathcal{A}$ (or the set with exponential dichotomy [Joh 86]) and $\mathcal{P} \setminus \mathcal{S}$ the nonuniform hyperbolic part. The elements in $\mathcal{P} \setminus int\mathcal{P}$ which will be constructed here are not uniform hyperbolic. But we will see, that we can choose such elements in the closure of $\mathcal{S}$.

A lot of unsuccessful efforts to find more powerful methods to prove positivity of the upper Lyapunov exponent of $Sl(2, R)-$ cocycles led us to believe that the question whether A is in $\mathcal{P}$ is difficult and subtle in general. We want to illustrate this by showing that the decision can be at least as hard as deciding whether a certain circle valued cocycle is a coboundary. The circle valued cocycles considered here have the range $\{1, -1\}$. The group $\mathcal{E}$ of such cocycles can be identified with the set of measurable subsets of X with group operation Δ. The elements in $Z \Delta T(Z)$ are called coboundaries and form a subgroup. We will prove that the positivity of the Lyapunov exponent of a cocycle can depend on the question whether a certain set is a coboundary or not. This question about coboundaries has been investigated in [Bag 88]. In the special case, when (X, T, m) is an irrational rotation on the circle and the sets considered are intervals the problem has been treated in ([Vee 69], [Mer 85]). Even in this reduced form, the coboundary problem is still not solved.

2 Preparations

A *dynamical system* (X, T, m) is a set X with a probability measure m and a measure preserving invertible map T on X. We assume that (X, m) is a Lebesgue space and that the dynamical system is ergodic. The later involves no loss of generality because the arguments can be applied to each ergodic fibre of the ergodic decomposition in general. The dynamical system is called *aperiodic* if the set of periodic points $\{x \in X | \exists n \in N \; with \; T^n(x) = x\}$ has measure zero.

Denote by $M(2, R)$ the vector space of all real 2×2 matrices equipped with the usual operator norm $|| \cdot ||$. In the Banach space

$$L^\infty(X, M(2, R)) = \{A : X \to M(2, R) | \; A_{ij} \in L^\infty(X)\}$$

with norm $|||A||| = | \, ||A(\cdot)|| \, |_{L^\infty(X)}$ lies the Banach manifold

$$\mathcal{A} = L^\infty(X, Sl(2, R))$$

where $Sl(2,R)$ is the group of 2×2 matrices with determinant 1. Take on $\mathcal{A}$ the induced topology from $L^\infty(X,M(2,R))$. Denote with $\circ$ matrix multiplication. With the multiplication $AB(x)=A(x)\circ B(x)$ the space $L^\infty(X,M(2,R))$ is a Banach algebra. Name A(T) the mapping $x\mapsto A(T(x))$. For $n>0$ we write

$$A^n = A(T^{n-1})\cdots A(T)A$$

and $A^0=1$ where $1(x)$ is the identity matrix. The mapping $(n,x)\mapsto A^n(x)$ is called a *matrix cocycle over the dynamical system (X,T,m).* With a slight abuse of language we will call the elements in $\mathcal{A}$ *matrix cocycles* or simply cocycles.

Denote with * matrix transposition. According to the *multiplicative ergodic theorem of Oseledec* (see [Rue 79]) the limit

$$M(A)(x) := \lim_{n\to\infty}((A^n)^*(x)A^n(x))^{1/2n}$$

exists almost everywhere for $A\in\mathcal{A}$. Let

$$\exp(\lambda^-(A,x)) \le \exp(\lambda^+(A,x))$$

be the eigenvalues of $M(A)(x)$. The numbers $\lambda^{+/-}(A,x)$ are called the *Lyapunov exponents* of A. Because T is ergodic we write $\lambda^{+/-}(A)$ for the value, $\lambda^{+/-}(A,x)$ takes almost everywhere. Because $M(A)(x)$ has determinant 1 one has $-\lambda^-(A)=\lambda^+(A)$. We call $\lambda(A)=\lambda^+(A)$ *the Lyapunov exponent* of A and define

$$\mathcal{P} = \{A\in\mathcal{A}|\ \lambda(A)>0\}.$$

For $A\in\mathcal{P}$ there exist two measurable mappings $W^{+/-}$ from X into the projective space P^1 of all one dimensional subspaces of R^2 which satisfy

$$A(x)W^{+/-}(x) = W^{+/-}(T(x)).$$

$W^{+/-}(x)$ are the eigenspaces of $M(A)(x)$. Given $A\in\mathcal{A}$ we can define the *skew product action* $T\times A$ on the space $X\times P^1$:

$$T\times A : (x,W)\mapsto (T(x),A(x)W).$$

The projection π from $X\times P^1$ onto X defines a projection π^* of probability measures. We say, a probability measure μ on $X\times P^1$ *projects down* to $\pi^*\mu$. Ledrappier [Led 82] has found an addendum of the multiplicative ergodic theorem. We report here only a special case. For $W\in P^1$ we will always denote with w a unit vector in W.

Proposition 2.1 *a) If $A\in\mathcal{P}$ there exist exactly two ergodic $T\times A$ invariant probability measures $\mu^{+/-}$ on $X\times P^1$ which project down to m and one has*

$$\lambda^{+/-}(A) = \int_{X\times P^1}\log|A(x)w|d\mu^{+/-}(x,W).$$

The measures $\mu^{+/-}$ have their support on

$$X^{+/-} = \{(x,W^{+/-}(x))|x\in X\}.$$

b) For every ergodic $T\times A$ invariant probability measure μ which projects down to m

$$\lambda(A) = |\int_{X\times P^1}\log|A(x)w|d\mu(x,W)|.$$

Remark: Part a) of proposition 2.1 has been stated also in [Her 81] and in the case of cocycles with random noise in [You 86].

Let $Z \subset X$ be a measurable set of positive measure. A new dynamical system (Z, T_Z, m_Z) can be defined as follows: Poincaré's recurrence theorem implies that the *return time* $n(x) = min\{n \geq 1 | T^n(x) \in Z\}$ is finite for almost all $x \in Z$. Now, $T_Z(x) = T^{n(x)}(x)$ is a measurable transformation of Z which preserves the probability measure $m_Z = m(Z)^{-1} \cdot m$. The system (Z, T_Z, m_Z) is called *the induced system* constructed from (X, T, m) and Z. It is ergodic if (X, T, m) is ergodic (see [Cor 82]).
The cocycle $A_Z(x) = A^{n(x)}(x)$ is called the *derived cocycle* of A over the system (Z, T_Z, m_Z). In the following lemma 2.2 we cite a formula which relates the Lyapunov exponent of an induced system $\lambda(A_Z)$ with $\lambda(A)$. This formula is analogues to the formula of Abramov (see [Den 76]) which gives the metric entropy of an induced system from the entropy of the system. Lemma 2.2 is also stated in a slightly different form by Wojtkowsky [Woj 85].

Lemma 2.2 (Wojtkowsky) *If $m(Z) > 0$ then $\lambda(A_Z) \cdot m(Z) = \lambda(A)$.*

Remark: Wojtkowsky gives the formula

$$\lambda(A_Z) = \int_Z n(x) dm_Z(x) \cdot \lambda(A).$$

The version given here follows with the recurrence lemma of Kac [Cor 82] which says $\int_Z n(x) dm_Z(x) = m(Z)^{-1}$.

3 Cocycles with values in $\{1, -1\}$

We denote with $\mathcal{E}$ the set of $\{1, -1\}-$*valued cocycles*

$$\mathcal{E} = \{A \in \mathcal{A} |\ A(x) \in \{1, -1\}\}.$$

To each $A \in \mathcal{E}$ we can associate a measurable set

$$\psi(A) = \{x \in X | A(x) = -1\}.$$

It is easy to see that

$$\psi(A) \Delta \psi(B) = \psi(AB)$$

where Δ denotes the symmetric difference. ψ is invertible. So the group $\mathcal{E}$ is isomorphic to the group of measurable sets in X with group operation Δ. We call a measurable set Y a *coboundary* if there exists a measurable set Z such that $Y = Z \Delta T(Z)$. Also $A \in \mathcal{E}$ is called a *coboundary* if $\psi(A)$ is a coboundary. We will use the notation $Y^c = X \setminus Y$. Given a cocycle $A \in \mathcal{E}$ we can build a skew product $T \times A$ on $X \times \{1, -1\}$ as follows:

$$T \times A : (x, u) \mapsto (T(x), A(x)u).$$

It leaves invariant the product measure $m \times \nu$ where ν is the measure $\nu(\{1\}) = \nu(\{-1\}) = 1/2$ on $\{1, -1\}$. One can see the skew product action $T \times A$ as follows: Take two copies of the dynamical system (X, T, m). The dynamics is then given on both copies as usual. Only when hitting the set $\psi(A)$ one jumps to the other system.
A necessary and sufficient condition under which the ergodicity of (X, T, m) implies the ergodicity of $(X \times \{1, -1\}, T \times A, m \times \nu)$ is given in the following result of Stepin [Ste 71]:

Proposition 3.1 (Stepin) *For $A \in \mathcal{E}$ the skew product $T \times A$ is ergodic if and only if A is not a coboundary.*

Proof: Assume $A = \psi^{-1}(Y)$ and $Y = Z \Delta T(Z)$. The set

$$Q = Z \times \{1\} \cup Z^c \times \{-1\}$$

is $T \times A$ invariant. Therefore $T \times A$ is not ergodic.
On the other hand, assume $A = \psi^{-1}(Y)$ and there exists a set $Q \subset X \times \{1, -1\}$, $0 < (m \times \nu)(Q) < 1$, which is $T \times A$ invariant. Let Z be defined by the equation

$$Q \cap (X \times \{1\}) = Z \times \{1\}.$$

One checks that $Y = Z \Delta T(Z)$. So, A is a coboundary. ■

Remark: There exists a generalization of the above result (as formulated in [Lem 89]): Take a compact abelian group G with Haar measure ν. A measurable map $A : X \to G$ is called a $G -$ *cocycle*. Such a cocycle defines a skew product $T \times A$ on $X \times G$:

$$(T \times A)(x, g) = (T(x), A(x)g)$$

which preserves the measure $m \times \nu$. The result is that $T \times A$ is ergodic if and only if for any nontrivial character $\chi \in \hat{G}$ the circle valued cocycle $x \mapsto \chi(A(x))$ is not a coboundary. The proof given in [Anz 51] in the case where G is the circle can be modified easily to prove the general result. As a special case, if G is the cyclic group of order 2, one gets the above result of Stepin.

Lemma 3.2 *A measurable set Y with $m(Y) > 0$ is a coboundary if and only if $(T_Y)^2$ is not ergodic.*

Proof: Assume first that Y is a coboundary $Y = Z \Delta T(Z)$. Call $Z_1 = Z \setminus T(Z)$ and $Z_2 = T(Z) \setminus Z$. Then $T_Y(Z_1) = Z_2$ and $T_Y(Z_2) = Z_1$ imply $(T_Y)^2(Z_1) = Z_1$. Therefore $(T_Y)^2$ is not ergodic because $0 < m(Z_1) < 1/2$.
If on the other hand $(T_Y)^2$ is not ergodic then $\exists Z \subset Y$ with $(T_Y)^2(Z) = Z$ and $0 < m(Z) < m(Y)$. We claim that $Y = Z \Delta T_Y(Z)$. Because

$$Z \cap T_Y(Z) = (T_Y)^2(Z) \cap T_Y(Z) = T_Y(T_Y(Z) \cap Z)$$

the ergodicity of T_Y implies that $Y = Z \cap T_Y(Z)$ or $m(Z \cap T_Y(Z)) = 0$. The first case implies $Y = Z$ which is not possible because of the assumption $m(Z) < m(Y)$. So $m(Z \cap T_Y(Z)) = 0$. The same argument with $Z' = Y \setminus Z$ implies

$$m(Z' \cap T_Y(Z')) = m(Y \setminus (Z \cup T_Y(Z))) = 0.$$

From $Y = Z \cup T_Y(Z)$ and $m(Z \cap T_Y(Z)) = 0$ we get $Y = Z \Delta T_Y(Z)$.
If $n(x)$ denotes the return time of a point $x \in Y$ to Y we define

$$U = \{T^k(x) | x \in Z, k = 0, \ldots, n(x) - 1\}.$$

Then

$$U \Delta T(U) = Z \Delta T_Y(Z) = Y$$

and Y is a coboundary. ■

We define on $\mathcal{E}$ the metric

$$d(A,B) = m(\{x \in X \mid A(x) \neq B(x)\}) = m(\psi(A)\Delta\psi(B))$$

which makes $\mathcal{E}$ to a topological group.

Proposition 3.3 *If the dynamical system is aperiodic then the set of coboundaries as well as its complement are both dense in $\mathcal{E}$ with respect to the metric d.*

Proof: It is known that the set of $A = \psi^{-1}(Y)$ such that $(T_Y)^2$ is not ergodic is dense in $\mathcal{E}$ (See [Fri 70] p. 125). Applying lemma 3.2 gives that coboundaries are dense.
It is known that the set of $A = \psi^{-1}(Y)$ such that T_Y is weakly mixing is dense in $\mathcal{E}$([Fri 70] p. 126). If T_Y is weakly mixing also $(T_Y)^2$ is weakly mixing ([Fur 81] p.83) and $(T_Y)^2$ must be ergodic. Apply again lemma 3.2. ∎

Remarks:

1) In proposition 3.3 has entered the assumption that the probability space (X,m) is a Lebesgue space. There exists an automorphism of a probability space such that each measurable set Y is a coboundary. (See [Akc 65].)
2) Proposition 3.3 gives some indication that the decision whether a set is a coboundary or not might be subtle, especially when trying to deal with the question numerically.
3) Let us mention that for an ergodic periodic dynamical system (X,T,m) a set $Z \subset X$ is a coboundary if and only if the cardinality of Z is even. This follows quickly from the above lemma 3.2. Proposition 3.3 is no more true in the periodic case.
4) Of course the construction of coboundaries is very easy: Take a measurable set $Z \subset X$ and form the coboundary $Y = Z\Delta T(Z)$. On the other hand, we don't know of an easy construction of sets which are not coboundaries.

Lemma 3.4 *Assume $Z \subset Y \subset X$ with $m(Z) > 0$. Then, Z is a coboundary for T if and only if Z is a coboundary for T_Y.*

Proof: Because $(T_Y)_Z = T_Z$ we have also $((T_Y)_Z)^2 = (T_Z)^2$. The claim follows with lemma 3.2. ∎

We will use the following corollary of the proposition 3.3:

Corollary 3.5 *For every $Y \subset X$ with $m(Y) > 0$ there exists $Z \subset Y$ which is not a coboundary.*

Proof: Look at the dynamical system (Y, T_Y, m_Y). If (X,T,m) is aperiodic the proposition 3.3 assures that there exists $Z \subset Y$ such that Z is not a coboundary for T_Y. This means with lemma 3.4 that Z is not a coboundary for T. If (X,T,m) is periodic, choose $Z \subset Y$ which consists of one element. This Z is not a coboundary because $(T_Z)^2$ is trivially ergodic. ∎

4 Continuity and Discontinuity of the Lyapunov exponent

Computer experiments indicate that the Lyapunov exponent λ is discontinuous. But from the topological point of view we have a big set were λ is continuous. Recall that a subset of a topological space is called *generic* if it contains a countable intersection of open dense sets. The complement of a generic set is called *meager*.

Theorem 4.1 *The set $\{A \in \mathcal{A} | \ \lambda$ is continuous in $A\}$ is generic in $\mathcal{A}$. $\mathcal{P} \setminus int\mathcal{P}$ is meager.*

Proof: We can write

$$\lambda(A) = \lim_{n \to \infty} \lambda_n(A)$$

with $\lambda_n(A) = n^{-1} \int \log ||A^n|| dm$. So, λ is the pointwise limit of continuous functions λ_n. A theorem of Baire (see [Hah 32] p.221) states that the set of continuity points of such a function is generic.
The set $\mathcal{P} \setminus int(\mathcal{P})$ is a subset of all the discontinuity points. It is therefore meager. ■

Definition: We say a cocycle $A \in \mathcal{P}$ *is weak on* $Y \subset X$ if the following three conditions are satisfied:
a) the return time to Y^c is unbounded,
b) $A(x) = 1$ for $x \in Y$,
c) $(1,0) \in W^+(x)$ and $(0,1) \in W^-(x)$ for $x \in Y$.
We call A *weak*, if $A \in \mathcal{P}$ and there exists $Y \subset X$ with $0 < m(Y) < 1$ such that A is weak on Y.

Lemma 4.2 *There exist weak cocycles if (X,T,m) is aperiodic.*

Proof: If the dynamical system is aperiodic there exists for every $n \in N, n > 0$ and every $\epsilon > 0$ a measurable set Z such that $Z, T(Z), \dots, T^{n-1}(Z)$ are pairwise disjoint and such that $m(\bigcup_{k=0}^{n-1} T^k(Z)) \geq 1 - \epsilon$. This is *Rohlin's lemma* (for a proof see [Hal 56]) and the set Z is called a $(n,\epsilon)-$ *Rohlin set.*
Define the set

$$Y = \bigcup_{n=1}^{\infty} \bigcup_{k=1}^{n} T^k(Z_n)$$

where Z_n is a $(n2^n, 1/2)$-Rohlin set. Then $m(Y) \leq 1/2$ and the return time to Y^c is not bounded. Take a diagonal cocycle $D(x) = Diag(\mu(x), \mu^{-1}(x))$ with $\mu(x) = 1$ for $x \in Y$ and $\mu(x) = 2$ for $x \in Y^c$. This cocycle D is weak. ■

The main result in this section is:

Theorem 4.3 *$\mathcal{P} \setminus int(\mathcal{P})$ is not empty if and only if (X,T,m) is aperiodic.*

For the proof we will need another lemma. Call

$$R(\phi) = \begin{pmatrix} cos(\phi) & sin(\phi) \\ -sin(\phi) & cos(\phi) \end{pmatrix}$$

and denote with $\mathcal{X}_Z$ the characteristic function of a measurable set $Z \subset X$.

Lemma 4.4 *If $A \in \mathcal{P}$ is weak on Y and $Z \subset Y$ is not a coboundary then the cocycle*

$$B(x) = R(\pi/2 \cdot \mathcal{X}_Z(x))A(x)$$

is in $\mathcal{A} \setminus \mathcal{P}$.

Proof: Given a cocycle A which is weak on $Y \subset X$. The two sets

$$X^{+/-} = \{(x, W^{+/-}(x)) | x \in X\} \subset X \times P^1$$

are invariant under the skew product action $T \times A$. We call $T^{+/-}$ the action of $T \times A$ restricted to $X^{+/-}$ and $\mu^{+/-}$ the two ergodic $T \times A$ invariant measures projecting down to m. The dynamical systems $(X^{+/-}, T^{+/-}, \mu^{+/-})$ are isomorphic to (X, T, m). Define

$$B(x) = R(\pi/2 \cdot \mathcal{X}_Z(x))A(x)$$

where $Z \subset Y$ is not a coboundary. The set $X^+ \cup X^-$ is invariant under $T \times B$ and $(\mu^+ + \mu^-)/2$ is an invariant measure of $T \times B$ which projects down to m. The system $(X^+ \cup X^-, T \times B, (\mu^+ + \mu^-)/2)$ is isomorphic to $(X \times \{1, -1\}, T \times \psi^{-1}(Z), m \times \nu)$ which we have met in the last section. Stepin's result implies that the measure $(\mu^+ + \mu^-)/2$ is an ergodic $T \times B$ invariant measure on $X \times P^1$. This gives then with proposition 2.1b)

$$\begin{aligned} \lambda(B) &= |\int \log |A(x)w| d\mu^+(x, W) + \int \log |A(x)w| d\mu^-(x, W)|/2 \\ &= |\lambda^+(A) + \lambda^-(A)|/2 = 0 \end{aligned}$$

■

Proof of theorem 4.3: Assume (X,T,m) is aperiodic. It is enough to show: If A is weak then $A \in \mathcal{P} \setminus int(\mathcal{P})$. With lemma 4.2 follows then that $\mathcal{P} \setminus int\mathcal{P}$ is not empty.
Let $A \in \mathcal{P}$ be weak on Y and let $\epsilon > 0$ be given. We will construct a $B \in \mathcal{A}$ such that $\lambda(B) = 0$ and $|||B - A||| \leq \epsilon$. Choose $V \subset Y^c$, such that $T(V), \dots, T^n(V)$ are disjoint from Y^c and $m(V) > 0$. This is possible because the return time to Y^c is not bounded. Then there exists with corollary 3.5 a set $Z \subset V$ which is not a coboundary. Define

$$U = X \setminus \bigcup_{k=1}^{n-1} T^k(Z)$$

and look at the induced system (U, T_U, m_U). Then A_U is weak over $Y \cap U$ and with lemma 3.4 follows that Z is not a coboundary for T_U, because it is not a coboundary for T. Application of lemma 4.4 gives that

$$C = R(\pi/2 \cdot \mathcal{X}_Z)A_U$$

has zero Lyapunov exponent. Define the cocycle

$$B(x) = R(\pi/(2n) \cdot \mathcal{X}_{U^c}(x))A(x).$$

We check that

$$B_U = C.$$

This gives with lemma 2.2 and $\lambda(C) = 0$ also $\lambda(B) = 0$. Further

$$|||B - A||| \leq |||A||| \cdot \pi/2n \leq \epsilon.$$

We have shown that a weak cocycle is in $\mathcal{P} \setminus int\mathcal{P}$.
If (X, T, m) is periodic then the Lyapunov exponent is continuous and so $\mathcal{P} = int\mathcal{P}$. ■

Remarks:
1) We say $A, B \in \mathcal{A}$ are *cohomologous in* $\mathcal{A}$ if there exists $C \in \mathcal{A}$, such that $C(T)AC^{-1} = A$. Cohomologous cocycles have the same Lyapunov exponents and if A is conjugated to a weak cocycle then it is also in $P \setminus int\mathcal{P}$.
2) It was surprising for us to find diagonal cocycles in $\mathcal{P} \setminus int\mathcal{P}$. We expected that the arbitrary closeness of stable and instable directions are responsible for the discontinuity of the Lyapunov exponent. This can also be the case as the following remark indicates.
3) Assume $A \in \mathcal{P} \setminus int(\mathcal{P})$ and $A_n \to A$ with $\lambda(A_n) = 0$. Because $\mathcal{P}$ is dense in $\mathcal{A}$ (see [Kni 90]) we can find $B_n \to A$ with $|||B_n - A_n||| \leq 1/n$ and $B_n \in \mathcal{P}$. If μ_n denotes a $T \times B_n$ invariant proability measure projecting down to m then μ_n converges weakly to $(\mu^+ + \mu^-)/2$ where $\mu^{+/-}$ are the $T \times A$ invariant ergodic measures projecting down to m. In some sense the stable and unstable directions of B_n come closer and closer together as n is increasing.

5 Difficulty of the decision whether the Lyapunov exponent is positive

There are only a few methods to decide whether $A \in \mathcal{P}$ or not. The only method which works for general dynamical systems is Wojtkowsky's cone method [Woj 85]. But there are many examples where one measures positive Lyapunov exponent numerically without being able to prove it. This suggests that the general problem is difficult. The next theorem could be one of the reasons for the subtlety.

Theorem 5.1 *Given a measurable set $Y \subset X$ with $0 < m(Y) < 1$. There exists $A \in \mathcal{A}$, such that $B = R(\pi/2 \cdot \mathcal{X}_Y)A \in \mathcal{P}$ if and only if Y is a coboundary.*

Proof: Given $Y \subset X$ with $0 < m(Y) < 1$ we build the Kakutani skyscraper over Y, which is a partition $X = \bigcup_{i \geq 1} Y_i$ where $Y_1 = Y$ and $Y_{n+1} = T(Y_n) \setminus Y$. We have $m(Y_2) > 0$ because $m(Y) < 1$. Define $U = Y_2$ and the diagonal cocycle $A(x) = Diag(2, 2^{-1})$ for $x \in U$ and $A(x) = 1$ else. The Lyapunov exponent of A is

$$\lambda(A) = \log(2) \cdot m(U) > 0.$$

Clearly $(1,0) \in W^+(x)$ and $(0,1) \in W^-(x)$. We denote with $\mu^{+/-}$ the two ergodic $T \times A$ invariant measures on $X \times P^1$ which project down to m and have their support on

$$X^{+/-} = \{(x, W^{+/-}(x)) | x \in X\}.$$

If Y is not a coboundary we conclude like in the proof of lemma 4.4 that $(\mu^+ + \mu^-)/2$ is an ergodic $T \times B$ invariant measure projecting down to m and so $\lambda(B) = 0$.
If Y is a coboundary, that is if $Y = Z \Delta T(Z)$, there is a $T \times A$ invariant set

$$Q = \{(x, W^+(x)) | \ x \in Z\} \cup \{(x, W^-(x)) | \ x \in Z^c\}.$$

This set Q carries an ergodic $T \times A$ invariant measure μ which projects onto the measure m. Because $U = Y_2$ is disjoint from Y either $U \subset Z \cap T(Z)$ or U is disjoint from $Z \cup T(Z)$. This implies $U \subset Z$ or $U \subset Z^c$ and we have either $\{(x, W^+(x)) | \ x \in U\} \subset Q$

or $\{(x, W^-(x))|\ x \in U\} \subset Q$. Because $A(x)$ is different from the identity matrix only on U and is there $Diag(2, 2^{-1})$ we have

$$\lambda(B) = |\int_Q \log|A(x)w| d\mu(x, W)| = \log(2) \cdot m(U) = \lambda(A) > 0.$$

■

If we would have an algorithm to find out whether a given cocycle $A \in \mathcal{A}$ is in $\mathcal{P}$ or not we could also find out for a measurable set $Z \subset X$ if Z is a coboundary or not. So, the cohomology problem in $\mathcal{E}$ exhibits already a difficulty for calculating or estimating the Lyapunov exponents.

Let us mention to the end some open questions:

1) Assume T is a homeomorphism of a compact metric space X leaving a Borel probability measure m invariant. Is the upper Lyapunov exponent continuous on

$$C(X, Sl(2,R)) = \{A : X \to Sl(2,R)|\ A\ continuous\}?$$

2) We believe that the cohomology problem in $\mathcal{E}$ is difficult. Is there a difficult mathematical problem which is embeddable in the cohomology problem for measurable sets ?
3) For which $r \geq 0$ is there a nonempty set in $\mathcal{A}$ such that the Lyapunov exponent is there r times but not $r+1$ times differentiable ?
4) Find discontinuities of the Lyapunov exponent λ on special curves through $\mathcal{A}$. In the theory of random Jacobi matrices [Cyc 87] one would like to know about regularity properties of λ on the curve

$$E \mapsto A_E = \begin{pmatrix} V+E & -1 \\ 1 & 0 \end{pmatrix}$$

where $V \in L^\infty(X, R)$. Johnson [Joh 84] has examples for discontinuities in the case of almost periodic Schroedinger operators. An other interesting curve would be the circle

$$\beta \mapsto AR(\beta) = A \begin{pmatrix} cos(\beta) & sin(\beta) \\ -sin(\beta) & cos(\beta) \end{pmatrix}.$$

Can we always find $A \in \mathcal{A}$ where $\beta \mapsto \lambda(AR(\beta))$ is not continuous ?
5) Is every $A \in \mathcal{P} \setminus int\mathcal{P}$ cohomologous to a weak cocycle ?

Bibliography

[Akc 65] M.A.Akcoglu,R.V.Chacon. *Generalized eigenvalues of automorphisms.* Proc.Amer.Math.Soc., 16:676-680, 1965.

[Anz 51] H.Anzai. *Ergodic Skew Product Transformations on the Torus.* Osaka Math. J., 3:83-99, 1951.

[Bag 88] L.Baggett. *On circle valued cocycles of an ergodic measure preserving transformation.* Israel J. Math., 61:29-38, 1988.

[Cor 82] I.P.Cornfeld,S.V.Fomin,Ya.G.Sinai. *Ergodic Theory.* Springer, New York, 1982

[Cyc 87] H.L.Cycon,R.G.Froese,W.Kirsch,B.Simon. *Schroedinger Operators.* Texts and Monographs in Physics, Springer, 1987

[Den 76] M.Denker,C.Grillenberg,K.Sigmund. *Ergodic Theory on Compact Spaces.* Lecture Notes in Math., No. 527, Springer 1976

[Fri 70] N.A.Friedman. *Introduction to Ergodic Theory.* Van Nostrand-Reinhold, Princeton, New York, 1970

[Fur 81] H.Fürstenberg. *Recurrence in Ergodic Theory and Combinatorial Number Theory.* Princeton University Press, Princeton, New Jersey, 1981

[Hah 32] H.Hahn. *Reelle Funktionen, 1.Teil, Punktfunktionen.* Akademische Verlagsgesellschaft, Leipzig, 1932

[Hal 56] P.Halmos. *Lectures on Ergodic Theory.* The mathematical society of Japan, 1956

[Her 81] M.R.Herman. *Construction d'un difféomorphisme minimal d'entropie topologique non nulle.* Ergodic Theory & Dynamical Systems,1:65-76, 1981

[Joh 84] R.A.Johnson. *Lyapunov numbers for the almost periodic Schroedinger equation.* Illinois J. Math., 28:397-419, 1984

[Joh 86] R.A.Johnson. *Exponential Dichotomy, Rotation Number and Linear Differential Operators with Bounded Coefficients.* J. Differential Equations, 61:54-78, 1986

[Kif 82] Y.Kifer. *Perturbations of Random Matrix Products.* Z.Wahrscheinlichkeitstheorie verw.Gebiete, 61:83-95, 1982

[Kni 90] O.Knill. *Positive Lyapunov exponents for a dense set of bounded measurable Sl(2,R) cocycles.* Submitted for publication

[Led 82] F.Ledrappier. *Quelques Proprietés des Exposants Caracteristiques.* Lecture Notes in Math., No. 1097, Springer, 1982

[Lem 89] M.Lemańczyk. *On the Weak Isomorphism of Stricly Ergodic Homeomorphisms.* Monatsh. Math., 108:39-46, 1989

[Mer 85] K.D.Merrill. *Cohomology of step functions under irrational rotations.* Israel J. Math.,52:320-340, 1985.

[Pag 89] E. Le Page. *Regularité du plus grand exposant caractéristique des produits de matrices aléatoires indépendantes et applications.* Ann. Inst. Henri Poincaré , 25:109-142, 1989

[Rue 79] D.Ruelle. *Ergodic theory of differentiable dynamical systems.* Publ.math. de l'IHES, 50:27-58, 1979

[Rue 79a] D.Ruelle. *Analycity Properties of the Characteristic Exponents of Random Matrix Products.* Advances in Mathematics, 32:68-80, 1979

[Ste 71] A.M.Stepin. *Cohomologies of automorphism groups of a Lebesque space.* Functional Anal. Appl., 5:167-168, 1971

[Vee 69] W.Veech. *Strict ergodicity in zero dimensional dynamical systems and the Kronecker-Weyl theorem mod 2.* Trans.Amer.Math.Soc.,140:1-33, 1969

[Woj 85] M.Wojtkowski. *Invariant families of cones and Lyapunov exponents.* Ergodic Theory & Dynamical Systems, 5:145-161, 1985.

[You 86] L.-S.Young. *Random perturbations of matrix cocycles.* Ergodic Theory & Dynamical Systems, 6:627-637, 1986.

Linear skew-product flows and semigroups of weighted composition operators

Yu. D. Latushkin
Hydrophysics Institute
Ukrainian Academy of Science
Odessa, USSR

A.M. Stepin
Moscow State University
Moscow, USSR

Abstract

The article contains the results on the relations between the spectral theory of linear skew-product flows, the multiplicative ergodic theorem and the spectral theory of the weighted composition operator semigroup. The latter is given by

$$(T_A^t f)(x) = \left(\frac{d\mu \circ \alpha^{-t}}{d\mu}(x)\right)^{1/2} A(\alpha^{-t}x, t) f(\alpha^{-t}x),$$

acting in the space $L_2(X, \mu; H)$ of H-valued functions f on the compact space X; where A is a cocycle over the flow $\{\alpha^t\}$ on X, $t \in \mathbb{R}$ or $\mathbb{Z}$.

Let $\{\alpha^t\}$ be a flow ($t \in \mathbb{R}$) or a cascade ($t \in \mathbb{Z}$) on a compact metric space $X, \mathcal{L}(\mathcal{H})$ – the algebra of bounded linear operators acting in a separable Hilbert space H, $A : X \times \mathbb{R}_+ \to \mathcal{L}(\mathcal{H})$ (or $A : X \times \mathbb{Z}_+ \to \mathcal{L}(\mathcal{H})$) — a cocycle over $\{\alpha^t\}$. We consider the semigroup of weighted composition operators (WCO)

$$(T_A^t f)(x) = \left(\frac{d\mu \circ \alpha^{-t}}{d\mu}(x)\right)^{1/2} A(\alpha^{-t}x, t) f(\alpha^{-t}x), \qquad x \in X, \tag{1}$$

in the space $L_2 = L_2(X, \mu; H)$ of square summable H–valued functions, where μ is a Borel quasiinvariant probability measure on X with supp $\mu = X$.

This paper contains results on relations between the spectral theory of WCO, in particular the generator of the continuous semigroup (1), and the spectral theory (due to R. Sacker and G.Sell) of linear extensions $\hat{\alpha} = \{\hat{\alpha}^t\}$, $\hat{\alpha}^t(x, v) = (\alpha^t x, A(x,t)v)$, $v \in H$; ([1], see also survey [2]). For compact-valued cocycles A we describe spectra of WCO (1) and the corresponding linear extension in terms of the multiplicative ergodic theorem. It is worth stressing that we consider general cocycles (without assumption to be invertible (cf.[1–6])) and also admit the case of infinite dimensional H. For simplicity we restrict ourselves to extensions acting on trivial linear bundles; besides this it is assumed that the set of $\{\alpha^t\}$–periodic points is μ–negligible.

1. Suppose that the semigroup $\{T_A^t\}_{t\in\mathbb{R}_+}$ is strongly continuous in L_2. This property is guaranteed by the following two conditions:

a) the group of unitary operators $\{T_{\mathbf{1}}^t\}_{t\in\mathbb{R}}$, where $\mathbf{1}$ is the unit cocycle, is strongly continuous in L_2,

b) for any $v \in H$, $\|A(x,t)v - v\|_H \to 0$ uniformly in $x \in X$ as $t \to 0$ (in particular, the function $(x,t) \to \|A(x,t)v\|_H$ is continuous for all $v \in H$).

Let $\mathcal{D} = \frac{d}{dt}T_A^t|_{t=0}$ be the generator of the semigroup $\{T_A^t\}_{t\in\mathbb{R}}$. If, for example, $\{\alpha^t\}$ is a smooth volume-preserving flow on a manifold X and $A(x,t)$ the Cauchy operator for the equation

$$\dot{v} = b(\alpha^t x)v, \qquad x \in X, \quad t \in \mathbb{R}, \tag{2}$$

where $b : X \to \mathcal{L}(\mathcal{H})$ is continuous, then the domain of $\mathcal{D}$ coincides with that of the operator

$$(\mathbf{d}f)(x) = \frac{d}{dt}f(\alpha^t x)|_{t=0}$$

and $(\mathcal{D}f)(x) = -(\mathbf{d}f)(x) + b(x)f(x)$.

Theorem 1 (Spectral mapping theorem). The spectrum $\sigma(\mathcal{D})$ is invariant under translations along the imaginary axis and is related to the spectrum $\sigma(T_A^t)$ by the formula

$$\sigma(T_A^t)\backslash\{0\} = \exp t\,\sigma(\mathcal{D}), \quad t > 0.$$

Proof. We need the following lemma.

Lemma. Let $m(Per\,\alpha) = 0$. For any $\epsilon > 0$ and $\xi \in \mathbb{R}$ there exists g, $|g(x)| = 1$ m–a.e., so that

$$\mathbf{d}g = \frac{d}{dt}g \circ \alpha^t|_{t=0}$$

exists in $L_\infty(X, m; \mathbb{R})$ and $\| \mathbf{d}g - i\xi g \|_{L_\infty} < \epsilon$.

The proof of this Lemma, which is omitted, is constructive and uses the isomorphism between $\{\alpha^t\}$ and a special quasiflow (see [7]).

To prove the theorem, we consider the flow $\{\alpha^{-t}\}$ and choose for given $\epsilon > 0$ and $\xi \in \mathbb{R}$ the function g, $|g(x)| = 1$ m–a.e., from the Lemma. Let D be a domain of $\mathcal{D}$. It is easy to see that for $f \in D$ we have $gf \in D$ and $\mathcal{D}(gf) = \mathbf{d}g \cdot f + g \cdot \mathcal{D}f$. For $f \in D$, $\| f \| = 1$, and $\lambda \in \mathbb{C}$ we obtain from the Lemma:

$$\| (\lambda + i\xi - \mathcal{D})fg \|_{L_2} \le \| g(\lambda f - \mathcal{D}f) \|_{L_2} + \| (\mathbf{d}g - i\xi g)f \|_{L_2} \le \| \lambda f - \mathcal{D}f \|_{L_2} + \epsilon.$$

Hence, $|||\lambda + i\xi - \mathcal{D}|||_* \le |||\lambda - \mathcal{D}|||_* + \epsilon$, where $|||\mathcal{D}|||_* = inf\{\| \mathcal{D}f \| : f \in D, \| f \| = 1\}$. Writing this inequality for $\lambda = \lambda' + i\xi'$, $\xi = -\xi'$ and letting $\epsilon \to 0$, we have that $|||\lambda - \mathcal{D}|||_*$ doesn't depend on Im λ, and the first part of the theorem is proved.

To prove the second part we need the spectral mapping theorem due to L. Gerhard (see e.g. [8]): For any c_0-semigroup $\{e^{t\mathcal{D}}\}_{t\ge 0}$, acting in a Hilbert space, the following formula is valid: $\sigma(e^{t\mathcal{D}})\backslash\{0\} = \{e^{\lambda t} : \omega_k = \lambda + 2\pi i k/t \in \sigma(\mathcal{D})$ for some $k \in \mathbb{Z}$, or the sequence $\{\| (\omega_k I - \mathcal{D})^{-1} \|\}_{k\in\mathbb{Z}}$ is unbounded$\}$. This together with the translational invariance of $\sigma(\mathcal{D})$ implies that for $e^{\lambda t} \in \sigma(T_A^t)\backslash\{0\}$ either $\omega_k \in \sigma(\mathcal{D})$ for all $k \in \mathbb{Z}$, or for all $k \in \mathbb{Z}$ $(\omega_k I - \mathcal{D})^{-1}$ exist and the sequence of their norms is unbounded. The latter is impossible, because $\|\omega_k I - \mathcal{D}\|_{\bullet}^{-1} = \| (\omega_k I - \mathcal{D})^{-1} \|$ doesn't depend on K as we have seen in the proof of the first part of the theorem. □

Let us agree to call an operator $T_A^{t_0}$ (or the generator $\mathcal{D}$ of a semigroup $\{T_A^t\}_{t\in\mathbb{R}_+}$) hyperbolic, if its spectrum is disjoint to the unit circle $\{z \in \mathbb{C} : |z| = 1\}$ (to the imaginary axis, respectively). According to Theorem 1 the operators T_A^t and $\mathcal{D}$ are hyperbolic simultaneously. This assertion is valid for an arbitrary strongly continuous semigroup of bounded WCO in L_2, generated by a flow on a Lebesgue measure space, having its period function bounded away from zero; (a related result for groups of WCO under the restriction $\dim H < \infty$ can be found in [4,5]).

Recall that an extension $\hat{\alpha}$ is said to be hyperbolic if there exist:

1. a continuous projection valued function $P : X \to \mathcal{L}(H)$ satisfying the condition $A(x,t)P(x) = P(\alpha^t x)A(x,t)$ and
2. constants $C, \beta > 0$ such that for any $x \in X$ the following estimates hold

$$\begin{aligned} \| A(x,t)v \| &\leq Ce^{-\beta t} \| v \|, v \in \mathbb{S}_x = \mathrm{Im}P(x), \\ \| A(x,t)v \| &\geq Ce^{\beta t} \| v \|, \; v \in \mathbb{U}_x = \mathrm{Im}(\mathbf{1} - P(x)), \; t > 0. \end{aligned} \tag{3}$$

Suppose that the mapping $a(\cdot) = A(\cdot, 1) : X \to \mathcal{L}(H)$ is continuous.

Lemma 2. [9] The hyperbolicity of the extension $\hat{\alpha}^1$ is equivalent to that of the operator $T := T_A^1$. The Riesz projection $\mathcal{P}$ for T corresponding to $\sigma(T) \cap \{z \in \mathbb{C} : |z| < 1\}$ is related to the function $P(\cdot)$ of (3) by the formula $(\mathcal{P}f)(x) = P(x)f(x)$.

This together with theorem 1 implies that hyperbolicity of $\hat{\alpha}$ is equivalent to invertibility of $\mathcal{D}$. It can be proved that the latter condition is also equivalent to the existence and uniqueness of a Green function for the extension $\hat{\alpha}$ (cf. [10]).

Now we will characterize hyperbolicity of $\hat{\alpha}, T$ and $\mathcal{D}$ in terms of "WCO along trajectories $\{\alpha^t x\}$".

In the discrete time case ($t \in \mathbb{Z}$) we define "discrete WCO" acting in $l_2 := l_2(\mathbb{Z}, H)$ by the formula

$$\theta_x : (v_t)_{t\in\mathbb{Z}} \to (a(\alpha^{t-1}x)v_{t-1})_{t\in\mathbb{Z}}.$$

Lemma 3. Hyperbolicity of T is equivalent to that of θ_x for all $x \in X$.

For proofs of Lemmas 2 und 3 cf. [8]; they are based on an application of C^*-algebra techniques to algebras generated by T_1^1 and operators of multiplication in L_2 by continuous operator functions (see [9] for details).

Regarding continuous time we consider the semigroup $\{\Theta_x^t\}_{t\in\mathbb{R}_+}$, $x \in X$, in $L_2(\mathbb{R}; H)$ acting according to the rule

$$(\Theta_x^t f)(s) = A(\alpha^{s-t}x, t)f(s-t),\ s \in \mathbb{R}.$$

Note that semigroups $\{\Theta_x^t\}_{t\in\mathbb{R}_+}$ are strongly continuous for any $x \in X$ if part (b) in the sufficient condition for strong continuity of $\{T_A^t\}_{t\in\mathbb{R}_+}$ above holds. For any $x \in X$ hyperbolicity of the generator $d_x = \frac{d}{dt}\Theta_x^t|_{t=0}$ is equivalent to that of the operator $\Theta_x^{t_0}, t_0 \neq 0$. Moreover $\sigma(\Theta_x^t)\backslash\{0\} = exp\, t\sigma(d_x)$. In case $A(x,t)$ is the Cauchy operator of equation (2) the generator d_x acts according to the formula $(d_x f)(s) = -\frac{df}{ds} + b(\alpha^s x)f(s)$. In particular: If $x : \mathbb{R} \to \mathcal{L}(\mathcal{H})$ is continuous, $X = cl\{x(\cdot + t) : t \in \mathbb{R}\}$ is compact and the flow $\{\alpha^t\}$ and function b on X are given by the formula

$$\alpha^t : X \to X,\ x(\cdot) \to x(\cdot + t),\ b : x(\cdot) \to x(0), \tag{4}$$

then $d_x = -\frac{d}{ds} + x(\cdot)$.

Using Lemma 3 we prove

Theorem 4. Hyperbolicity of the extension $\hat{\alpha}$ is equivalent to hyperbolicity of Θ_x^t, $t > 0$, for all $x \in X$ and consequently to invertibility of d_x for all $x \in X$.

Proof. We may restrict ourselves to the case $t \in \mathbb{Z}$. According to Lemma 2 we have to prove only that the hyperbolicity of $T = T_a$ is equivalent to that of the operators $\Theta_x = \Theta_x^1$ for all $x \in X$, where $(\Theta_x f)(s) = a(\alpha^{s-1}x)f(s-1)$, $s \in \mathbb{R}, f \in L_2(\mathbb{R}; H)$.

Using the group of unitary operators $\{T_1^t\}_{t\geq 0}$, we construct the family of operators $\mathcal{T}(s) = T_{a\circ\alpha^s}$, $0 \leq s \leq 1$. Let us consider the operator

$$\mathcal{T} = \int_0^1 \mathcal{T}(s)ds, (\mathcal{T}f)(s,x) = \left[\frac{dm\alpha^{-1}}{dm}(x)\right]^{\frac{1}{2}} a(\alpha^{s-1}x)f(s,\alpha^{-1}x),$$

acting in the space $L_2([0,1]; L_2(X,m;H)) = L_2([0,1] \times X; H) = L_2(X,m; L_2([0,1];H))$. Hyperbolicity of T_a is equivalent to that of $\mathcal{T}$. But in the space $L_2(X,m;L_2([0,1];H))$ the operator $\mathcal{T} = T_1\tilde{a}$ acts as a weighted composition operator, generated by α and $\tilde{a} : X \to L(L_2([0,1];H))$, $(\tilde{a}(x)v)(s) = a(\alpha^s x)v(s)$, $s \in [0,1]$. Applying Lemma 3 to $\mathcal{T}$ we obtain that hyperbolicity of T is equivalent to the following operators acting in $l_2(\mathbb{Z}; L_2([0,1];H))$ being hyperbolic for all $x \in X$: $\tilde{\theta}_x : (v_t(s))_{t\in\mathbb{Z}} \to (a(\alpha^{s+t-1}x)v_{t-1}(s))_{t\in\mathbb{Z}}, s \in [0,1]$, $x \in X$. If $\tau : (v_t(\cdot))_{t\in\mathbb{Z}} \to f(\cdot) = v_{[\cdot]}(\{\cdot\}) \in L_2(\mathbb{R}; H)$, where $\{\cdot\}$ and $[\cdot]$ are the fractional and entire parts of $(\cdot)$, then $\tau\tilde{\theta}_x\tau^{-1} = \Theta_x$, and we are done. □

2. Recall (cf. [1,2]) that for a cocycle A with invertible values the extension $\hat{\alpha}$ is called exponentially dichotomous at $x \in X$ if there exist a projection P and constants $C, \beta > 0$ such that

$$\| A(x,t)PA^{-1}(x,s) \| \leq Ce^{-\beta(t-s)},\quad s \leq t;$$
$$\| A(x,t)(I-P)A^{-1}(x,s) \| \leq Ce^{-\beta(s-t)},\quad s \geq t.$$

Lemma 5. $\hat{\alpha}^1$ is exponentially dichotomous at x iff θ_x is hyperbolic.

Proof. Let $\hat{\alpha}^1$ be exponentially dichotomous at x. Let us define the projection $\mathcal{P} = \text{diag}\,\{P(\alpha^t x)\}_{t\in\mathbb{Z}}$ in l_2 where $P(\alpha^t x) = A(x,t)PA^{-1}(x,t)$. It is easy to check that the spectral radius is

$$R(\theta_x) = \lim_{\mathbb{Z}\ni s\to\infty} \sup_{t\in\mathbb{Z}} \| A(\alpha^t x, s) \|^{\frac{1}{s}}$$

and that $\theta_x\mathcal{P} = \mathcal{P}\theta_x$. Due to the cocycle property of A we have:

$$\begin{aligned} R(\theta_x\mathcal{P}) &= \lim_{s\to\infty} \sup_{t\in\mathbb{Z}} \| A(\alpha^s x, t)P(\alpha^t x) \|^{\frac{1}{s}} \\ &= \lim_{s\to\infty} \sup_{t\in\mathbb{Z}} \| A(x, t+s)PA^{-1}(x,t) \|^{\frac{1}{s}} \le e^{-\beta}, \end{aligned}$$

and $R((\theta_x(I-\mathcal{P}))^{-1}) < e^{-\beta}$ analogously.

For the hyperbolic operator θ_x one can prove, as in Lemma 2 (see e.g. [9], p.726), that the Riesz projection is $\mathcal{P} = \text{diag}\,\{P_t\}_{t\in\mathbb{Z}}, P_t^2 = P_t \in \mathrm{L}(H)$. Putting $P(\alpha^t x) = P_t$, $t \in \mathbb{Z}$, the equality $\theta_x\mathcal{P} = \mathcal{P}\theta_x$ yields that $A(x,t)P(x) = P(\alpha^t x)A(x,t)$. $\| A(x,t)PA^{-1}(x,s) \| = \| A(\alpha^t x, t-s)P(\alpha^s x) \| \le CR(\theta_x\mathcal{P})^{t-s}$ for some constant C and $t \ge s$, and the first estimate in the definition of an exponential dichotomy follows. The second one is also true because $R\left((\theta_x(I-\mathcal{P}))^{-1}\right) < 1$. □

This and Lemma 3 imply:

Corollary 6. Exponential dichotomy of the linear extension $\hat{\alpha}$ at any point $x \in X$ is equivalent to hyperbolicity of $\hat{\alpha}$.

The following proposition can be considered as a generalization of the result (in case (4) above) concerning the equivalence of exponential dichotomy for the equation $\dot{v} = x(t)v$ and invertibility of the operator $\frac{d}{dt} - x(\cdot)$ (see [12], cf. also [13]).

Theorem 7. For an exponential dichotomy of $\hat{\alpha}$ at a point $x_0 \in X$ invertibility of d_{x_0} is sufficient and hyperbolicity of d_x for all $x \in cl\{\alpha^t x_0 : t \in \mathbb{R}\}$ is necessary.

Proof. If d_{x_0} is invertible, then $\Theta^1_{x_0}$ is hyperbolic by the spectral mapping theorem for the semigroup $\{\Theta^t_{x_0}\}_{t\ge 0}$. Hence, for some $C, \beta > 0$ and all $t \in \mathbb{R}_+$ we have

$$\| \Theta^t_{x_0}\mathcal{P} \| \le Ce^{-\beta t};\ \| \Theta^{-t}_{x_0}(I-\mathcal{P}) \| \le Ce^{-\beta t}, \tag{5}$$

where $\mathcal{P}$ is the Riesz projection for $\Theta^1_{x_0}$. As in Lemma 2, one can prove (cf. [9], p.726) that $(\mathcal{P}f)(s) = P_s f(s)$, where the projection-valued function $s \to P_s$ is continuous and bounded on $\mathbb{R}$. But $\Theta^t_{x_0}\mathcal{P} = \mathcal{P}\Theta^t_{x_0}$; hence $P_s = A(x,s)P_0A^{-1}(x,s)$. Thus we get the exponential estimate for $P = P(x_0) : \| A(x_0,t)PA^{-1}(x_0,s) \| \le \sup_{\tau\in\mathbb{R}} \| A(\alpha^\tau x_0, t-s)A(x_0,\tau)\cdot PA^{-1}(x_0,\tau) \| = \| \Theta^{t-s}_{x_0}\mathcal{P} \| \le Ce^{-\beta(t-s)}$, for $t \ge s$ (and the same for $t \le s$).

Now let $\{\hat{\alpha}^t\}$ be exponentially dichotomous at $x_0 \in X$ with projection $P = P(x_0)$. Putting $P_s = A(x,s)PA^{-1}(x,s)$, $s \in \mathbb{R}$, we consider the projection $(\mathcal{P}f)(s) = P_s f(s)$ in $L_2(\mathbb{R}; H)$. Then (5) is a consequence of the exponential dichotomy, and $\Theta_{x_0} = \Theta^1_{x_0}$ is hyperbolic. To prove the hyperbolicity of Θ_x for any $x \in cl\{\alpha^t x_0 : t \in \mathbb{R}\}$ let us consider the hyperbolic operators $\Theta_{\alpha^{n_t}x_0} = U^{-n_t}\Theta_{x_0}U^{n_t}$, $(U^n f)(s) = f(s-n)$, where $\{n_t\} \subset \mathbb{Z}$ and $\alpha^{n_t}x_0 \to x$ when $\mathbb{Z} \ni t \to \infty$. It is clear, that $\Theta_{\alpha^{n_t}x_0} \xrightarrow{t} \Theta_x$, $\Theta^*_{\alpha^{n_t}x_0} \xrightarrow{t} \Theta^*_n$ strongly, and $\| (z - \Theta_{\alpha^{n_t}x_0})^{-1} \| = \| (z - \Theta_{x_0})^{-1} \| < \infty$ for $|z| = 1$. That is why (see, e.g. [14], p.112) Θ^1_x is hyperbolic, so that d_x is hyperbolic by the spectral mapping theorem for the semigroup $\{\Theta^t_x\}_{t\geq 0}$. □

3. The dynamical spectrum $\sum = \sum(\hat{\alpha})$ of the linear extension $\hat{\alpha}$ is the set of $\omega \in \mathbb{R}$ for which the extension

$$\hat{\alpha}^t_\omega : (x,v) \to (\alpha^t x, e^{-\omega t}A(x,t)v)$$

is nonhyperbolic (see [1,2]).

From Lemma 2 and Theorem 1 we deduce

Corollary 8. $\sum = ln|\sigma(T)| = \sigma(\mathcal{D}) \cap \mathbb{R}$.

For $A(x,t) \in GL(\mathbb{R}^m)$ this result is contained in [4,5].

Let F^1 be an $\hat{\alpha}$-invariant subbundle in $X \times H$, F^2 its linear complement and $Q(\cdot)$ a continuous projection valued function such that

$$F^1_x = \mathrm{Ker}\, Q(x), \quad F^2_x = \mathrm{Im}\, Q(x).$$

We define the linear extension $\hat{\alpha}_2$ acting in F^2 by the formula

$$\hat{\alpha}^t_2(x,v) = (\alpha^t x, Q(\alpha^t x)A(x,t)v), \quad (x,v) \in F^2.$$

Let L^1_2 and L^2_2 be the spaces of square-summable sections of subbundles F^1 and F^2, respectively. From the triangular form of the operator T w.r. to the splitting $L_2 = L^1_2 + L^2_2$ we deduce

Corollary 9. The spectrum $\sum(\hat{\alpha}_2)$ of the linear extension $\hat{\alpha}_2$ does not depend on the choice of the linear complement F^2. The following inclusion holds: $\sum(\hat{\alpha}) \subset \sum(\hat{\alpha}|_{F^1}) \cup \sum(\hat{\alpha}_2)$; it turns into an equality if

a) $\sum(\hat{\alpha}_2) \cap \sum(\hat{\alpha}|_{F^1}) = \emptyset$ or

b) F^1 has an $\hat{\alpha}$-invariant linear complement.

This assertion enables one to correctly define the so-called normal spectrum of an extension $\hat{\alpha}$ w.r. to an $\hat{\alpha}$-invariant subbundle. A finite dimensional version of Corollary 9 is proved in [6].

Now we turn to the construction of the spectral decomposition for $\hat{\alpha}$. Choose points $\gamma, \gamma', \gamma < \gamma'$, from different connected components of $\mathbb{R} \setminus \sum$ and denote by $P_\gamma, P_\gamma^{\gamma'}, P^{\gamma'}$ the Riesz projections for T corresponding to parts of $\sigma(T)$ contained in $\{z \in \mathbb{C} : |z| < e^\gamma\}$, $\{z \in \mathbb{C} : e^\gamma \leq |z| \leq e^{\gamma'}\}$ and $\{z \in \mathbb{C} : |z| \geq e^{\gamma'}\}$, resp. According to Lemma 2 these projections are multiplication operators by continuous projection valued functions. Putting

$$\mathbb{S}_x^\gamma = \operatorname{Im} P_\gamma(x), \;\; \mathbb{E}_x^{\gamma\gamma'} = \operatorname{Im} P_\gamma^{\gamma'}(x), \;\; \mathbb{U}_x^{\gamma'} = \operatorname{Im} P^{\gamma'}(x),$$

we obtain continuous spectral subbundles for $\hat{\alpha}$ (cf. [1]):

$$X \times H = \mathbb{S}^\gamma \oplus \mathbb{E}^{\gamma\gamma'} \oplus \mathbb{U}^{\gamma'}, \;\; \sum(\hat{\alpha}|_{\mathbb{S}^\gamma}) = \sum \cap [-\infty, \gamma),$$
$$\sum(\hat{\alpha}|_{\mathbb{E}^{\gamma\gamma'}}) = \sum \cap (\gamma, \gamma'), \;\; \sum(\hat{\alpha}|_{\mathbb{U}^{\gamma'}}) = \sum \cap (\gamma', \infty).$$

These subbundles can be characterized (in a similar way as it is done in [2]) by Lyapunov numbers $\underline{\lambda}^\pm$ and $\overline{\lambda}^\pm$ defined by the expressions

$$\begin{aligned}
\overline{\lambda}^+(x,v) &= \limsup_{t\to\infty} t^{-1}\, ln\|A(x,t)v\|,\\
\underline{\lambda}^+(x,v &= \liminf_{t\to\infty} t^{-1}\, ln\|A(x,t)v\|;\\
\overline{\lambda}^-(x,v) &= \limsup_{t\to\infty}(-t^{-1}\, ln\|v_t\|),\\
\underline{\lambda}^-(x,v &= \liminf_{t\to\infty}(-t^{-1}\, ln\|v_t\|,
\end{aligned}$$

if for $v = v_0$ there exists a sequence $\{v_t\}_{t=1}^\infty$ such that

$$a(\alpha^{-1}x)v_t = v_{t-1}, \;\; t = 1, 2, \ldots$$

(otherwise we set $\underline{\lambda}^- = \overline{\lambda}^- = -\infty$).

Theorem 10 [9]. The subspaces $\mathbb{E}_x^{\gamma\gamma'}$ are closed linear spans of vectors $v \neq 0$ such that $\lambda_-^\pm, \overline{\lambda}^\pm \in \sum \cap (\gamma, \gamma')$.

4. We consider now the case of discrete time and suppose that the cocycle A takes values in the set of compact operators in H. According to the multiplicative ergodic theorem (we follow [15]), for any ergodic probability measure ν on X and some subset X_ν, $\nu(X_\nu) = 1$, subspaces

$$V_x^1 = H \supset V_x^2 \supset \cdots, \;\; V_x^0 = \{0\} \subset V_x^{-1} \subset \cdots,$$

are defined on X_ν so that there exist exact Lyapunov exponents

$$\begin{aligned}
\lambda_\nu^1 &> \lambda_\nu^2 > \cdots \geq -\infty,\\
\lambda_\nu^j &= \lim_{t\to\infty} t^{-1}\, ln\|A(x,t)v\|, \; v \in V_x^j \setminus V_x^{j+1}.
\end{aligned}$$

Furthermore

$$H = V_x^{j+1} + W_x^j + \cdots + W_x^1,$$

where

$$W_x^j = V_x^j \cap V_x^{-j}, \ j = 1, 2, \cdots (< s_\nu + 1), \ s_\nu \leq \infty.$$

Notice that dim $W_{x,\nu}^j$ and λ_ν^j do not depend on $x \in X_\nu$ due to ergodicity of ν.

Lemma 11. The spectral radius $R(T)$ can be calculated by the formula $lnR(T) = \sup\{\lambda_\nu^1 : \nu \in \mathcal{E}\}$ (here $\mathcal{E}$ is the set of ergodic invariant Borel probability measures on X).

Proof. Since the function $a(\cdot) = A(\cdot, 1) : X \to G(H)$ (the set of compact operators in H) is continuous, it is easy to check that $R(T) = \lim_{\mathbb{Z} \ni t \to \infty} \max_{x \in X} \|A(x,t)\|^{1/t}$, and the inequality "$\geq$" of Lemma 11 is evident. To prove the inequality "$\leq$" we will find for any $\epsilon > 0$ the measure $\nu = \nu(\epsilon) \in \mathcal{E}$, so that $\lambda_\nu^1 \geq lnR(T) - \epsilon$. Let S, B be the unit sphere and the unit ball of H endowed with the weak topology, $X_B = X \times B, X_S = X \times S$ with the product-topology, $\overset{o}{X}_B = \{(x,v) \in X_B : a(x)v \neq 0\}$, $\tilde{a}(x,v) = ln\|a(x)v\|, (x,v) \in \overset{o}{X}_B$,

$$\tilde{\alpha} : X_B \to X_B : (x,v) \to \begin{cases} \left(\alpha x, \frac{a(x)v}{\|a(x)v\|}\right) & , \ (x,v) \in \overset{o}{X}_B \\ (\alpha x, 0) & , \ (x,v) \in X_B \setminus \overset{o}{X}_B \, . \end{cases}$$

Since $a(x) \in G(H)$, the function $\tilde{a}$ is continuous on $\overset{o}{X}_B$, $\tilde{\alpha}$ is a Borel endomorphism X_B, and $\tilde{\alpha}(X_B)$ is a Borel subset of X_B. If $\tilde{\alpha}^j(x,v) \in \overset{o}{X}_B$, $j = 0, \ldots, t-1 \in \mathbb{Z}$, then $\|A(x,t)v\|^{1/t} = \exp \frac{1}{t} \sum_{j=0}^{t-1} \tilde{a}(\tilde{\alpha}^j(x,v))$.

The measure ν will be defined as a projection to X of some Borel measure μ. Now we construct μ. To do this, let us fix $(x_t, v_t) \in X_s$ so that $\|A(x_t,t)v_t\| = \max_{x \in X} \|A(x,t)\|$ and define Borel measure μ_t on X_B by the rule: $\mu_t(f) = \frac{1}{t} \sum_{j=0}^{t-1} f(\tilde{\alpha}^j(x_t, v_t))$, $f \in C(X_B)$. Let us choose a subsequence $\{\mu_{n_t}\}$, so that $\{\mu_{n_t}\}$ converges to some Borel measure μ. Although $\tilde{a} \notin C(X_B)$, $\mu_t(\tilde{a})$ are well-defined and due to the choice of (x_t, v_t) we have: $\lim_{t \to \infty} \mu_{n_t}(\tilde{a}) = ln\, R(Ta)$.

Now we will prove that:

a) $\mu(\overset{o}{X}_B) = 1$;

b) $\tilde{a}$ is $\tilde{\mu}$-integrable;

c) $\mu(\tilde{a}) \geq ln\, R(Ta)$.

To do this, let us define the sets $F_N = \{(x,v) : \tilde{a}(x,v) \leq -N\} \supset X_B \setminus \overset{o}{X}_B$ for $N = 1, 2, \ldots$ and consider the functions $\tilde{a}_N \in C(X_B)$, $\tilde{a}_N(x,v) = \tilde{a}(x,v)$, $(x,v) \in X_B \setminus F_N$, $\tilde{a}_N(x,v) = -N$, $(x,v) \in F_N$. The inequalities $\tilde{a}(x,v) \leq \tilde{a}_N(x,v) \leq \max_{x \in X} ln\|a(x)\| \overset{def}{=} M$, $(x,v) \in X_B$, yield $\mu_{n_t}(\tilde{a}) \leq \mu_{n_t}(\tilde{a}_N)$, and for all N we have:

$$ln\, R(T) = \lim_{t \to \infty} \mu_{n_t}(\tilde{a}) \leq \lim_{t \to \infty} \mu_{n_t}(\tilde{a}_N) = \mu(\tilde{a}_N). \tag{6}$$

This and the inequality $\mu(\tilde{a}_N) \leq -N\mu(F_N) + M$ imply that $\mu(F_N) \to 0$ for $N \to \infty$, and a) is proved. But $\tilde{a}_N \searrow \tilde{a}$ pointwise on $\overset{o}{X}_B$ (that is μ-a.e.). This yields b) and, moreover, $\mu(\tilde{a}) = \lim_{N \to \infty} \mu(\tilde{a}_N) \geq ln\, R(T)$ because of (6), so c) is also proved.

One can check that for any function f which is bounded on X_B and continuous on $\overset{o}{X}_B$ (in particular, for $f = h \circ \tilde{\alpha}$, $h \in C(X_B)$), $\mu(f) = \lim_{t\to\infty} \mu_{n_t}(f)$ holds. Hence for any $h \in C(X_B)$ we have

$$\begin{aligned} \mu(h \circ \tilde{\alpha}) - \mu(h) &= \lim_{t\to\infty} [\mu_{n_t}(h \circ \tilde{\alpha}) - \mu_{n_t}(h)] \\ &= \lim_{t\to\infty} \frac{1}{t} \left[h(\tilde{\alpha}^t(x_t, v_t)) - h(x_t, v_t)\right] = 0. \end{aligned}$$

This means that μ is $\tilde{\alpha}$-invariant, or $\tilde{\alpha}$ is a μ-mod 0 endomorphism of X_B.

According to the Birkhoff-Hinchin ergodic theorem for μ-integrable $\tilde{a}$ we can find a set $Y \subset X_B$, $\mu(Y) = 1$, so that $\overline{a}(x, v) = \lim_{t\to\infty} \frac{1}{t} \sum_{j=0}^{t-1} \tilde{a}(\tilde{\alpha}^j(x, v))$ exists for every $(x, v) \in Y$ and $\mu(\overline{a}) = \mu(\tilde{a}) \geq ln\, R(T)$.

To construct the measure ν let us consider the set $Y^\epsilon = \{(x, v) \in Y : \overline{a}(x, v) \geq \mu(\tilde{a}) - \epsilon\}$, $\mu(Y^\epsilon) > 0$, for a given $\epsilon > 0$ and denote $X^\epsilon = pr\, Y^\epsilon$, $\nu = pr\, \mu$, where $pr(x, v) = x$. Using ergodic components, if necessary, we may assume ν to be ergodic. Applying the multiplicative ergodic theorem, we find a set X_ν and a point $\overline{x} \in X_\nu \cap X^\epsilon$, so that $\lim_{t\to\infty} t^{-1} ln\|A(\overline{x}, t)\| = \lambda_\nu^1$. Thus we have for $(\overline{x}, \overline{v}) \in Y^\epsilon$:

$$\begin{aligned} ln\, R(T) - \epsilon \leq \mu(\tilde{a}) - \epsilon &\leq \overline{a}(\overline{x}, \overline{v}) = \lim_{t\to\infty} \sum_{j=0}^{t-1} \tilde{a}(\tilde{\alpha}^j(\overline{x}, \overline{v})) \\ &\leq \lim_{t\to\infty} \frac{1}{t} ln\|A(\overline{x}, t)\| = \lambda_\nu^1, \end{aligned}$$

and the proof is finished. □

Theorem 12. The spectrum $\sum$ consists of a finite or countably infinite collection of intervals: $\sum = \cup_{k=1}^N [r_k^-, r_k^+]$, $N \leq \infty$. If $\gamma_0 > r_1^+ \geq r_1^- > \gamma_1 > r_2^+ \geq \ldots$ then dim $\mathbb{E}_x^{\gamma_k \gamma_k} < \infty$. If dim $H = \infty$ then $-\infty \in \sum$; if $N = \infty$ then $\lim_{k\to\infty} e^{r_k^\pm} = 0$.

The spectrum can be calculated similarly in case $a(x)$ is a sum of unitary and compact operators and dim $H = \infty$ (see [9] for proofs).

Theorem 13 [9]. For any measure $\nu \in \mathcal{E}$ there exist numbers $i_1 = 1 < i_2 < \ldots < i_{N+1} = s_\nu + 1$ such that

$$\begin{aligned} \mathbb{S}_x^{\gamma_{k-1}} &= V_{x,\nu}^{i_k},\ k = 1, 2, \ldots (< N+1), \\ \mathbb{E}_x^{\gamma_k \gamma_{k-1}} &= \sum_{j=i_k}^{i_{k+1}-1} W_{x,\nu}^j, \lambda_\nu^j \in [r_k^-, r_k^+],\ i_k \leq j < i_{k+1},\ k = 1, 2, \ldots (< N+1). \end{aligned}$$

Furthermore,

$$r_k^- = \inf\{\lambda_\nu^{i_{k+1}-1} : \nu \in \mathcal{E}\},\ r_k^+ = \sup\{\lambda_\nu^{i_k} : \nu \in \mathcal{E}\}.$$

A similar statement for $a : X \to GL(\mathbb{R}^m)$ can be found in [1,2].

Theorems 12 and 13 suggest an inductive procedure to calculate the spectrum that consists in successive "chopping off" of those unstable spectral subbundles which extend the subbundles

$$V_{x,\nu}^{-j} = W_{x,\nu}^{j} \oplus \ldots \oplus W_x^1, \quad x \in X_\nu, \quad \nu \in \mathcal{E}.$$

We are going to describe this procedure in detail:

1. Calculate $r_1^+ = \ln R(T) = \sup\{\lambda_\nu^1 : \nu \in \mathcal{E}\}$.
2. For $\nu \in \mathcal{E}$ calculate the sums $m_\nu(k) = \sum_{j=1}^{k} \dim W_{x,\nu}^j$, $k = 1, 2, \ldots$, and choose $k = k(\nu)$ so that $m = m_\nu(k(\nu))$ does not depend on $\nu \in \mathcal{E}$.
3. For $\nu \in \mathcal{E}$ check the existence of $\hat{\alpha}$-invariant continuous subbundles $\mathbb{S}$ and $\mathbb{U}$ over X that extend the subbundles $V_{x,\nu}^{k(\nu)+1}$ and $V_{x,\nu}^{-k(\nu)}$, respectively from X_ν. Notice that such an extension exists for any $\nu \in \mathcal{E}$ if it exists for some $\nu_0 \in \mathcal{E}$. If there is no such extension we turn back to step 2) and increase m.
4. Calculate $r_1^- = \inf\{\lambda_\nu^{k(\nu)} : \nu \in \mathcal{E}\}$, $r_2^+ = \sup\{\lambda_\nu^{k(\nu)+1} : \nu \in \mathcal{E}\}$ and fix γ_1 so that $r_2^+ < \gamma_1 < r_1^-$. If $r_1^- \le r_2^+$ come back to step 2).
5. Check hyperbolicity of the linear extension $\hat{\alpha}_{\gamma_1}$ w.r. to the splitting $H = \mathbb{S}_x \oplus \mathbb{U}_x$ (if the restriction $a_u(x) = a(x)|_{\mathbb{U}_x}$ is invertible for all $x \in X$ then hyperbolicity automatically holds); if there is no hyperbolicity come back to step 2).
6. Build a cocycle $a_s(x) = a(x)|_{\mathbb{S}_x}$ and apply to it the procedure described above.

A simple modification of this scheme enables one to describe subbundles with the property of exponential separation (cf. [16]). In fact: Steps 1) – 3) proceed without changes. If step 4) can be accomplished then we gain the subbundles sought for. If $r_1^- \le r_2^+$ check the condition $0 \notin cl\{\lambda_\nu^j + \int_X g d\nu : \nu \in \mathcal{E}\}$, where $g(x) = (\|a_s(x)\|_\bullet \cdot \|a_u(x)\|_\bullet)_{-1/2}$, $\|a\|_\bullet = \inf_{\|v\|=1} \|av\|$. If this condition is fulfilled then $\hat{\alpha}$ is exponentially separated w.r. to the decomposition $X \times H = \mathbb{S} \oplus \mathbb{U}$; otherwise come back to step 2) and increase m.

To conclude we formulate two results concerning the case $a : X \to GL(\mathbb{R}^m)$.

Theorem 14 [9]. $\hat{\alpha}$ is hyperbolic iff it is quasihyperbolic (i.e. $\sup\{\|A(x,t)v\| : t \in \mathbb{Z}\} < \infty \Rightarrow v = 0$) and the sum of multiplicities

$$m_\nu^+ = \sum_{j, \lambda_\nu^j > 0} \dim W_{x,\nu}^j$$

does not depend on $\nu \in \mathcal{E}$.

The spectral radius $R(Q)$ of the WCO $(Qf)(x) = \left(\frac{d\mu\circ\alpha}{d\mu}(x)\right)^{1/2} a(x) f(\alpha x)$ generated by a "finite leaf" endomorphism α of a compact metric space satisfies the following variational principle.

Theorem 15. Let $h_\nu(\alpha)$ denote the entropy of the endomorphism α, χ^1_ν the greatest Lyapunov exponent for the cocycle $a^* \circ \alpha^{t-1} \cdot \ldots \cdot a^*$ w.r. to an α-invariant probability measure ν.
Then $\ln R(Q) \leq \sup\{\frac{1}{2} h_\nu(\alpha) + \chi^1_\nu : \nu\text{-invariant probability on} X\}$. This inequality turns out to be an equality when α is a one-sided topological Markov shift.

Proof. At first we introduce the matrix analogue L_a of the Ruelle-Frobenius-Perron operator (cf.[17]). Let $\mathcal{L}(X)$ be the space of continuous functions $\psi : X \to \mathrm{L}(\mathbb{R}^m)$, with self-adjoint values, endowed with the sup-norm. We define L_a by a rule:

$$(L_a\psi)(x) = \sum_{y\in\alpha^{-1}(x)} a^*(y)\psi(y)a(y).$$

It is clear, that $(L^t_a\psi)(x) = \sum_{y\in\alpha^{-t}x} \hat{A}(y,t)\psi(y)\hat{A}^*(y,t)$, where $\hat{A}(x,t) = a^*(\alpha^{t-1}x)\cdot \ldots \cdot a^*(x)$. Using the Radon-Nikodym theorem and the partition $X = \cup_j X_j$, $\alpha : X_j \to X$ is one-to-one, we have: $\|Q^t_a f\|^2_{L_2} =< L^t_a \mathbf{1} \cdot f, f >_{L_2}$, $f \in L_2$. Since a is a continuous function and supp $m = X$, we obtain: $R(Q) = \lim_{t\to\infty} \|L^t_a \mathbf{1}\|^{\frac{1}{2t}}_{\mathcal{L}(X)}$. The definition of L_a also yields a chain of equalities:

$$\|L^t_a\|_{\mathrm{L}(\mathcal{L}(X))} = \|L^t_a\mathbf{1}\|_{\mathcal{L}(X)} = \max_{(x,v)\in X_s} \sum_{y\in\alpha^{-t}x} \|a(y)\cdot\ldots\cdot a(\alpha^{t-1}y)v\|^2. \tag{7}$$

Hence

$$R(Q) = R^{\frac{1}{2}}(L_a) = \lim_{t\to\infty} \max_{(x,v)\in X_s} \left(\sum_{y\in\alpha^{-t}x} \|a(y)\cdot\ldots\cdot a(\alpha^{t-1}y)v\|^2\right)^{\frac{1}{2t}}. \tag{8}$$

In particular, we get for the scalar situation $m = 1$

$$R(Q) = \lim_{t\to\infty} \max_{x\in X} \left(\sum_{y\in\alpha^{-t}x} |a(y)\cdots a(\alpha^{t-1}y)|^2\right)^{\frac{1}{2t}}.$$

That is why (as it was shown in [18], see also [19]) we have the following inequality for $m = 1$:

$$\ln R(Q) \leq \frac{1}{2} P(\alpha, 2\ln|a|) = \sup\{\frac{1}{2} h_\nu(\alpha) + \int_X \ln|a|d\nu : \nu \in \mathcal{M}(\alpha)\}. \tag{9}$$

This inequality turns out to be an equality when (α, X) is a one-sided subshift of finite type. Here $P(\cdot,\cdot)$ is the topological pressure [17,20], $\mathcal{M}(\alpha)$ is the set of α-invariant measures on X (we use the variational principle [17,20]).

To deal with the case $m > 1$ let us denote: $\tilde{\alpha} : X_s \to X_s$: $(x,v) \to (\alpha x, a^{-1}(x)v/\|a^{-1}(x)v\|$, $\tilde{a}(x,v) = \|a^{-1}(x)v\|^{-1}$. It is easy to check that

$$\sum_{y\in\alpha^{-t}x} \|a(y)\cdot\ldots\cdot a(\alpha^{t-1}y)v\|^2 = \sum_{(y,u)\in\tilde{\alpha}^{-t}(x,v)} \prod_{j=0}^{t-1} |\tilde{a}(\tilde{\alpha}^j(y,u))|^2.$$

Applying (8) and the preceeding reasoning (see (9)) to $\tilde{a}$ and $\tilde{\alpha}$ instead of a and α, we conclude:

$$ln\, R(Q) \leq \sup\{\frac{1}{2}h_\mu(\tilde{\alpha}) + \int_{X_S} ln|\tilde{a}|d\mu : \mu \in \mathcal{M}(\tilde{\alpha})\}.$$

According to the Birkhoff-Hinchin theorem, for any $\mu \in \mathcal{M}(\tilde{\alpha})$ there exists a set of positive μ-measure, and for each of its points (x,v) we have:

$$\begin{aligned}\int_{X_S} ln|\tilde{a}|d\mu &\leq \lim_{t\to\infty} t^{-t}\sum_{j=0}^{t-1} ln\, \tilde{a}(\tilde{\alpha}^j(x,v)) \\ &= -\lim_{t\to\infty}\frac{1}{t} ln\|a^{-1}(\alpha^{t-1}x)\cdot\ldots\cdot a^{-1}(x)v\|.\end{aligned}$$

Using the multiplicative ergodic theorem for $\nu = pr\mu$ and the cocycles generated by a^{-1}, α and a^*, α, we conclude that the last term is less than or equal χ^1_ν. The desired inequality follows now from

Lemma. If $\mu = pr\mu$, $\mu \in \mathcal{M}(\tilde{\alpha})$, then $h_\mu(\tilde{\alpha}) = h_\nu(\alpha)$.

We omit the proof of the Lemma, which uses the Abramov-Rohlin formula.

Let now (α, X) be a one-sided subshift of finite type. To prove that the inequality in the theorem turns out to be an equality, we need at first the triangularization procedure (cf. [21]). Let $SO(m)$ be the set of orthogonal matrices with unit determinant, $\tilde{X} = X \times SO(m)$, $\pi_1 : \tilde{X} \to X$, $\pi_2 : \tilde{X} \to SO(m)$ the projections, $g(x,u) = u^T a(x)\tilde{u}(x,u)$, where $\tilde{u} = \tilde{u}(x,u) \in SO(m)$ is chosen such that $\tilde{u}^T a^*(x)u$ is a lower triangular matrix with positive entries on the main diagonal. Let us note that $g : \tilde{X} \to GL(\mathbb{R}^m)$ is continuous, and define the extension β of α by the rule $\beta : \tilde{X} \to \tilde{X} : (x,u) \to (\alpha x, \tilde{u}(x,u))$. In the space $\mathcal{L}(\tilde{X})$ for β we consider Ruelle's operators, corresponding to $a \circ \pi_1$ and g:

$$\begin{aligned}(\tilde{\mathrm{L}}\psi)(x,u) &= \sum_{(y,w)\in\beta^{-1}(x,u)} [a \circ \pi_1(y,w)]^* \psi(y,w)\, a \circ \pi_1(y,w); \\ (\tilde{\mathrm{L}}_g\psi)(x,u) &= \sum_{(y,w)\in\beta^{-1}(x,u)} g^*(y,w)\psi(y,w)g(y,w).\end{aligned}$$

If $J : \mathcal{L}(\tilde{X}) \to \mathcal{L}(\tilde{X}) : \psi \to \pi_2\psi\pi_2^T$, then $\tilde{\mathrm{L}}_g = J^{-1}\mathrm{L}J$. Combining this with (7) one can note that $R(\tilde{\mathrm{L}}_g) = R(\tilde{\mathrm{L}}) = R(L_a)$. The Lyapunov spectrum of the cocycle $g^* \circ \beta^{-1} \cdot \ldots \cdot g^*$ consists of the numbers $\int_{\tilde{X}} ln\, g_i\, d\mu$, $i = 1, ..., m$, where g_i are the elements of the main diagonal of the lower triangular matrix g^*. To prove that

$$R(\tilde{\mathrm{L}}_g) \geq \exp\{h_\mu(\beta) + 2\int_{\tilde{X}} ln\, g_i(x,u)d\mu\},\ \mu \in \mathcal{M}(\beta),\ i = 1,\ldots,m, \tag{10}$$

we need induction on m.

For $m = 1$ the assertion follows from the scalar result of [18], which was cited above. If (10) is valid for $m-1$, then we consider g in the form

$$g = \begin{bmatrix} g_1 & * \\ 0 & \tilde{g}\end{bmatrix};\ \tilde{g} : \tilde{X} \to GL(\mathbb{R}^{m-1}),\ v = (v_1, \overline{v})^T \in \mathbb{R}^m,\ \overline{v}^T \in \mathbb{R}^{m-1}.$$

To calculate $R(\tilde{L}_g)$ we will use (8). We note, that

$$\sup_{\|v\|_{\mathbb{R}^m}=1} \sum_{(y,w)\in\beta^{-t}(x,u)} \|g(y,w)\cdot\ldots\cdot g\circ\beta^{t-1}(y,w)v\|^2 \geq$$

$$\max\left\{\sum_{(y,w)} |g_1(y,w)\cdot\ldots\cdot g_1\circ\beta^{t-1}(y,w)|^2, \sup_{\|\bar{v}\|_{\mathbb{R}^{m-1}}=1} \sum_{(y,w)} \|\tilde{g}(y,w)\cdot\ldots\cdot\tilde{g}\circ\beta^{t-1}(y,w)\bar{v}\|^2\right\},$$

and (10) follows from the induction assumption.

To finish the proof we have to deduce the inequality $R(L_a) = R(\tilde{L}_g) \geq \exp\{h_\nu(\alpha) + 2\chi_\nu^1\}$ from (10) for any measure $\nu \in \mathcal{M}(\alpha)$. Choosing the Borel probability measure $\tilde{\nu}$ on $SO(m)$ and starting from the measure $\mu_0 = \nu \times \tilde{\nu}$ on $\tilde{X}$, let us construct the β-invariant measure μ with the help of the Bogolubov-Krylov construction. Since $\nu = pr\,\mu$, the Abramov-Rohlin formula implies that $h_\mu(\beta) = h_\nu(\alpha) + h_\mu(\beta|\alpha) \geq h_\nu(\alpha)$. The cocycles $a^*\circ\alpha^{t-1}\cdot\ldots\cdot a^*$ and $g^*\circ\beta^{t-1}\cdot\ldots\cdot g^*$ are homological, hence

$$\chi_\nu^1 = \lim_{t\to\infty} t^{-1}ln\|\hat{A}(x,t)\| = \lim_{t\to\infty} t^{-1}ln\|g^*\circ\beta^{t-1}(x,u)\cdot\ldots\cdot g^*(x,u)\| = \chi_\mu^1(g^*,\beta),$$

if $x \in X_\nu$, $(x,u) \in \tilde{X}_\mu$. Estimating the maximum over $i = 1,\ldots,m$ of the right hand side of (10) by $h_\nu(\alpha) + 2\chi_\nu^1$, we get the desired inequality. □

References

[1] Sacker, R.T. and G.R. Sell, A spectral theory for linear differential systems, J. Diff. Eqns. 27(1978) 320-338.

[2] Johnson, R.A., K. Palmer and G. Sell, Ergodic properties of linear dynamical systems, SIAM J. Math. Anal. 1(1987) 1-33.

[3] Sacker, R. and G. Sell, Existence of dichotomies and invariant splitting for linear differential systems, J. Diff. Eqns. 15(1974) 429-455.

[4] Johnson, R., Analyticity of spectral subbundles, J. Diff. Eqns. 35(1980) 366-387.

[5] Chicone, C. and R. Swanson, Spectral theory for linearization of dynamical systems, J. Diff. Eqns. 40(1981) 155-167.

[6] Sacker, R. and G. Sell, The spectrum of an invariant submanifold, J. Diff. Eqns. 38(1980) 135-160.

[7] Kubo, I., Quasi-flows, Nagoya Math. J. 35(1969) 1-30.

[8] Greiner, G., Some applications of Fejer's theorem to one-parameter semigroups, Semesterbericht Funktionalanalysis Wintersemester 1984/85 Tübingen (1985) 33-50.

[9] Latushkin, Y.D. and A.M. Stepin, Weighted composition operators, spectral theory of linear skew-product flows and multiplicative ergodic theorem, Matem. sbornik 181(1990), no. 6, 723-742.

[10] Samojlenko, A.M., The elements of the mathematical theory of many frequency oscillation (Russian), Kiev, 1987.

[11] Antonevich, A.B., Linear functional equations. Operators approach (Russian), Minsk, 1988.

[12] Coppel, W.A., Dichotomies in stability theory, Lect. Notes Math. 629(1978).

[13] Palmer, K., Exponential dichotomies and Fredholm operators Proc. AMS 104(1988) 149-156.

[14] Gohberg, I. and I. Feldman, Convolution equations and projective methods of their solving, (Russian), Moscow (1971).

[15] Ruelle, D., Ergodic theory of differentiable dynamical systems, Publ. Math. 50(1979) 275-306.

[16] Bronshtejn, I.U., Nonautonomous dynamical systems (Russian), Kishinev (1984).

[17] Bowen, R., Equilibrium states and ergodic theory of Anosov diffeomorphisms, Lect. Notes Math. 470(1975).

[18] Latushkin, Y.D. and A.M. Stepin, Weighted composition operator on topological Markov chain, Funkc.anal.priloz. 22(1988) 86-87 (in Russian).

[19] Latushkin Y.D., On weighted composition operators, Col.: "Boundary value problems", Perm (1982) 148-150 (in Russian).

[20] Walters P., An introduction to ergodic theory, Springer-Verlag, New York 1982.

[21] Oseledec, V.I., A multiplicative ergodic theorem. Lyapunov characteristic numbers for dynamical systems. Trans. Moscow Math. Soc. 19(1968) 197-231.

FILTRE DE KALMAN BUCY
ET EXPOSANTS DE LYAPOUNOV

Philippe BOUGEROL
Département de Mathématiques, Université Nancy 1
54506 Vandoeuvre les Nancy, BP 239. France

ABSTRACT: In this paper, we consider linear filtering with random stationary coefficients. We show that generalizations of Kalman's results on the asymptotic behavior of the filter, previously obtained using contraction properties, can also be proved using Osseledets's theorem and a result of M. Wojtkowski. The filter is exponentially stable with a rate given by the smallest positive Lyapunov exponent of some associated symplectic matrices.

RESUME: On considère la situation du filtrage linéaire à coefficients aléatoires stationnaires. On montre que les généralisations des résultats de Kalman sur le comportement asymptotique du filtre, que nous avions obtenues à partir de propriétés de contraction, peuvent aussi être montrées en utilisant le théorème d'Osseledets et un résultat de M. Wojtkowski. Le filtre est exponentiellement stable avec un taux déterminé par le plus petit exposant de Lyapounov positif d'un produit de matrices symplectiques.

1. Introduction.

Nous considérons une situation de filtrage linéaire dans un environnement aléatoire stationnaire. Nous avons établi les propriétés asymptotiques du filtre dans [Bo] de façon élémentaire. Ici, nous mettons en évidence les liens entre ces propriétés et les exposants de Lyapounov.

Nous étudions le système linéaire suivant :

$$\begin{aligned} X_n &= A_n X_{n-1} + F_n \varepsilon_n , \quad n \geq 1, \\ Y_n &= C_n X_n + \eta_n, \end{aligned} \tag{1}$$

où $X_n \in \mathbb{R}^d$, $\varepsilon_n \in \mathbb{R}^p$, η_n et $Y_n \in \mathbb{R}^q$. Les coefficients A_n, F_n et C_n sont des matrices de taille respective $d \times d$, $d \times p$ et $q \times d$, indexées par n dans $\mathbb{Z}$. Toutes ces quantités sont des variables aléatoires définies sur un espace de probabilité $(\Omega, \mathcal{F}, \mathbb{P})$.

Dans tout l'article, nous supposons qu'il y a une sous tribu $\mathcal{F}_0$ de $\mathcal{F}$ telle que, si $\mathcal{F}_n$ est la tribu engendrée par $\mathcal{F}_0$ et les variables aléatoires $Y_1, Y_2, ..., Y_n$, alors, pour tout $n \geq 1$,

(i) A_n, F_n et C_n sont $\mathcal{F}_{n-1}$ mesurables;

(ii) Les vecteurs aléatoires ε_n et η_n sont indépendants de loi gaussienne centrée de matrice de covariance égale à l'identité et ils sont indépendants de la tribu $\sigma(\mathcal{F}_{n-1}, X_{n-1})$.

(iii) Conditionnellement à $\mathcal{F}_0$, le vecteur X_0 a une loi gaussienne d'espérance $\hat{X}_0$ et de matrice de covariance P_0.

De plus nous supposons que $\{(A_n, F_n, C_n), n \in \mathbb{Z}\}$ est un processus stationnaire ergodique et que les matrices A_n sont inversibles.

Nous sommes dans la situation dite "conditionnellement gaussienne". Les équations classiques du filtre de Kalman (cf., e.g., [Ba], [Wh]) nous indiquent alors que les vecteurs $\hat{X}_n$ et les matrices symétriques P_n définies, pour $n \geq 1$, par

$$\hat{X}_n := \mathbb{E}(X_n / \mathcal{F}_n) \quad \text{et} \quad P_n := \mathbb{E}((X_n - \hat{X}_n)(X_n - \hat{X}_n)^* / \mathcal{F}_n)$$

peuvent être obtenus par la récurrence suivante : Si $R_n = C_n^* C_n$ et $S_n = F_n F_n^*$,

$$P_n = (A_n P_{n-1} A_n^* + S_n)\,(I + R_n S_n + R_n A_n P_{n-1} A_n^*)^{-1} \tag{2}$$

$$\hat{X}_n = (A_n - P_n R_n A_n)\;\hat{X}_{n-1} + P_n C_n^* Y_n\,. \tag{3}$$

Dans [Bo], nous avons donné des exemples concrets où cette modélisation intervient. Nous y avons aussi décrit les propriétés asymptotiques de la suite P_n et de ce filtre, généralisant ainsi les résultats classiques de Kalman (qui correspondent au cas des coefficients constants). Avant d'énoncer ces propriétés, introduisons quelques définitions.

Définition 1: *On dit que le système* (1) *satisfait à la condition* (C) *si les conditions* (C_1) *et* (C_2) *suivantes sont vérifiées :*

(C_1) *Le système est faiblement observable et controlable au sens où, avec une probabilité non nulle, il existe un entier* $n \geq 1$ *tel que les deux matrices*

$$A_1^* R_1 A_1 + A_1^* A_2^* R_2 A_2 A_1 + \dots + A_1^* \,..\, A_n^* R_n A_n \,..\, A_1$$

et

$$S_n + A_n S_{n-1} A_n^* + \dots + A_n \,..\, A_2 S_1 A_2^* \,..\, A_n^*$$

soient inversibles.

(C_2) *Les variables aléatoires* $\mathrm{Log}^+\|A_1\|$, $\mathrm{Log}^+\|A_1^{-1}\|$, $\mathrm{Log}^+\|C_1\|$ *et* $\mathrm{Log}^+\|F_1\|$ *sont intégrables.*

Notation. On note $\mathcal{P}$ l'ensemble des matrices $d \times d$ symétriques semi-définies positives et $\mathcal{P}_0$ l'ensemble des matrices symétriques définies positives. On munit $\mathcal{P}_0$ de la distance riemannienne δ définie ainsi: si $P, Q \in \mathcal{P}_0$,

$$\delta(P, Q) = \Big\{ \sum_{i=1}^{d} \mathrm{Log}^2 \lambda_i \Big\}^{1/2},$$

où $\lambda_1, \dots, \lambda_d$ sont les valeurs propres de PQ^{-1}.

Le théorème suivant est montré dans [Bo].

Théorème 2: *Supposons que le système* (1) *satisfait à la condition* (C). *Alors*:
(i) *Il existe un processus stationnaire* $\{\bar{P}_n,\ n\in\mathbb{Z}\}$ *à valeurs dans* $\mathcal{P}_0$, *solution de* (2), *et un réel* $\alpha>0$, *tels que,*

$$\overline{\lim_{n\to+\infty}}\ \frac{1}{n}\ \mathrm{Log}\ \delta(P_n,\bar{P}_n)\leq-\alpha<0, \qquad p.s.,$$

pour toute solution $\{P_n,\ n\in\mathbb{N}\}$ *de* (2) *pour laquelle* $P_0\in\mathcal{P}_0$.
(ii) *Le filtre* (3) *est exponentiellement stable au sens où,*

$$\overline{\lim_{n\to+\infty}}\ \frac{1}{n}\ \mathrm{Log}\ \|(A_n-P_n\,R_n\,A_n)\dots(A_2-P_2\,R_2\,A_2)(A_1-P_1\,R_1\,A_1)\|<0, \quad p.s..$$

pour toute solution $\{P_n,\ n\in\mathbb{N}\}$ *de* (2) *à valeurs dans* $\mathcal{P}$.

La preuve simple que nous avons donnée de ce théorème reposait sur les propriétés de contraction de la récurrence (2) vis à vis de la distance δ. Il peut aussi s'interpréter à l'aide des exposants de Lyapounov. Nous allons ici expliciter ce point de vue en montrant une variante de ce théorème à partir du théorème d'Osseledets ([Os]) et d'un résultat de M. Wojtkowski ([Wo1]). Nous montrerons en particulier que la limite apparaisant au (ii) est déterminée par le plus petit exposant de Lyapounov positif d'un produit de matrices hamiltoniennes (Théorème 8).

2. Matrices hamiltoniennes et théorème de Wojtkowski.

Rappelons d'abord que le groupe symplectique Sp(d,$\mathbb{R}$) est le groupe des matrices carrées d'ordre $2d$ s'écrivant

$$M=\begin{pmatrix}A & B\\ C & D\end{pmatrix}$$

où A, B, C, D sont des matrices carrées d'ordre d telles que

$$AD^*-BC^*=I\ ,\quad AB^*=BA^*\ ,\quad CD^*=DC^*.$$

Au système linéaire (1) on associe les matrices

$$M_n=\begin{pmatrix}A_n & S_nA_n^{*-1}\\ R_nA_n & (I+R_nS_n)A_n^{*-1}\end{pmatrix}$$

parfois appelées matrices hamiltoniennes. Ces matrices sont dans l'ensemble $\mathcal{H}$ défini par :

$$\mathcal{H} = \left\{ \begin{pmatrix} A & B \\ C & D \end{pmatrix} \in \mathrm{Sp}(d,\mathbb{R}) \,;\, A \text{ est inversible}, AB^* \in \mathcal{P} \text{ et } CD^* \in \mathcal{P} \right\}.$$

Posons

$$\mathcal{H}_0 = \left\{ \begin{pmatrix} A & B \\ C & D \end{pmatrix} \in \mathrm{Sp}(d,\mathbb{R}) \,;\, AB^* \in \mathcal{P}_0 \text{ et } CD^* \in \mathcal{P}_0 \right\}.$$

Le produit de deux matrices de $\mathcal{H}$ est dans $\mathcal{H}$, et le produit d'une matrice de $\mathcal{H}$ par une matrice de $\mathcal{H}_0$ est dans $\mathcal{H}_0$, (cf. [Wo1], [Wo2] ou [Bo]). Autrement dit, $\mathcal{H}_0$ est un ideal bilatère du semi groupe $\mathcal{H}$.

On appelle sous espace lagrangien de $\mathbb{R}^{2d}$, tout sous espace vectoriel L de dimension d tel que, si u_1, $u_2 \in L$ alors $u_1^* J u_2 = 0$ où $J = \begin{pmatrix} 0 & I \\ -I & 0 \end{pmatrix}$. Si S est une matrice d'ordre d symétrique, on pose

$$L(S) := \left\{ \begin{pmatrix} Sx \\ x \end{pmatrix}, x \in \mathbb{R}^d \right\}.$$

Alors $L(S)$ est un sous espace lagrangien.

Définition 3: *On appelera sous espace lagrangien non négatif (resp. positif) tout sous espace lagrangien de la forme $L(P)$ où $P \in \mathcal{P}$ (resp. $P \in \mathcal{P}_0$).*

L'image par une matrice symplectique d'un sous espace lagrangien est encore lagrangien. Le groupe symplectique opère donc sur l'espace des sous espaces lagrangiens. De plus, si $M = \begin{pmatrix} A & B \\ C & D \end{pmatrix}$ est symplectique et si S est une matrice symétrique d'ordre d pour laquelle $CS + D$ est inversible, alors

$$M\,(L(S)) = L(M \bullet S) \tag{4}$$

où l'on a posé

$$M \bullet S = (AS + B)\,(CS + D)^{-1} \tag{5}$$

Lorsque M est une matrice de $\mathcal{H}$ et lorsque P est dans $\mathcal{P}$, resp. $\mathcal{P}_0$, alors $CP + D$ est inversible et $M \bullet P$ est dans $\mathcal{P}$, resp. $\mathcal{P}_0$. Si de plus M est dans $\mathcal{H}_0$, alors $M \bullet P$ est dans $\mathcal{P}_0$ (cf. [Wo1], p. 141). La formule de récurrence (2) s'écrit aussi

$$P_n = (A_n P_{n-1} + S_n A_n^{*-1})\,(R_n A_n P_{n-1} + (I + R_n S_n) A_n^{*-1})$$

c'est à dire

$$P_n = M_n \bullet P_{n-1}\,. \tag{6}$$

Elle correspond donc à l'action des matrices hamiltoniennes sur les sous espaces lagrangiens. Introduisons les matrices Π_n définies pour $n \geq 1$, par

$$\Pi_n = M_n\, M_{n-1} \ldots M_1 .$$

Le lemme suivant est montré dans [Bo] :

Lemme 4: *Sous la condition* (C_1), *il existe p.s. un entier* $n \geq 1$ *pour lequel* Π_n *est dans* $\mathcal{H}_0$.

La condition de moment (C_2) entraine que $\mathrm{Log}^+\|M_n\|$ est intégrable. On peut donc appliquer le théorème d'Osseledets aux produits de matrices $\Pi_n = M_n\, M_{n-1} \ldots M_1$. Comme ces matrices sont symplectiques les $2d$ exposants de Lyapounov vont par paires et on peut les écrire

$$\gamma_1 \geq \gamma_2 \geq \ldots \geq \gamma_d \geq -\gamma_d \geq \ldots \geq -\gamma_1.$$

Remarquons que γ_d est le plus petit exposant non négatif. Notre étude va reposer sur le théorème suivant :

Théorème 5 (Wojtkowski): *Sous la condition* (C), γ_d *est strictement positif.*

Preuve : Dans le cas particulier où M_1 est presque sûrement dans $\mathcal{H}_0$ ce résultat est le théorème 5.1 de [Wo1]. Dans le cas général il résulte de ce théorème et du lemme précédent exactement comme le théorème 2.2 de Wojtkowski résulte de son théorème 2.1.

Nous aurons besoin du lemme suivant qui est implicite dans [Wo1]. Nous utilisons les notations classiques sur les produits extérieurs (cf. [Lo, St] par exemple).

Lemme 6: *Supposons la condition* (C) *vérifiée. Alors, si* $u_1, u_2, \ldots, u_d$ *engendrent un sous espace lagrangien non négatif* L,

$$\lim_{n\to+\infty} \frac{1}{n} \mathrm{Log}\, \|\Pi_n u_1 \wedge .. \wedge \Pi_n u_d\| = \gamma_1 + \gamma_2 + \ldots + \gamma_d, \qquad p.s..$$

Preuve: Pour la commodité du lecteur nous explicitons la preuve. Soit u un élément de $\mathbb{R}^{2d}$ que nous écrivons $u = \binom{x}{y}$ avec $x, y \in \mathbb{R}^d$. On pose

$$Q(u) = \langle x, y \rangle = \sum_{i=1}^{d} x_i\, y_i.$$

M. Wojtkowski a montré que pour tout M de $\mathcal{H}$, il existe un réel $\rho(M) \geq 1$ tel que, si u est de la forme $u = \binom{Py}{y}$ avec P dans $\mathcal{P}$, alors

$$Q(Mu) \geq \rho(M) Q(u) .$$

De plus si M est dans $\mathcal{H}_0$, $\rho(M) > 1$, (cf [Wo1], Proposition 5.1). D'après le lemme 4 il existe un entier k tel que $\mathbb{P}(\Pi_k \in \mathcal{H}_0) \neq 0$, ce qui entraîne que $\mathbb{E}(\mathrm{Log}\ \rho(\Pi_k)) > 0$. Quitte à remplacer la suite (Π_n) par la suite (Π_{nk}), nous pouvons sans perte de généralité supposer que $k = 1$ pour établir le lemme. Soit alors u un élément du lagrangien non négatif L. Presque sûrement il existe un indice p tel que Π_p est dans $\mathcal{H}_0$, ce qui assure que $Q(\Pi_p\, u) > 0$. Alors en utilisant l'inégalité, si $n \geq p$,

$$Q(\Pi_n\, u) \geq \rho(M_n)\, \rho(M_{n-1}) \dots \rho(M_{p+1})\, Q(\Pi_p\, u),$$

et le théorème ergodique de Birkhoff, on voit que

$$\lim_{n\to+\infty} \frac{1}{n}\ \mathrm{Log}\ Q(\Pi_n\, u) \geq \mathbb{E}(\mathrm{Log}\ \ \rho(M_1)) > 0, \quad \text{p.s..}$$

Comme, par ailleurs, $\|\Pi_n\, u\|^2 \geq 2Q(\Pi_n\, u)$, on obtient que pour tout u de L,

$$\lim_{n\to+\infty} \frac{1}{n}\ \mathrm{Log}\ \|\Pi_n\, u\| > 0, \quad \text{p.s..}$$

En d'autres termes, l'intersection de L et du sous espace contractant d'Osseledets $L^s(\omega)$ défini par

$$L^s(\omega) = \{u \in \mathbb{R}^{2d}\ ;\ \lim_{n\to+\infty} \frac{1}{n}\ \mathrm{Log}\ \|\Pi_n(\omega) u\| < 0\}$$

est réduite à $\{0\}$, p.s.. Il en résulte comme on voulait le montrer, que si $u_1, \dots, u_d$ est une base de L,

$$\lim_{n\to+\infty} \frac{1}{n}\ \mathrm{Log}\ \|\Pi_n\, u_1 \wedge .. \wedge \Pi_n\, u_d\| = \gamma_1 + \dots + \gamma_d\,, \qquad \text{p.s..}$$

(en effet, dans le cas contraire, si $f_1, \dots, f_d$ est une base de L^s , pour tout élément $\vec{w}$ de la base $\{u_{i_1} \wedge .. \wedge u_{i_k} \wedge f_{i_{k+1}} \wedge .. \wedge f_{i_d}, 1 \leq i_1, \dots, i_d \leq \mathrm{d}\}$ de $\wedge^d \mathbb{R}^{2d}$, on aurait $\lim_{n\to+\infty} \frac{1}{n} \mathrm{Log}\ \|\wedge^d \Pi_n\, \vec{w}\|$ strictement inférieur à $\gamma_1 + \dots + \gamma_d$; cette somme ne pourrait donc pas être un exposant de $(\wedge^d \Pi_n)$).

3. Comportement asymptotique de l'erreur quadratique.

Dans cette section, nous montrons directement une variante affaiblie de la partie (i) du théorème 2 et nous interprétons $L(\overline{P}_n)$ comme le sous espace dilatant d'Osseledets. En fait nous établissons l'analogue de cet énoncé pour la distance sur $\mathcal{P}$ induite par la distance naturelle sur la grassmannienne des sous espaces de dimension d de $\mathbb{R}^{2d}$. Pour définir cette distance, munissons $W = \wedge^d\ \mathbb{R}^{2d}$ du produit scalaire pour lequel

$$< u_1 \wedge .. \wedge u_d, \mathrm{v}_1 \wedge .. \wedge \mathrm{v}_d > = \det\, \{< u_i, \mathrm{v}_j >\}_{i,j} \quad ,$$

puis $\wedge^2 W$ du produit scalaire défini de façon analogue. Alors si V_1 (resp. V_2) est le sous espace engendré par $u_1, \ldots, u_d$ (resp. $v_1, \ldots, v_d$), on pose

$$\delta_2(V_1,V_2) = \frac{<\vec{u}, \vec{v}>}{\|\vec{u}\| \, \|\vec{v}\|} = \frac{\|\vec{u} \wedge \vec{v}\|}{\|\vec{u}\| \, \|\vec{v}\|} \tag{7}$$

où $\vec{u} = u_1 \wedge .. \wedge u_d$ et $\vec{v} = v_1 \wedge .. \wedge v_d$. Si $P, Q \in \mathcal{P}$, on pose $\delta_2(P, Q) = \delta_2(L(P), L(Q))$.

On introduit les sous espaces dilatants d'Osseledets $\{L_p^u(\omega), p \in \mathbb{Z}, \omega \in \Omega\}$ définis, pour presque tout ω de Ω, par

$$L_p^u(\omega) = \{u \in \mathbb{R}^{2d} \, ; \lim_{n\to+\infty} \frac{1}{n} \text{ Log } \|M_{p-n}^{-1}(\omega) \, .. \, M_{p-1}^{-1}(\omega) \, M_p^{-1}(\omega) \, u\| < 0\} \, .$$

Sous la condition (C), γ_d est non nul (théorème 5), et ces sous espaces sont de dimension d.

Proposition 7 : *Sous la condition* (C), *il existe un processus stationnaire* $\{\bar{P}_n, n \in \mathbb{Z}\}$, *à valeurs dans* $\mathcal{P}_0$, *solution de* (2), *vérifiant* $L_n^u = \bar{P}_n$, *tel que*

$$\overline{\lim_{n\to+\infty}} \, \frac{1}{n} \text{ Log } \delta_2(P_n, \bar{P}_n) \le -2\gamma_d < 0, \qquad p.s.,$$

pour toute solution $\{P_n, n \in \mathbb{N}\}$ *de* (2) *pour laquelle* $P_0 \in \mathcal{P}$.

Démonstration: D'après le théorème d'Osseledets, L_0^u est complémentaire de l'espace contractant L^s. Ceci entraine (voir la fin de la preuve du lemme 6 par exemple) que, si $v_1, \ldots, v_d$ est une base de L_0^u alors

$$\lim_{n\to+\infty} \frac{1}{n} \text{ Log } \|\Pi_n v_1 \wedge .. \wedge \Pi_n v_d\| = \gamma_1 + \ldots + \gamma_d \, . \tag{8}$$

Considérons une solution $\{P_n, n \in \mathbb{N}\}$ de (2), avec $P_0 \in \mathcal{P}$. Soit $u_1, \ldots, u_d$ une base du sous espace lagrangien non-négatif $L(P_0)$. Par le lemme 6,

$$\lim_{n\to+\infty} \frac{1}{n} \text{ Log } \|\Pi_n u_1 \wedge .. \wedge \Pi_n u_d\| = \gamma_1 + \ldots + \gamma_d \, . \tag{9}$$

Considérons les éléments $\vec{u}$ et $\vec{v}$ de $\wedge^d \mathbb{R}^{2d}$ définis par $\vec{u} = u_1 \wedge .. \wedge u_d$ et $\vec{v} = v_1 \wedge .. \wedge v_d$. Les deux plus grands exposants de Lyapounov des produits $\wedge^d \Pi_n$ sont $\sum_{i=1}^{d-1} \gamma_i + \gamma_d$ et $\sum_{i=1}^{d-1} \gamma_i - \gamma_d$. On a donc

$$\lim_{n\to+\infty} \frac{1}{n} \text{ Log} \|(\wedge^d \Pi_n)(\vec{u}) \wedge (\wedge^d \Pi_n)(\vec{v})\| \le 2(\gamma_1 + \ldots + \gamma_{d-1}). \tag{10}$$

Comme

$$\delta_2(\Pi_n(L_0^u), \Pi_n(L(P_0))) = \frac{\|(\wedge^d \Pi_n)(\vec{u}) \wedge (\wedge^d \Pi_n)(\vec{v})\|}{\|(\wedge^d \Pi_n)\vec{u}\| \, \|(\wedge^d \Pi_n)\vec{v}\|} ,$$

il résulte de (8), (9) et (10) que

$$\overline{\lim_{n\to+\infty}} \frac{1}{n} \operatorname{Log} \delta_2(\Pi_n(L_0^u), \Pi_n(L(P_0))) \leq -2\,\gamma_d .$$

Or, d'une part

$$L_n^u = M_n(L_{n-1}^u) ,$$

d'où $\Pi_n(L_0^u) = L_n^u$ et d'autre part d'après (6), $\Pi_n(L(P_0)) = L(P_n)$. On obtient donc que

$$\overline{\lim_{n\to+\infty}} \frac{1}{n} \operatorname{Log} \delta_2(L_n^u, L(P_n)) \leq -2\gamma_d .$$

Il est clair que $\{L_n^u, n \in \mathbb{Z}\}$ est un processus stationnaire. Pour terminer la preuve, il suffit de montrer que ce processus est à valeurs dans l'ensemble des sous espaces lagrangiens positifs. La suite $\delta_2(L_n^u, L(P_n))$ tend vers 0. En utilisant la stationarité de la suite (L_n^u) et le fait que l'ensemble des lagrangiens non négatifs est compact, on en déduit que L_n^u est p.s. un lagrangien non-négatif. Par ailleurs, d'après le lemme 4, presque sûrement, pour k assez grand Π_k est dans $\mathcal{H}_0$. L'image d'un lagrangien non-négatif par un élément de $\mathcal{H}_0$ est un lagrangien positif. Comme $L_k^u = \Pi_k(L_0^u)$, on déduit de la stationarité que L_n^u est un sous espace lagrangien positif.

4. Stabilité exponentielle du filtre.

Le but de ce paragraphe est de montrer que l'équation (3) du filtre de Kalman est exponentiellement stable avec un taux controlé par le plus petit exposant de Lyapounov positif γ_d. Nous précisons ainsi le point (ii) du théorème 2. On peut penser que dans la plupart des cas la limite supérieure apparaissant dans cet énoncé est une limite et que l'inégalité est une égalité (c'est par exemple clair quand $d = 1$).

Théorème 8: *Supposons que la condition* (C) *est vérifiée. Alors, pour toute solution* $\{P_n, n \in \mathbb{N}\}$ *de* (2) *telle que* $P_0 \in \mathcal{P}$, *presque sûrement,*

$$\overline{\lim_{n\to+\infty}} \frac{1}{n} \operatorname{Log} \|(A_n - P_n R_n A_n) \dots (A_1 - P_1 R_1 A_1)\| \leq -\gamma_d < 0.$$

Pour la preuve, nous utiliserons les deux lemmes suivants. Le premier est montré dans [Bo].

Lemme 9: *Pour toute solution* $\{P_n, n \in \mathbb{N}\}$ *de* (2) *à valeurs dans* $\mathcal{P}$, *p.s.,*

$$\lim_{n\to+\infty} \frac{1}{n} \operatorname{Log}^+ \|P_n\| = 0.$$

Lemme 10: *Considérons une suite* $N_n = \begin{pmatrix} A_n & B_n \\ C_n & D_n \end{pmatrix}$, $n = 1, 2, \ldots,$ *formée de matrices symplectiques. Posons* $Z_n = N_n \ldots N_1$ *que l'on écrit* $Z_n = \begin{pmatrix} A^{(n)} & B^{(n)} \\ C^{(n)} & D^{(n)} \end{pmatrix}$. *Pour toute matrice symétrique* P *on a, si* $P_n = Z_n \cdot P$,

$$(A_n - P_n C_n)^* = (C^{(n-1)} P + D^{(n-1)} (C^{(n)} P + D^{(n)})^{-1}. \tag{11}$$

Preuve : Par définition, la relation $P_n = Z_n \cdot P$ s'écrit

$$P_n = (A^{(n)} P + B^{(n)}) (C^{(n)} P + D^{(n)})^{-1}.$$

Comme P_n est symétrique, il en résulte que

$$(A_n - P_n C_n)^* (C^{(n)} P + D^{(n)}) = A_n^* (C^{(n)} P + D^{(n)}) - C_n^* (A^{(n)} P + B^{(n)}).$$

En écrivant que $Z_n = N_n Z_{n-1}$ on obtient immédiatement que l'expression précédente est égale à

$$A_n^* (C_n A^{(n-1)} + D_n C^{(n-1)}) P + A_n^* (C_n B^{(n-1)} + D_n D^{(n-1)}) \\ - C_n^* (A_n A^{(n-1)} + B_n C^{(n-1)}) P - C_n^* (A_n B^{(n-1)} + B_n D^{(n-1)}).$$

En utilisant alors que $A_n^* C_n = C_n^* A_n$ on voit que ceci est égal à

$$(A_n^* D_n - C_n^* B_n) (C^{(n-1)} P + D^{(n-1)}).$$

La matrice N_n est symplectique, donc $A_n^* D_n - C_n^* B_n = I$. Finalement, on a prouvé que

$$(A_n - P_n C_n)^* (C^{(n)} P + D^{(n)}) = C^{(n-1)} P + D^{(n-1)}$$

et le lemme est établi.

Démonstration du Théorème 8. Soit P une matrice semidéfinie positive et $\{P_n, n \in \mathbb{N}\}$ la solution de (2) vérifiant $P_0 = P$. Il résulte immédiatement du lemme 10 et de la relation (6) que si l'on pose $\Pi_n ==$

$\begin{pmatrix} A^{(n)} & B^{(n)} \\ C^{(n)} & D^{(n)} \end{pmatrix}$, alors

$$(A_n - P_n R_n A_n) \ldots (A_1 - P_1 R_1 A_1) = (C^{(n)} P + D^{(n)})^{*-1}.$$

Il nous faut donc montrer que

$$\overline{\lim_{n\to+\infty}} \frac{1}{n} \operatorname{Log} \|(C^{(n)}P + D^{(n)})^{-1}\| \leq -\gamma_d . \tag{12}$$

Notons I la matrice identité d'ordre d et posons $Q = \begin{pmatrix} P & 0 \\ I & I \end{pmatrix}$. On remarque que chaque cofacteur de $C^{(n)} P + D^{(n)}$ est un coefficient de la matrice $\wedge^{d-1} (\Pi_n Q)$. Puisque

$$\lim_{n\to+\infty} \frac{1}{n} \operatorname{Log} \|\wedge^{d-1} \Pi_n Q\| = \lim_{n\to+\infty} \frac{1}{n} \operatorname{Log} \|\wedge^{d-1} \Pi_n\| = \gamma_1 + \gamma_2 + \ldots + \gamma_{d-1},$$

il suffit donc, pour établir (12), de montrer que, p.s.,

$$\lim_{n\to+\infty} \frac{1}{n} \operatorname{Log} |\det (C^{(n)} P + D^{(n)})| \geq \gamma_1 + \gamma_2 + \ldots + \gamma_d. \tag{13}$$

Posons, pour $n \geq 0$,

$$K_n = \begin{pmatrix} P_n & 0 \\ I & 0 \end{pmatrix} \quad \text{et} \quad H_n = \begin{pmatrix} C^{(n)}P + D^{(n)} & 0 \\ 0 & I \end{pmatrix}.$$

On vérifie immédiatement que $\Pi_n K_0 = K_n H_n$. Si $\{e_1, \ldots, e_{2d}\}$ est la base canonique de $\mathbb{R}^{2d}$, on a

$$\begin{aligned} \Pi_n K_0 e_1 \wedge \ldots \wedge \Pi_n K_0 e_d &= K_n H_n e_1 \wedge \ldots \wedge K_n H_n e_d \\ &= \{\det (C^{(n)} P + D^{(n)})\} (K_n e_1 \wedge \ldots \wedge K_n e_d) \end{aligned}$$

donc

$$|\det (C^{(n)} P + D^{(n)})| = \frac{\|\Pi_n K_0 e_1 \wedge \ldots \wedge \Pi_n K_0 e_d\|}{\|K_n e_1 \wedge \ldots \wedge K_n e_d\|}.$$

D'une part,

$$\lim_{n\to+\infty} \frac{1}{n} \operatorname{Log} \|\Pi_n K_0 e_1 \wedge \ldots \wedge \Pi_n K_0 e_d\| = \gamma_1 + \ldots + \gamma_d, \qquad \text{p.s.},$$

d'après le lemme 6, car $K_0 e_1, \ldots, K_0 e_d$ engendrent le sous espace lagrangien non négatif $L(P)$. D'autre part, il existe une constante C telle que

$$\text{Log}\, \|K_n\, e_1 \wedge \ldots \wedge K_n\, e_d\| \leq C\, \text{Log}\, \|K_n\| \leq C\, (1 + \text{Log}^+ \|P_n\|)\ ,$$

d'où, par le lemme 9,

$$\lim_{n \to +\infty} \frac{1}{n}\, \text{Log}\, \|K_n\, e_1 \wedge \ldots \wedge K_n\, e_d\| \leq 0.$$

Ces relations prouvent (14).

Références

[Ba]. Balakrishnan, A.V. (1984) : Kalman Filtering Theory. Optimization Software, New York.

[Bo]. Bougerol, P. (1990) : Kalman filtering with random coefficients and contractions (Preprint).

[Lo, St]. Loomis, L.H. et Sternberg, S. : Advanced Calculus. Addison Wesley, Reading, Ma.

[Os]. Osseledets, V.I. (1968) : The multiplicative ergodic theorem. The Lyapunov characteristic numbers of a dynamical system. Trans. Mosc. Math. Soc., 19, 197-231.

[Wh]. Whittle, P. (1982) : Optimization over time. Vol. 1. Wiley, Chichester, New-York.

[Wo1]. Wojtkowski, M. (1985) : Invariant families of cones and Lyapunov exponents. Ergod. Th. & Dynam. Sys., 5, 145-161.

[Wo2]. Wojtkowski, M. (1988) : Measure theoretic entropy of the system of hard spheres. Ergod. Th. & Dynam. Sys., 8, 133-153.

INVARIANT MEASURES FOR NONLINEAR STOCHASTIC DIFFERENTIAL EQUATIONS

Peter H. Baxendale
Department of Mathematics
University of Southern California
Los Angeles, CA 90089-1113, USA

1 Introduction

Consider the (Stratonovich) stochastic differential equation in $\mathbf{R}^d$

$$dx_t = V_0(x_t)dt + \sum_{\alpha=1}^{r} V_\alpha(x_t) \circ dW_t^\alpha \tag{1.1}$$

where $V_0, V_1, \ldots, V_r$ are smooth vector fields on $\mathbf{R}^d$ and $\{(W_t^1, \ldots, W_t^r) : t \geq 0\}$ is a standard $\mathbf{R}^r$-valued Brownian motion on some probability space $(\Omega, \mathcal{F}, \mathbf{P})$. The corresponding Itô stochastic differential equation is

$$dx_t = \overline{V}_0(x_t)dt + \sum_{\alpha=1}^{r} V_\alpha(x_t)dW_t^\alpha \tag{1.2}$$

where

$$\overline{V}_0 = V_0 + \frac{1}{2}\sum_{\alpha=1}^{r} D_{V_\alpha}V_\alpha.$$

The resulting (possibly explosive) diffusion process $\{x_t : t \geq 0\}$ in $\mathbf{R}^d$ has generator

$$L = \frac{1}{2}\sum_{i,j=1}^{d} a^{ij}(x)\frac{\partial^2}{\partial x_i \partial x_j} + \sum_{i=1}^{d} b^i \frac{\partial}{\partial x_i} \tag{1.3}$$

where

$$a^{ij}(x) = \sum_{\alpha=1}^{r} V_\alpha^i(x)V_\alpha^j(x) \tag{1.4}$$

and

$$b^i(x) = \overline{V}_0^i(x) = V_0^i(x) + \frac{1}{2}\sum_{\alpha=1}^{r}\sum_{j=1}^{d} V_\alpha^j(x)\frac{\partial V_\alpha^i}{\partial x_j}(x). \tag{1.5}$$

Questions concerning recurrence, transience, and the existence of invariant measures for $\{x_t : t \geq 0\}$ have been studied by many authors. Maruyama and Tanaka [14] and Khas'minskii [10] constructed invariant measures for non-degenerate recurrent processes. Criteria for recurrence for non-degenerate processes (in terms of the coefficients of L near infinity) have been given by Khas'minskii [10], Friedman [9], Azencott [3], and Bhattacharya [7]. For degenerate processes the support theorem of Stroock and Varadhan [18] provides a connection between recurrence and control theory. Results in this area have been obtained by Arnold and Kliemann [1] and Kliemann [13].

In this paper we shall assume that

$$V_0(0) = V_1(0) = \cdots = V_r(0) = 0 \tag{1.6}$$

so that 0 is a fixed point of the diffusion, and the non-trivial behavior of the diffusion takes place in $\mathbf{R}^d \setminus \{0\}$. The statement that $x_t \to 0$ as $t \to \infty$ can now be interpreted both as stability of the fixed point 0 and also as transience of the process on $\mathbf{R}^d \setminus \{0\}$. In fact transience now becomes a matter of whether $x_t \to 0$ as well as whether $\|x_t\| \to \infty$. In this paper we shall be concerned mostly with the behavior of $\{x_t : t \geq 0\}$ near 0, and we shall use the theory of Lyapunov exponents to investigate the relationship between stability or instability of the fixed point 0, transience or recurrence of the process on $\mathbf{R}^d \setminus \{0\}$, and existence of invariant measures on $\mathbf{R}^d \setminus \{0\}$. For these purposes we shall impose conditions on the diffusion process to ensure that it does not diverge to infinity (assumption (2.3)), and also that it is sufficiently non-degenerate that it has positive probability of hitting arbitrarily small neighborhoods of 0 (assumption (2.4)). These two conditions together will ensure that the behavior of $\{x_t : t \geq 0\}$ near 0 affects the overall behavior on $\mathbf{R}^d \setminus \{0\}$.

In order to analyze the behavior of $\{x_t : t \geq 0\}$ near 0, we will first linearize (1.1) at 0 to obtain the linear stochastic differential equation

$$dv_t = A_0 v_t dt + \sum_{\alpha=1}^{r} A_\alpha v_t \circ dW_t^\alpha \tag{1.7}$$

where $A_\alpha = DV_\alpha(0) \in L(\mathbf{R}^d)$ for $0 \leq \alpha \leq r$. The diffusion process $\{v_t : t \geq 0\}$ has generator TL, say, given by

$$TL = \frac{1}{2} \sum_{i,j=1}^{d} \left(\sum_{k,l=1}^{d} \frac{\partial^2 a^{ij}}{\partial x_k \partial x_l}(0) v_k v_l \right) \frac{\partial^2}{\partial v_i \partial v_j} + \sum_{i=1}^{d} \left(\sum_{k=1}^{d} \frac{\partial b^i}{\partial x_k}(0) v_k \right) \frac{\partial}{\partial v_i}. \tag{1.8}$$

Under appropriate non-degeneracy conditions (assumption (2.5)) the *Lyapunov exponent*

$$\lambda = \lim_{t \to \infty} \frac{1}{t} \log \|v_t\| \qquad \text{w.p.1} \tag{1.9}$$

and the *Lyapunov moment function*

$$\tilde{\Lambda}(p) = \lim_{t \to \infty} \frac{1}{t} \log \mathbf{E}\left(\|v_t\|^p\right) \qquad \text{for } p \in \mathbf{R} \tag{1.10}$$

are well-defined (i.e. the limits exist and do not depend on $v_0 \neq 0$). It is immediate from their definitions that the values of λ and $\tilde{\Lambda}(p)$ control the almost-sure stability and p^{th}-moment stability of the linearized process $\{v_t : t \geq 0\}$. We will give results in Sect. 5 showing how they control the behavior near 0 of the original non-linear process $\{x_t : t \geq 0\}$. For example, in the case $\lambda > 0$, the value of λ controls the expected exit time of $\{x_t : t \geq 0\}$ from small balls $B(0,r)$, and $\tilde{\Lambda}$ is used in the estimate of the expected time that $\{x_t : t \geq 0\}$ spends inside $B(0,\varepsilon)$ before exiting $B(0,r)$, for $0 < \varepsilon \leq \|x_0\| < r$. See Corollary 5.9 for details. The material in Sect. 5 is based closely on similar results in Baxendale and Stroock [6] and Baxendale [5].

The information on the behavior of $\{x_t : t \geq 0\}$ near 0 is then combined with the assumed behavior of $\{x_t : t \geq 0\}$ near infinity to yield Theorems 2.12, 2.13 and 2.14. The construction of invariant measures for the cases $\lambda > 0$ and $\lambda = 0$ is the one used by Maruyama and Tanaka [14] and Khas'minskii [10]. Our results say, roughly, that the process $\{x_t : t \geq 0\}$ on $\mathbf{R}^d \setminus \{0\}$ is transient, or null-recurrent, or recurrent according as $\lambda < 0$, or $\lambda = 0$, or $\lambda > 0$.

2 Statement of Results

Let $\{x_t : t \geq 0\}$ be the (possibly explosive) diffusion process on $\mathbf{R}^d \cup \{\infty\}$ given by (1.1) and (1.6), with generator L given by (1.3). Let $P(t,x,A) = \mathbf{P}\{x_t \in A \mid x_0 = x\}$ denote the corresponding transition probability. We write $\mathbf{P}^x$ to denote the law of $\{x_t : t \geq 0\}$ conditioned so that $x_0 = x$; thus $P(t,x,A) = \mathbf{P}^x\{x_t \in A\}$. We write $B(0,r) = \{x \in \mathbf{R}^d : \|x\| < r\}$ and $B'(0,r) = \{x \in \mathbf{R}^d : 0 < \|x\| < r\}$.

(2.1) Definition. For $0 < R < \infty$ let $\tau_R = \inf\{t \geq 0 : \|x_t\| = R\}$, with the usual convention that $\inf(\emptyset) = \infty$.

(2.2) Definition. The diffusion process $\{x_t : t \geq 0\}$ is said to be *regular* (or *non-explosive*, or *conservative*, or *complete*) if $\mathbf{P}^x\{\lim_{n\to\infty} \tau_n < \infty\} = 0$ for all $x \in \mathbf{R}^d$.

At various places in the paper we shall make one or more of the following assumptions.

(2.3) $\{x_t : t \geq 0\}$ is regular and there exist $f \in C^2(\mathbf{R}^d)$, $g \in C(\mathbf{R}^d)$ and $R < \infty$ such that $f \geq 0$, $g \geq 1$ and $Lf(x) \leq -g(x)$ whenever $\|x\| \geq R$.

(2.4) For all $r > 0$ and $x \neq 0$ there exists $T < \infty$ such that $P(T,x,B(0,r)) > 0$.

(2.5) $\mathrm{Lie}(A_1, A_2, \ldots, A_r)(v) = \mathbf{R}^d$ for all $v \neq 0$.

(2.6) Remark. The assumptions (2.3), (2.4), and (2.5) concern the behavior of $\{x_t : t \geq 0\}$ near infinity, between infinity and 0, and near 0 respectively. We study the diffusion in each of these ranges in the next three sections. An equivalent formulation of assumption (2.4) is given in Proposition 4.3.

(2.7) Remark. Although assumption (2.5) is easy to state and to check it is stronger than is necessary for our results. For information about a weaker replacement assumption see Remark 6.4.

Let $\{\Phi_t : t \geq 0\}$ in $L(\mathbf{R}^d)$ denote the fundamental matrix solution to the linear stochastic differential equation (1.7), so that $v_t = \Phi_t v_0$ for all $t \geq 0$ and $v_0 \in \mathbf{R}^d$. With no extra assumptions the multiplicative ergodic theorem (Oseledec [15]) applies to $\{\Phi_t : t \geq 0\}$ to give the following.

(2.8) Proposition. *There exist $\lambda_1 > \lambda_2 > \cdots > \lambda_N$ $(N \leq d)$ and $\Omega_0 \in \mathcal{F}$ with $\mathbf{P}(\Omega_0) = 1$ such that for each $\omega \in \Omega_0$ there exists a strictly decreasing family of subspaces*

$$\mathbf{R}^d = E_1(\omega) \supset E_2(\omega) \supset \cdots \supset E_N(\omega) \supset E_{N+1}(\omega) = \{0\}$$

with the property that

$$\lim_{t\to\infty} \frac{1}{t} \log \|v_t\| = \lambda_i$$

whenever $v_0 \in E_i(\omega) \setminus E_{i+1}(\omega)$. In particular if $v_0 \neq 0$ then

$$\mathbf{P}\{\lim_{t\to\infty} \frac{1}{t} \log \|v_t\| \leq \lambda_1\} = 1, \tag{2.9}$$

and for Lebesgue-almost all $v \in \mathbf{R}^d$

$$\mathbf{P}\{\lim_{t\to\infty} \frac{1}{t} \log \|v_t\| = \lambda_1\} = 1. \tag{2.10}$$

We write $\lambda_1 = \lambda$ and call λ the (top) Lyapunov exponent. Under the assumption (2.5) (or the weaker assumption (5.5), see later) it can be shown that (1.9) and (1.10) are good definitions and that $\tilde{\Lambda}$ is a convex analytic function with $\tilde{\Lambda}(0) = 0$ and $\tilde{\Lambda}'(0) = \lambda$ (see Arnold, Oeljeklaus and Pardoux [2]).

(2.11) Remark. λ and $\tilde{\Lambda}$ are determined by the law of $\{v_t : t \geq 0\}$, which is determined by TL and hence by L. However the lower Lyapunov exponents $\lambda_2, \ldots, \lambda_N$ are determined by $\{\Phi_t : t \geq 0\}$ given by (1.7), and this contains the extra information about the way in which the solutions for different initial positions and the same noise are correlated. The distinction between the law of $\{v_t : t \geq 0\}$ and the law of $\{\Phi_t : t \geq 0\}$ is essentially the same as the distinction between the law of $\{x_t : t \geq 0\}$, given by L, and the law of the stochastic flow of diffeomorphisms generated by (1.1). In this paper our results deal with the behavior of the 'one-point motions' $\{x_t : t \geq 0\}$ and $\{v_t : t \geq 0\}$.

(2.12) Theorem. *Assume* (2.3), (2.4) *and $\lambda < 0$. Then*

$$\mathbf{P}^x \left\{ \limsup_{t\to\infty} \frac{1}{t} \log \|x_t\| \leq \lambda \right\} = 1$$

for all $x \neq 0$.

(2.13) Theorem. *Assume* (2.3), (2.4), (2.5) *and $\lambda > 0$. Then there exists a unique probability measure μ on $\mathbf{R}^d \setminus \{0\}$ such that*

$$\mathbf{P}^x \left\{ \frac{1}{t} \int_0^t \phi(x_s)\, ds \to \int \phi\, d\mu \text{ as } t \to \infty \right\} = 1$$

for all bounded measurable $\phi : \mathbf{R}^d \setminus \{0\} \to \mathbf{R}$ *and all* $x \neq 0$. *In particular* μ *is the unique invariant measure for* $\{x_t : t \geq 0\}$ *on* $\mathbf{R}^d \setminus \{0\}$. *Moreover there exist* $\gamma > 0$, $\delta > 0$ *and* $K < \infty$ *such that* $\bar{\Lambda}(-\gamma) = 0$ *and*

$$\frac{1}{K} r^\gamma \leq \mu\{x \in \mathbf{R}^d \setminus \{0\} : \|x\| < r\} \leq K r^\gamma$$

for $0 < r < \delta$, *and*

$$\int g \, d\mu < \infty.$$

(2.14) Theorem. *Assume* (2.3), (2.4), (2.5) *and* $\lambda = 0$. *Then there exists a* σ*-finite measure* μ *on* $\mathbf{R}^d \setminus \{0\}$, *unique up to a multiplicative constant, such that*

$$\mathbf{P}^x \left\{ \frac{\int_0^t \phi(x_s)\, ds}{\int_0^t \psi(x_s)\, ds} \to \frac{\int \phi \, d\mu}{\int \psi \, d\mu} \text{ as } t \to \infty \right\} = 1$$

for all bounded measurable μ*-integrable* $\phi, \psi : \mathbf{R}^d \setminus \{0\} \to \mathbf{R}$ *with* $\int \psi \, d\mu \neq 0$ *and all* $x \neq 0$. *In particular* μ *is the unique, up to multiplicative constant, invariant measure for* $\{x_t : t \geq 0\}$ *on* $\mathbf{R}^d \setminus \{0\}$. *Moreover there exists* $a \in (0, \infty)$ *such that*

$$\mu\{x \in \mathbf{R}^d \setminus \{0\} : \|x\| > \varepsilon\} / |\log \varepsilon| \to a \text{ as } \varepsilon \to 0,$$

and

$$\int_{\|x\| > \varepsilon} g \, d\mu < \infty \text{ for all } \varepsilon > 0.$$

(2.15) Remark. Under assumption (2.5) the linearized process $\{v_t : t \geq 0\}$ satisfies $\mathbf{P}\{\|v_t\| \to \infty \text{ as } t \to \infty\} = 1$ if $\lambda > 0$ and $\mathbf{E}(\inf\{t \geq 0 : \|v_t\| = r\}) = \infty$ for sufficiently large $\|v_t\|$ if $\lambda = 0$. Thus the results of Theorems 2.13 and 2.14 depend strongly on the non-linearity of equation (1.1).

(2.16) Remark. Suppose in addition to the assumptions (2.3), (2.4) and (2.5) that $\{x_t : t \geq 0\}$ is non-degenerate on $\mathbf{R}^d \setminus \{0\}$. Then, using recurrence in the sense of hitting open sets from arbitrary starting points (see e.g. Khas'minskii [12] or Bhattacharya [7]), the theorems above give a complete classification of $\{x_t : t \geq 0\}$ on $\mathbf{R}^d \setminus \{0\}$ as positive recurrent, null recurrent, or transient according as λ is positive, zero, or negative. With just the assumptions (2.3), (2.4) and (2.5) a similar remark can be made in terms of hitting probabilities and expected hitting times of annuli $\{x \in \mathbf{R}^d : r_1 < \|x\| < r_2\}$ for $0 < r_1 < r_2$ and sufficiently small r_2.

Conditions (2.3) and (2.4) are used together to show that $\{x_t : t \geq 0\}$ visits arbitrarily small neighborhoods of 0 almost surely. This is carried out in Sections 3 and 4. Theorem 2.12 ($\lambda < 0$) will then follow easily from the local stable manifold theorem (see Theorem 5.1). The cases $\lambda > 0$ and $\lambda = 0$ are more complicated and use essentially the same techniques as were used in Baxendale and Stroock [6] and Baxendale [5] to analyze the two point motion of a stochastic flow of diffeomorphisms on a compact manifold. In particular Theorems 2.13 and 2.14 are similar to [6, Theorem 4.6] and [5, Theorem 5.8]. The similarity arises because studying the behavior of $\{x_t : t \geq 0\}$ relative to the fixed point 0 is a particular case of studying the motion of a pair of points under the flow generated by a stochastic differential equation. The construction of invariant measures in [6] and [5] was in turn based on the method used for recurrent diffusion processes by Maruyama and Tanaka [14] and Khas'minskii [10]. The construction is based on expected occupation times of sets; as a result we will be much concerned with estimates of expected exit times from open balls and complements of open balls. Results concerning the behavior of the linearized process $\{v_t : t \geq 0\}$ and the original process $\{x_t : t \geq 0\}$ near 0 are given in Section 5, and the proofs of Theorems 2.12 to 2.14 are given in Section 6.

3 Behavior near infinity

The results in this section do not use (1.6). We state some known results concerning regularity and recurrence for the diffusion process $\{x_t : t \geq 0\}$ with generator L as given by (1.1)–(1.5).

(3.1) Proposition. *Assume there exists $f \in C^2(\mathbf{R}^d)$ and $c < \infty$ such that $f(x) \geq 0$ for all $x \in \mathbf{R}^d$, $Lf(x) \leq cf(x)$ for all $x \in \mathbf{R}^d$, and $f(x) \to \infty$ as $x \to \infty$. Then $\{x_t : t \geq 0\}$ is regular. Moreover for each $t > 0$, $\mathbf{P}^x\{\tau_n \leq t\} \to 0$ as $n \to \infty$ uniformly for x in compact subsets of $\mathbf{R}^d$.*

Proof. See Khas'minskii [12, Theorem III.4.1]. □

(3.2) Remark. The well-known result that a linear growth condition on $\overline{V}_0, V_1, \ldots, V_r$ implies regularity can be verified by taking $f(x) = \|x\|^2$.

(3.3) Proposition. *Assume $\{x_t : t \geq 0\}$ is regular and there exists $f \in C^2(\mathbf{R}^d)$ and $R < \infty$ such that $f(x) \geq 0$ whenever $\|x\| \geq R$, and $Lf(x) \leq -1$ whenever $\|x\| \geq R$. Then $\mathbf{E}^x(\tau_R) \leq f(x)$ whenever $\|x\| \geq R$.*

Proof. See Khas'minskii [12, Theorem III.7.1]. □

(3.4) Remark. Proposition 3.3 applies whenever assumption (2.3) is satisfied. In fact (2.3) implies the stronger result:

$$\mathbf{E}^x\left(\int_0^{\tau_R} g(x_s)\, ds\right) \leq f(x)$$

whenever $\|x\| \geq R$. The proof of this is contained within the proof of Corollary 4.8. Notice also that (2.3) is close to being a necessary condition for $\mathbf{E}^x(\tau_R) < \infty$ whenever $\|x\| \geq R$, in the sense that $h(x) \equiv \mathbf{E}^x(\tau_R)$ satisfies $Lh(x) = -1$ for $\|x\| \geq R$ (see Khas'minskii [12, Theorem III.7.3]).

(3.5) Remark. Khas'minskii [10], Friedman [9], Azencott [3] and Bhattacharya [7] all give similar collections of sufficient conditions, in terms of the coefficients of L, for the existence of functions f satisfying the requirements for Propositions 3.1 and 3.3. (Khas'minskii's results were stated without proof.) In [9], [3], and [7] the stated conditions were used to construct functions of the form $f(x) = F(\|x-z\|)$ for some fixed $z \in \mathbf{R}^d$, satisfying the requirements of the corresponding proposition; the results of Propositions 3.1 and 3.3 were essentially due to these earlier authors.

(3.6) Remark. Propositions (3.1) and (3.3) often apply when the (Itô) drift $\overline{V}_0$ has a sufficiently large inward pointing component relative to the size of the noise vector fields $V_1, \ldots, V_r$. More precisely suppose there exist $\beta > \alpha$, $K < \infty$, $\varepsilon > 0$ and $R_0 < \infty$ such that

$$\sum_{i=1}^{d} b^i(x)x_i \leq -\varepsilon\|x\|^\beta \quad \text{and} \quad |a^{ij}(x)| \leq K\|x\|^\alpha,\ 1 \leq i,j \leq d,$$

whenever $\|x\| \geq R_0$. Then a direct calculation shows the existence of $c > 0$, $\gamma > 0$, and $R \geq R_0$ such that

$$L\left(\exp(\gamma\|\cdot\|^{\beta-\alpha})\right)(x) \leq -c\|x\|^{2\beta-\alpha-2}\exp(\gamma\|x\|^{\beta-\alpha})$$

whenever $\|x\| \geq R$. It follows (by Proposition 3.1) that $\{x_t : t \geq 0\}$ is regular and that (2.3) holds with both f and g of the form $\exp(\gamma\|x\|^{\beta-\alpha})$ (with possibly different γ for f and g).

4 Hitting times of small neighborhoods of 0

The results in this section do not use (1.6). We first give an equivalent condition to (2.4) in terms of the associated control problem. (Connections between control theory and recurrence are examined in detail in Arnold and Kliemann [1] and Kliemann [13].) We then obtain estimates on $\mathbf{E}^x(\tau_r)$ for arbitrary $r > 0$ and $\|x\| > r$.

(4.1) Definition. For $T > 0$ let $\mathcal{U}_T = C([0,T]; \mathbf{R}^r)$. For $u \in \mathcal{U}_T$ let $\{\xi(t,x;u) : 0 \leq t \leq T\}$ denote the solution of the associated control problem

$$\frac{\partial \xi}{\partial t}(t,x;u) = V_0(\xi(t,x;u)) + \sum_{\alpha=1}^{r} V_\alpha(\xi(t,x;u))u_\alpha(t) \tag{4.2}$$

with $\xi(0,x;u) = x$, if it exists.

(4.3) Proposition. *Condition* (2.4) *is equivalent to the following condition:*

(4.4) *For all $r > 0$ and $x \neq 0$ there exists $T > 0$ and $u \in \mathcal{U}_T$ such that $\{\xi(t,x;u) : 0 \leq t \leq T\}$ exists and $\xi(T,x;u) \in B(0,r)$.*

Proof. If $V_0, V_1, \ldots, V_r$ are all bounded with bounded derivatives, the result follows immediately from the support theorem of Stroock and Varadhan [18] (which describes the topological support of the distribution of the diffusion $\{x_t : 0 \leq t \leq T\}$ given by (1.1) in terms of the control paths $\{\xi(t,x;u) : 0 \leq t \leq T\}$). In the general case we use a localization argument as follows. Let $r > 0$ and $x \neq 0$ be given and assume (2.4), so that $\mathbf{P}^x\{\|x_T\| < r\} > 0$ for some $T > 0$. Then there exists $n > 0$ such that

$$\mathbf{P}^x\{\|x_T\| < r \ \text{and}\ \|x_t\| < n \text{ for all } t \in [0,T]\} > 0.$$

Choose compactly supported vector fields $V_0^{(n)}, V_1^{(n)}, \ldots, V_r^{(n)}$ agreeing with $V_0, V_1, \ldots, V_r$ on $B(0,n)$ and let $\{x_t^{(n)} : t \geq 0\}$ and $\{\xi^{(n)}(t,x;u) : 0 \leq t \leq T\}$ denote the diffusion and control processes obtained by replacing V_α with $V_\alpha^{(n)}$, $0 \leq \alpha \leq r$, in (1.1) and (4.2). Let $\mathbf{P}^{x,(n)}$ denote the law of $\{x_t^{(n)} : t \geq 0\}$ given $x_0^{(n)} = x$. Since $\{x_t : t \geq 0\}$ and $\{x_t^{(n)} : t \geq 0\}$ agree up to the time of first exit from $B(0,n)$ then

$$\begin{aligned}\mathbf{P}^{x,(n)}&\{\|x_T^{(n)}\| < r \text{ and } \|x_t^{(n)}\| < n \text{ for all } t \in [0,T]\}\\ &= \mathbf{P}^x\{\|x_T\| < r \text{ and } \|x_t\| < n \text{ for all } t \in [0,T]\}\\ &> 0.\end{aligned}$$

The support theorem applies to $\{x_t^{(n)} : t \geq 0\}$, and so there exists $u \in \mathcal{U}_T$ such that $\{\xi^{(n)}(t,x;u) : 0 \leq t \leq T\}$ exists with $\|\xi^{(n)}(T,x;u)\| < r$ and $\|\xi^{(n)}(t,x;u)\| < n$ for all $t \in [0,T]$. The last assertion implies that $\xi^{(n)}(t,x;u) = \xi(t,x;u)$ for all $t \in [0,T]$, and so (4.4) is valid. The converse argument is similar and we omit the details. □

(4.5) Lemma. *Assume* (2.4). *For all* $r > 0$ *and* $x \neq 0$ *there exist* $T > 0$, $\varepsilon > 0$, *and a neighborhood* U *of* x *such that* $P(T,y,B(0,r)) > 0$ *whenever* $y \in U$.

Proof. If the vector fields $V_0, V_1, \ldots, V_r$ are such that, for every $T > 0$, $\mathbf{P}^x\{\tau_n \leq T\} \to 0$ as $n \to \infty$ uniformly for x in compact subsets of $\mathbf{R}^d$, then $\{x_t : t \geq 0\}$ is a Feller process and the result follows from the lower-semi-continuity of $y \mapsto P(T,y,B(0,r))$. More generally for $r > 0$ and $x \neq 0$ let T, n, and $\{x_t^{(n)} : t \geq 0\}$ be as in the proof of Proposition 4.3; thus

$$\mathbf{P}^{x,(n)}\{\|x_T^{(n)}\| < r \text{ and } \|x_t^{(n)}\| < n \text{ for all } t \in [0,T]\} > 0.$$

Since the vector fields $V_0^{(n)}, V_1^{(n)}, \ldots, V_r^{(n)}$ are all compactly supported then the mapping $x \mapsto \mathbf{P}^{x,(n)}$ is weakly continuous (see Stroock and Varadhan [19, Cor 6.3.3]). In particular

$$\begin{aligned}\liminf_{y\to x} &\mathbf{P}^y\{\|x_T\| < r\}\\ &\geq \liminf_{y\to x} \mathbf{P}^y\{\|x_T\| < r \text{ and } \|x_t\| < n \text{ for all } t \in [0,T]\}\\ &= \liminf_{y\to x} \mathbf{P}^{y,(n)}\{\|x_T^{(n)}\| < r \text{ and } \|x_t^{(n)}\| < n \text{ for all } t \in [0,T]\}\\ &\geq \mathbf{P}^{x,(n)}\{\|x_T^{(n)}\| < r \text{ and } \|x_t^{(n)}\| < n \text{ for all } t \in [0,T]\}\\ &> 0\end{aligned}$$

as required. □

(4.6) Proposition. *Assume* (2.4). *For all* $0 < r < R < \infty$ *there exist* $T > 0$ *and* $\varepsilon > 0$ *such that*

$$\mathbf{P}^x\{\tau_r < T\} \geq \varepsilon$$

whenever $r < \|x\| < R$.

Proof. The lemma above implies that whenever $r < \|x\| < R$ there exist $T_x > 0$, $\varepsilon_x > 0$ and a neighborhood U_x of x such that $P(T_x,y,B(0,r)) \geq \varepsilon_x$ for $y \in U_x$. By compactness there exist $N \geq 1$ and $x_1, \ldots, x_N$ such that $\{x \in \mathbf{R}^d : r \leq \|x\| \leq R\} \subset \bigcup_{i=1}^N U_{x_i}$. Take $T = \max\{T_{x_i} : 1 \leq i \leq N\}$ and $\varepsilon = \min\{\varepsilon_{x_i} : 1 \leq i \leq N\}$. Then, for $r < \|x\| < R$, if $x \in U_{x_i}$ we have

$$\begin{aligned} \mathbf{P}^x\{\tau_r < T\} &\geq \mathbf{P}^x\{\tau_r < T_{x_i}\} \\ &\geq P(T_{x_i}, x, B(0,r)) \\ &\geq \varepsilon \end{aligned}$$

as required. □

(4.7) Theorem. *Assume* (2.4). *For all* $0 < r < R < \infty$ *there exists* $K < \infty$ *such that*

$$\mathbf{E}^x\left(\int_0^{\tau_r} \chi_{(0,R]}(\|x_s\|)\, ds\right) \leq K$$

whenever $\|x\| \geq r$.

Proof. Let T and ε be as in Proposition 4.6, and define

$$f_n(x) = \mathbf{P}^x\left\{\int_0^{\tau_r} \chi_{(0,R]}(\|x_s\|)\, ds \geq nT\right\}$$

for $n \geq 1$. We will prove by induction that $f_n(x) < (1-\varepsilon)^n$ for all $\|x\| \geq r$ and all $n \geq 1$. For $n = 1$ and $r \leq \|x\| \leq R$ the result is immediate from Proposition 4.6; and if $\|x\| > R$ then $\tau_R < \tau_r$ and so

$$f_1(x) = \mathbf{E}^x\left(f_1(x_{\tau_R})\chi_{\tau_R<\infty}\right) < 1 - \varepsilon.$$

The inductive step goes as follows. Assume $f_n(x) < (1-\varepsilon)^n$ for all $\|x\| \geq r$. By conditioning on $\mathcal{F}_T \equiv \sigma\{x_t : 0 \leq t \leq T\}$ we obtain

$$\begin{aligned} f_{n+1}(x) &\leq \mathbf{P}^x\left\{\tau_r \geq T \text{ and } \int_T^{\tau_r} \chi_{(0,R]}(\|x_s\|)\, ds \geq nT\right\} \\ &= \mathbf{E}^x\left(f_n(x_T)\chi_{\tau_r \geq T}\right) \\ &< (1-\varepsilon)^n \mathbf{P}^x\{\tau_r \geq T\}. \end{aligned}$$

So, by Proposition 4.6, $f_{n+1}(x) < (1-\varepsilon)^{n+1}$ for $r \leq \|x\| \leq R$. For $\|x\| > R$ then, arguing as for f_1 above,

$$f_{n+1}(x) = \mathbf{E}^x\left(f_{n+1}(x_{\tau_R})\chi_{\tau_R<\infty}\right) < (1-\varepsilon)^{n+1}$$

and the inductive step is complete. Finally

$$\mathbf{E}^x\left(\int_0^{\tau_r} \chi_{(0,R]}(\|x_s\|)\, ds\right) \leq T + T\sum_{n=1}^{\infty} f_n(x) < \frac{T}{\varepsilon}$$

and we are done. □

(4.8) Corollary. *Assume (2.3) and (2.4). For all $r > 0$ there exists $K < \infty$ such that*

$$\mathbf{E}^x\left(\int_0^{\tau_r} g(x_s)\,ds\right) \le K + f(x)$$

whenever $\|x\| \ge r$. In particular $\mathbf{E}^x(\tau_r) < \infty$ whenever $\|x\| \ge r$.

Proof. Let f, g and R be as in (2.3), and then let K be as in Theorem 4.7. Define $c \equiv \sup\{Lf(x) + g(x) : \|x\| \le R\} < \infty$ and notice that $Lf(x) + g(x) \le 0$ when $\|x\| \ge R$. For $\|x\| \ge r$ and $n \ge \|x\|$ we have, writing $\sigma = t \wedge \tau_r \wedge \tau_n$,

$$\begin{aligned}\mathbf{E}^x(f(x_\sigma)) - f(x) &= \mathbf{E}^x\left(\int_0^\sigma Lf(x_s)\,ds\right)\\ &= -\mathbf{E}^x\left(\int_0^\sigma g(x_s)\,ds\right) + \mathbf{E}^x\left(\int_0^\sigma (Lf+g)(x_s)\,ds\right)\\ &\le -\mathbf{E}^x\left(\int_0^\sigma g(x_s)\,ds\right) + c\mathbf{E}^x\left(\int_0^\sigma \chi_{(0,R]}(\|x_s\|)\,ds\right).\end{aligned}$$

Since $f \ge 0$,

$$\begin{aligned}\mathbf{E}^x\left(\int_0^\sigma g(x_s)\,ds\right) &\le f(x) + c\mathbf{E}^x\left(\int_0^\sigma \chi_{(0,R]}(\|x_s\|)\,ds\right)\\ &\le f(x) + cK.\end{aligned}$$

Now (2.3) implies that $\mathbf{P}^x\{\lim_{n\to\infty} \tau_n = \infty\} = 1$, so that $t \wedge \tau_r \wedge \tau_n \to \tau_r$ as $t \to \infty$ and $n \to \infty$, and the result follows from the monotone convergence theorem. □

5 Behavior near 0

This section will use only assumption (2.5) which is an assumption about the linearized process $\{v_t : t \ge 0\}$. The first result, for the case $\lambda < 0$, does not even require (2.5). We shall write $\{x_t(x,\omega) : t \ge 0\}$ to denote the dependence of the strong solution of (1.1) on its initial position x and on the noise ω.

(5.1) Theorem. *Assume $\lambda < 0$. For each $\mu \in (\lambda, 0)$ there exist random variables $r(\omega)$ and $C(\omega)$, positive and finite with probability 1, such that*

$$\{x \in \mathbf{R}^d : \|x_t(x,\omega)\| \le C(\omega)\|x\|e^{\mu t} \text{ for all } t \ge 0\} \subset B(0, r(\omega)).$$

In particular, for all $\varepsilon > 0$ there exist constants $r > 0$ and $C < \infty$ such that

$$\mathbf{P}^x\{\|x_t\| \le C\|x\|e^{\mu t} \text{ for all } t \ge 0\} \ge 1 - \varepsilon \tag{5.2}$$

whenever $\|x\| \le r$.

Proof. This a partial statement of the local stable manifold theorem for stochastic flows (see Carverhill [8] or Ruelle [16]), using if necessary a localization argument similar to the one in Proposition 4.3 in order to ensure that the stochastic flow exists. □

In order to deal with the cases $\lambda > 0$ and $\lambda = 0$ we need to make more assumptions than were needed for Theorem 5.1 above, although (2.5) turns out to be much stronger than necessary. Recall the linearized process $\{v_t : t \geq 0\}$ given by (1.7). As soon as $\lambda_1 \geq 0$ the distinction between the statements (2.9) and (2.10) becomes more important, and we need an assumption which will ensure that the initial direction of v_0 does not influence too greatly the large time behavior of $\{\|v_t\| : t \geq 0\}$. Writing $\theta_t = v_t/\|v_t\|$ we obtain a diffusion process $\{\theta_t : t \geq 0\}$ on the unit sphere S^{d-1} given by the (Stratonovich) stochastic differential equation

$$d\theta_t = \tilde{A}_0(\theta_t)dt + \sum_{\alpha=1}^{r} \tilde{A}_\alpha(\theta_t) \circ dW_t^\alpha \tag{5.3}$$

where the vector fields $\tilde{A}_\alpha$ on S^{d-1} are given by

$$\tilde{A}_\alpha(\theta) = A_\alpha\theta - \langle A_\alpha\theta, \theta\rangle\theta.$$

Equation (5.3) describes the angular part of $\{v_t : t \geq 0\}$; its norm is given by

$$d(\log\|v_t\|) = \langle A_0\theta_t, \theta_t\rangle dt + \sum_{\alpha=1}^{r} \langle A_\alpha\theta_t, \theta_t\rangle \circ dW_t^\alpha \tag{5.4}$$

This skew product decomposition, due to Khas'minskii [11], is of major importance in the study of linear stochastic differential equations. In order to ensure that the diffusion $\{\theta_t : t \geq 0\}$ is sufficiently non-degenerate, we make the assumption

(5.5)

(i) $\mathrm{Lie}(\tilde{A}_0, \tilde{A}_1, \ldots, \tilde{A}_r)(\theta) = T_\theta S^{d-1}$ for all $\theta \in S^{d-1}$.

(ii) For $u \in \mathcal{U}_T$ let $\{\eta(t,\theta;u) : 0 \leq t \leq T\}$ be the corresponding control path, given by $\eta(0,\theta;u) = \theta$ and $\frac{\partial}{\partial t}\eta(t,\theta;u) = \tilde{A}_0(\eta(t,\theta;u)) + \sum_{\alpha=1}^{r} \tilde{A}_\alpha(\eta(t,\theta;u))u_\alpha(t)$; then $\{\eta(T,\theta;u) : T > 0, u \in \mathcal{U}_T\}$ is dense in S^{d-1} for all $\theta \in S^{d-1}$.

Notice that (2.5) implies $\mathrm{Lie}(\tilde{A}_1, \ldots, \tilde{A}_r)(\theta) = T_\theta S^{d-1}$ for all $\theta \in S^{d-1}$. This in turn implies both (5.5)(i) and also that the set of control paths $\{\eta(t,\theta;u) : 0 \leq t \leq T\}$ for all $u \in \mathcal{U}_T$ is dense in the set of all continuous paths from $[0,T]$ to S^{d-1} starting at θ. Thus (2.5) implies (5.5). Notice also that (5.5)(i) implies that the infinitesimal generator for $\{\theta_t : t \geq 0\}$ is hypoelliptic and so, for example, any invariant probability measure on S^{d-1} has a smooth density; the extra strength of (5.5)(ii) ensures that the density is strictly positive. In the language of geometric control theory, given (5.5)(i) then (5.5)(ii) implies that the unique invariant control set on S^{d-1} is S^{d-1} itself.

Under assumption (5.5) the definitions (1.9) and (1.10) for the Lyapunov exponent λ and the Lyapunov moment function $\tilde{\Lambda}$ are well-defined. The interpretation of $\tilde{\Lambda}(p)$ in terms of pth-moment stability is obvious. In addition $\tilde{\Lambda}$ is the Legendre transform of the rate function for large deviations of $(1/t)\log\|v_t\|$ away from λ; and it is the largest eigenvalue for an associated partial differential equation on S^{d-1}. For details of these and other properties see Arnold, Oeljeklaus and Pardoux [2], Stroock [17], Baxendale [4] and Baxendale and Stroock [6]. The eigenfunctions corresponding to $\tilde{\Lambda}(p)$ give rise to the following collection of Lyapunov type functions for the linearized process $\{v_t : t \geq 0\}$.

(5.6) Theorem. *Assume* (5.5).

(i) *For each* $p \in \mathbf{R}$ *there exists a smooth* $\phi_p : S^{d-1} \to (0,\infty)$ *such that*

$$\left(TL + \frac{\partial}{\partial t}\right)\left(\phi_p(\frac{v}{\|v\|})\|v\|^p e^{-\tilde{\Lambda}(p)t}\right) = 0.$$

(ii) *There exists a smooth* $\psi : S^{d-1} \to \mathbf{R}$ *such that*

$$\left(TL + \frac{\partial}{\partial t}\right)\left(\psi(\frac{v}{\|v\|}) + \log\|v\| - \lambda t\right) = 0.$$

(iii) *If* $\lambda = 0$ *there exists a smooth* $\eta : S^{d-1} \to \mathbf{R}$ *such that*

$$\left(TL + \frac{\partial}{\partial t}\right)\left((\log\|v\|)^2 + 2\psi(\frac{v}{\|v\|})\log\|v\| + \eta(\frac{v}{\|v\|}) - Vt\right) = 0$$

where $V = \frac{d^2\tilde{\Lambda}}{dp^2}(0) \geq 0$.

Proof. For (i) and (ii) see the preliminaries to Theorem 3.18 of Baxendale and Stroock [6]. For (iii) see the proof of Prop. 5.2 of Baxendale [5]. □

The result above refers to TL, the generator of the linearized process $\{v_t : t \geq 0\}$, and it allows many estimates on the behavior of $\{\|v_t\| : t \geq 0\}$. See for example Theorems 4.1, 4.2, and 4.5 of Baxendale [4]. The following result will allow similar estimates for the original process $\{x_t : t \geq 0\}$ while it is close to 0.

(5.7) Theorem. *Assume* (5.5). *In case* $\lambda = 0$ *assume also* $V > 0$. *For each choice of* $-\infty < a < b < \infty$ *there exist* $\delta > 0$ *and* $K < \infty$ *for which the following assertions are valid.*

(i) *For each* $p \in [a,b]$ *there exist smooth functions* $\phi_p^{\pm} : B'(0,\delta) \to (0,\infty)$ *such that*

$$\left(L + \frac{\partial}{\partial t}\right)\left(\phi_p^+(x)e^{-\tilde{\Lambda}(p)t}\right) \geq 0 \geq \left(L + \frac{\partial}{\partial t}\right)\left(\phi_p^-(x)e^{-\tilde{\Lambda}(p)t}\right)$$

and

$$\frac{1}{K}\|x\|^p \leq \phi_p^{\pm}(x) \leq K\|x\|^p$$

whenever $0 < \|x\| < \delta$.

(ii) *There exist smooth functions* $\psi^{\pm} : B'(0,\delta) \to \mathbf{R}$ *such that*

$$\left(L + \frac{\partial}{\partial t}\right)\left(\psi^+(x) - \lambda t\right) \geq 0 \geq \left(L + \frac{\partial}{\partial t}\right)\left(\psi^-(x) - \lambda t\right)$$

and

$$\left|\psi^{\pm}(x) - \log\|x\|\right| \leq K$$

whenever $0 < \|x\| < \delta$.

(iii) *In the case $\lambda = 0$ and $V > 0$, there exist smooth functions $\eta^\pm : B'(0,\delta) \to \mathbf{R}$ such that*

$$\left(L + \frac{\partial}{\partial t}\right)\left(\eta^+(x) - Vt\right) \geq 0 \geq \left(L + \frac{\partial}{\partial t}\right)\left(\eta^-(x) - Vt\right)$$

and

$$\left|\eta^\pm(x) - (\log\|x\|)^2\right| \leq K\left|\log\|x\|\right|$$

whenever $0 < \|x\| < \delta$.

Proof. The proofs of Theorem 3.18 of [6] and Proposition 5.2 of [5] remain valid in this much simpler situation where the unit sphere bundle SM is replaced by a single unit sphere S^{d-1}. □

(5.8) Remark. Under (5.5), if $\lambda = 0$ and $V = 0$ then $\tilde{\Lambda}(p) \equiv 0$ (see Proposition 5.1 of [5]). This implies there exists an invertible $Q \in L(\mathbf{R}^d)$ such that $QA_0Q^{-1}, QA_1Q^{-1}, \ldots, QA_rQ^{-1}$ are all skew-symmetric (see Arnold, Oeljeklaus and Pardoux [2, Theorem 3.2]), and so $\|Q^{-1}v_t\| \equiv \|Q^{-1}v_0\|$ almost surely. This cannot happen when (2.5) is satisfied.

Theorems 5.6(i–ii) and 5.7(i–ii) are valid for all $\lambda \in \mathbf{R}$. We now specialize and study the implications of Theorem 5.7 for the two cases $\lambda > 0$ and $\lambda = 0$. Of course Theorem 5.7 also has information to give about the case $\lambda < 0$, but we have no need for the extra information here.

(5.9) Corollary. *Assume (5.5) and $\lambda > 0$.*

(i) *There exist $\delta > 0$ and $K < \infty$ such that*

$$\frac{1}{\lambda}\left(\log(\frac{R}{\|x\|}) - K\right) \leq \mathbf{E}^x(\tau_R) \leq \frac{1}{\lambda}\left(\log(\frac{R}{\|x\|}) + K\right)$$

whenever $0 < \|x\| < R < \delta$.

(ii) *Suppose $\gamma > 0$ satisfies $\tilde{\Lambda}(-\gamma) = 0$. Then there exist $\delta > 0$, $K < \infty$, and $k \in (0,1)$ such that*

$$\frac{1}{K}\left(\frac{\varepsilon}{\|x\|}\right)^\gamma \leq \mathbf{E}^x\left(\int_0^{\tau_R} \chi_{(0,\varepsilon)}(\|x\|)\,ds\right) \leq K\left(\frac{\varepsilon}{\|x\|}\right)^\gamma$$

whenever $0 < \varepsilon < \|x\| < kR < k\delta$.

(iii) *Suppose $\tilde{\Lambda}(-p) < 0$ for all $p > 0$. Then for all $p > 0$ there exist $\delta > 0$, $K < \infty$, and $k \in (0,1)$ such that*

$$\mathbf{E}^x\left(\int_0^{\tau_R} \chi_{(0,\varepsilon)}(\|x\|)\,ds\right) \leq K\left(\frac{\varepsilon}{\|x\|}\right)^p$$

whenever $0 < \varepsilon < \|x\| < kR < k\delta$.

Proof. See Theorem 3.19 and Corollary 3.24 of [6] for (i) and (ii). The proof of (iii) is the same as the proof of the upper bound in (ii). □

(5.10) Remark. Recall $\tilde{\Lambda}(0) = 0$ and $\tilde{\Lambda}'(0) = \lambda > 0$. The convexity of $\tilde{\Lambda}$ implies that if $\tilde{\Lambda}$ has a second zero then the second zero is unique and of the form $-\gamma$ for some $\gamma > 0$. If γ does not exist then $\tilde{\Lambda}(p) < 0$ for all $p < 0$, so that $\tilde{\Lambda}(p)/p$ has a finite non-negative limit as $p \to \infty$. This can happen only if there exists an invertible $Q \in L(\mathbf{R}^d)$ such that $QA_1Q^{-1}, \ldots, QA_rQ^{-1}$ are all skew-symmetric and $\langle QA_0Q^{-1}\theta, \theta\rangle \geq 0$ for all $\theta \in S^{d-1}$ ([2, Theorem 3.2 and Remark 3.2]). In particular it cannot happen under the stronger assumption (2.5).

(5.11) Corollary. *Assume* (5.5) *and* $\lambda = 0$ *and* $V > 0$. *There exist* $\delta > 0$, $K < \infty$, *and* $k \in (0,1)$ *for which the following assertions are valid.*

(i) $\mathbf{P}^x\{\tau_R < \infty\} = 1$ *whenever* $0 < \|x\| < R < \delta$.

(ii) $\mathbf{E}^x(\tau_R) = \infty$ *whenever* $0 < \|x\| < kR < k\delta$.

(iii)

$$\begin{aligned}
\frac{2}{V}\left[\log\left(\frac{R}{\|x\|}\right) - K\right] &\leq \liminf_{\varepsilon\to 0} \frac{1}{|\log\varepsilon|}\mathbf{E}^x\left(\int_0^{\tau_R} \chi_{[\varepsilon,\infty)}(\|x_s\|)\,ds\right) \\
&\leq \limsup_{\varepsilon\to 0} \frac{1}{|\log\varepsilon|}\mathbf{E}^x\left(\int_0^{\tau_R} \chi_{[\varepsilon,\infty)}(\|x_s\|)\,ds\right) \\
&\leq \frac{2}{V}\left[\log\left(\frac{R}{\|x\|}\right) + K\right]
\end{aligned}$$

whenever $0 < \|x\| < R < \delta$.

Proof. See Corollary 5.4 and Proposition 5.6 of [5]. □

(5.12) Remark. The papers [10, 9, 3, 7] mentioned in Remark 3.5 give sufficient conditions for transience or recurrence of $\{x_t : t \geq 0\}$ near infinity. By applying these criteria to the process $\{x_t/\|x_t\|^2 : t \geq 0\}$, the results will describe the behavior of $\{x_t : t \geq 0\}$ near 0. However, unless the generator L has sufficient rotational symmetry, there may be a gap between the sufficient condition for $\mathbf{P}^x\{\|x_t\| \to 0 \text{ as } t \to \infty\} > 0$ and the sufficient condition for recurrence to sets of the form $\mathbf{R}^d \backslash B(0,r)$. The results (5.1), (5.9), and (5.11) presented here do not have any such gap. This is due partly to the fact that the evaluation of λ involves an averaging over different directions (see the formula of Khas'minskii [11]). The advantage of these methods is in the precision of (5.1), (5.9), and (5.11) once λ is known; the disadvantage is that the computation of λ can be difficult.

6 Proofs of main results

We deal first with the case $\lambda < 0$.

Proof of Theorem 2.12. Given $\varepsilon > 0$ and $\mu \in (\lambda, 0)$ there exist, by Theorem 5.1, $r > 0$ and $C < \infty$ such that

$$\mathbf{P}^x\left\{\|x_t\| \leq c\|x\|e^{\mu t} \text{ for all } t \geq 0\right\} \geq 1 - \varepsilon$$

whenever $0 < \|x\| \le r$. For such x

$$\mathbf{P}^x\left\{\limsup_{t\to\infty}\frac{1}{t}\log\|x_t\|\le\mu\right\}\ge 1-\varepsilon$$

If $\|x\| > r$ then $\mathbf{P}^x\{\tau_r < \infty\} = 1$ (by Corollary 4.8) and so

$$\begin{aligned}\mathbf{P}^x&\left\{\limsup_{t\to\infty}\frac{1}{t}\log\|x_t\|\le\mu\right\}\\ &\ge \mathbf{P}^x\left\{\tau_r<\infty \text{ and } \|x_{t+\tau_r}\|\le c\|x_{\tau_r}\|e^{\mu t} \text{ for all } t\ge 0\right\}\\ &\ge 1-\varepsilon.\end{aligned}$$

In either case we let $\varepsilon \to 0$ and $\mu \to \lambda$ to obtain the required result. □

The proofs for the cases $\lambda > 0$ and $\lambda = 0$ will use the construction of an invariant measure developed by Maruyama and Tanaka [14] and Khas'minskii [10]. Fixing $0 < r < R < \infty$, define times inductively

$$\begin{aligned}\sigma_0 &= \inf\{t\ge 0 : \|x_t\| = r\}\\ \sigma_n' &= \inf\{t\ge\sigma_n : \|x_t\| = R\}\\ \sigma_{n+1} &= \inf\{t\ge\sigma_n' : \|x_t\| = r\}.\end{aligned}$$

Corollaries 4.8, 5.9, and 5.11 imply that for sufficiently small R,

$$\mathbf{P}^x\{\sigma_n < \infty \text{ for all } n\ge 0\} = 1$$

for all $x \ne 0$. Consider the induced Markov chain $\{Z_n : n \ge 0\}$ on $\Gamma \equiv \{x\in\mathbf{R}^d : \|x\| = r\}$ given by $Z_n = x_{\sigma_n}$. We denote its transition function

$$\begin{aligned}\Pi(x,A) &= \mathbf{P}^x\{x_{\sigma_1}\in A\}\\ &= \int_{\|y\|=R}\mathbf{P}^y\{x_{\tau_r}\in A\}\mathbf{P}^x\{x_{\tau_R}\in dy\}\end{aligned}$$

for $x \in \Gamma$, $A \in \mathcal{B}(\Gamma)$. The following lemma is based on Lemma 4.4 of Baxendale and Stroock [6].

(6.1) Lemma. *Assume* (2.5). *Assume also there exists* $\delta_0 > 0$ *such that* $\mathbf{P}^x\{\sigma_1 < \infty\} = 1$ *whenever* $0 < r < R < \delta_0$ *and* $\|x\| = r$. *Then there exists a positive* $\delta \le \delta_0$ *such that for* $0 < r < R < \delta$ *and* $A \in \mathcal{B}(\Gamma)$

(i) $x \mapsto \Pi(x,A)$ *is a continuous function on* Γ; *and*

(ii) *either* $\Pi(\cdot,A)\equiv 0$ *or else* $\Pi(x,A) > 0$ *for all* $x\in\Gamma$.

Proof. The assumption (2.5) implies the existence of $\delta_1 > 0$ such that

(6.2) $$\mathrm{Lie}(V_1,\ldots,V_r)(x) = \mathbf{R}^d \quad\text{for all}\quad x\in B'(0,\delta_1)$$

(see [6, Lemma 4.4]). Then L is hypoelliptic and satisfies the strong maximum principle on $B'(0,\delta_1)$. Let $\delta = \min(\delta_0,\delta_1)$. For fixed $A\in\mathcal{B}(\Gamma)$ define $f : B'(0,\delta)\to[0,1]$ by

$$f(x) = 1 - \int_{\|y\|=R}\mathbf{P}^y\{x_{\tau_r}\in A\}\mathbf{P}^x\{x_{\tau_R}\in dy\}.$$

Then $Lf = 0$ and $f|_\Gamma = 1 - \Pi(\cdot, A)$. Now hypoellipticity implies (i) and the strong maximum principle implies (ii). □

(6.3) Remark. The assumption (2.5), and its consequence (6.2), is stronger than is needed here. The hypoellipticity of L is true under the weaker assumption that $\text{Lie}(V_0, V_1, \ldots, V_r)(x) = \mathbf{R}^d$ for all $x \in B'(0, \delta_1)$ and Bony's strong maximum principle can be replaced by the strong maximum principle of Stroock and Varadhan [18] whenever there is a sufficiently large set of control paths inside $B'(0, R)$ corresponding to the diffusion $\{x_t : t \geq 0\}$.

Whenever the conclusions of Lemma 6.1 are true then the Markov chain $\{Z_n : n \geq 0\}$ satisfies Doeblin's condition (see Maruyama and Tanaka [14, Prop. 4.1]) and so has a unique invariant probability measure $\tilde{\mu}$, say, which is equivalent to $\Pi(x, \cdot)$ for each $x \in \Gamma$.

Proof of Theorem 2.13. Choose $0 < r < kR < k\delta$ so that the conclusions of Corollary 5.9 and Lemma 6.1 are valid, and let $\tilde{\mu}$ denote the invariant measure for $\{Z_n : n \geq 0\}$ in $\Gamma \equiv \{x \in \mathbf{R}^d : \|x\| = r\}$. Define μ by

$$\mu(A) = \frac{\int_\Gamma \left(\mathbf{E}^x \left(\int_0^{\sigma_1} \chi_A(x_s)\, ds\right)\right) d\tilde{\mu}(x)}{\int_\Gamma \mathbf{E}^x(\sigma_1)\, d\tilde{\mu}(x)}$$

for $A \in \mathcal{B}(\mathbf{R}^d \setminus \{0\})$. The fact that the denominator is finite comes from the estimates of $\mathbf{E}^x(\tau_R)$ for $\|x\| = r$ and $\mathbf{E}^x(\tau_r)$ for $\|x\| = R$ given in Corollaries 5.9(i) and 4.8. The invariance of μ and the ergodic properties are proved in [14] and [10]. The estimate of $\mu(B'(0, r))$ comes from Corollary 5.9(ii); and the integrability of g with respect to μ follows from Corollary 4.8. □

Proof of Theorem 2.14. Choose $0 < r < kR < k\delta$ so that the conclusions of Corollary 5.11 and Lemma 6.1 are valid, and let $\tilde{\mu}$ and Γ be as above. Define μ by

$$\mu(A) = \int_\Gamma \left(\mathbf{E}^x \left(\int_0^{\sigma_1} \chi_A(x_s)\, ds\right)\right) d\tilde{\mu}(x)$$

for $A \in \mathcal{B}(\mathbf{R}^d \setminus \{0\})$. The estimate of Corollary 5.11(ii) shows that μ is an infinite measure, and so we do not attempt to normalize. The invariance of μ and the ergodic properties are proved as in [14] and [10]. The estimate on the limiting behavior of $\mu(\mathbf{R}^d \setminus B(0, \varepsilon))$ comes from Corollary 4.8 and Corollary 5.11(iii), see the proof of [5, Theorem 5.8] for more details of a similar estimate; and the integrability of g over $\mathbf{R}^d \setminus B(0, \varepsilon))$ comes from Corollary 4.8. □

(6.4) Remark. We summarize here our earlier comments concerning the extent to which Theorems 2.13 and 2.14 remain valid under weakening of assumption (2.5).

Theorem 2.13 remains valid when (2.5) is replaced by

(a) assumption (5.5); and
(b) the conclusion of Lemma 6.1 is valid (see Remark 6.3); and
(c) γ exists (see Remark 5.10).

If (a) and (b) hold but γ does not exist (i.e. $\tilde{\Lambda}(-p) < 0$ for all $p > 0$) then Theorem 2.13 remains valid except that the estimate on $\mu(B'(0, \varepsilon))$ must be replaced by: for all $p > 0$ there exist $K < \infty$ and $\delta > 0$ such that $\mu(B'(0, \varepsilon)) \leq K\varepsilon^p$ for $0 < \varepsilon < \delta$.

Theorem 2.14 remains valid when (2.5) is replaced by (a) and (b) as above together with

(c′) $V > 0$ (see Remark 5.8).

References

[1] L. Arnold and W. Kliemann (1983). Qualitative theory of stochastic systems. In *Probabilistic Analysis and Related Topics* (A. T. Bharucha-Reid, ed.) **3** 1–79. Academic Press, New York.

[2] L. Arnold, E. Oeljeklaus and E. Pardoux (1986). Almost sure and moment stability for linear Itô equations. In *Lyapunov exponents* (L. Arnold, V. Wihstutz, eds) Lect. Notes Math. **1186** 129–159. Springer, Berlin Heidelberg New York.

[3] R. Azencott (1974). Behavior of diffusion semi-groups at infinity. *Bull. Soc. Math. France* **102** 193–240.

[4] P. H. Baxendale (1987). Moment stability and large deviations for linear stochastic differential equations. In *Proc. Taniguchi Symposium on Probabilistic Methods in Mathematical Physics. Katata and Kyoto 1985.* (N. Ikeda, ed.) 31–54. Kinokuniya, Tokyo.

[5] P. H. Baxendale (1990). Statistical equilibrium and two-point motion for a stochastic flow of diffeomorphisms. In *Spatial Stochastic Processes: Festschrift for T. E. Harris* (K. Alexander, J. Watkins, eds) Birkhauser, Boston Basel Stuttgart (in press).

[6] P. H. Baxendale and D. W. Stroock (1988). Large deviations and stochastic flows of diffeomorphisms. *Probab. Th. Rel. Fields* **80** 169–215.

[7] R. N. Bhattacharya (1978). Criteria for recurrence and existence of invariant measures for multidimensional diffusions. *Ann. Probab.* **6** 541–553.

[8] A. P. Carverhill (1985). Flows of stochastic dynamical systems: ergodic theory. *Stochastics* **14** 273–317.

[9] A. Friedman (1973). Wandering out to infinity of diffusion processes. *Trans. Am. Math. Soc.* **184** 185–203.

[10] R. Z. Khas'minskii (1960). Ergodic properties of recurrent diffusion processes and stabilization of the solution of the Cauchy problem for parabolic equations. *Theory Probab. Appl.* **5** 179–196.

[11] R. Z. Khas'minskii (1967). Necessary and sufficient conditions for the asymptotic stability of linear stochastic systems. *Theory Probab. Appl.* **12** 144–147.

[12] R. Z. Khas'minskii (1980). *Stochastic stability of differential equations.* Sijthoff and Noordhoff, Alphen aan den Rijn.

[13] W. Kliemann (1987). Recurrence and invariant measures for degenerate diffusions. *Ann. Probab.* **15** 690–707.

[14] G. Maruyama and H. Tanaka (1959). Ergodic property of N-dimensional recurrent Markov processes. *Mem. Fac. Sci. Kyushu Univ.* **A-13** 157–172.

[15] V. I. Oseledec (1968). A multiplicative ergodic theorem. Lyapunov characteristic numbers for dynamical systems. *Trans. Moscow Math. Soc.* **19** 197–231.

[16] D. Ruelle (1979). Ergodic theory of differential dynamical systems. *Publ. Math. IHES* **50** 275–306.

[17] D. W. Stroock (1986). On the rate at which a homogeneous diffusion approaches a limit, an application of the large deviation theory of certain stochastic integrals. *Ann. Probab.* **14** 840–859.

[18] D. W. Stroock and S. R. S. Varadhan (1972). On the support of diffusion processes with applications to the strong maximum principle. *Proc. Sixth Berkeley Symp. Math. Statist. Probab.* **3** 333–359. Univ. California Press.

[19] D. W. Stroock and S. R. S. Varadhan (1979). *Multidimensional diffusion processes.* Springer, Berlin Heidelberg New York.

HOW TO CONSTRUCT STOCHASTIC CENTER MANIFOLDS ON THE LEVEL OF VECTOR FIELDS

Petra Boxler

Institut für Dynamische Systeme, Universität Bremen
2800 Bremen 33, Fed. Rep. of Germany

Abstract

It is well-known by now that in a nonlinear ordinary differential equation with random coefficients the existence of a stochastic center manifold can be shown (see Boxler [3], [4]) if one of the Lyapunov exponents of the linearization vanishes. So far this was proved on the level of the random dynamical system (cocycle, "flow") generated by the equation. From the point of view of applications this is a disadvantage because a statement in terms of the original vector field would be preferable. For this reason we will present a different proof here which entirely stays on the level of vector fields. In these terms we will also derive an approximation result which is thus particularly useful for applications. It is illustrated by an example.

1 Introduction

Consider the following ordinary differential equation

$$\dot{x} = F(x), \quad x \in \mathbb{R}^d, \qquad F(x_0) = 0, \tag{1.1}$$

where $F: \mathbb{R}^d \to \mathbb{R}^d$ is assumed to be sufficiently smooth and to have a steady state at x_0. For convenience we take $x_0 = 0$.

We ask whether the asymptotic behavior of (1.1) and in particular the stability of the zero solution may be derived from a lower dimensional system. For this the eigenvalues of the Jacobian matrix $DF\big|_{x_0=0}$ have to be investigated: If all their real parts are $\neq 0$ then it is well-known that the stability properties of the linearized system carry over to the nonlinear system. However, this does no longer hold true if one of the real parts vanishes, the situation typically encountered in bifurcation theory.

In this case center manifold theory applies and allows to decouple the system and to separate that part of the system which corresponds to the vanishing real part. In order to examine the stability properties of the entire system it is sufficient to investigate the asymptotic behavior of the lower dimensional system obtained as the restriction of the original equation to the center manifold. In many applications this system will be of dimension 1 or 2, no matter what the dimension of the original system was. For more details see e. g. Carr [6], Vanderbauwhede [12] or Iooss [11].

In this paper we are going to study what happens to these schemes if a system like (1.1) is influenced by noise, i. e. if we have to deal with an ordinary differential equation with random coefficients. Although these questions have already been answered in the author's thesis [4] (see [3] for a published version) it seems to be worthwhile to address them again. The advantages of the rather general method developed there with which for example discrete time systems and continuous time systems disturbed by real as well as by white noise may be treated simultaneously have to be paid for by the drawback that one has to work on the level of random dynamical systems or cocycles generated by the solution of the original system.

After several discussions with physicists and others mainly interested in concrete applications we have thus decided to present a different proof of existence of a stochastic center manifold in the real noise case. The new procedure enables us to stay on the level of vector fields which is a considerable advantage for applications. However, this method does not carry over immediately to systems disturbed by white noise because serious problems with the non-adaptedness of a certain integrand arise.

A combination of these tools with an approximation result for stochastic center manifolds proved for random dynamical systems in Boxler [3] will also lead to an approximation theorem on the level of vector fields. This result, the use of which will be illustrated by an example, makes stochastic center manifolds an appropriate tool for applications, for example in a stochastic bifurcation theory.

2 Stochastic framework

2.1 Set up and the Multiplicative Ergodic Theorem

Suppose $(\Omega, \mathcal{F}, P)$ is an abstract probability space and $\vartheta_t: \Omega \to \Omega$, $t \in \mathbb{R}$ is a flow of P-preserving maps, i. e. $\vartheta_0 = \mathrm{id}$ and $\vartheta_{t+s} = \vartheta_t \circ \vartheta_s$ for all t, $s \in \mathbb{R}$. Assume $\{\vartheta_t | t \in \mathbb{R}\}$ to be ergodic and measurable, i. e.$(t,\omega) \to \vartheta_t\omega$ is jointly measurable.

Consider the equation

$$\dot{z}_t = X(\vartheta_t\omega, z_t),\ z_0 = z \in \mathbb{R}^d,\ t \in \mathbb{R} \tag{2.1}$$

which may be understood pathwise as an ordinary differential equation with random coefficients. Let us suppose that for almost all $\omega \in \Omega$, $X(\omega,\cdot)$ is $\mathcal{C}^2$ and that $X(\cdot, z)$ is measurable.

Assume that $X(\omega,0) = 0$ P-a.s., i.e. we assume that the vector field has 0 as an equilibrium point. This can be done without loss of generality, as shown in Boxler [3], Prop. 4.1. After having linearized (differentiated) at this zero solution we obtain

$$\dot{z}_t = A(\vartheta_t\omega)z_t + N(\vartheta_t\omega, z_t),\quad z_0 \doteq z, \tag{2.2}$$

where $A: \Omega \to \mathbb{R}^{d\times d}$, $N: \Omega \to \mathcal{C}^1(\mathbb{R}^d)$.

Furthermore, N describes the nonlinear part of X, i. e. $N(\omega,0) = D_z N(\omega,0) = 0$ P-a.s. As a consequence of this, $N(\omega,\cdot)$ satisfies P-a.s. a local Lipschitz condition with Lipschitz constant $L_N(\omega)$. We suppose that $L_N \in L^1(\Omega, \mathcal{F}, P)$. This implies that, by Fubini's theorem,

$L_N(\vartheta,\omega)$ is almost surely locally integrable w. r. t. Lebesgue measure.

Under these conditions a local solution $\varphi(t,\omega)$ exists. It is absolutely continuous in t and is unique (see e.g. Has'minskii [9], p. 10).

Let us assume that $A \in L^1(\Omega, \mathcal{F}, P)$. Then we may apply the Multiplicative Ergodic Theorem (see e. g. Arnold, Kliemann and Oeljeklaus [2]) to the linear random dynamical system $\Psi(t,\omega)$ obtained as the solution of

$$\dot{\Psi}(t,\omega) = A(\vartheta_t\omega)\Psi(t,\omega), \qquad \Psi(0,\omega) = \mathrm{id}.$$

Thus, $\Psi(t,\omega)$ is nothing but the corresponding "fundamental matrix". In particular it has the cocycle property:

$$\Psi(t+s,\omega) = \Psi(t,\vartheta_s\omega) \circ \Psi(s,\omega) \text{ for all } t, s \in \mathbb{R}, \text{ P-a.s.}$$

We will therefore obtain *Lyapunov exponents* $\lambda_1 > \lambda_2 > \ldots > \lambda_r$, $1 \le r \le d$, and a decomposition of $\mathbb{R}^d$ into the direct sum of the corresponding r Oseledec spaces $E_i(\omega)$:

$$\mathbb{R}^d = E_1(\omega) \oplus E_2(\omega) \oplus \ldots \oplus E_r(\omega).$$

These linear spaces depend measurably on ω and satisfy $\Psi(t,\omega)E_i(\omega) = E_i(\vartheta_t\omega)$ for all $t \in \mathbb{R}$, P-a.s.

Henceforth we will assume that all the Lyapunov exponents are different and that one of them vanishes. As explained in the introduction the situation would be much simpler if all exponents were different from 0.

We introduce the *stable* and the *unstable* subspace of Ψ, resp.:

$$E^s(\omega) := \bigoplus_{\lambda_i < 0} E_i(\omega), \quad E^u(\omega) := \bigoplus_{\lambda_i > 0} E_i(\omega).$$

Furthermore, let $E^c(\omega)$ be the center subspace, i.e. the (one-dimensional) Oseledec space corresponding to the Lyapunov exponent 0.

We put $\lambda^s := \max_{\lambda_i < 0} \lambda_i$, $\lambda^u := \min_{\lambda_i > 0} \lambda_i$.

Fix $\varepsilon > 0$ such that $\lambda^s + \varepsilon < 0$ and $\lambda^u - \varepsilon > 0$.

Then the Multiplicative Ergodic Theorem yields the following estimate where $\|.\|$ denotes the norm induced on the subspaces by the usual Euclidean norm.

$$\|\Psi_s(t,\omega)z\| \le C_s(\omega)\, e^{(\lambda^s+\varepsilon)t} \|z\| \text{ for all } z \in E^s(\omega) \text{ and all } t \ge 0 \text{ P-a.s.}$$

Here C_s is a random variable which takes values in $[1,\infty[$. Analogous estimates hold for the center and the unstable part. See Boxler [3], Lemma 4.1., for details.

For later use we note that $\lim_{t\to\infty} \frac{1}{t} \log \|\Psi(t,\omega)\| = \lambda_1$ and $\lim_{t\to-\infty} \frac{1}{t} \log \|\Psi(t,\omega)\| = \lambda_r$ P-a.s.

(see e.g. Boxler [4], Cor. 2.1., for a proof). Thus we obtain:

$$\| \Psi(t,\omega) \| \le C(\varepsilon,\omega)\, e^{(\lambda_1+\varepsilon)t} \text{ for all } t \ge 0, \quad \| \Psi(t,\omega) \| \le \tilde{C}(\varepsilon,\omega)\, e^{(\lambda_r+\varepsilon)t} \text{ for all } t \le 0,$$

where C, $\tilde{C}$ are measurable functions on Ω with values in $[1,\infty[$.

2.2 Transformation of the original system

Since we have assumed the Lyapunov exponents to be different the intersection of the Oseledec spaces $E_i(\omega)$, which are thus one-dimensional, and the projective space $\mathbb{P}^{d-1}$ consists of exactly one point. Following Crauel [7] they may be used to construct a matrix $S(\omega) = \mathrm{col}(s_1(\omega), ..., s_d(\omega))$ consisting of stationary solutions $s_i(\vartheta_t\omega) = s(t; s_i(\omega))$ which are the projections of solutions of (2.2) onto $\mathbb{S}^{d-1}$.

Using this as a transformation matrix the "fundamental matrix" becomes diagonal, namely

$$\Phi(t,\omega) := S^{-1}(\vartheta_t\omega)\Psi(t,\omega)S(\omega) = e^{\left(\mathrm{diag}\left\{ \int_0^t q_i(\tau,\omega)\, d\tau : i = 1, ..., d \right\}\right)},$$

where $q_i = s_i{}'As_i$. If we put $Q = \mathrm{diag}(q_1, ..., q_d)$ then Φ will be the "fundamental matrix" of the equation

$$\dot{v}_t = Q(\vartheta_t\omega)\, v_t + S^{-1}(t,\omega)\, N(\vartheta_t\omega, S(t,\omega)v_t). \tag{2.3}$$

which has the same Lyapunov exponents as (2.2).

Remarks.

(i) The s_i may also be interpreted as the solutions of the angular equation

$$\dot{s}_i = (A(\vartheta_t\omega) - q_i[s_i, A(\vartheta_t\omega)]\ \mathrm{id})\, s_i.$$

(ii) If there were $r < d$ different Lyapunov exponents only the existence of at least r stationary columns in the matrix S would be guaranteed. It is for this reason that the assumption about the exponents being different is made. However, the linear part can always be block diagonalized w. r. t. E^c, E^s and E^u. This is all that is really needed here although one will have to make sure in this case that the Lyapunov exponents are not altered by the transformation. For the sake of simplicity we will not elaborate this in detail.

Note that the spaces E^c, E^s and E^u describe a random coordinate system. Projecting (2.3) onto E^s along E^c and E^u and so on and denoting the projections by indices s, c, u we obtain:

$$\dot{x}_c = Q_c(\vartheta_t\omega)\, x_c + F_c(\vartheta_t\omega, x_c, x_s, x_u), \tag{2.4a}$$

$$\dot{x}_s = Q_s(\vartheta_t\omega)\, x_s + F_s(\vartheta_t\omega, x_c, x_s, x_u), \tag{2.4b}$$

$$\dot{x}_u = Q_u(\vartheta_t\omega)\, x_u + F_u(\vartheta_t\omega, x_c, x_s, x_u), \tag{2.4c}$$

where we have denoted $S^{-1} N(\vartheta_t\omega,Sv)$ by F. Furthermore, the influences of ϑ and S are jointly described by a new noise source which we have again denoted by ϑ and which takes values in a higher dimensional space.

For later use we introduce

$$\|p(\omega)\| := \max\{\|p_c(\omega)\|, \|p_s(\omega)\|, \|p_u(\omega)\|\}$$

where $p_{c,s,u}(\omega)$: $\mathbb{R}^d \to E^{c,s,u}(\omega)$ denotes the projection maps.

3 Definition of stochastic center manifolds

As described in detail in Boxler [3], Lemma 4.2., we introduce a random norm in $\mathbb{R}^d$ by incorporating into its definition the knowledge about the long term behavior of the system. This knowledge may then already be used e. g. at time t = 0. For this we put for example:

$$\|x_s\|^s_\omega := \int_0^\infty e^{-(\lambda^s+\varepsilon)t} \|\Phi_s(t,\omega)x\| \, dt.$$

We will make extensive use of the estimates of Lemma 4.2. in Boxler [3].We obtain for example:

$$\|\Phi_s(t,\omega)x\|^s_{\vartheta_t\omega} \le e^{(\lambda^s+\varepsilon)t} \|x\|_\omega \quad \text{for all } t \ge 0 \text{ P-a.s.}$$

The norms $\|\cdot\|$ and $\|\cdot\|_\omega$ are related as follows: For each $\gamma > 0$ there is a random variable $C(\omega,\gamma)$ such that

$$C^{-1}(\omega,\gamma)\|x\| \le \|x\|_\omega \le C(\omega,\gamma)\|x\|, \tag{3.1a}$$

$$C(\omega,\gamma)e^{-\gamma|t|} \le C(\vartheta_t\omega,\gamma) \le C(\omega,\gamma)e^{\gamma|t|} \tag{3.1b}$$

for all $t \in \mathbb{R}$ (see Dahlke [8] , Lemma 2.1.2).

As in the deterministic case we wish to describe stochastic center manifolds as graphs in appropriate spaces. For this we define:

$$E := \{(\omega,x) \in \Omega \times \mathbb{R}^d \mid x \in E^c(\omega)\},$$

$$\mathcal{X} := \{h: E \to \mathbb{R}^d \text{ meas.} \mid h(\omega,\cdot): E^c(\omega) \to E^s(\omega) \oplus E^u(\omega) \text{ a.s. continous and bounded}\}.$$

Boundedness of $h(\omega,\cdot)$ is understood with respect to the norm

$$\|h(\omega.\cdot)\|_{\infty,\omega} := \sup_{x\in E^c(\omega)} \|h(\omega,x)\|_\omega = \sup_{x\in E^c(\omega)} \left[\max(\|h_s(\omega,x)\|^s_\omega, \|h_u(\omega,x)\|^u_\omega)\right].$$

If we endow $\mathcal{X}$ with the pseudometric d of convergence in probability and denote by X the space of equivalence classes w. r. t. almost sure equality then it follows from Lemma 4.3. in Boxler [3] that (X, d) is a complete metric space.

We are now ready to define local stochastic center manifolds:

Definition 3.1. A set

$$M(\omega) = \{(x, h(\omega,x)): x \in E^c(\omega), \|x\|_\omega^c \le \delta(\omega)\},$$

δ a positive valued random variable and $h \in X$, is called a *local stochastic center manifold* for the equation (2.1) if the following conditions hold:
(i) $M(\omega)$ is P-a.s. invariant under the solution $\varphi(t,\omega)$, i.e. for all t for which $\|\varphi(t,\omega)x\|_{\vartheta_t\omega} \le \delta(\vartheta_t\omega)$ we have

$$\varphi(t,\omega)M(\omega) \subset M(\vartheta_t\omega) \qquad \text{P-a.s.} \tag{3.2}$$

(ii) $h(\omega,0) = 0$ P-a.s.
(iii) For all $x, \tilde{x} \in E^c(\omega)$, $\|x\|_\omega^c, \|\tilde{x}\|_\omega^c \le 2\delta$:

$$\|h_{s,u}(\omega,x)\|_\omega^{s,u} \le L \quad \text{P-a.s.}$$

$$\|h_{s,u}(\omega,x) - h_{s,u}(\omega,\tilde{x})\|_\omega^{s,u} \le L\|x - \tilde{x}\|_\omega^c \quad \text{P-a.s.}$$

Remark.
For a more geometrical definition see Boxler [3], Def. 6.2.

4 Existence of local stochastic center manifolds

Theorem 3.1. *Assume the situation described in section 2. Then there is a local stochastic center manifold* $M(\omega) = \{(x, h(\omega,x)): x \in E^c(\omega), \|x\|_\omega^c \le \delta(\omega)\}$ *for the system (2.1) where* δ *is a positive valued random varible and* $h \in X$.

Remark. We omit tackling questions of differentiability here because there is no difference to the case of random dynamical systems treated in Boxler [3].

Proof. Instead of proving existence of a *local* stochastic center manifold for (2.1) or, equivalently, for (2.4) we are going to show existence of a *global* stochastic center manifold for a system obtained by "cutting-off" the vector field. For this let $\Gamma: \mathbb{R} \to [0,1]$ be a $\mathfrak{C}^\infty$-function such that $\Gamma(u) = 1$ if $|u| \le 1$ and $= 0$ if $|u| \ge 2$.

For any positive valued random variable δ for which $\delta(\omega) \le \delta$ for an appropriate constant $\delta > 0$ we put:

$$f^\delta_{c,s,u}(\omega,x_c,x_s,x_u) := F_{c,s,u}\left(\omega, \Gamma\left(\frac{\|x_c\|_\omega^c}{\delta(\omega)}\right)x_c, \Gamma\left(\frac{\|x_s\|_\omega^s}{\delta(\omega)}\right)x_s, \Gamma\left(\frac{\|x_u\|_\omega^u}{\delta(\omega)}\right)x_u\right).$$

Then it is obvious that f^δ and F will coincide in a δ-neighborhood of the origin. Hence we consider the following system:

$$\dot{x}_c = Q_c(\vartheta_t\omega)\, x_c + f_c^{\delta}(\vartheta_t\omega, x_c, x_s, x_u), \tag{4.1a}$$

$$\dot{x}_s = Q_s(\vartheta_t\omega)\, x_s + f_s^{\delta}(\vartheta_t\omega, x_c, x_s, x_u), \tag{4.1b}$$

$$\dot{x}_u = Q_u(\vartheta_t\omega)\, x_u + f_u^{\delta}(\vartheta_t\omega, x_c, x_s, x_u). \tag{4.1c}$$

As a consequence of the assumptions and the cut-off procedure $f^{\delta}_{c,s,u}$ satisfies global Lipschitz conditions, and by shrinking the range of the random variable δ the Lipschitz constant η, which can be chosen independently of ω because we accept a random neighborhood, will become arbitrarily small.

Thus for all $x, y \in \mathbb{R}^d$ we obtain:

$$\| f^{\delta}_{c,s,u}(\vartheta_t\omega, x) - f^{\delta}_{c,s,u}(\vartheta_t\omega, y) \|^{c,s,u}_{\vartheta_t\omega} \le \eta \| x - y \|_{\omega}. \tag{4.2}$$

To prove the theorem we proceed in several steps:

1) For a given $h \in X$ we consider the equation

$$\dot{x}_c = Q_c(\vartheta_t\omega)\, x_c + f_c^{\delta}(\vartheta_t\omega, x_c, h_s(\vartheta_t\omega, x_c), h_u(\vartheta_t\omega, x_c)), \tag{4.3}$$

and show that it has a P-a.s. unique solution.

2) Incorporating the solution of (4.3) we define an operator T which acts on a closed subset A of X.

3) We show that $T(A) \subset A$.

4) We prove that T is a contraction. Thus it has a unique fixed point h by the contraction mapping theorem.

5) It is shown that h has the required invariance property.

1) This will be achieved by applying Theorem I. 3.1. in Has'minskii [9]. The first condition required there is satisfied because in the new norm a global Lipschitz condition with a Lipschitz constant that does not depend on ω holds, as we have seen in (4.2). The second condition is trivial in our case since 0 is an equilibrium point.

2) For any $h \in X$ let $\varphi_c(t,\omega,x_c,h(\omega,x_c))$ be the unique solution of (4.3). Denote by $\Phi_{s,u}(t,\omega)$ the projection onto $E^{s,u}(\vartheta_t\omega)$ of the "fundamental matrix" Φ which we have seen to be diagonal. Integrating (4.1b) yields the "solution" at time t of that equation, starting at time t_0:

$$\varphi_s(t_0,t,\omega,x_c,h(\omega,x_c)) = \Phi_s(t,\omega)\Phi_s^{-1}(t_0,\omega)h_s(\omega,x_c) +$$

$$\int_{t_0}^{t} \Phi_s(t,\omega)\Phi_s^{-1}(\tau,\omega) f^{\delta}(\vartheta_\tau\omega, \varphi_c(\tau,\omega,x_c,h(\omega,x_c)), h(\vartheta_\tau\omega, \varphi_c(\tau,\omega,x_c,h(\omega,x_c))))d\tau.$$

We put $t = 0$ and examine the behavior of this solution for $t_0 \to -\infty$. We take into account that $\Phi(0,\omega) = 0$ and argue exactly as in Arnold and Crauel [1], proof of theorem 4, and obtain:

$$\lim_{t_0 \to -\infty} \| \Phi_s^{-1}(t_0,\omega) \|^{s}_{\omega} = 0.$$

The "solution" of (4.1c) may be treated similarly, and we obtain: $\lim_{t_0 \to \infty} \| \Phi_u^{-1}(t_0,\omega) \|_\omega^u = 0.$

Let us now introduce the operator T by

$$Th(\omega,x_c) := \int_{-\infty}^{0} \Phi_s^{-1}(\tau,\omega) f^\delta(\vartheta_\tau\omega,\varphi_c(\tau,\omega,x_c,h(\omega,x_c)),h(\vartheta_\tau\omega,\varphi_c(\tau,\omega,x_c,h(\omega,x_c))))d\tau$$

$$- \int_{0}^{\infty} \Phi_u^{-1}(\tau,\omega) f^\delta(\vartheta_\tau\omega,\varphi_c(\tau,\omega,x_c,h(\omega,x_c)),h(\vartheta_\tau\omega,\varphi_c(\tau,\omega,x_c,h(\omega,x_c))))d\tau.$$

Since this may also be written as

$$Th(\omega,x_c) = \lim_{t \to -\infty} \varphi_s(t,0,\omega,x_c,h(\omega,x_c)) + \lim_{t \to \infty} \varphi_u(t,0,\omega,x_c,h(\omega,x_c))$$

this operator enables us to read off what will remain of the system "in the long run". However, this is nothing but the stochastic center manifold.

We claim that T is an operator on X. For this let $h \in X$.

a) We prove that for almost all $\omega \in \Omega$, $Th(\omega,\cdot)$ is continuous:

Obviously Theorem 1.6, p. 21 in Bunke [5] applies and ensures that the solution of (4.2) depends continuously on the initial value. Since h and f^δ are continuous functions the assertion follows.

b) We show that Th is measurable:

The solution of (4.3) is absolutely continuous and thus there is a measurable process which is equivalent to it (see e. g. Ikeda and Watanabe [10]). Hence we may suppose without loss of generality that for all x_c, h, $\varphi_c(\cdot,\cdot,x_c,h(\omega,x_c))$ is measurable.

By assumption, $\vartheta: (t,\omega) \to \vartheta_t\omega$ is jointly measurable. Since h and $f^\delta(\cdot,z)$ are also measurable by assumption, $f^\delta(\omega,\cdot)$ is continuous and Φ_c is measurable as a "fundamental matrix", the integrand is measurable in τ and ω.

Thus the integral is measurable w.r.t. ω. Since we know from a) that Th depends continuously on the second variable we may conclude the desired measurability of Th (see e.g. Ikeda and Watanabe [10]).

The proof of boundedness is postponed because it will follow from a more restrictive condition which we will show in 3).

In a second step we single out a subset A of X which reflects the properties a stochastic center manifold is required to have (except invariance). For this let

$A_L := \{h \in X: h(\omega,\cdot)$ satisfies (ii) and (iii) of Def. 3.1. for all $x \in E^c(\omega)$ P–a.s.$\}$.
Then it is immediately checked that A_L is a closed subset of X.

3) We show that $T(A_L) \subset A_L$.

Let $h \in A_L$. If we write down the "solution" of (4.2) with 0 as an initial value then it is evident that $\varphi_c(t,\omega,0,h(\omega,0)) = 0$ because $f^\delta(\omega,0) = 0$ by definition. If we insert this into the definition of T and argue the same way then (ii) follows at once.

To prove (iii) we restrict ourselves to the stable part because the estimates for the unstable part are completely analogous.

For any $x_c \in E^c(\omega)$ we obtain:

$$\|(Th)_s(\omega,x_c)\|^s_\omega \leq \int_{-\infty}^0 \|\Phi_s^{-1}(\tau,\omega)f^\delta(\vartheta_\tau\omega,\varphi_c(\tau,\omega,x_c,h(\omega,x_c)),h(\vartheta_\tau\omega,\varphi_c(\tau,\omega,x_c,h(\omega,x_c))))\|_{\vartheta_\tau\omega}d\tau$$

We make use of the estimate mentioned in Section 3 (see Lemma 4.2. in Boxler [3]), artificially insert $f^\delta(\vartheta_\tau\omega,0,h(\vartheta_\tau\omega,0))$, which is 0, and apply estimate (4.2) to get:

$$\|(Th)_s(\omega,x_c)\|^s_\omega \leq \eta \int_{-\infty}^0 e^{-(\lambda^s+\varepsilon)\tau}(\|\varphi_c(\tau,\omega,x_c,h(\omega,x_c))\|_{\vartheta_\tau\omega} + \|h(\vartheta_\tau\omega,\varphi_c(\tau,\omega,x_c,h(\omega,x_c)))\|_{\vartheta_\tau\omega})d\tau.$$

By definition of f^δ and our choice of δ the first term in this sum has to be $< 2\delta$. Since $h \in A_L$ the second term is bounded by L, and thus we obtain:

$$\|(Th)_s(\omega,x_c)\|^s_\omega \leq \frac{2\delta + L}{-(\lambda^s+\varepsilon)}\eta,$$

which is $\leq L$ if we choose η sufficiently small (which can be achieved by shrinking $\delta(\omega)$).

Finally we have to prove that Th satisfies a Lipschitz condition with Lipschitz constant L.

We first derive an estimate for the solution of (4.3) for $t \leq 0$. For this we make use of the cocycle property, the estimate mentioned in Section 3 (cf. Boxler [3], Lemma 4.2.(ii)) and the Lipschitz condition for f^δ to obtain:

$$\|\varphi_c(t,\omega,x_c,h(\omega,x_c)) - \varphi_c(t,\omega,\tilde{x}_c,h(\omega,\tilde{x}_c))\|^c_{\vartheta_t\omega} \leq \|\Phi_c(t,\omega)(x_c-\tilde{x}_c)\|^c_{\vartheta_t\omega} +$$

$$\int_t^0 \|\Phi_c(t-\tau,\vartheta_{-\tau}\omega)[f^\delta(\vartheta_\tau\omega,\varphi_c(\tau,\omega,x_c,h(\omega,x_c)),h(\vartheta_\tau\omega,\varphi_c(\tau,\omega,x_c,h(\omega,x_c)))) -$$

$$f^\delta(\vartheta_\tau\omega,\varphi_c(\tau,\omega,\tilde{x}_c,h(\omega,\tilde{x}_c)),h(\vartheta_\tau\omega,\varphi_c(\tau,\omega,\tilde{x}_c,h(\omega,\tilde{x}_c))))]\|_{\vartheta_\tau\omega}\,d\tau$$

$$\leq e^{-\varepsilon t}\|x_c-\tilde{x}_c\|^c_\omega + e^{-\varepsilon t}\eta\int_t^0 e^{\varepsilon\tau}(\|\varphi_c(\tau,\omega,x_c,h(\omega,x_c)) - \varphi_c(\tau,\omega,\tilde{x}_c,h(\omega,\tilde{x}_c))\|_{\vartheta_\tau\omega} +$$

$$\|h(\vartheta_\tau\omega,\varphi_c(\tau,\omega,x_c,h(\omega,x_c))) - h(\vartheta_\tau\omega,\varphi_c(\tau,\omega,\tilde{x}_c,h(\omega,\tilde{x}_c)))\|_{\vartheta_\tau\omega})\,d\tau$$

$$\le e^{-\varepsilon t}\|x_c-\tilde{x}_c\|^c_\omega + e^{-\varepsilon t}\eta(1+L)\int_t^0 e^{\varepsilon\tau}\|\varphi_c(\tau,\omega,x_c,h(\omega,x_c)) - \varphi_c(\tau,\omega,\tilde{x}_c,h(\omega,\tilde{x}_c))\|_{\vartheta_\tau\omega}\,d\tau,$$

where the last inequality is a consequence of the Lipschitz condition for h.

Multiplying both sides by $e^{\varepsilon t}$ and applying a stochastic version of Gronwall's lemma (see e.g. Bunke [5], p. 21) yields:

$$\|\varphi_c(t,\omega,x_c,h(\omega,x_c)) - \varphi_c(t,\omega,\tilde{x}_c,h(\omega,\tilde{x}_c))\|^c_{\vartheta_t\omega} \le \|x_c-\tilde{x}_c\|^c_\omega\, e^{-(1+L)\eta t}. \tag{4.4}$$

Now the desired Lipschitz condition follows by the same arguments employed to prove boundedness. The only difference is that we use the Lipschitz condition for h and then combine this with estimate (4.4). This leads to:

$$\|(Th)_s(\omega,x_c) - (Th)_s(\omega,\tilde{x}_c)\|^s_\omega \le (1+L)\eta\int_{-\infty}^0 e^{-(\lambda^s+\varepsilon)-(1+L)\eta\tau}\,d\tau\,\|x_c-\tilde{x}_c\|^c_\omega.$$

By shrinking η (and thus $\delta(\omega)$) if necessary, we make sure that $(1+L)\eta < -(\lambda^s+\varepsilon)$. Then we obtain:

$$\|(Th)_s(\omega,x_c) - (Th)_s(\omega,\tilde{x}_c)\|^s_\omega \le \frac{(1+L)\eta}{-(\lambda^s+\varepsilon)-(1+L)\eta}\,\|x_c-\tilde{x}_c\|^c_\omega,$$

and the last constant is $< L$ for η sufficiently small.

4) We show that T is a contraction on A_L.
For this we have to prove that there is a constant c, $0 < c < 1$, such that for any $h, \tilde{h} \in A_L$, $d(Th, T\tilde{h}) \le c\, d(h, \tilde{h})$.

In a first step we estimate $\|Th(\omega,\cdot) - T\tilde{h}(\omega,\cdot)\|_{\infty,\omega} = \sup_{x_c\in E^c(\omega)} \|Th(\omega,x_c) - T\tilde{h}(\omega,x_c)\|_\omega$.

Such an estimate can be derived using the same reasoning as in the case of the Lipschitz condition. This means that one first estimates $\|\varphi_c(t,\omega,x_c,h(\omega,x_c)) - \varphi_c(t,\omega,x_c,\tilde{h}(\omega,x_c))\|^c_{\vartheta_t\omega}$. This is done as before, the only difference being that one will have to make use of the following estimate:

$$\|h(\vartheta_t\omega,\varphi_c(t,\omega,x_c,h(\omega,x_c))) - \tilde{h}(\vartheta_t\omega,\varphi_c(t,\omega,x_c,\tilde{h}(\omega,x_c)))\|_{\vartheta_t\omega} \le$$

$$\|h(\vartheta_t\omega,\varphi_c(t,\omega,x_c,h(\omega,x_c))) - h(\vartheta_t\omega,\varphi_c(t,\omega,x_c,\tilde{h}(\omega,x_c)))\|_{\vartheta_t\omega} +$$

$$\|h(\vartheta_t\omega,\varphi_c(t,\omega,x_c,\tilde{h}(\omega,x_c))) - \tilde{h}(\vartheta_t\omega,\varphi_c(t,\omega,x_c,\tilde{h}(\omega,x_c)))\|_{\vartheta_t\omega}$$

$$\le L\|\varphi_c(t,\omega,x_c,h(\omega,x_c)) - \varphi_c(t,\omega,x_c,\tilde{h}(\omega,x_c))\|^c_{\vartheta_t\omega} + \sup_{y\in E^c(\vartheta_t\omega)}\|h(\vartheta_t\omega,y) - \tilde{h}(\vartheta_t\omega,y)\|_{\vartheta_t\omega}.$$

Proceeding as in the Lipschitz case this yields the estimate for $\| Th(\omega,\cdot) - T\tilde{h}(\omega,\cdot) \|_{\infty,\omega}$. We will thus omit further details.

After this the assertion follows by the same arguments as employed in Boxler [3], p. 527. We will therefore not repeat the proof here.

Now the contraction mapping theorem applies and provides us with a unique fixed point $h \in A_L$ which has properties (ii) and (iii) of Definition 3.1. by construction.

5) We show that (3.1) holds for the fixed point h obtained in 4), i.e. that as long as the δ-neighborhood is not left, $\varphi(t,\omega)M(\omega) \subset M(\vartheta_t\omega)$ P-a.s. where $M(\omega) = \{(x, h(\omega,x)): x \in E^c(\omega), \|x\|^c_\omega \leq \delta(\omega)\}$, δ as determined in the course of the proof.

Thus let us choose a point $(x_c, h(\omega,x_c)) \in M(\omega)$.

Provided that at time t, $\|\varphi(t,\omega,x_c,h(\omega,x_c))\|_{\vartheta_t\omega} \leq \delta(\vartheta_t\omega)$, its image under the solution will be on $M(\vartheta_t\omega)$ if $\varphi_{s,u}(t,\omega,x_c,h(\omega,x_c)) = h_{s,u}(\vartheta_t\omega,\varphi_c(t,\omega,x_c,h(\omega,x_c)))$ P-a.s. As usual we restrict ourselves to the stable part.

Since h satisfies Th = h we may write

$$h_s(\omega,x) = \int_{-\infty}^{0} \Phi_s^{-1}(\tau,\omega)\, f^\delta(\vartheta_\tau\omega,\varphi_c(\tau,\omega,x_c,h(\omega,x_c)),h(\vartheta_\tau\omega,\varphi_c(\tau,\omega,x_c,h(\omega,x_c))))\, d\tau\,, \quad (4.5)$$

which implies

$$h_s(\vartheta_t\omega,\varphi_c(t,\omega,x_c,h(\omega,x_c))) =$$

$$\int_{-\infty}^{0} \Phi_s^{-1}(\tau,\vartheta_t\omega) f^\delta(\vartheta_\tau\vartheta_t\omega,\varphi_c(\tau,\vartheta_t\omega,\varphi_c(t,\omega,x_c,h(\omega,x_c)),h(\vartheta_t\omega,\varphi_c(t,\omega,x_c,h(\omega,x_c)))),$$

$$h(\vartheta_\tau\vartheta_t\omega,\varphi_c(\tau,\vartheta_t\omega,\varphi_c(t,\omega,x_c,h(\omega,x_c)),h(\vartheta_t\omega,\varphi_c(t,\omega,x_c,h(\omega,x_c)))))))\, d\tau$$

$$= \int_{-\infty}^{0} \Phi_s(t,\omega)\Phi_s^{-1}(\tau+t,\omega) f^\delta(\vartheta_{\tau+t}\omega,\varphi_c(\tau+t,\omega,x_c,h(\omega,x_c)),h(\vartheta_{\tau+t}\omega,\varphi_c(\tau+t,\omega,x_c,h(\omega,x_c))))\, d\tau$$

where the last equality is a consequence of the cocycle property. Finally a substitution yields:

$$h_s(\vartheta_t\omega,\varphi_c(t,\omega,x_c,h(\omega,x_c))) =$$

$$\Phi_s(t,\omega) \int_{-\infty}^{t} \Phi_s^{-1}(\tau,\omega) f^\delta(\vartheta_\tau\omega,\varphi_c(\tau,\omega,x_c,h(\omega,x_c)),h(\vartheta_\tau\omega,\varphi_c(\tau,\omega,x_c,h(\omega,x_c))))\, d\tau\,.$$

On the other hand the stable part of the "solution" at time t which starts at $(x_c, h(\omega,x_c))$ at time 0 may be written as

$$\varphi_s(t,\omega,x_c,h(\omega,x_c)) = \Phi_s(t,\omega)h_s(\omega,x_c) +$$

$$\Phi_s(t,\omega)\int_0^t \Phi_s^{-1}(\tau,\omega)f^\delta(\vartheta_\tau\omega,\varphi(\tau,\omega,x_c,h(\omega,x_c)))\, d\tau .$$

We first treat the case $t \geq 0$. Inserting (4.5) we obtain:

$$\| \varphi_s(t,\omega,x_c,h(\omega,x_c)) - h_s(\vartheta_t\omega,\varphi_c(t,\omega,x_c,h(\omega,x_c))) \|^s_{\vartheta_t\omega} =$$

$$\| \Phi_s(t,\omega)\int_{-\infty}^0 \Phi_s^{-1}(\tau,\omega)f^\delta(\vartheta_\tau\omega,\varphi_c(\tau,\omega,x_c,h(\omega,x_c)),h(\vartheta_\tau\omega,\varphi_c(\tau,\omega,x_c,h(\omega,x_c))))\, d\tau +$$

$$\Phi_s(t,\omega)\int_0^t \Phi_s^{-1}(\tau,\omega)f^\delta(\vartheta_\tau\omega,\varphi_c(\tau,\omega,x_c,h(\omega,x_c)),\varphi_s(\tau,\omega,x_c,h(\omega,x_c)),\varphi_u(\tau,\omega,x_c,h(\omega,x_c)))\, d\tau \; -$$

$$\Phi_s(t,\omega)\int_{-\infty}^t \Phi_s^{-1}(\tau,\omega)f^\delta(\vartheta_\tau\omega,\varphi_c(\tau,\omega,x_c,h(\omega,x_c)),h(\vartheta_\tau\omega,\varphi_c(\tau,\omega,x_c,h(\omega,x_c))))\, d\tau \, \|^s_{\vartheta_t\omega}$$

$$\leq \eta\int_0^t \| \Phi(t-\tau,\vartheta_\tau\omega) \|_{\vartheta_t\omega} \max\{ \| \varphi_s(\tau,\omega,x_c,h(\omega,x_c)) - h_s(\vartheta_\tau\omega,\varphi_c(\tau,\omega,x_c,h(\omega,x_c))) \|^s_{\vartheta_\tau\omega},$$

$$\| \varphi_u(\tau,\omega,x_c,h(\omega,x_c) - h_u(\vartheta_\tau\omega,\varphi_c(\tau,\omega,x_c,h(\omega,x_c))) \|^u_{\vartheta_\tau\omega}\}\, d\tau,$$

where we have made use of (4.2), the cocycle property and the fact that the definition of the norms implies that $\| \Phi_s(t,\omega) \|^s_{\vartheta_t\omega} \leq \| \Phi(t,\omega) \|_{\vartheta_t\omega}$.

Combining inequality (3.1a) relating the random and the Euclidean norm with the consequence of the Multiplicative Ergodic Theorem mentioned at the end of Section 2.1. we obtain that $\| \Phi(t,\omega) \|_{\vartheta_t\omega} \leq K(\omega)e^{(\lambda_1+\varepsilon)t}$ where $K(\omega)$ depends on ε.

Exactly the same reasoning as above holds true for the unstable part. After having multiplied both sides with $e^{-(\lambda_1+\varepsilon)t}$ this yields:

$$e^{-(\lambda_1+\varepsilon)t} \max\{ \| \varphi_s(t,\omega,x_c,h(\omega,x_c)) - h_s(\vartheta_t\omega,\varphi_c(t,\omega,x_c,h(\omega,x_c))) \|^s_{\vartheta_t\omega},$$

$$\| \varphi_u(t,\omega,x_c,h(\omega,x_c) - h_u(\vartheta_t\omega,\varphi_c(t,\omega,x_c,h(\omega,x_c))) \|^u_{\vartheta_t\omega}\} \leq$$

$$\leq \eta\int_0^t K(\vartheta_\tau\omega)\, e^{-(\lambda_1+\varepsilon)\tau} \max\{ \| \varphi_s(\tau,\omega,x_c,h(\omega,x_c)) - h_s(\vartheta_\tau\omega,\varphi_c(\tau,\omega,x_c,h(\omega,x_c))) \|^s_{\vartheta_\tau\omega},$$

$$\| \varphi_u(\tau,\omega,x_c,h(\omega,x_c) - h_u(\vartheta_\tau\omega,\varphi_c(\tau,\omega,x_c,h(\omega,x_c))) \|^u_{\vartheta_\tau\omega}\}\, d\tau.$$

Taking into account where K comes from we see that an estimate like (3.1b) holds true for K as well. This implies the integrability of $K(\vartheta_t\omega)$ over any finite time interval, P-a.s. Thus a stochastic version of the Gronwall lemma (see Bunke [5], Lemma 1.3.) may be applied and yields that

$$\max\{ \| \varphi_s(t,\omega,x_c,h(\omega,x_c)) - h_s(\vartheta_t\omega,\varphi_c(t,\omega,x_c,h(\omega,x_c))) \|^s_{\vartheta_t\omega},$$

$$\| \varphi_u(t,\omega,x_c,h(\omega,x_c)) - h_u(\vartheta_t\omega,\varphi_c(t,\omega,x_c,h(\omega,x_c))) \|^u_{\vartheta_t\omega}\} = 0.$$

This finishes the proof in case $t \geq 0$. If $t \leq 0$ then one only has to use the corresponding estimate in Section 2.1. for $t \leq 0$.

Thus we have shown that the stochastic center manifold is invariant for all $t \in \mathbb{R}$. As explained earlier, the theorem is now entirely proved. □

5 Approximation of the stochastic center manifold

After having seen in the last section how to combine techniques stemming from the deterministic situation with stochastic tools we could now continue and derive various properties of stochastic center manifolds. However, we will not do this here because there are no new ideas involved. One may for example adapt Section 2.4. in Carr [6] with such a procedure and learn that stochastic center manifolds are attracting and that it is sufficient to examine the system restricted to the center manifold in order to deduce the stability properties of the entire system.

The stochastic center manifold being invariant it is sufficient to study the equation

$$\dot{x}_c = Q_c(\vartheta_t\omega)\, x_c \; + \; F_c(\vartheta_t\omega,x_c,h_s(\vartheta_t\omega,x_c),h_u(\vartheta_t\omega,x_c)). \tag{5.1}$$

Typically (5.1) will be one or two dimensional even if the original system was of a very high dimension. Thus stochastic center manifolds allow a considerable reduction of the equation to be analyzed.

One may also show in the case of a stable system that the error which is made by replacing the original system by its restriction to the center manifold decays exponentially. For statements (and proofs on the level of random dynamical systems) of this and the results mentioned above see Boxler [3].

Next one might try to carry over Carr's proof of existence of an approximation of the center manifold. It turns out, however, that this is not possible because one would have to differentiate $\vartheta_t\omega$ with respect to t which can only be done if we assume differentiable noise. Therefore it is the aim of the present section to make use of the approximation result derived in Boxler [3], Th. 8.1., to show how stochastic center manifolds constructed on the level of vector fields can be approximated.

5.1 Existence of an approximation

First we note that, despite our remark above, one can give a meaning to the derivative of

$h(\vartheta_t\omega,\varphi_c)$ with respect to t, where φ_c denotes $\varphi_c(t,\omega,x_c,h(\omega,x_c))$. This is simply due to the fact that the invariance of the stochastic center manifold implies that

$$h_{s,u}(\vartheta_t\omega,\varphi_c(t,\omega,x_c,h(\omega,x_c))) = \varphi_{s,u}(t,\omega,x_c,h(\omega,x_c)).$$

The right-hand side being the solution of a differential equation it can be differentiated for Lebesgue-almost all t. Thus we may write:

$$\dot{h}_{s,u}(\vartheta_t\omega,\varphi_c) = Q_{s,u}(\vartheta_t\omega)\ \varphi_{s,u}(t,\omega,x_c,h(\omega,x_c)) + F_{s,u}(\vartheta_t\omega,\varphi_c,h(\vartheta_t\omega,\varphi_c))$$

$$= Q_{s,u}(\vartheta_t\omega)\ h_{s,u}(\vartheta_t\omega,\varphi_c) + F_{s,u}(\vartheta_t\omega,\varphi_c,h(\vartheta_t\omega,\varphi_c))$$

On the other hand we obtain as a consequence of the chain rule and of (5.1):

$$\dot{h}_{s,u}(\vartheta_t\omega,\varphi_c) = \frac{\partial h_{s,u}(\vartheta_t\omega,\cdot)}{\partial\varphi_c}[\,Q_c(\vartheta_t\omega)\ \varphi_c + F_c(\vartheta_t\omega,\varphi_c,h(\vartheta_t\omega,\varphi_c))].$$

Thus, in order to obtain the stochastic center manifold, one would have to solve the equation

$$h'_{s,u}(\omega,u)[Q_c(\omega)u + F_c(\omega,u,h(\omega,u))] - Q_{s,u}(\omega)\ h_{s,u}(\omega,u) - F_{s,u}(\omega,u,h(\omega,u)) = 0, \quad (5.2)$$

where the prime denotes differentiation w. r. t. the second variable.

In general it will be impossible to solve (5.2) (as it is in the deterministic case) but one may use it as a starting point for an approximation result. For this we define an operator U on A_L by

$$(U\rho)_s(\omega,u) := \rho'(\omega,u)[Q_c(\omega)u + F_c(\omega,u,\rho(\omega,u))] - Q_s(\omega)\ \rho_s(\omega,u) - F_s(\omega,u,\rho(\omega,u))$$

and analogously for the unstable part. Obviously, $Uh(\omega,x) = 0$.

Denote by $g(\omega,y) = O(\|y\|)$ the fact that $\|g(\omega,y)\| \le K(\omega)\|y\|$ a.s. for a suitable positive valued random variable K and y sufficiently small. Then we can show:

Theorem 5.1. *For a given* $\rho \in A_L$ *assume that* $(U\rho)(\omega,y) = O(\|y\|^q)$ *for some* $q > 1$, *almost all* $\omega \in \Omega$ *and all* $y \in E^c(\omega)$ *sufficiently small. Then we obtain:*

$$\|h(\omega,y) - \rho(\omega,y)\| = O(\|y\|^q).$$

Remark. At first glance this result looks the same as the one given in Boxler [3], Th. 8.1. It is different and more useful for applications, however, because, in contrast to the latter, it will enable us to derive approximations in terms of random Taylor series on the level of vector fields. We will illustrate this by an example.

Proof. For the sake of notational simplicity let us assume that all the Lyapunov exponents are ≤ 0, i.e. we will not have to deal with an unstable part.

According to Theorem 8.1. in Boxler [3] the assertion will follow once we have shown that $(V\rho)(\omega,y) = O(\|y\|^q)$ for $q > 1$ where $(V\rho)(\vartheta_1\omega,y) := \varphi_s(1,\omega,x_c,\rho(\omega,x_c)) - \rho(\vartheta_1\omega,y)$ and $y = \varphi_c(1,\omega,x_c,\rho(\omega,x_c))$.

As explained above for h we obtain:

$$\dot{\rho}(\omega,u) = \rho'(\omega,u)[Q_c(\omega)u + F_c(\omega,u,\rho(\omega,u))] = (U\rho)(\omega,u) + Q_s(\omega)\rho_s(\omega,u) + F_s(\omega,u,\rho(\omega,u)).$$

Interpreting the time derivative of $\rho(\vartheta_\tau\omega)$ as explained above for h and denoting $\rho(\omega,x_c)$ by ρ we may write:

$$\| \varphi_s(1,\omega,x_c,\rho(\omega,x_c)) - \rho(\vartheta_1\omega,y) \|^s_{\vartheta_1\omega} =$$

$$\| \int_0^1 \dot{\varphi}_s(\tau,\omega,x_c,\rho(\omega,x_c))\, d\tau - \int_0^1 \dot{\rho}(\vartheta_\tau\omega,\varphi_c(\tau,\omega,x_c,\rho(\omega,x_c)))\, d\tau \|^s_{\vartheta_1\omega} =$$

$$\| \int_0^1 [Q_s(\vartheta_\tau\omega)\varphi_s(\tau,\omega,x_c,\rho) + F_s(\vartheta_\tau\omega,\varphi_c(\tau,\omega,x_c,\rho),\varphi_s(\tau,\omega,x_c,\rho))]\, d\tau -$$

$$\Big[\int_0^1 [Q_s(\vartheta_\tau\omega)\rho(\vartheta_\tau\omega,\varphi_c(\tau,\omega,x_c,\rho)) + F_s(\vartheta_\tau\omega,\varphi_c(\tau,\omega,x_c,\rho),\rho(\vartheta_\tau\omega,\varphi_c(\tau,\omega,x_c,\rho))) +$$

$$(U\rho)(\vartheta_\tau\omega,\varphi_c(\tau,\omega,x_c,\rho))]\, d\tau \Big] \|^s_{\vartheta_1\omega}$$

$$\leq \int_0^1 (\| Q_s(\vartheta_\tau\omega) \|^s_{\vartheta_\tau\omega} + \eta) \| \varphi_s(\tau,\omega,x_c,\rho) - \rho(\vartheta_\tau\omega,\varphi_c(\tau,\omega,x_c,\rho)) \|^s_{\vartheta_\tau\omega}\, d\tau +$$

$$\int_0^1 \| (U\rho)(\vartheta_\tau\omega,\varphi_c(\tau,\omega,x_c,\rho))\, d\tau \|^s_{\vartheta_\tau\omega},$$

where we have made use of equations (2.4b) and (4.2). The latter was possible because we are only interested in small initial values whence we can replace F by f^δ.

Combining the assumption about $U\rho$ with inequality (3.1a) ($\gamma > 0$ fixed), which relates the Euclidean and the random norm, we easily check that

$$\int_0^1 \| (U\rho)(\vartheta_\tau\omega,\varphi_c(\tau,\omega,x_c,\rho))\, d\tau \|^s_{\vartheta_\tau\omega} \leq$$

$$\sup_{0\leq\tau\leq 1} [C(\vartheta_\tau\omega,\gamma)K(\vartheta_\tau\omega)] \| \varphi_c(1,\omega,x_c,\rho) \|^q \int_0^1 [\| \varphi_c(\tau,\omega,x_c,\rho) \| \, \| \varphi_c^{-1}(1,\omega,x_c,\rho) \|^q]\, d\tau$$

$$\leq \kappa(\omega) \| \varphi_c(1,\omega,x_c,\rho) \|^q$$

for an appropriate (positive valued) random variable κ. Here we have estimated the integral by the supremum over $\tau \in [0,1]$ and initial values belonging to a small neighborhood of 0.

After having applied (3.1a) again we deduce by means of the stochastic Gronwall lemma (see e. g. Bunke [5], Lemma 1.3.):

$$\| (V\rho)(\vartheta_1\omega,\varphi_c(1,\omega,x_c,\rho) \| \leq \kappa(\omega)C(\vartheta_1\omega,\gamma)\, e^{\int_0^1 (|Q_s(\vartheta_\tau\omega)|^s_{\vartheta_\tau\omega}+\eta)d\tau} \| \varphi_c(1,\omega,x_c,\rho) \|^q.$$

Finally we take into account that the exponential is nothing but $\| \Phi_s(1,\omega) \|^s_{\vartheta_1\omega}$ and that we may thus use the estimates derived earlier. This leads to:

$$\| (V\rho)(\vartheta_1\omega,\varphi_c(1,\omega,x_c,\rho) \| \leq \tilde{\kappa}(\omega)\ \| \varphi_c(1,\omega,x_c,\rho) \|^q$$

with $\tilde{\kappa}(\omega) = \kappa(\omega)\ C(\vartheta_1\omega,\gamma)\ e^{(\lambda^s+\varepsilon+\eta)}$.

As explained earlier this yields the assertion. □

5.2 How to approximate stochastic center manifolds in concrete examples

In this section we are going to summarize the procedure that leads to an approximation of the stochastic center manifold. This will then be illustrated by an example.

In a first step we linearize the vector field at the zero solution. As explained earlier we have to require that all the Lyapunov exponents are different (or that the linear part is in block diagonal form already).

Next we diagonalize the linear part as outlined in Section 2.2. This leads to a system of the form (2.4).

Since the stochastic center manifold is to be tangent to $E^c(\omega)$ we use the ansatz

$$\rho(\omega,u) = a_2(\omega)u^2 + a_3(\omega)u^3 + \ldots \tag{5.3}$$

which will be inserted into the equation defining the operator U.

Then we equate cefficients in the equation

$$\rho'(\omega,u)[Q_c(\omega)u + F_c(\omega,u,\rho(\omega,u))] - Q_{s,u}(\omega)\ \rho_{s,u}(\omega,u) - F_{s,u}(\omega,u,\rho(\omega,u)) = 0 \tag{5.4}$$

until $(U\rho)(\omega,u)$ is of the desired order. Theorem 5.1. will then guarantee that we have in fact found an approximation for h and thus for the stochastic center manifold.

Example

We consider the following system which has already been discussed in Boxler [3], p. 542.

$$\dot{x}_c = a(\vartheta_t\omega)x_c,$$

$$\dot{x}_s = b(\vartheta_t\omega)x_s + c(\vartheta_t\omega)x_c^2,$$

where a, b and c are integrable functions with $Ea = 0$, $Eb < 0$. We will see that the expression obtained for h on the vector field level is much handier and thus more useful for approximations than the one derived in [3].

It is easy to see that $\lambda^c = Ea$, $\lambda^s = Eb$. Let $\rho(\omega,u) := a_2(\omega)u^2 + a_3(\omega)u^3$. Then (5.4) implies:

$$(2a_2(\omega)u + 3a_3(\omega)u^2)\, a(\omega)u - b(\omega)a_2(\omega)u^2 - b(\omega)a_3(\omega)u^3 - c(\omega)u^2 = 0.$$

Thus, equating coefficients yields:

$$\begin{cases} 2a(\omega)a_2(\omega) - b(\omega)a_2(\omega) - c(\omega) = 0, \\ 3a(\omega)a_3(\omega) - b(\omega)a_3(\omega) = 0, \end{cases}$$

which implies that then one obtains:

$$a_2(\omega) = \frac{c(\omega)}{2a(\omega) - b(\omega)},\ a_3(\omega) = 0$$

if $b(\omega) \neq 2a(\omega)$ and $\neq 3a(\omega)$ P-a.s. Otherwise we will have to take into account higher order terms. Thus an approximation of the stochastic center manifold is for P-almost all $\omega \in \Omega$ given by

$$h(\omega,u) = \frac{c(\omega)}{2a(\omega) - b(\omega)}u^2 + O(\|u\|^4),\quad u \in E^c(\omega) = \mathbb{R} \times \{0\}.$$

References

[1] *Arnold, L., Crauel, H.:* Iterated function systems and multiplicative ergodic theory. M. Pinsky, V. Wihstutz (eds.): Stochastic flows. Birkhäuser (in press).

[2] *Arnold, L., Kliemann, W., Oeljeklaus, E.:* Lyapunov exponents of linear stochastic systems. Proceedings of a workshop Bremen 1984. Lecture Notes in Mathematics vol. 1186. Springer Berlin-Heidelberg-New York 1986.

[3] *Boxler, P.:* A stochastic version of center manifold theory. Probab. Th. Rel. Fields 83 (1989), 509 - 545.

[4] *Boxler, P.:* Stochastische Zentrumsmannigfaltigkeiten. Ph. D. thesis, Institut für Dynamische Systeme, Universität Bremen 1988.

[5] *Bunke, H.:* Gewöhnliche Differentialgleichungen mit zufälligen Parametern. Akademie-Verlag, Berlin 1972.

[6] *Carr, J.:* Applications of Centre Manifold Theory. Springer, Berlin-Heidelberg-New York 1981.

[7] *Crauel, H.:* Lyapunov exponents and invariant measures of stochastic systems on manifolds. Proceedings of a workshop Bremen 1984. Lecture Notes in Mathematics vol. 1186. Springer Berlin-Heidelberg-New York 1986.

[8] *Dahlke, S.:* Invariante Mannigfaltigkeiten für Produkte zufälliger Diffeomorphismen. Ph. D. thesis, Institut für Dynamische Systeme, Universität Bremen 1989.

[9] *Has'minskii, R. Z.:* Stochastic Stability of Differential Equations. Sijthoff and Noordhoff, Alphen 1980.

[10] *Ikeda, N., Watanabe, S.:* Stochastic differential equations and diffusion processes. North-Holland, Amsterdam 1981.

[11] *Iooss, G.:* Bifurcation of Maps and Applications. North-Holland, Amsterdam 1979.

[12] *Vanderbauwhede , A.:* Center manifolds, normal forms and elementary bifurcations. In: U. Kirchgraber, H. O. Walther (eds.): Dynamics Reported, Vol. 2, Teubner and Wiley 1989, 89 - 169.

Additive noise turns a hyperbolic fixed point into a stationary solution

Ludwig Arnold and Petra Boxler
Institut für Dynamische Systeme
Universität Bremen
2800 Bremen 33, Germany

Abstract

Suppose (*) $\dot{x} = A(\vartheta_t\omega)x$ is hyperbolic, i.e. all of its Lyapunov exponents are different from zero. Then $\dot{x} = A(\vartheta_t\omega)x + f(\vartheta_t\omega, x) + b(\vartheta_t\omega)$ with $f(\omega, \cdot)$ locally Lipschitz and $f(\omega, 0) = 0$ has a (unique) stationary solution in a neighborhood of $x = 0$ provided f and b are 'small'. 'Smallness' is being described in terms of a random norm measuring the non-uniformity of the hyperbolicity of (*).

1 Introduction. The linear case

If a differential equation $\dot{x} = f(t, x)$ has a bounded solution $x_0(t)$ whose variational equation $\dot{v} = f_x(t, x_0(t))v$ has an exponential dichotomy (e.g. a hyperbolic fixed point or a periodic solution with non-zero Floquet exponents), and if $g(t, x)$ is 'small enough' then $\dot{x} = f(t, x) + g(t, x)$ has a unique bounded solution close to $x_0(t)$. Recent statements of this fact can be found in Meyer and Sell ([7], Theorem 0), and in Scheurle ([9], Lemma 2.4), or for discrete time in Palmer ([8], Proposition 2.8), based on classical formulations in Coppel ([4],p.137) and Hale [5].

The aim of this paper is to present a stochastic version of the above fact, with the result summarized by the title of the paper.

This generalizes earlier results by Bunke ([3], Satz 5.16) and others for a stable fixed point and uses a recent result by Arnold and Crauel [1] for affine differential or difference equations.

Let $T = \mathbb{R}$ (continuous time) or $\mathbb{Z}$ (discrete time) and let $\vartheta_{\cdot}(\cdot) : T \times \Omega \to \Omega$ be measurable such that $(\vartheta_t(\cdot))_{t\in T}$ is an ergodic flow (i.e. $\vartheta_t \circ \vartheta_s = \vartheta_{t+s}$ for all $t, s \in T$) of measurable bijections of a probability space $(\Omega, \mathcal{F}, P)$ leaving P invariant. The dynamical system $(\Omega, \mathcal{F}, P, (\vartheta_t)_{t\in T})$ serves as our model for stationary noise entering into differential or difference equations.

Assume for the measurable $A : \Omega \to \mathbb{R}^{d\times d}$ the integrability condition $A \in L^1(\Omega, P)$. Then for the linear random differential equation in $\mathbb{R}^d$

$$\dot{x} = A(\vartheta_t\omega)x \tag{1.1}$$

we have Oseledec's *Multiplicative Ergodic Theorem* [1]: Let $\phi(t,\omega)$ be the fundamental matrix of (1.1). Then there is an invariant set $\Omega_0 \in \mathcal{F}$ of full measure, such that for all $\omega \in \Omega_0$ the following holds: There are (nonrandom) numbers $\lambda_1 > \ldots > \lambda_r$ (Lyapunov exponents of (1.1)) and random subspaces $E_1(\omega), \ldots, E_r(\omega)$ of $\mathbb{R}^d$ with nonrandom dimensions $d_1 + \ldots + d_r = d$ splitting $\mathbb{R}^d$,

$$\mathbb{R}^d = E_1(\omega) \oplus \ldots \oplus E_r(\omega),$$

such that

$$\begin{aligned}
&(i) && \phi(t,\omega)E_i(\omega) = E_i(\vartheta_t\omega), \text{ all } i = 1, \ldots, r,\\
&(ii) && \lim_{t\to\pm\infty} \frac{1}{t}\log|\phi(t,\omega)x| = \lambda_i \text{ iff } x \in E_i(\omega),\\
&(iii) && \text{for all } x \in \mathbb{R}^d \lim_{t\to\infty} \frac{1}{t}\log|\phi(t,\omega)x| = \lambda_{i(\omega,x)},\\
& && i(\omega,x) = \min\{i : x_i \neq 0, x = x_1 \oplus \ldots \oplus x_r\}.
\end{aligned}$$

The system (1.1) is said to be *hyperbolic* if all $\lambda_i \neq 0$. In this case let

$$E^s(\omega) := \bigoplus_{\lambda_i<0} E_i(\omega), \quad E^u(\omega) := \bigoplus_{\lambda_i>0} E_i(\omega)$$

be the stable and unstable subspaces, resp., with the corresponding projections

$$\pi^{s,u}(\omega) : \mathbb{R}^d \to E^{s,u}(\omega), \quad x = x^s \oplus x^u \mapsto x^{s,u}.$$

The main result of Arnold and Crauel [1] is

Theorem 1.1 *Let $A : \Omega \to \mathbb{R}^{d\times d}$, $b : \Omega \to \mathbb{R}^d$ satisfy $A, b \to L^1(\Omega, P)$. Assume (1.1) is hyperbolic. Then the affine random differential equation*

$$\dot{x} = A(\vartheta_t\omega)x + b(\vartheta_t\omega) \tag{1.2}$$

has a uniqe stationary solution whose initial value is given by

$$\kappa(\omega) = \pi^s(\omega)\int_{-\infty}^0 \phi(t,\omega)^{-1}b(\vartheta_t\omega)dt \bigoplus -\pi^u(\omega)\int_0^\infty \phi(t,\omega)^{-1}b(\vartheta_t\omega)dt. \tag{1.3}$$

There are discrete time and white noise versions of the above statement, cf. [1]. Theorem 1.1 says that additive noise turns a hyperbolic fixed point of a *linear* system (1.1) into a (unique) stationary solution.

The proof of Theorem 1.1 uses a random norm $|\cdot|_\omega$ in $\mathbb{R}^d$ equivalent to $|\cdot|$ (cf. Boxler [2], Lemma 4.2): Put

$$\lambda^s := \max_{\lambda_i<0} \lambda_i, \ \lambda^u := \min_{\lambda_i>0} \lambda_i,$$

fix an $\epsilon > 0$ such that $\lambda^s + \epsilon < 0$, $\lambda^u - \epsilon > 0$ and put

$$|x|_\omega := \max(|\pi^s(\omega)x|_\omega, |\pi^u(\omega)x|_\omega),$$

where

$$|\pi^s(\omega)x|_\omega := \int_0^\infty e^{-t(\lambda^s+\epsilon)}|\phi(t,\omega)\pi^s(\omega)x|dt,$$
$$|\pi^u(\omega)x|_\omega := \int_0^\infty e^{t(\lambda^u-\epsilon)}|\phi(-t,\omega)\pi^u(\omega)x|dt.$$

The crucial advantage of this norm is that for alle $x \in \mathbb{R}^d$

$$|\phi(t,\omega)\pi^s(\omega)x|_{\vartheta_t\omega} \leq e^{t(\lambda^s+\epsilon)}|\pi^s(\omega)x|_\omega \text{ for } t \geq 0, \tag{1.4}$$

$$|\phi(t,\omega)\pi^u(\omega)x|_{\vartheta_t\omega} \leq e^{t(\lambda^u-\epsilon)}|\pi^u(\omega)x|_\omega \text{ for } t \leq 0, \tag{1.5}$$

i.e. on the stable (unstable) subspace $E^s(\omega)$ $(E^u(\omega))$ the mapping $\phi(t,\omega)$ is a contraction for all $t \geq 0$ $(t \leq 0)$. In other words: The new norm $|\cdot|_\omega$ compensates for the nonuniformity of the hyperbolic situation.

Note that (1.3) certainly makes sense for those b's for which

$$\beta := \text{ess sup } |b(\omega)|_\omega < \infty$$

2 The nonlinear case

Now consider

$$\dot{x} = A(\vartheta_t\omega)x + f(\vartheta_t\omega, x) + b(\vartheta_t\omega) \tag{2.1}$$

with $A, b \in L^1(\Omega, P)$ and $f(\omega, \cdot)$ locally Lipschitz with Lipschitz constant $C \in L^1(\Omega, P)$ and with $f(\omega, 0) = 0$. Then a local solution of (2.1) exists and is unique, see e.g. Hasminskii [6], p.10.

Theorem 2.1 *Assume*

(i) $x = 0$ *is a huperbolic fixed point for* $\dot{x} = A(\vartheta_t\omega)x$.

(ii) $\beta := \text{ess sup } |b(\omega)|_\omega < \infty$.

(iii) For given $R > 0$

$$|f(\omega,x) - f(\omega,y)|_\omega \leq C(\omega)|x-y|_\omega \text{ for } |x|_\omega, |y|_\omega \leq R,$$

with

$$\gamma := \text{ess sup } C(\omega) < \Lambda := \min(\lambda^u - \epsilon, -(\lambda^s + \epsilon)).$$

(iv) $\beta \leq (\Lambda - \gamma)R$.

Then (2.1) has a stationary solution whose initial value $\kappa(\omega)$ *satisfies ess sup* $|\kappa(\omega)|_\omega \leq R$ *(and is unique with this property).*

Proof. (i) The assertion is true in the affine case $f = 0$. We want to extend it to the case $f \neq 0$ by a fixed point argument for which we now define the appropriate space. Let

$$\mathcal{L}^\infty := \{x : \Omega \to \mathbb{R}^d \text{ measurable with } \|x\| := \sup_{t\in\mathbb{R}} |x(\vartheta_t\omega)|_{\vartheta_t\omega} < \infty \text{ P-a.s.}\}.$$

Note that[1] $\sup_{t\in\mathbb{R}} |x(\vartheta_t\omega)|_{\vartheta_t\omega}$ is an invariant random variable and thus, by the ergodicity of ϑ_t, P-a.s. equal to a constant, which is denoted by $\|x\|$. Also,

$$\text{ess sup}\, |x(\omega)|_\omega = \|x\|.$$

$\mathcal{L}^\infty$ is a Banach space. The subset

$$D_R := \{x \in \mathcal{L}^\infty : \|x\| \leq R\}$$

is closed.

(ii) We now define an operator $T : D_R \to \mathcal{L}^\infty$ by

$$Ty(\omega) = \pi^s(\omega) \int_{-\infty}^0 \phi(t,\omega)^{-1}\tilde{b}(\vartheta_t\omega)dt \bigoplus -\pi^u(\omega) \int_0^\infty \phi(t,\omega)^{-1}\tilde{b}(\vartheta_t\omega)dt, \tag{2.2}$$

where

$$\tilde{b}(\omega) = f(\omega, y(\omega)) + b(\omega).$$

Comparing (2.2) with (1.3) shows that Ty is nothing but the initial value of the unique stationary solution of the affine equation

$$\dot{x} = A(\vartheta_t\omega)x + \tilde{b}(\vartheta_t\omega).$$

This solution exists and (2.2) makes sense since

$$\begin{aligned} |\tilde{b}(\omega)|_\omega &\leq |f(\omega, y(\omega)) - f(\omega, 0)|_\omega + |b(\omega)|_\omega \\ &\leq \gamma R + \beta \end{aligned}$$

by assumptions (ii) and (iii).
We now prove that T maps D_R to D_R and is a contraction. The unique fixed point of T will then be the initial value of the unique stationary solution of (2.1) in D_R.

(iii) T maps D_R to D_R: For the stable part of Ty, using (1.4), we obtain

$$\begin{aligned} |\pi^s(\omega)Ty(\omega)|_\omega &\leq |\pi^s(\omega)|_\omega \int_{-\infty}^0 e^{-t(\lambda^s+\epsilon)}(\gamma R + \beta)dt \\ &\leq \frac{\gamma R + \beta}{-(\lambda^s + \epsilon)}. \end{aligned}$$

Similarly for the unstable part

$$|\pi^u(\omega)Ty(\omega)|_\omega \leq \frac{\gamma R + \beta}{\lambda^u - \epsilon},$$

[1]All processes are measurable and can w.l.o.g. be assumed to be separable.

thus

$$|Ty(\omega)|_\omega \le \frac{\gamma R + \beta}{\Lambda} \le R$$

by condition (iv).

(iv) T is contracting on D_R:

Using the Lipschitz condition and estimating the stable and unstable parts of $Ty - T\overline{y}$ separately gives

$$|Ty(\omega) - T\overline{y}(\omega)|_\omega \le \frac{\gamma}{\Lambda}\|y - \overline{y}\|$$

and thus $\|Ty - T\overline{y}\| \le \frac{\gamma}{\Lambda}\|y - \overline{y}\|$ with $\frac{\gamma}{\Lambda} < 1$ by assumption (iii). □

Remarks. (i) There is obviously a discrete time version of Theorem 2.1: Let $A : \Omega \to Gl(d, \mathbb{R})$, $b : \Omega \to \mathbb{R}^d$ with $log^+\|A\| + \log^+\|A^{-1}\| + \log^+|b| \in L^1(\Omega, P)$, and let $x_{n+1} = A(\vartheta^n\omega)x_n$ be hyperbolic. Then

$$x_{n+1} = A(\vartheta^n\omega)x_n + b(\vartheta^n\omega), \quad n \in \mathbb{Z},$$

has a unique stationary solution, by the discrete time version of Theorem 1.1. For

$$x_{n+1} = A(\vartheta^n\omega)x_n + f(\vartheta^n\omega, x_n) + b(\vartheta^n\omega), \quad n \in \mathbb{Z},$$

Theorem 2.1 holds unchanged with

$$\Lambda := \min\left(\frac{e^{\lambda^u - \epsilon} - 1}{e^{\lambda^u - \epsilon}}, \frac{1 - e^{\lambda^s + \epsilon}}{e^{\lambda^s + \epsilon}}\right).$$

The random norms $|\cdot|_\omega$ are now defined with sums instead of integrals.

(ii) Although a continuous time white noise version of Theorem 1.1 exists (see [1]), the analogue of Theorem 2.1 (even for an f globally Lipschitz) cannot be obtained by the above method.

References

[1] Arnold, L. and H. Crauel: Iterated function systems and multiplicative ergodic theory. M. Pinsky, V. Wihstutz (eds): Stochastic flows. Birkhäuser 1991.

[2] Boxler, P.: A stochastic version of center manifold theory. Probab. Th. Rel. Fields 83, 509-545 (1989).

[3] Bunke, H.: Gewöhnliche Differentialgleichungen mit zufälligen Parametern. Akademie-Verlag, Berlin 1972.

[4] Coppel, W.A.: Stability and asymptotic behaviour of differential equations. Heath, Boston 1965.

[5] Hale, J.K.: Ordinary differential equations. Wiley, New York 1969.

[6] Hasminskii, R.Z.: Stochastic stability of differential equations. Sijthoff and Noordhoff, Alphen 1980.

[7] Meyer, K.R and G.R. Sell: Melnikov transforms, Bernoulli bundles, and almost periodic perturbations. Trans. Amer. Math. Soc. 314, 63-105 (1989).

[8] Palmer, K.: Exponential dichotomies, the shadowing lemma and transversal homoclinic points. Dynamics Reported, Vol. 1. Wiley, New York 1988, 265-306.

[9] Scheurle, J.: Chaotic solutions of systems with almost periodic forcing. J. Applied Math. and Phys. (ZAMP) 37, 12-26 (1986).

LYAPUNOV FUNCTIONS
AND ALMOST SURE EXPONENTIAL STABILITY

Xuerong Mao
Mathematics Institute, University of Warwick, Coventry CV4 7AL, England

1. INTRODUCTION AND NOTATIONS

There exists an extensive literature in the exponential stability of stochastic systems, in particular, we mention Arnold [1], Arnold and Kliemann [2], Arnold, Oeljeklaus and Pardoux [3], Carverhill [4], Chappell [5], Crauel [7], Curtain [7], Elworthy [8], Has'minskii [9] among others. However there were few papers using the Lyapunov function for studying the almost sure exponential stability of stochastic systems. Recently we showed in [12] the relation between Lyapunov function and Lyapunov exponent for a general stochastic differential equation based on a random field in the Kunita sense. This is the continuation of the previous paper.

Let $(\Omega, \mathcal{F}, \{\mathcal{F}_t\}_{t\geq 0}, P)$ be a complete probability space with a right continuous filtration $\{\mathcal{F}_t\}$ containing all P-null sets of $\mathcal{F}$. Let $F(x, t) = (F^1(x, t), \dots, F^n(x, t))^T$, $(x, t) \in R^n \times R_+$, be a continuous C-semimartingale with spatial parameters, i.e., $F(x, t)$ is a continuous semimartingale for any $x \in R^n$ and is continuous in x for any t almost surely. Let $F^i(x, t) = M^i(x, t) + B^i(x, t)$ be the decomposition such that $M^i(x, t)$ is a continuous local martingale and $B^i(x, t)$ a continuous process of bounded variation. Set

$$A^{ij}(x, y, t) = \langle M^i(x, t), M^j(y, t)\rangle, \quad 1 \leq i, j \leq n.$$

Then there exists a continuous strictly increasing process A_t with $A_0 = 0$ such that all $A^{ij}(x, y, t)$ and $B^i(x, t)$ are absolutely continuous with respect to A_t almost surely for any $x, y \in R^n$. Therefore there exist predictable processes $a^{ij}(x, y, t)$ and $b^i(x, t)$ with parameters x, y such that

$$A^{ij}(x, y, t) = \int_0^t a^{ij}(x, y, s)\, dA_s$$

and

$$B^i(x, t) = \int_0^t b^i(x, s)\, dA_s.$$

Set $a(x, y, t) = (a^{ij}(x, y, t))_{n\times n}$ and $b(x, t) = (b^1(x, t), \dots, b^n(x, t))^T$. The triple $(a(x, y, t), b(x, t), A_t)$ is called the characteristic of $F(x, t)$ (cf. Kunita [10]).

We shall continue to investigate the almost sure exponential stability for a stochastic differential equation

$$\varphi_t = x_0 + \int_0^t F(\varphi_s, ds) \tag{1.1}$$

via the Lyapunov function, where $x_0 \in R^n$.

Let $C^{2,1}(R^n \times R_+)$ denote the family of all functions $V(x, t) : R^n \times R_+ \to R$ with continuous second partial derivatives in x and first partial derivative in t. If $V \in C^{2,1}(R^n \times R_+)$, we define an operator L by

$$LV(x, A_t) := \frac{\partial}{\partial t} V(x, A_t) + \sum_{i=1}^{n} \frac{\partial}{\partial x_i} V(x, A_t)\, b^i(x, t)$$

$$+ \frac{1}{2} \sum_{i,j=1}^{n} \frac{\partial^2}{\partial x_i \partial x_j} V(x, A_t)\, a^{ij}(x, x, t). \tag{1.2}$$

2. MAIN RESULTS

Theorem 2.1. Assume there exist a function $V \in C^{2,1}(R^n \times R_+)$, a nonnegative predictable process $\psi(t)$ on $t \geq 0$, and constants $p, \lambda, \eta, d, \sigma > 0$, $0 \leq \alpha < 1$ such that

(1) $e^{\lambda t} |x|^p \leq V(x, t)$ on $(x, t) \in R^n \times R_+$;

(2) $LV(x, A_t) + \eta(1+t)^{-d} \sum_{i,j=1}^{n} \frac{\partial}{\partial x_i} V(x, A_t) \frac{\partial}{\partial x_j} V(x, A_t)\, a^{ij}(x, x, t)$

$$\leq \psi(t) \left(1 + \left[V(x, A_t)\right]^{\alpha}\right) \quad \text{a.s. on } (x, t) \in R^n \times R_+;$$

(3) $\liminf_{t \to \infty} \frac{A_t}{t} \geq \sigma$ a.s.;

(4) $\limsup_{t \to \infty} \frac{1}{t} \log \int_0^t \psi(s)\, dA_s \leq 0$ a.s.

Then the solution of equation (1.1) satisfies

$$\limsup_{t \to \infty} \frac{1}{t} \log |\varphi_t| \leq -\lambda\, \sigma / p \quad \text{a.s.} \tag{2.1}$$

In order to prove the theorem we need to prepare an inequality first.

Lemma 2.2. Let $T > 0$ and $0 \leq \alpha < 1$. Let $\mu(t)$ be a right continuous nondecreasing function on $[0, T]$ with $\mu(0) = 0$. Let $u(t)$ be a Borel measurable, left limit, bounded and nonnegative function on $[0, T]$. Let $C \geq 0$. Assume

$$u(t) \leq C + \int_0^t \left[u(s-)\right]^{\alpha} d\mu(s) \quad \text{on } 0 \leq t \leq T.$$

Then

$$u(t) \le \left(C^{1-\alpha} + (1-\alpha)\mu(t)\right)^{1/(1-\alpha)} \quad \text{on } 0 \le t \le T. \tag{2.2}$$

Proof. By Theorem 3.1 of Mao [13] we see that

$$u(t) \le G^{-1}(G(C) + \mu(t)) \quad \text{on } 0 \le t \le T, \tag{2.3}$$

where

$$G(r) = \int_0^r \frac{ds}{s^\alpha} = \frac{1}{1-\alpha} r^{1-\alpha} \quad \text{on } r \ge 0,$$

and G^{-1} is the inverse function of G that has the form

$$G^{-1}(r) = \left[(1-\alpha)\, r \right]^{1/(1-\alpha)} \quad \text{on } r \ge 0.$$

Substituting these into (2.3) we get the required inequality (2.2).

Proof of Theorem 2.1. Itô's formula gives

$$V(\varphi_t, A_t) = V(x, 0) + \int_0^t LV(\varphi_s, A_s)\, dA_s + \int_0^t \sum_{i=1}^n \frac{\partial}{\partial x_i} V(\varphi_s, A_s)\, M^i(\varphi_s, ds). \tag{2.4}$$

Thanks to the exponential martingale inequality, we have

$$P\Big[\omega : \sup_{0 \le t \le \tau} \Big\{ \int_0^t \sum_{i=1}^n \frac{\partial}{\partial x_i} V(\varphi_s, A_s)\, M^i(\varphi_s, ds)$$

$$- \frac{\gamma}{2} \int_0^t \sum_{i,j=1}^n \frac{\partial}{\partial x_i} V(\varphi_s, A_s) \frac{\partial}{\partial x_j} V(\varphi_s, A_s)\, a^{ij}(\varphi_s, \varphi_s, s)\, dA_s \Big\} > \delta \Big] \le e^{-\gamma\delta} \tag{2.5}$$

for any positive constants γ, δ and τ. In particular, we take

$$\gamma = 2\eta(1+2^k)^{-d}, \quad \delta = \frac{1}{\eta}(1+2^k)^d \log k, \quad \tau = 2^k$$

for each $k = 1, 2, \ldots$. Applying the Borel-Cantelli lemma we deduce that there exists an integer $k_0(\omega)$ for almost sure $\omega \in \Omega$ such that

$$\int_0^t \sum_{i=1}^n \frac{\partial}{\partial x_i} V(\varphi_s, A_s)\, M^i(\varphi_s, ds) \le \frac{1}{\eta}(1+2^k)^d \log k \; +$$

$$+ \eta(1+2^k)^{-d} \int_0^t \sum_{i,j=1}^n \frac{\partial}{\partial x_i} V(\varphi_s, A_s) \frac{\partial}{\partial x_j} V(\varphi_s, A_s) a^{ij}(\varphi_s, \varphi_s, s)\, dA_s$$

for all $0 \le t \le 2^k$, $k \ge k_0$. Substituting this into (2.4) gives

$$V(\varphi_t, A_t) \le V(x, 0) + \frac{1}{\eta}(1+2^k)^d \log k + \int_0^t LV(\varphi_s, A_s)\, dA_s +$$

$$+ \eta(1+2^k)^{-d} \int_0^t \sum_{i,j=1}^n \frac{\partial}{\partial x_i} V(\varphi_s, A_s) \frac{\partial}{\partial x_j} V(\varphi_s, A_s) a^{ij}(\varphi_s, \varphi_s, s)\, dA_s$$

$$\leq V(x, 0) + \frac{1}{\eta}(1+2^k)^d \log k +$$

$$+ \int_0^t \Big(LV(\varphi_s, A_s) + \eta(1+s)^{-d} \sum_{i,j=1}^n \frac{\partial}{\partial x_i} V(\varphi_s, A_s) \frac{\partial}{\partial x_j} V(\varphi_s, A_s) a^{ij}(\varphi_s, \varphi_s, s) \Big)\, dA_s$$

$$\leq V(x, 0) + \frac{1}{\eta}(1+2^k)^d \log k + \int_0^t \psi(s)\Big(1 + \big[V(\varphi_s, A_s)\big]^\alpha \Big)\, dA_s$$

$$\leq V(x, 0) + \frac{1}{\eta}(1+2^k)^d \log k + \int_0^{2^k} \psi(s)\, dA_s + \int_0^t \psi(s) \big[V(\varphi_s, A_s)\big]^\alpha\, dA_s$$

for all $0 \leq t \leq 2^k$, $k \geq k_o$ almost surely. Applying Lemma 2.2 gives

$$V(\varphi_t, A_t) \leq \Big\{ \Big(V(x, 0) + \frac{1}{\eta}(1+2^k)^d \log k + \int_0^{2^k} \psi(s)\, dA_s \Big)^{1-\alpha} +$$

$$+ (1-\alpha) \int_0^t \psi(s)\, dA_s \Big\}^{1/(1-\alpha)}$$

for all $0 \leq t \leq 2^k$, $k \geq k_o$ almost surely. Therefore, if $2^{k-1} \leq t \leq 2^k$, $k \geq k_o$,

$$\frac{1}{t} \log\big[e^{\lambda A_t} |\varphi_t|^p \big] \leq \frac{1}{t} \log V(\varphi_t, A_t)$$

$$\leq \frac{1}{2^{k-1}} \log\Big\{ \Big(V(x, 0) + \frac{1}{\eta}(1+2^k)^d \log k + \int_0^{2^k} \psi(s)\, dA_s \Big)^{1-\alpha} + (1-\alpha) \int_0^{2^k} \psi(s)\, dA_s \Big\}^{1/(1-\alpha)}$$

which, together with condition (4), implies immediately that

$$\limsup_{t\to\infty} \frac{1}{t} \log\big[e^{\lambda A_t} |\varphi_t|^p \big] \leq 0 \quad \text{a.s.}$$

Finally

$$\limsup_{t\to\infty} \frac{1}{t} \log|\varphi_t| = \limsup_{t\to\infty} \frac{1}{pt} \log\big[e^{-\lambda A_t} e^{\lambda A_t} |\varphi_t|^p \big]$$

$$\leq -\frac{\lambda}{p} \liminf_{t\to\infty} \frac{A_t}{t} \leq -\lambda\sigma/p \qquad \text{a.s.}$$

as required. The proof is complete.

Theorem 2.3. Assume there exist a function $V \in C^{2,1}(R^n \times R_+)$, two nonnegative predictable processes $\psi_1(t)$ and $\psi_2(t)$ on $t \geq 0$, and constants $p, \lambda, \eta, d, \sigma > 0$, $0 \leq \alpha < 1$ such that

(1) $$e^{\lambda t}|x|^p \le V(x,t) \quad \text{on } (x,t)\in R^n\times R_+;$$

(2) $$LV(x,A_t) + \eta e^{-dt}\sum_{i,j=1}^{n}\frac{\partial}{\partial x_i}V(x,A_t)\frac{\partial}{\partial x_j}V(x,A_t)\,a^{ij}(x,x,t)$$
$$\le \psi_1(t) + \psi_2(t)\left[V(x,A_t)\right]^{\alpha} \quad \text{a.s. on } (x,t)\in R^n\times R_+;$$

(3) $$\liminf_{t\to\infty}\frac{A_t}{t} \ge \sigma \quad \text{a.s.};$$

(4) $$\limsup_{t\to\infty}\frac{1}{t}\log\int_0^t \psi_1(s)\,dA_s \le d \quad \text{a.s.};$$

$$\limsup_{t\to\infty}\frac{1}{t}\log\int_0^t \psi_2(s)\,dA_s \le (1-\alpha)d \quad \text{a.s.}$$

Then the solution of equation (1.1) satisfies

$$\limsup_{t\to\infty}\frac{1}{t}\log|\varphi_t| \le \frac{1}{p}(\lambda\sigma - d) \quad \text{a.s.}$$

Proof. We take

$$\gamma = 2\eta e^{-dk}, \quad \delta = \frac{1}{\eta}e^{-dk}\log k, \quad \tau = k$$

for each $k = 1, 2, \ldots$ in (2.5). Applying the Borel-Cantelli lemma we deduce similarly that there exists an integer $k_0(\omega)$ for almost sure $\omega\in\Omega$ such that

$$\int_0^t\sum_{i=1}^{n}\frac{\partial}{\partial x_i}V(\varphi_s,A_s)\,M^i(\varphi_s,ds) \le \frac{1}{\eta}e^{dk}\log k +$$

$$+\eta e^{-dk}\int_0^t\sum_{i,j=1}^{n}\frac{\partial}{\partial x_i}V(\varphi_s,A_s)\frac{\partial}{\partial x_j}V(\varphi_s,A_s)a^{ij}(\varphi_s,\varphi_s,s)\,dA_s$$

for all $0\le t\le k$, $k\ge k_0$. Substituting this into (2.4) and applying condition (2) give

$$V(\varphi_t,A_t) \le V(x,0) + \frac{1}{\eta}e^{dk}\log k + \int_0^t\left(\psi_1(s) + \psi_2(s)\left[V(\varphi_s,A_s)\right]^{\alpha}\right)dA_s$$

for all $0\le t\le k$, $k\ge k_o$ almost surely. By Lemma 2.2 we obtain

$$V(\varphi_t,A_t) \le \Big\{\Big(V(x,0) + \frac{1}{\eta}e^{dk}\log k + \int_0^k \psi_1(s)\,dA_s\Big)^{1-\alpha} +$$

$$+ (1-\alpha)\int_0^t \psi_2(s)\,dA_s\Big\}^{1/(1-\alpha)}$$

for all $0\le t\le k$, $k\ge k_o$ almost surely. Let $\varepsilon > 0$ be arbitrary. By condition (4) there exists $k_1 = k_1(\omega)$ such that

$$\int_0^k \psi_1(s)\, dA_s \le e^{(d+\varepsilon)k} \quad \text{and} \quad \int_0^k \psi_2(s)\, dA_s \le e^{(1-\alpha)(d+\varepsilon)k}$$

for all $k \ge k_1$ almost surely. Therefore, if $(k-1) \le t \le k$, $k \ge k_o \vee k_1$,

$$e^{\lambda A_t} |\varphi_t|^p \le \left\{ \left(V(x, 0) + \frac{1}{\eta} e^{dk} \log k + e^{(d+\varepsilon)k} \right)^{1-\alpha} + (1-\alpha)\, e^{(1-\alpha)(d+\varepsilon)k} \right\}^{1/(1-\alpha)}$$

almost surely, which gives immediately that

$$\limsup_{t\to\infty} \frac{1}{t} \log\left[e^{\lambda A_t} |\varphi_t|^p \right] \le d+\varepsilon \quad \text{a.s.}$$

Since ε is arbitrary, we obtain

$$\limsup_{t\to\infty} \frac{1}{t} \log\left[e^{\lambda A_t} |\varphi_t|^p \right] \le d \quad \text{a.s.}$$

Finally

$$\limsup_{t\to\infty} \frac{1}{t} \log|\varphi_t| = \limsup_{t\to\infty} \frac{1}{pt} \log\left[e^{-\lambda A_t} e^{\lambda A_t} |\varphi_t|^p \right]$$

$$\le \frac{1}{p}\left(d - \lambda \liminf_{t\to\infty} \frac{A_t}{t} \right) \le -\frac{1}{p}(\lambda\sigma - d) \qquad \text{a.s.}$$

as required. The proof is complete.

3. USEFUL COROLLARIES

Let $N_t = (N_t^1, \ldots, N_t^m)^T$, $t \ge 0$, be an m-dimensional continuous martingale such that $N_0 = 0$ and

$$\langle N^i, N^j \rangle_t = \int_0^t K^{ij}(s)\, dA_s, \quad t \ge 0, \ 1 \le i, j \le m,$$

where K^{ij}, $1 \le i, j \le m$ are all predictable processes. Let $f(x, t) = (f^{ij})_{n\times m}$ and $b(x, t) = (b^1, \ldots, b^n)^T$, $t \ge 0$ be a predictable matrix and a predictable vector respectively for each $x \in R^n$. Consider a stochastic differentialequation

$$\varphi_t = x_0 + \int_0^t f(\varphi_s, s)\, dN_s + \int_0^t b(\varphi_s, s)\, dA_s \quad \text{on } t \ge 0. \tag{3.1}$$

Assume the equation satisfies the condition of existence and uniqueness of the solution (cf. Mao and Wu [14]). Note equation (3.1) is equivalent to equation (1.1) if we set

$$F(x, t) = \int_0^t f(x, s)\, dN_s + \int_0^t b(x, s)\, dA_s \quad \text{on } t \ge 0,$$

and the operator L has the form

$$LV(x, A_t) = \frac{\partial}{\partial t} V(x, A_t) + \sum_{i=1}^n \frac{\partial}{\partial x_i} V(x, A_t)\, b^i(x, t) +$$

$$+\frac{1}{2}\sum_{i,j=1}^{n}\sum_{l,k=1}^{m}\frac{\partial^2}{\partial x_i \partial x_j}V(x, A_t)\, f^{il}(x, t)\, K^{lk}(t)\, f^{jk}(x, t).$$

Therefore we have the following corollaries.

Corollary 3.1. Assume there exist a function $V \in C^{2,1}(R^n \times R_+)$, a nonnegative predictable process $\psi(t)$ on $t \geq 0$, and constants $p, \lambda, \eta, d, \sigma > 0$, $0 \leq \alpha < 1$ such that

(1) $e^{\lambda t}|x|^p \leq V(x, t)$ on $(x, t) \in R^n \times R_+$;

(2) $$LV(x, A_t) + \eta(1+t)^{-d}\sum_{i,j=1}^{n}\sum_{l,k=1}^{m}\frac{\partial}{\partial x_i}V(x, A_t)\frac{\partial}{\partial x_j}V(x, A_t)\, f^{il}(x, t)\, K^{lk}(t)\, f^{jk}(x, t)$$

$$\leq \psi(t)\left(1 + \left[V(x, A_t)\right]^{\alpha}\right) \quad \text{a.s. on } (x, t) \in R^n \times R_+;$$

(3) $\liminf_{t\to\infty}\frac{A_t}{t} \geq \sigma$ a.s.;

(4) $\limsup_{t\to\infty}\frac{1}{t}\log\int_0^t \psi(s)\, dA_s \leq 0$ a.s.

Then the solution of equation (3.1) satisfies

$$\limsup_{t\to\infty}\frac{1}{t}\log|\varphi_t| \leq -\lambda\sigma/p \quad \text{a.s.}$$

Corollary 3.2. If conditions (2) and (4) in Corollary 3.1 are replaced by

$$LV(x, A_t) + \eta e^{-dt}\sum_{i,j=1}^{n}\sum_{l,k=1}^{m}\frac{\partial}{\partial x_i}V(x, A_t)\frac{\partial}{\partial x_j}V(x, A_t)\, f^{il}(x, t)\, K^{lk}(t)\, f^{jk}(x, t)$$

$$\leq \psi_1(t) + \psi_2(t)\left[V(x, A_t)\right]^{\alpha} \quad \text{a.s. on } (x, t) \in R^n \times R_+,$$

where ψ_1 and ψ_2 are nonnegative predictable processes such that

$$\limsup_{t\to\infty}\frac{1}{t}\log\int_0^t \psi_1(s)\, dA_s \leq d \quad \text{a.s.}$$

and

$$\limsup_{t\to\infty}\frac{1}{t}\log\int_0^t \psi_2(s)\, dA_s \leq (1-\alpha)d \quad \text{a.s.}$$

Then the solution of (3.1) satisfies

$$\limsup_{t\to\infty}\frac{1}{t}\log|\varphi_t| \leq -(\lambda\sigma - d)/p \quad \text{a.s.}$$

More specially we consider an Itô equation

$$\varphi_t = x_0 + \int_0^t f(\varphi_s, s)\, dW_s + \int_0^t b(\varphi_s, s)\, ds, \quad t \geq 0, \tag{3.2}$$

where W is an m-dimensional Wiener process. In this case the operator L takes the form

$$LV(x, t) = \frac{\partial}{\partial t} V(x, t) + \sum_{i=1}^{n} \frac{\partial}{\partial x_i} V(x, t)\, b^i(x, t)$$

$$+ \frac{1}{2} \sum_{i,j=1}^{n} \sum_{k=1}^{m} \frac{\partial^2}{\partial x_i \partial x_j} V(x, t)\, f^{ik}(x, t)\, f^{jk}(x, t).$$

We then have another corollary.

Corollary 3.3. Assume there exist a function $V \in C^{2,1}(R^n \times R_+)$, a polynomial $\mu(t)$ $(t \geq 0)$ with positive coefficients, and constants $p, \lambda, \eta, d > 0$, $0 \leq \alpha < 1$ such that

(1) $e^{\lambda t} |x|^p \leq V(x, t)$ on $(x, t) \in R^n \times R_+$;

(2) $LV(x, t) + \eta(1+t)^{-d} \sum_{i,j=1}^{n} \sum_{k=1}^{m} \frac{\partial}{\partial x_i} V(x, t) \frac{\partial}{\partial x_j} V(x, t)\, f^{ik}(x, t)\, f^{jk}(x, t)$

$$\leq \mu(t) \left(1 + \left[V(x, A_t) \right]^{\alpha} \right) \qquad \text{a.s. on } (x, t) \in R^n \times R_+.$$

Then the solution of equation (3.2) satisfies

$$\limsup_{t \to \infty} \frac{1}{t} \log |\varphi_t| \leq -\lambda/p \quad \text{a.s.}$$

Furthermore, if condition (2) is replaced by

$$LV(x, t) + \eta e^{-dt} \sum_{i,j=1}^{n} \sum_{k=1}^{m} \frac{\partial}{\partial x_i} V(x, t) \frac{\partial}{\partial x_j} V(x, t)\, f^{ik}(x, t)\, f^{jk}(x, t)$$

$$\leq c_1 e^{dt} + c_2 e^{d(1-\alpha)t} \left[V(x, A_t) \right]^{\alpha} \qquad \text{a.s. on } (x, t) \in R^n \times R_+$$

for some constants $c_1, c_2 \geq 0$, then the solution satisfies

$$\limsup_{t \to \infty} \frac{1}{t} \log |\varphi_t| \leq -(\lambda - d)/p \quad \text{a.s.}$$

4. EXAMPLES

Before applying our results to study the bound of Lyapunov exponents of stochastic flows, we give here some interesting examples to illustrate the theorems.

Example 4.1. Let w_t be a one-dimensional Wiener process. Consider a semilinear Itô equation

$$d\varphi_t = -\varphi_t\, dt + e^{-\gamma t} |\varphi_t|^{1/2}\, dw_t \quad \text{on } t \geq 0 \tag{4.1}$$

with initial data $\varphi_0 = x_0 \in R$, where $0 < \gamma < 1/2$. Let $1/2 < p < 1$ and introduce a Lyapunov function

$$V(x, t) = e^{pt}|x|^p \quad \text{on } (x, t) \in R \times R_+.$$

It is easy to check that

$$LV(x, t) + e^{-(1-2\gamma)pt}\left[V_x(x, t)\right]^2 e^{-2\gamma t}|x|$$

$$\leq e^{-(1-2\gamma)pt}\, p^2\, e^{2(p-\gamma)t}|x|^{2p-1}$$

$$\leq e^{(1-2\gamma)(1-p)t}\left[V(x, t)\right]^{(2p-1)/p}.$$

Note $0 < (2p-1)/p < 1$ and $\left(1-(2p-1)/p\right)(1-2\gamma)p = (1-p)(1-2\gamma)$. We can then apply Corollary 3.3 to derive that the solution of equation (4.1) satisfies

$$\limsup_{t\to\infty} \frac{1}{t}\log|\varphi_t| \leq -(p-(1-2\gamma)p)/p = -2\gamma \quad \text{a.s.}$$

Example 4.2. Let us consider an n-dimensional nonlinear stochastic oscillator

$$\ddot{x} + b(x, \dot{x}) + \nabla G(x) = \sigma(x, \dot{x}, t)\,\dot{w}_t + e(t) \quad \text{on } t \geq 0, \tag{4.2}$$

where $\nabla G(x) = (G_{x_1}, ..., G_{x_n})^T$, and w_t, $t \geq 0$ is an n-dimensional Wiener process. Let $y = \dot{x}$ and the corresponding 2n-dimensional Itô stochastic differential equation is

$$\begin{aligned} dx &= y\,dt \\ dy &= \left[-b(x, y) - \nabla G(x) + e(t)\right]dt + \sigma(x, y, t)\,dw_t. \end{aligned} \tag{4.3}$$

We assume

(H1) $b(x, y) = (b_1, ..., b_n)^T$ is locally Lipschitzian; $G(x)$ is a C^2-function such that $\nabla G(x)$ is locally Lipschitzian; $\sigma(x, y, t) = (\sigma_{ij})_{n\times n}$ is continuous in $(x, y, t) \in R^{2n}\times R_+$ and also locally Lipschitzian in (x, y) at each fixed $t \geq 0$; and $e(t)$, $t \geq 0$ is an n-dimensional adapted continuous process.

(H2) $$G(x) > 0 \text{ except } G(0) = 0, \quad \langle y, b(x, y)\rangle \geq \lambda\, U(x, y),$$

$$|\sigma|^2 = \text{trace}(\sigma\sigma^T) \leq e^{-\lambda t} \quad \text{and} \quad |e(t)| \leq e^{-\lambda t/2} \quad \text{a.s.}$$

for some constant $\lambda > 0$, where U is the energy of the system defined by

$$U(x, y) = \tfrac{1}{2}|y|^2 + G(x).$$

We now assign $0 < \varepsilon < \lambda$ arbitrarily and introduce a Lyapunov function

$$V(x, y, t) = e^{(\lambda-\varepsilon)t}\, U(x, y).$$

(Note we only require V has continuous first derivatives in x in this special case). We then deduce that

$$LV(x, y, t) + \varepsilon \left(e^{2(\lambda-\varepsilon)t} y^T \sigma\sigma^T y \right)$$

$$= e^{(\lambda-\varepsilon)t} \left((\lambda-\varepsilon)U(x, y) - \langle y, b(x, y)\rangle + \langle y, e(t)\rangle + |\sigma|^2 \right) + \varepsilon\, e^{2(\lambda-\varepsilon)t} |\sigma|^2 |y|^2$$

$$\leq -\varepsilon\, V(x, y, t) + e^{(\lambda-\varepsilon)t} \left(|e(t)|\, |y| + |\sigma|^2 \right) + \varepsilon\, e^{-2\varepsilon t}\, V(x, y, t) \leq$$

$$\leq 1 + 2\left[V(x, y, t)\right]^{1/2}.$$

Therefore, by Corollary 3.3 (with careful reading of the proof of Theorem 2.1) we derive that the energy of the system satisfies

$$\limsup_{t\to\infty} \frac{1}{t} \log U(x, y) \leq -(\lambda-\varepsilon) \quad \text{a.s.}$$

Since ε is arbitrary, we obtain that

$$\limsup_{t\to\infty} \frac{1}{t} \log U(x, y) \leq -\lambda \quad \text{a.s.}$$

5. BOUND FOR LYAPUNOV EXPONENTS OF STOCHASTIC FLOWS

In this section we will apply our results to study the bound for Lyapunov exponents of stochastic flows. For the readers' convenience, let us first give the definition of the stochastic flow of homomorphisms and the Brownian flow (cf. Kunita [10).

Let $\varphi_{s,t}(x)$, $s, t \in R_+$, $x \in R^n$ be a continuous R^n-valued random field defined on the probability space $(\Omega, \mathcal{F}, P)$. It is called a *stochastic flow of homomorphisms* if it satisfies the following properties:

(1) $\varphi_{s,u} = \varphi_{t,u} \circ \varphi_{s,t}$ for any s, t ,u a.s., where $\circ$ denotes the composition of maps;

(2) $\varphi_{s,s}$ = identity map for any s a.s.;

(3) $\varphi_{s,t} : R^n \to R^n$ is an onto homomorphism for any s, t a.s..

It is called a *Brownian flow* if it still satisfies

(4) for any $0 \leq t_0 < t_1 < \dots < t_k$, $\varphi_{t_i, t_{i+1}}$, $i = 0, \dots, k-1$ are independent.

If $\varphi_{s,t}$ is only defined on $0 \leq s \leq t < \infty$, we call it a *forward stochastic flow of homomorphisms* or *forward Brownian flow* respectively.

We shall assume the following conditions.

(A.1) $\varphi_{s,t}(x)$ is square integrable for each s, t, x, and there exist the infinitesimal mean b(x, t) and the infinitesimal covariance a(x, y, t) for any t, x, y:

$$b(x, t) := \lim_{h\to 0+} \frac{1}{h}\, E\{ \varphi_{t,t+h}(x) - x \}$$

$$a(x, y, t) := \lim_{h\to 0+} \frac{1}{h} E\{(\varphi_{t,t+h}(x) - x)\, (\varphi_{t,t+h}(y) - y)^T\}$$

(A.2) There exists a positive constant K such that for any s, t, x, y

$$| E\{ \varphi_{s,t}(x) - x \} | \le K(1 + | x |) | t-s |,$$

$$| E\{(\varphi_{s,t}(x) - x) (\varphi_{s,t}(y) - y)^T\} | \le K(1 + | x |)(1 + | y |) | t-s |.$$

(A.3) a(x, y, t) and b(x, t) are continuous in (x, y, t) and (x, t) respectively. Moreover they are locally δ-Hölder continuous ($\delta > 0$): for any compact subset C of R^n, there exists a positive constant K_c such that for any $x, y \in C$ and $t \ge 0$

$$\| a(x, x, t) - 2a(x,y, t) + a(y, y, t) \| \le K_c | x-y |^{2\delta},$$

$$| b(x, t) - b(y, t) | \le K_c | x - y |^{\delta}.$$

We shall need the following theorem due to Kunita [10] (Theorem 4.2.8).

Theorem 5.1 (Kunita [10]). Let $\varphi_{s,t}(x)$, $0 \le s \le t < \infty$, $x \in R^n$ be a forward Brownian flow. Suppose that the pair of infinitesimal covariance and infinitesimal mean (a(x, y, t), b(x, t)) satisfies (A.1)-(A.3). Then there exists a Brownian motion F(x, t) with the characteristic (a(x, y, t), b(x, t), t) such that the flow is governed by Itô's stochastic differential equation based on F(x, t), i.e.,

$$\varphi_{s,t}(x) = x + \int_s^t F(\varphi_{s,r}(x), dr) \quad \text{a.s. on } 0 \le s \le t < \infty, x \in R^n.$$

We now have the following theorem immediately which shows one can use the Lyapunov function to estimate the Lyapunov exponent of the stochastic flow.

Theorem 5.2. Let $\varphi_{s,t}(x)$, $0 \le s \le t < \infty$, $x \in R^n$ be a forward Brownian flow. Suppose that the pair of infinitesimal covariance and infinitesimal mean (a(x, y, t), b(x, t)) satisfies (A.1)-(A.3). Assume there exist a function $V \in C^{2,1}(R^n \times R_+)$, a polynomial $\mu(t)$ of t (≥ 0) with positive coefficients, and constants p, λ, η, $d > 0$, $0 \le \alpha < 1$ such that

(1) $e^{\lambda t} |x|^p \le V(x, t)$ on $(x, t) \in R^n \times R_+$;

(2) $$LV(x, t) + \eta(1+t)^{-d} \sum_{i,j=1}^{n} \frac{\partial}{\partial x_i} V(x, t) \frac{\partial}{\partial x_j} V(x, t)\, a^{ij}(x, x, t)$$
$$\le \mu(t) \Big(1 + \big[V(x, t) \big]^{\alpha} \Big) \qquad \text{a.s. on } (x, t) \in R^n \times R_+.$$

Then the flow satisfies

$$\limsup_{t \to \infty} \frac{1}{t} \log|\varphi_{s,t}(x)| \le -\lambda/p \quad \text{a.s.}$$

for any $s \ge 0$ and $x \in R^n$. Furthermore, if condition (2) is replaced by

$$LV(x,t) + \eta e^{-dt} \sum_{i,j=1}^{n} \frac{\partial}{\partial x_i} V(x,t) \frac{\partial}{\partial x_j} V(x,t)\, a^{ij}(x,x,t)$$

$$\leq c_1 e^{dt} + c_2 e^{d(1-\alpha)t} \left[V(x, A_t) \right]^{\alpha} \quad \text{a.s. on } (x,t) \in R^n \times R_+$$

for some constants $c_1, c_2 \geq 0$, then the flow satisfies

$$\limsup_{t\to\infty} \frac{1}{t} \log|\varphi_{s,t}(x)| \leq -(\lambda-d)/p \quad \text{a.s.}$$

For a stochastic flow of homomorphisms one can have a similar result but we leave the details to the readers. We finally give one more example to close our paper.

Example 5.3. Let $\varphi_{s,t}(x)$, $0 \leq s \leq t < \infty$, $x \in R^2$ be a forward Brownian flow satisfying (A.1)–(A.3) with

$$b(x,t) = \begin{bmatrix} x_1 + 3x_2 \\ -4x_1 - 6x_2 \end{bmatrix}, \quad a(x,y,t) = e^{-2t} \begin{bmatrix} \sin x_1 \sin y_1, & \sin x_1 \cos(y_1 - y_2) \\ \cos(x_1 - x_2) \sin y_1, & \cos(x_1 - x_2)\cos(y_1 - y_2) \end{bmatrix}.$$

Introduce a Lyapunov function

$$V(x,t) = 100 e^{3t} \left(17x_1^2 + 26x_1x_2 + 10x_2^2 \right).$$

Then we have

$$e^{3t}|x|^2 \leq V(x,t).$$

Note

$$LV(x,t) \leq 100e^{3t}\left(3\left[17x_1^2 + 26x_1x_2 + 10x_2^2\right] - 70x_1^2 - 108x_1x_2 - 42x_2^2 \right) + 5300e^t \leq -e^{3t}|x|^2 + 5300e^t$$

and

$$\sum_{i,j=1}^{2} \frac{\partial}{\partial x_i} V(x,t) \frac{\partial}{\partial x_j} V(x,t)\, a^{ij}(x,x,t) \leq C e^{4t} |x|^2,$$

where C is a positive constant. We also have

$$LV(x,t) + \frac{1}{C} e^{-t} \sum_{i,j=1}^{2} \frac{\partial}{\partial x_i} V(x,t) \frac{\partial}{\partial x_j} V(x,t)\, a^{ij}(x,x,t) \leq 5300e^t.$$

Hence, by Theorem 5.2, the flow satisfies

$$\limsup_{t\to\infty} \frac{1}{t} \log|\varphi_{s,t}(x)| \leq -(3-1)/2 = -1 \quad \text{a.s}$$

for all $s \geq 0$ and $x \in R^2$.

ACKNOWLEDGEMENTS

This research is supported by grant GR/F51241 of SERC. The author also wish to thank Professors K. D. Elworthy and L. Markus for their kind assistance and to the referee for his useful suggestions.

REFERENCES

[1] Arnold, L., A formula connecting sample and moment stability of linear stochastic systems, SIAM J. Appl. Math. 44 (1984), 793-802.

[2] Arnold, L. and Kliemann, W., Qualitative theory of stochastic systems, Probabilistic Analysis and Related Topics, Vol.3 (Bharucha-Reid, A. T., ed.), 1-79, New York, Academic Press 1983.

[3] Arnold, L., Oeljeklaus, E. and Pardoux, E., Almost sure and moment stability for linear Ito equations, Lecture Notes Math. 1186(1984), 129-159.

[4] Carverhill, A., Flows of stochastic dynamical systems: ergodic theory, Stochastics 14 (1985), 273-317.

[5] Chappell, M., Bounds for average Lyapunov exponents of gradient stochastic systems, Lecture Notes Math. 1186(1984), 292-307.

[6] Crauel, H., Lyapunov numbers of Markov solutions of linear stochastic systems, Stochastics 14 (1984), 11-28.

[7] Curtain, R. (ed.), Stability of stochastic dynamical systems, Lecture Notes Math. 294, Berlin-Heidelberg-New York, Springer 1972.

[8] Elworthy, K. D., Stochastic Differential Equations on Manifolds, London Math. Society Lecture Note 70 , C.U.P, 1982.

[9] Has'minskii, R. Z., Necessary and sufficient conditions for the asymptotic stability of linear stochastic systems, Theory Probability Appl. 12 (1967), 144-147.

[10] Kunita, H., Stochastic Flows and Stochastic Differential Equations, Cambridge University Press, 1990.

[11] Mao, X., Lebesgue-Stieltjes integral inequality and stochastic stability, Quarterly J. Math. Oxford (2), 40 (1989), 301-311.

[12] Mao, X., Lyapunov functions and almost sure exponential stability of stochastic differential equations based on semimartingales with spatial parameters, SIAM J. Control and Optimization 28(6) (1990), 1481-1490.

[13] Mao, X., Lebesgue-Stieltjes integral inequalities of the Wendroff type in n-independent variables, Chinese J. Math. 17(1) (1989), 29-50.

[14] Mao, X. and Wu, R., Existence and uniqueness of the solutions of stochastic differential equations, Stochastics 11 (1983), 19-32.

[15] Ladde, G. S., Stochastic stability analysis of model ecosystems with time-delay, in "Differential Equations and Applications in Biology, Epidemics, and Population Problems" edited by S. N. Busenberg and K. Cook, Academic Press(1981), 215-228.

Large Deviations for Random Expanding Maps

by
YURI KIFER

Institute of Mathematics
Hebrew University, Jerusalem, Israel

Abstract

Employing a general theorem on large deviations together with Walters' result on uniqueness of equilibrium states for some mappings which expand distances I derive large deviation bounds for random expanding maps which involve the entropy similarly to large deviations for dynamical systems. Relativized large deviations are discussed, as well.

1. A large deviations theorem.

Suppose that X is a compact metric space, $P(X)$ is the space of probability measures on X endowed with the topology of weak convergence, and $(\Omega_\lambda, \mathcal{F}_\lambda, P_\lambda)$ is a family of probability spaces indexed by λ from a directed set Λ. Let $\varsigma^\lambda : \Omega_\lambda \to P(X), \varsigma^\lambda : \omega \to \varsigma^\lambda_\omega, \lambda \in \Lambda$ be a family of measurable maps.

Theorem 1 ([K]) *Suppose that the limit*

$$Q(V) = \lim_{\lambda \in \Lambda} (\frac{1}{r(\lambda)}) \log \int \exp(r(\lambda) \int V(x) d\varsigma^\lambda_\omega(x)) dP_\lambda(\omega) \tag{1}$$

exists for any $V \in C(X)$ where $r(\lambda) > 0$, $\lim_{\lambda \in \Lambda} r(\lambda) = \infty$ is a scaling function on Λ. Then for any closed $K \subset P(X)$,

$$\limsup_{\lambda \in \Lambda} (\frac{1}{r(\lambda)}) \log P_\lambda\{\varsigma^\lambda \in K\} \le - \inf\{I(\nu) : \nu \in K\} \tag{2}$$

where the functional I is the convex conjugate of Q, i.e.

$$I(\mu) = \sup_{V \in C(X)} (\int V d\mu - Q(V)) \tag{3}$$

if $\mu \in P(X)$ and $I(\mu) = \infty$ for other signed measures μ, and also

$$Q(V) = \sup_{\mu \in P(X)} (\int V d\mu - I(\mu)). \tag{4}$$

If, furthermore, there exists a countable set of functions $V_1, V_2, \ldots \in C(X)$ such that their span is dense in $C(X)$ with respect to $\|\cdot\|, \|V_i\| = 1$ for all $i = 1, 2, \ldots$, and for each $n = 1, 2, \ldots$ and all numbers $\beta_1, \ldots \beta_n$ the function $V = \beta_1 V_1 + \ldots + \beta_n V_n$ has a unique measure μ_V satisfying

$$Q(V) = \int V d\mu_V - I(\mu_V) \tag{5}$$

then for any open $G \subset \mathcal{P}(X)$,

$$\liminf_{\lambda \in \Lambda} (1/r(\lambda))) \log P_\lambda\{\varsigma^\lambda \in G\} \geq -\inf\{I(\nu) : \nu \in G\}. \tag{6}$$

This theorem was employed in [K2] to obtain as corollaries large deviations results á la Donsker-Varadhan [DV] for diffusions and Markov chains having transition densities and also large deviations bounds for hyperbolic dynamical systems which generalize results of Orey and Pelikan [OP].

I will be mainly concerned here with an application of this theorem to the case where one has an X-valued Markov chain $Y_n = Y_n(\omega)(\Lambda = \mathbb{Z}_+, \lambda = r(\lambda) = n \in \mathbb{Z}_+)$ generated by random transformations and the measures ς_ω^n are defined by $\varsigma_\omega^n = \frac{1}{n}\sum_{\ell=0}^{n-1} \delta_{Y_\ell(\omega)}$ where δ_x denotes the unit mass at x.

2. Random expanding maps.

Let M be a compact connected Riemannian manifold and $f_1, \cdots, f_k$ be C^2 expanding maps of M onto itself, i.e. there are real numbers $c > 0$ and $\gamma > 1$ such that

$$\|Df_i^n \xi\| \geq c\gamma^n \|\xi\| \tag{7}$$

for each $n \geq 0$ and all tangent vectors ξ. The randomness of the system is described by a collection of continuous functions $p_{ij}(x) > 0$, $i, j = 1, \cdots, k$, $x \in M$ such that $\sum_j P_{ij}(x) = 1$. These define a Markov chain Y_n on $\mathcal{R} = K \times M$ with $K = \{1, \cdots, k\}$ by the transition probability

$$P((i, x), (j, \Gamma)) = p_{ij}(x)\chi_\Gamma(f_i x) \tag{8}$$

for any Borel set $\Gamma \subset M$. In particular, if $p_{ij}(x) = p_j$ is independent of i and x then we arrive at the scheme of independent random transformations. If we write $Y_n = (\nu(n), X_n)$ then this Markov chain starting at x can be written as

$$X_n = f_{\nu(n-1)} \circ \cdots \circ f_{\nu(0)} x, \quad X_0 = x$$

From the dynamical systems point of view one has here a generalized skew product system acting on $\Omega \times M$ where $\Omega = K^{\mathbb{N}}$ by the formula

$$\tau(\omega, x) = (\theta\omega, f_{\omega_0} x)$$

where $\theta : \Omega \to \Omega$ is the shift and $\omega_n = (\theta^n\omega)_0 = \nu(n)$. One can show that the Markov chain Y_n and the transformation τ have invariant measures whose conditional measures on M are equivalent to the Riemannian volume m on M.

Put $\varphi_i(x) = -\log|detD_x f_i|, \varphi(\omega,x) = \varphi_{\omega_0}(x), V(\omega,x) = V_{\omega_0}(x), p(\omega,x) = p_{\omega_0\omega_1}(x)$, and remark that

$$\tau^n(\omega,x) = (\theta^n\omega, f_{\omega_{n-1}} \circ \cdots \circ f_{\omega_0}x), \tau^0(\omega,x) = (\omega,x).$$

For each $(i,x) \in K \times M$ the transition probabilities generate a probability measure $\tilde{P}_{i,x}$ on $\Omega \times M^{\mathbb{N}}$ corresponding to the condition $Y_0 = (i,x)$. Since given (ω,x) the whole sequence $\{x_\ell\} \in M^{\mathbb{N}}, x_\ell = f^\ell x, \ell = 0,1,\ldots$ is determined uniquely the Markov chain Y_n is already described by the projection $P_{i,x}$ of $\tilde{P}_{i,x}$ to Ω. The expectation corresponding to $P_{i,x}$ is denoted by $E_{i,x}$. Let also mP_i denotes the measure on $\Omega \times M$ defined by $dmP_i(\omega,x) = dP_{i,x}(\omega)dm(x)$.

Theorem 2. *For any collection $\bar{V} = (V_1,\cdots,V_k)$ of continuous functions on M and $i \in K$ the limit*

$$Q(\bar{V}) = \lim_{n\to\infty}\frac{1}{n}\log\int_M E_{i,x}\exp\Big(\sum_{\ell=0}^{n-1} V_{\nu(\ell)}(f_{\nu(\ell-1)} \circ \cdots \circ f_{\nu(0)}x)\Big)dm(x) \tag{9}$$

$$= \lim_{n\to\infty}\frac{1}{n}\log\sum_{i=1}^{k}\int_M E_{i,x}\exp\Big(\sum_{\ell=0}^{n-1} V_{\nu(\ell)}(f_{\nu(\ell-1)} \circ \cdots \circ f_{\nu(0)}x)\Big)dm(x)$$

exists. Moreover

$$Q(\bar{V}) = \sup_{\mu\in P(\Omega\times M),\mu-\tau-\text{invariant}}\Big(\int (V+\varphi+\log p)d\mu + h_\mu(\tau)\Big) \tag{10}$$

where $h_\mu(\tau)$ is the entropy of τ with respect to the $\tau-$invariant measure μ.

Sketch of the proof

Define the metric on Ω by

$$d(\omega,\omega') = \sum_{n=0}^{\infty} 2^{-n}|\omega_n - \omega'_n|$$

and the metric on $\Omega \times M$ by

$$d((\omega,x),(\omega',x')) = d(\omega,\omega') + d(x,x').$$

Put $d_n((\omega,x),(\omega',x')) = \max_{0\le\ell\le n-1} d(\tau^\ell(\omega,x),\tau^\ell(\omega',x'))$.

If $d_n((\omega,x),(\omega',x')) \le \delta < 1$ then $\omega_i = \omega'_i$ for all $i = 0,\cdots,n-1$.

On the other hand, if $\omega_i = \omega'_i$ for all $i = 0,\cdots,n+n_\delta, n_\delta = 2\log\frac{k}{\delta}$ then

$$d_n((\omega,x),(\omega',x')) \le \delta + \max_{1\le\ell\le n-1} d(f^\ell x, f^\ell x') \tag{11}$$

$$\text{where} \quad f^\ell = f_{\nu(\ell-1)} \circ \cdots \circ f_0, f^0 = id.$$

It is easy to see that the following volume estimates hold true. The ratios

$$m\{y : d_n((\omega,x),(\omega,y)) \le \delta\} / \exp(\sum_{\ell=0}^{n-1} \varphi(\tau^\ell(\omega,x))) \tag{12}$$

and

$$\frac{mP_i\{(\omega',y) : d_n((\omega,x),(\omega',y)) \le \delta\}}{p(\omega,x)p(\tau(\omega,x)) \cdots p(\tau^{\ell-1}(\omega,x)) \exp(\sum_{\ell=0}^{n-1} \varphi(\tau^\ell(\omega,x)))} \tag{13}$$

are sandwiched between two constants depending only on δ.

From these one derives in the same way as in [K2] that the limit in (9) exists and it is the topological pressure for τ corresponding to the function $V + \varphi + \log p$. □

It is easy to see that τ is an expansive transformation and so $h_\mu(\tau)$ is upper-semicontinuous. So we can write

$$Q(\bar{V}) = \sup_{\mu \in P(\Omega \times M)} (\int V d\mu - I(\mu)) \tag{14}$$

where

$$I(\mu) = \begin{cases} -\int(\varphi + \log p)d\mu - h_\mu(\tau) & \text{if } \mu - \tau \text{ invariant} \\ \infty \text{ if } \mu \text{ is not } \tau - & \text{invariant} \end{cases}$$

is lower semicontinuous.

By Theorem 1 we obtain the upper large deviation bound.

Corollary 1. For any closed set $K \subset \mathcal{P}(M)$, and $i \in K$,

$$\limsup_{n\to\infty} \frac{1}{n} \log mP_i\{(\omega,x) : \frac{1}{n} \sum_{\ell=0}^{n-1} \delta_{f^\ell(\omega)x} \in K\} \le -\inf_{\nu \in K} I(\nu). \tag{15}$$

To obtain this corollary one has to take in Theorem 1 $P_i \times m$ in place of $P_n, \Omega \times M$ in place of Ω, and M in place of X. For this corollary one needs Theorem 2 applied to functions V depending on x only. Taking functions $V(\omega,x) = V_{\omega_0}(x)$ we can derive also large deviations bounds for occupational measures

$$\frac{1}{n} \sum_{\ell=0}^{n-1} \delta_{(\nu(\ell), f^\ell x)} \quad \text{on} \quad K \times M. \tag{16}$$

Since $f_i, i = 1, \cdots, k$ are C^2 then φ_i are C^1 functions. Assume that $p_{ij}(x)$ are Hölder continuous. Then the conditions I and (i) on p. 123 and (iii) on p. 125 from Walters [W] are clearly satisfied. The condition II on p. 125 holds true, as well, with the same proof as in Lemma 17 in [W]. Thus by Theorem 16 in [W] for each Hölder continuous vector-function $\bar{V}$ there exists a unique measure μ_V (equilibrium state) such that

$$Q(\bar{V}) = \int V d\mu_V - I(\mu_V). \tag{17}$$

Thus by Theorem 1 we obtain

Corollary 2. For any open $G \subset \mathcal{P}(X)$,

$$\liminf_{n\to\infty}\frac{1}{n}\log mP_i\{(\omega,x):\frac{1}{n}\sum_{\ell=0}^{n-1}\delta_{f^\ell(\omega)x}\in G\}\geq -\inf_{\eta\in G}I(\eta) \tag{18}$$

Remark 1. There exists a unique τ-invariant measure μ on $\Omega\times M$ such that its projection to $\{(\omega,x):(\omega_0,x)\in K\times M\}$ is an invariant measure of the Markov chain Y_n with conditional measures on M being equivalent to m. Then it is easy to see that $h_\mu(\tau)=-\int(\varphi+\log p)d\mu$ and this is the only measure for which $I(\mu)=0$. Remark that one can consider this model as a random bundle map and the sum of corresponding Lyapunov exponents will be $-\int\varphi d\mu$.

Remark 2. Similar results can be obtained when $f_1,\cdots,f_k$ are small C^1 perturbations of a diffeomorphism f having a hyperbolic set. Then in a neighborhood of this hyperbolic set one will have volume estimates similar to (12)-(13) for the corresponding random transformation, and so Theorem 2 and the upper large deviation bound can be derived again. Nevertheless Walters' condition II does not hold here and it is not clear how to prove uniqueness of equilibrium states which implies lower large deviation bounds for this case. Remark that usually the uniqueness is proved via Markov partitions and symbolic dynamics which is not clear how to employ for random transformations. The Walter's method is one way to avoid symbolic dynamics.

Remark 3. If one wants to generalize Theorem 2 to the case when expanding transformations are chosen from an arbitrary set not necessarily finite then the main problem is how to obtain the volume estimate of the type (13). One can do this if, for instance, $\nu(n)$ takes values on a compact manifold K and the transition probability of the corresponding Markov chain Y_n is

$$P((y,x),(Q,\Gamma))=\int_Q p_{yz}(x)\chi_\Gamma(f_zx)dm(z)$$

where $y\in K, Q\subset K$, $x\in M,\Gamma\subset M$, and m is the Riemannian volume. Then the result completely similar to (9) and (10) holds true.

3. Relativized large deviations

By relativized large deviations I mean bounds

$$\limsup_{n\to\infty}\frac{1}{n}\log m\{x:\frac{1}{n}\sum_{\ell=0}^{n-1}\delta_{f^\ell(\omega)x}\in K\}\leq -\inf_{\nu\in K}I(\nu) \tag{19}$$

for any closed $K\subset\mathcal{P}(M)$, and

$$\liminf\frac{1}{n}\log m\{x:\frac{1}{n}\sum_{\ell=0}^{n-1}\delta_{f^\ell(\omega)x}\in G\}\geq -\inf_{\nu\in G}I(\nu) \tag{20}$$

for any open $G\subset\mathcal{P}(M)$ where $f^\ell(\omega)=f_{\theta^\ell\omega}\cdots f_{\theta\omega}f_\omega$. Here the limits and the inequalities may hold only almost surely.

Theorem 3. *For any collection $\bar{V} = (V_1, \cdots, V_k)$ of continuous functions on M the limit*

$$Q^{(r)}(\bar{V}) = \lim_{n\to\infty} \frac{1}{n} \log \int_M \exp\Big(\sum_{\ell=0}^{n-1} V_{\nu(\ell)}(f_{\nu(\ell-1)} \circ \cdots \circ f_{\nu(0)} x)\Big) dm(x) \tag{21}$$

exist almost surely. Moreover if $p_{ij}(x) = p_{ij}$ are independent of x and p is the corresponding probability measure on Ω invariant under θ then

$$Q^{(r)}(\bar{V}) = \sup_{\substack{\mu \in \mathcal{P}(\Omega\times M),\\ proj_\Omega \mu = p, \mu-\tau-\text{invariant}}} \Big(\int (V + \varphi) d\mu + h_\mu^{(r)}(\tau)\Big) \tag{22}$$

where $Q^{(r)}$ and $h_\mu^{(r)}(\tau)$ are relativized topological pressure and relativized entropy, respectively and $\tau(\omega, x) = (\theta\omega, f_{\omega_0} x)$.

Sketch of the proof.

To prove (21) one uses the volume estimates (12) and the arguments connected with the relativized topological pressure similar to Section 2.3 in [K1]. The relation (22) is the Ledrappier-Walters relativized variational principle [LW]. □

Now (19) follows from (21) and Theorem 1 with m in place of P_n and M in place of Ω and X. If $p_{ij}(x) = p_{ij}$ are independent of x then $I(\nu)$ in (19) is given by

$$I(\nu) = \begin{cases} -\int \varphi d\nu - h_\nu^{(r)}(\tau) \text{ if } \nu \text{ is } \tau - \text{invariant and } \mathrm{proj}_\Omega \nu = p \\ \infty \text{ for otherwise.} \end{cases} \tag{23}$$

Note that $h_\nu^{(r)}(\tau)$ will be upper semicontinuous in our situation (using the definition of $h_\nu^{(r)}(\tau)$ via partitions and the expansiveness of τ), and so $I(\nu)$ defined by (23) will be lower semicontinuous. Remark also that $-\int \varphi d\nu$ is the sum of Lyapunov exponents for corresponding random transformations so the formula for $I(\nu)$ is similar to the deterministic case.

The uniqueness of relativized equilibrium states in (22) for Hölder continuous V can be proved similarly to [W] and so the lower bound (20) holds true.

In fact, Theorem 3 can be extended to the situation of a general stationary sequence of expanding maps. Let M be a compact Riemannian manifold, $\mathcal{F}$ be the space of C^2 expanding maps on M, and $(\Omega, \mathbf{P})$ be a probability space with a measure $\mathbf{P}$ invariant with respect to a measurable transformation $\theta : \Omega \to \Omega$ (the shift). Suppose that $\Psi : \Omega \to \mathcal{F}$ is a measurable map, $\Psi(\omega) = f_\omega$, so that $\Psi\mathbf{P}$ has compact support.

Theorem 4. *For any continuous function $V(x)$ on M the limit*

$$Q^{(r)}(V) = \lim_{n\to\infty} \frac{1}{n} \log \int_M \exp\Big(\sum_{\ell=0}^{n-1} V(f_{\theta^{\ell-1}\omega} \circ \cdots \circ f_\omega x)\Big) dm(x) \tag{24}$$

exists almost surely. Moreover

$$Q^{(r)}(V) = \sup_{\substack{\mu \in \mathcal{P}(\Omega\times M)\\ proj_\Omega \mu = \mathbf{P}, \mu-\tau-\text{invariant}}} \Big(\int (V + \varphi) d\mu + h_\mu^{(r)}(\tau)\Big) \tag{25}$$

where $\varphi_\omega(x) = -\log|detD_x f_\omega|$, $\tau(\omega,x) = (\theta\omega, f_\omega x)$ is the skew-product transformation and $Q^{(r)}$ and $h^{(r)}$ are relativized topological pressure and relativized entropy, respectively. Furthermore, if V is Hölder continuous then there exists a unique equilibrium state in the variational principle (25). By Theorem 1 the above means that we have both upper and lower relativized large deviations bounds.

Sketch of the proof

The existence of the limit (24) follows from the same volume estimates as in Theorem 3. The main point here is the proof of uniqueness of equilibrium states which proceeds similarly to [W] but employs additional arguments. First, one defines random Perron-Frobenius operator

$$(\mathcal{L}^\omega_{r_\omega} q_\omega)(x) = \sum_{y \in f_\omega^{-1}x} e^{r_\omega(y)} q_\omega(y) \tag{26}$$

acting on measurable functions q_ω which are continuous in x where $\{r_\omega\}$ is a equi Hölder continuous family of functions on M. The main point here is to construct a family of probability measures $\{\nu^\omega\}$ on M and a family of functions $\{h^\omega\}$ such that

$$(\mathcal{L}^\omega_{r_\omega})^* \nu^{\theta\omega} = \lambda^\omega \nu^\omega \text{ and } \mathcal{L}^\omega_{r_\omega} h^\omega = \lambda^\omega h^{\theta\omega} \tag{27}$$

where $\lambda^\omega > 0$ is a family of numbers. This construction can be done separately for each sequence $\{\theta^i\omega, -\infty < i < \infty\}$ which simplifies the proof and enables one to go along the lines of [W]. If ν^ω and h^ω satisfy (27) then it is easy to check that $\mu^\omega = h^\omega \cdot \nu^\omega$ satisfies $f_\omega \mu^\omega = \mu^{\theta\omega}$ a.s and so if $\Gamma^\omega = \{(x,\omega) \in \Gamma : x \in M\}$ then $\eta(\Gamma) = \int_\Omega \mu^\omega(\Gamma^\omega)dP(\omega)$ denoted $\eta = (P,\mu)$ is a τ-invariant measure. It turns out that the pair $\{\nu^\omega, h^\omega\}$ satisfying (27) is unique,

$$\begin{aligned}
\int \log \lambda^\omega dP(\omega) &= \int r_\omega(x) d\eta(\omega,x) + h^{(r)}_\eta(\tau) \\
&= \sup_{\substack{\{m^\omega, \omega\in\Omega\}, m^\omega \in \mathcal{P}(M) \\ f_\omega m^\omega = m^{\theta\omega}}} \left(\int r_\omega(x) dm^\omega(x) dP(\omega) + h^{(r)}_{P,m}(\tau)\right) \\
&= \sup_{\substack{\nu \in \mathcal{P}(\Omega\times M) \\ proj_\Omega \nu = P, \nu-\tau-\text{invariant}}} \left(\int r_\omega(x) d\nu(\omega,x) + h^{(r)}_\nu(\tau)\right),
\end{aligned} \tag{28}$$

and this supremum is achieved on η only. Since we take V independent of ω and φ_ω is equi Hölder continuous (we need only f_ω from the compact $supp\Psi P$) then there exists a unique equilibrium state in the variational principle (25) and so both upper and lower relativized large deviation bounds are valid. □

Remark 4. A more detailed proof of Theorem 4 including the case of random expanding only in average maps will appear elsewhere.

Remark 5. The limits (21) and (24) exist for the example of Remark 2, as well, and so the upper bound (19) holds true for it.

Remark 6. The results of this paper can be obtained also for some random maps of the interval such as expanding, piece-wise expanding, and some others.

Remark 7. If μ is a $\tau-$invariant measure whose conditional measures on M are equivalent to m then using, say, a relativized version of the Brin-Katok's local entropy theorem one can prove that $h_\nu^{(r)}(\tau) = -\int \varphi d\nu$ if ν is $\tau-$invariant, $proj_\Omega \nu = P$ and conditional measures of ν on M are equivalent to the Lebesque measure.

4. Stochastic flow.

Let f^t be a stochastic flow generated by a stochastic differential equation on M. Again we may be interested in "full" and relativized large deviations bounds:

$$\limsup_{t\to\infty} \frac{1}{t} \log Em\{x : \frac{1}{t}\int_0^t \delta_{f^s x} \in K\} \le - \inf_{\nu\in K} I(\nu), \tag{29}$$
$K \subset \mathcal{P}(M)$ is closed,

$$\liminf_{t\to\infty} \frac{1}{t} \log Em\{x : \frac{1}{t}\int_0^t \delta_{f^s x} \in G\} \ge - \inf_{\nu\in G} I(\nu) \tag{30}$$
$G \subset \mathcal{P}(M)$ is open, and

$$\limsup_{t\to\infty} \frac{1}{t} \log m\{x : \frac{1}{t}\int_0^t \delta_{f^s x} \in K\} \le - \inf_{\nu\in K} I^{(r)}(\nu), \tag{31}$$
$K \subset \mathcal{P}(M)$ is closed, and

$$\liminf_{t\to\infty} \frac{1}{t} \log m\{x : \frac{1}{t}\int_0^t \delta_{f^s x} \in G\} \ge - \inf_{\nu\in G} I^{(r)}(\nu) \tag{32}$$
$G \subset \mathcal{P}(M)$ is open.

The bounds (29) and (30) are, essentially, the standard large deviations bounds for diffusion processes which correspond to the limit

$$\begin{aligned} Q(V) &= \lim_{t\to\infty} \frac{1}{t} \log E\int_M \exp(\int_0^t V(f^s x) ds) dm(x) \\ &= \lim_{t\to} \frac{1}{t} \log \int_M E_x \exp(\int_0^t V(Y_s) ds) dm(x) \end{aligned} \tag{33}$$

where Y_s is the corresponding diffusion. It is easy to see that $Q(V)$ is the principal eigenvalue of the corresponding generator L and by Donsker-Varadhan's formula

$$Q(V) = \sup_\mu (\int V d\mu - I(\mu)) \tag{34}$$

where $I(\mu) = -\inf_{u>0, u\in D_L} \int \frac{Lu}{u} d\mu$ and the conditions of Theorem 1 hold true here (see [K2]). So (29) and (30) are satisfied. On the other hand the bounds (31), (32) are connected with the limit

$$Q^{(r)} = \lim_{t\to\infty} \frac{1}{t} \log \int_M \exp\left(\int_0^t V(f^s x) ds\right) dm(x)$$

whose existence was not studied, as far as I know.

The large deviations results for diffusions of (29) - (30) type can be generalized to processes of random evolutions which are essentially when one applies several stochastic flows switching from one to another at random moments of time. Such evolutions are governed by stochastic differential equations

$$dY_t = \sigma_{\nu_t}(Y_t) dw_t + b_{\nu_t}(Y_t) dt$$

where ν_t is the random process with states $\{1, \cdots, n\}$ satisfying

$$P_{(y,k)}\{\nu^\varepsilon_{t+\Delta} = m | \nu^\varepsilon_t = \ell, Y_t = x\} = d_{\ell m}(x)\Delta + o(\Delta)$$

for $\Delta \downarrow 0, \ell \neq m, 1 \le \ell, m \le n, d_{\ell m}(x) > 0$ are continuous functions. Then one obtains a two component Markov process $Z^\varepsilon_t = (X^\varepsilon_t, \nu^\varepsilon_t)$ and one can show that the probabilities of large deviations for occupational measures $\frac{1}{t}\int_0^t \delta_{Z^\varepsilon_s} ds$ have bounds of the type (29) - (30) where the functional $I(\nu)$ is defined on vector-measures $\nu = (\nu_1, \cdots, \nu_n)$, $\nu_i \in \mathcal{P}(M), \sum_{i=1}^n \nu_i(M) = 1$ by the formula

$$I(\nu) = -\inf_{u>0} \sum_{k=1}^n \int_M \frac{(Lu)_k}{u_k} d\nu,$$

$$u = (u_1, \cdots, u_n), u_i \in C^2,$$

$$(Lu)_k = \frac{1}{2}(a_k \nabla, \nabla) u_k + (f_k, \nabla) u_k + (b_k, \nabla) u_k + \sum_{\ell=1}^n q_{k\ell}(u_\ell - u_k),$$

$$a_k = \sigma_k \sigma_k^*.$$

References

[DV] M. D. Donsher and S. R. S. Varadhan, Asymptotic evaluation of certain Markov processes for large time I, Comm. Pure Appl. Math 28 (1975), 1-47.

[K1] Y. Kifer, Ergodic Theory of Random Transformations, Birkhäuser, 1986.

[K2] Y. Kifer, Large deviations for dynamical systems and stochastic processes, Trans. Amer. Math. Soc., 231 (1990), 505-524.

[LW] F. Ledrappier and P. Walters, A relativized variational principle for continuous transformations, J. London Math. Soc. (2), 16 (1977), 568-576.

[OP] S. Orey and S. Pelikan, Deviations of trajectories and the defect in Pesin's formula for Anosov diffeomorphisms, Trans. Amer. Math. Soc. 315 (1989), 741-753.

[W] P. Walters, Invariant measures and equilibrium states for some mappings which expand distances, Trans. Amer. Math. Soc. 236, (1978), 121 - 153.

Multiplicative Ergodic Theorems in Infinite Dimensions

Kay-Uwe Schaumlöffel
Institut für Dynamische Systeme, Universität Bremen
Postfach 330 440, D-2800 Bremen 33

1 Introduction

If one thinks of extending Oseledec's Theorem to an infinite-dimensional context (for cocycles on a Banach space X, say) one could naively expect that the only change will consist of the possibility of infinitely many (isolated) Lyapunov exponents. However, a look at the most simple example — the iterates of a single bounded operator — reveals that things are not that simple.

An example due to T. Körner [8] shows that there is a shift operator (denoted by T) on $X = \ell^2$ such that for all elements x_k of an orthonormal basis $\liminf_{n\to\infty} \frac{1}{n} \log \|T^n x_k\| < \limsup_{n\to\infty} \frac{1}{n} \log \|T^n x_k\|$, so the orbits of a linear operator can be quite complicated (cf. also Beauzamy [2, Chapter 3]).

This phenomenon reflects the fact that in infinite dimensions the spectral properties of an operator are more involved than in finite dimensions (recall that in the case of matrices Oseledec's theorem generalizes the fact that $\mathbb{R}^d$ decomposes into the direct sum of generalized eigenspaces). The relationship between the spectrum of a bounded linear operator T and the exponential growth rates of vectors under iterates of T is less straightforward (of course, the logarithm of the spectral radius is the Lyapunov exponent of the norm and eigenvectors grow as they should). Only in the case of a self-adjoint operator (on a Hilbert space X) more can be said in general. Let E_λ denote the associated resolution of the identity and for $x \in X$ let P_x denote the measure induced by $\langle E_\lambda x, x\rangle$. Then= $\lim_{n\to\infty} \frac{1}{n} \log \|T^n x\| = \log \max\{|\inf \operatorname{supp} P_x|, |\sup \operatorname{supp} P_x|\}$ and, if T is invertible, $\lim_{n\to-\infty} \frac{1}{n} \log \|T^n x\| = \log(\sup\{\alpha : P_x(|\lambda| < \alpha) = 0\})$ (these facts were pointed out to the author by I. Lindemann [6]). This shows that even if all vectors have exponential growth rates these rates may not be isolated, see also the example at the end of Section 3.

To summarize, a general cocycle of bounded linear operators (to be denoted by $T(n, \omega)$) over a dynamical system $(\Omega, \mathcal{F}, P, \vartheta)$ may exhibit strange features if compared with the finite-dimensional case: there may be vectors with no exact exponential growth rates and the Lyapunov spectrum (if it exists) may be quite complicated.

To obtain infinite-dimensional versions of Oseledec's Theorem one has to assume additional properties of the cocycle. The most satisfactory situation is when the cocycle consists of *compact* operators. In that case there is a sequence of Lyapunov exponents λ_k (possibly finitely many, otherwise tending to $-\infty$) and an associated decomposition of the state space into a flag $X = V_1(\omega) \supset V_2(\omega) \supset \cdots$ such that $\lim_{n\to\infty} \frac{1}{n} \log \|T(n,\omega)x\| = \lambda_k$ iff $x \in V_k(\omega) \setminus V_{k+1}(\omega)$ where the V_k are subspaces of finite codimension.

This was proved in 1982 by Ruelle [9] in the case X = separable Hilbert space and in 1983 by Mañé [7] under topological assumptions on $(\Omega, \mathcal{F}, P, \vartheta)$ for the case of a general Banach space X. If ϑ is invertible and $T(n,\omega)$ is injective, X splits into finite dimensional Oseledec spaces, i.e. $X = E_1(\omega) \oplus E_2(\omega) \oplus \cdots$ with $\lim_{n\to\pm\infty} \frac{1}{n} \log \|T(n,\omega)x\| = \lambda_k$ for $x \in E_k(\omega) \setminus \{0\}$ (where $T(n,\omega)$ has to be defined in an appropriate way for $n < 0$).

The result of Mañé was extended in 1987 by Thieullen [13] for the case where the cocycle consists of operators which asymptotically behave like compact operators (see Section 3 for precise definitions). Ruelle also treats the case where the operators are compact perturbations of isometries.

The proofs of Ruelle on the one hand and those of Mañé and Thieullen on the other are quite different. Ruelle's proof is based on the spectral theory of non-negative symmetric operators and hence is limited to the Hilbert space setting. The Lyapunov exponents are identified using the norms of the operators in exterior products of X and the flag spaces are constructed as orthogonal complements of eigenspaces of a random operator. In the approach of Mañé and Thieullen the system has to be enlarged to make it invertible. Then the Oseledec spaces for the enlarged system are constructed by collecting all vectors with maximal growth backwards in time, and the flag spaces of the original system arise as images under a projection.

We will describe both approaches in some detail in Sections 2 and 3. Section 4 is devoted to a certain extension of Thieullen's result to a measurable set-up, which requires, however, some extra conditions on the cocycle.

In the sequel, we will mostly treat the case of discrete time. The extension to continuous time under some uniformity in $T(t,\omega)$, $t \in [0,1]$, $\omega \in \Omega$ is straightforward, cf. e.g. Ruelle [9, section 7.5] or [11, section 2.2].

The most relevant applications of the theory of Lyapunov exponents lie in the study of the geometry of attractors of dissipative nonlinear systems in Hilbert spaces. Suppose that $S(t)$ is a nonlinear semigroup on a Hilbert space X which has a compact attractor A. A supports the invariant measures of $S(t)$. If $S(t)$ is differentiable and the linearization on A is compact, Ruelle's theorem applies: for every invariant measure there is an associated Lyapunov spectrum. In general, if there is more than one invariant measure, these spectra will differ and reflect different dynamics of the system on the attractor. To get information about the structure of the attractor itself, one studies *uniform Lyapunov exponents* which exist for all $x \in A$ (not only almost everywhere w.r.t. an invariant measure) and which give e.g. bounds for the dimension of the attractor. See Temam [12, chapter V] for an overview.

2 The Approach of Ruelle

Let $(\Omega, \mathcal{F}, P)$ be a probability space, $\vartheta: \Omega \to \Omega$ a measurable P-preserving transformation, and X a separable Hilbert space. A discrete-time linear cocycle over ϑ is given by a strongly measurable[1] map $T: \Omega \to \mathcal{L}(X)$ (the space of bounded linear operators) via $T(n,\omega) := T(\vartheta^{n-1}\omega) \circ \cdots \circ T(\vartheta\omega) \circ T(\omega)$. Since X is separable, $\log^+ \|T(\cdot)\|$ is measurable.

[1] $T: \Omega \to \mathcal{L}(X)$ is strongly measurable if, for each $x \in X$, $T(\cdot)x$ is an X-valued random variable.

Assume

(H1) $\log^+ \|T(\cdot)\| \in L^1(\Omega, \mathcal{F}, P)$.

Then Kingman's subadditive ergodic theorem implies that the limit

$$l_1(\omega) := \lim_{n\to\infty} \frac{1}{n} \log \|T(n,\omega)\|$$

exists on a ϑ-invariant subset $\Omega_1 \subset \Omega$ of full measure.

Let $\bigwedge^m X$ denote the m-th exterior product of X completed with respect to the canonical scalar product. $T(n,\omega)$ acts as a bounded linear operator on $\bigwedge^m X$ to be denoted by $T(n,\omega)^{\wedge m}$. Since $\|T(n,\omega)^{\wedge m}\| \le \|T(n,\omega)\|^m$, the limits

$$l_m(\omega) := \lim_{n\to\infty} \frac{1}{n} \log \|T(n,\omega)^{\wedge m}\|$$

exist on sets Ω_m of full measure and are ϑ-invariant random variables.
Let $\Omega_0 \subset \bigcap_{m=1}^{\infty} \Omega_m$ be an ϑ-invariant set of full measure such that on Ω_0 also

$$\limsup_{n\to\infty} \frac{1}{n} \log \|T(\vartheta^n\omega)\| \le 0$$

holds.

Define $\bar{l}_m(\omega) := l_m(\omega) - \sum_{k=1}^{m-1} l_k(\omega)$. For general cocycles it is not clear whether the random variables $\bar{l}_m$ are distinct. Assume (for $\omega \in \Omega_0$ fixed)

(H2) There is a $q \in \mathbb{N}$ such that $\{\bar{l}_m(\omega) \ : \ m \in \mathbb{N}\}$ contains $q+1$ different values $\bar{l}_{m_1}(\omega), \dots, \bar{l}_{m_{q+1}}(\omega)$.

Define $\lambda_1(\omega) > \dots > \lambda_{s+1}(\omega)$ by $\lambda_k(\omega) = \bar{l}_{m_k}(\omega)$.

Consider the non-negative definite operator $S(n,\omega) := T(n,\omega)^* T(n,\omega)$. If n is large enough, $S(n)$ has m_q dominating eigenvalues (counting multiplicities). Let $[S(n,\omega)]_q$ denote the composition of $S(n,\omega)$ with the orthogonal projection onto the sum of the corresponding eigenspaces.

THEOREM 1 (Ruelle) *Assume (H1) and (H2).*

(a) The sequence $([S(n,\omega)]_q)^{1/2n}$ converges in norm to a non-negative definite operator $\Lambda_q(\omega)$ with nonzero eigenvalues $\exp(\lambda_1), \dots, \exp(\lambda_q)$.

(b) Denote by $U_1(\omega), \dots, U_q(\omega)$ the corresponding eigenspaces of $\Lambda_q(\omega)$ and define $V_k(\omega)$ $:= (U_1(\omega) \oplus \cdots \oplus U_{k-1}(\omega))^{\perp}$. Then $\lim_{n\to\infty} \frac{1}{n} \log \|T(n,\omega)x\| = \lambda_k$ if $x \in V_k(\omega) \setminus V_{k+1}(\omega)$.

The key step in the proof (cf. Ruelle [9, p. 249]) is to show that the eigenspaces of $[S(n,\omega)]_q$ corresponding to the leading eigenvalues converge. The subspaces $V_k(\omega)$ obviously form a flag decomposition of X: $X = V_1(\omega) \supset V_2(\omega) \supset \cdots$.

If ϑ has a measurable inverse one can consider the cocycle (over ϑ^{-1}) $\tilde{T}(n,\omega) :=$ $T(\vartheta^{-n}\omega)^* \circ \cdots \circ T(\vartheta^{-1}\omega)^*$. The (a.e. defined) limits of $\frac{1}{n} \log \|\tilde{T}(n,\omega)^{\wedge m}\|$ coincide with the invariant random variables l_m and Theorem 1 applies (under (H2)) to $\tilde{T}(n,\omega)$ yielding an operator $\tilde{\Lambda}(\omega)$ with eigenvalues λ_k and corresponding eigenspaces $\tilde{U}_k(\omega)$. Ruelle shows ([9, p. 261]) that $\tilde{U}_1(\omega) \oplus \cdots \oplus \tilde{U}_k(\omega)$ is complementary to $V_k(\omega)$.

Under what conditions is (H2) true a.s.? A first case is given by cocycles consisting of compact operators since in that case $l_m(\omega)/m \to -\infty$ a.s. ([9, p. 254]) and if l_1 is finite, (H2) holds. The limiting operator $\Lambda(\omega)$ is compact and, if there are infinitely many Lyapunov exponents λ_k, they tend to $-\infty$. If there are only finitely many, i.e., there is a maximal q in condition (H2), the kernel of $\Lambda(\omega)$ is nontrivial. The elements in this kernel have growth rate $-\infty$.

A second case is given by operators of the form $T(\omega) =$ isometry + compact such that $T^{-1}(\omega)$ exists and satisfies the integrability condition (H1). Then $\Lambda(\omega)$–identity is compact and either there are infinitely many Lyapunov exponents converging to 0 or there is a finite-codimensional subspace $V_0(\omega)$ such that $\lim_{n\to\infty} \frac{1}{n} \log \|T(n,\omega)x\| = 0$ for the nonzero vectors in $V_0(\omega)$ ([9, p. 254]).

3 The Approach of Mañé and Thieullen

This approach is based on some topological properties of the underlying dynamical system $(\Omega, \mathcal{F}, P, \vartheta)$ and on the cocycle. Let Ω be a Suslin space, $\mathcal{F}$ its Borel σ-field, P a probability measure on $\mathcal{F}$, $\vartheta\colon \Omega \to \Omega$ a continuous P-preserving map, and X an arbitrary (real) Banach space.

$T\colon \Omega \to \mathcal{L}(X)$ is assumed to be P-continuous, i.e. there is a sequence (K_k) of compact subsets of Ω such that $P(\bigcup_{k=1}^{\infty} K_k) = 1$ and $T|_{K_k}$ is continuous. On $\mathcal{L}(X)$ we take the topology induced by the operator norm. This implies that T is uniformly measurable, an assumption which is too strong for certain applications to stochastic dynamical systems, see Section 4. The cocycle induced by $T(\omega)$ will be denoted by $T(n,\omega)$ as before.

Assume

(H3) ϑ is a homeomorphism and $T(\omega)$ is a.s. injective.

This condition will be relaxed (under additional assumptions on Ω).

We have to introduce a few more notations. Let (for a bounded linear operator T) $\varrho(T)$ denote the infimum of all reals $\varepsilon > 0$ such that the image of the unit ball under T can be covered by a *finite* number of balls of radius ε. We have $\varrho(T) = 0$ for a compact operator and $\varrho(T) = 1$ for an isometry (if $\dim X = \infty$).

Obviously $\varrho(T) \le \|T\|$ and, since ϱ is a submultiplivative functional,

$$\lim_{n\to\infty} \frac{1}{n} \log \varrho(T(n,\omega)) =: \kappa(\omega)$$

exists a.s. for every cocycle satisfying (H1).

If $\kappa(\omega) \equiv -\infty$, the cocycle is called *asymptotically compact.*

Assumption (H3) allows to define orbits going to $-\infty$ (in time): if $x \in X_\infty(\omega) := \bigcap_{n=1}^{\infty} T(n, \vartheta^{-n}\omega)X$, x has a unique inverse image under $T(n, \vartheta^{-n}\omega)$ which will be called $T(-n,\omega)x$.

If $\lambda(\omega)$, the Lyapunov exponent of the norm, is bigger than $\kappa(\omega)$ the following version of the Multiplicative Ergodic Theorem is true.

THEOREM 2 (Thieullen) *Assume (H1) and (H3).*
Then there is a $p \in \mathbb{N} \cup \infty$, *a* ϑ*-invariant subset* $\Omega_0 \subset \{\lambda > \kappa\}$ *of full measure (i.e.* $P(\Omega_0) = P(\lambda > \kappa)$*),* ϑ*-invariant P-continuous functions* $\lambda_1(\omega) > \lambda_2(\omega) > \cdots > \lambda_p(\omega) > \kappa(\omega)$ *(if* $p = \infty$, $\lambda_i(\omega) \downarrow \kappa(\omega)$*), a family of T-invariant P-continuous subspaces* $E_i(\omega)$ *of finite dimension, and a family of T-invariant P-continuous subspaces* $F_i(\omega)$ *of finite codimension such that for all* $i = 1, \ldots, p$

1. $F_1(\omega) = X$, $F_i(\omega) = E_i(\omega) \oplus F_{i+1}(\omega)$.

2. *If* $x \in E_i(\omega) \setminus \{0\}$ *then* $x \in X_\infty(\omega)$ *and* $\lim_{n\to\pm\infty} \frac{1}{n} \log \|T(n,\omega)x\| = \lambda_i(\omega)$.

3. $\lim_{n\to\infty} \frac{1}{n} \log \|T(n,\omega)|_{F_{i+1}(\omega)}\| = \lambda_{i+1}(\omega)$ *(if p is finite,* $\lambda_{p+1} := \kappa$*).*

4. *If* $x \in X_\infty(\omega) \cap F_{i+1}(\omega)$ *then* $\limsup_{n\to-\infty} \frac{1}{n} \log \|T(n,\omega)x\| < \lambda_i(\omega)$.

REMARK. The spaces $E_i(\omega)$ are of course the Oseledec spaces of the cocycle. T-invariance mean that $T(\omega)F_i(\omega) \subset F_i(\vartheta\omega)$. Note that if $x \in F_{p+1}(\omega)$ if p is finite or $x \in \bigcap_{i=1}^{\infty} F_i(\omega)$ otherwise the theorem gives no information about the asymptotic behavior of $\|T(n,\omega)x\|$. If, a.s., $\lambda(\omega) = \kappa(\omega)$, then this applies to all $x \in X \setminus \{0\}$. This case, however, is excluded for asymptotically compact cocycles except for the case $\lambda = -\infty$.

The idea of the proof (cf. [7, 13]) is as follows. Define

$$E^\lambda(\omega) := \left\{ x \in X_\infty(\omega) \ : \ \limsup_{n\to-\infty} \frac{1}{n} \log \|T(n,\omega)x\| \geq \lambda(\omega) \right\}.$$

Then on $\{\lambda > \kappa\}$

- $E^\lambda(\omega)$ is a T-invariant subspace,
- $\dim E^\lambda(\omega) < \infty$,
- $\omega \to E^\lambda(\omega) \in \mathcal{G}(X)$ is P-continuous (where $\mathcal{G}(X)$ denotes the Graßmann-manifold of split subspaces of X, see e.g. [1, Example 3.1.8G]),
- $\dim E^\lambda(\omega) \geq 1$,
- $\lim_{n\to\pm\infty} \frac{1}{n} \log \|T(n,\omega)x\| = \lambda(\omega)$ if $x \in E^\lambda(\omega) \setminus \{0\}$.

Define $\lambda_1(\omega) := \lambda(\omega)$ and $E_1(\omega) := E^\lambda(\omega)$. The main step in the proof is to establish the P-continuity of the random subspace $E^\lambda(\omega)$. This allows the construction of a T-invariant P-continuous complementary space $F^\lambda(\omega)$. This is done as follows. Since $E^\lambda(\omega)$ is finite-dimensional and P-continuous, there exists a random family $G(\omega)$ of complementary subspaces such that the projection $p(\omega)$ associated with $X = E^\lambda(\omega) \oplus G(\omega)$ with kernel $E^\lambda(\omega)$ is bounded by $\dim E^\lambda(\omega) + 1$. Consider the following functional equation for a projection-valued measurable map U

$$T(\omega) \circ U(\omega)x = (\mathrm{id} - p(\vartheta\omega)) \circ T(\omega)x + U(\vartheta\omega) \circ p(\vartheta\omega) \circ T(\omega) \circ p(\omega)x$$

for all $x \in G(\omega)$. Since $p(\omega)$ has the mentioned bound, this equation can be solved. Then $F^\lambda(\omega) := \ker U(\omega)$ is T-invariant and complementary to $E^\lambda(\omega)$.

Let $F_2(\omega) := F^\lambda(\omega)$ and $\lambda_2(\omega) := \lim_{n\to\infty} \frac{1}{n} \log \|T(n,\omega)|_{F_2(\omega)}\|$. To proceed one has to show that on $\{\lambda > \kappa\}$

- $\lambda_1(\omega) > \lambda_2(\omega)$.

If $\lambda_2(\omega) = \kappa(\omega)$ a.s., the theorem is proved (with $p = 1$). If not, repeat the steps above with X replaced by $F_2(\omega)$. This procedure either stops (i.e. $\lambda_{p+1}(\omega) = \kappa(\omega)$ for some p) or yields a sequence of Lyapunov exponents and Oseledec spaces. In the latter case, it remains to show that $\inf_{i\geq 1} \lambda_i(\omega) = \kappa(\omega)$.

Assumption (H3) was needed in this construction to consider complete orbits $\{T(n,\omega)x \, : \, n \in \mathbb{Z}\}$. As in the finite dimensional case, if the system is not invertible one can only hope for a flag decomposition of X. This can be shown if Ω has more structure.

THEOREM 3 (Thieullen) *Assume that Ω is a compact metric space and (H1) holds. Then there is a $p \in \mathbb{N} \cup \infty$, a ϑ-invariant subset $\Omega_0 \subset \{\lambda > \kappa\}$ of full measure (i.e. $P(\Omega_0) = P(\lambda > \kappa)$), ϑ-invariant functions $\lambda_1(\omega) > \lambda_2(\omega) > \cdots > \lambda_p(\omega) > \kappa(\omega)$ (if $p = \infty$, $\lambda_i(\omega) \downarrow \kappa(\omega)$), and a family of T-invariant measurable subspaces $V_i(\omega)$ of finite codimension such that for all $i = 1, \ldots, p$*

1. $X = V_1(\omega) \supset V_2(\omega) \supset \cdots \supset V_{p+1}(\omega)$.

2. $\lim_{n\to\infty} \frac{1}{n} \log \|T(n,\omega)x\| = \lambda_i(\omega)$ *if* $x \in V_i(\omega) \setminus V_{i+1}(\omega)$.

3. $\lim_{n\to\infty} \frac{1}{n} \log \|T(n,\omega)|_{F_i(\omega)}\| = \lambda_i(\omega)$
   and, if p is finite, $\lim_{n\to\infty} \frac{1}{n} \log \|T(n,\omega)|_{F_{p+1}(\omega)}\| = \kappa(\omega)$.

The proof of this result is based on two constructions:
1) enlarge $(\Omega, \mathcal{F}, P, \vartheta)$ to obtain an invertible dynamical system $(\Omega', \mathcal{F}', P', \vartheta')$, where $\Omega' = \{(\omega_p)_{p\in\mathbb{Z}} \, : \, \omega_p \in \bigcap_{k=0}^\infty \vartheta^k \Omega \quad \forall p, \vartheta\omega_p = \omega_{p+1}\}$. $(\Omega', \mathcal{F}', P', \vartheta')$ is called the *natural extension* of the original system (see Cornfeld, Fomin, Sinai [3, p. 239]). The projection $\pi_1 : \Omega' \ni (\omega_p)_{p\in\mathbb{Z}} \to \omega_0 \in \Omega$ has an image of full measure.
2) the state space X is enlarged by chosing $X' := \{x' := (x_k)_{k\in\mathbb{N}} \, : \, \sup_{k\geq 0} \|x_k\| < \infty\}$. Pick a sequence (α_k) of positive reals and define a random bounded linear operator on X' by $T'(\omega')x' = (T(\pi_1\omega')x_1, \alpha_1 x_1, \alpha_2 x_2, \ldots)$. $T'(\omega')$ is injective and if $\pi_2 : X' \ni x' = (x_k) \to x_1 \in X$, then $\pi_2 \circ T' = T(\pi_1 \cdot) \circ \pi_2$. The sequence (α_k) can be choosen in such a way that the Lyapunov exponents of T' and T agree (see Thieullen [13, Lemma 4.2]). Theorem 2 applies to the system $(\Omega', \vartheta', X', T')$, so we have on a set Ω_0' of full measure in $\{\lambda(\omega') > \kappa(\omega')\}$ the flag decomposition $X = F_1(\omega') \supset F_2(\omega') \supset \cdots$. Since Ω was compact, $\Omega_0 := \pi_1(\Omega_0')$ is a Borel set of the same measure and Theorem 3 follows with the choice $V_i(\pi_1\omega') := \pi_2 F_i(\omega')$.

We conclude this section with a deterministic example in continuous time showing that if $\lambda = \kappa$, there may exist a continuum of exponential growth rates.

Let $X = L^p(0,2\pi)$ and $T(t)$ be the flow induced by the partial differential equation

$$\begin{aligned}\frac{\partial x}{\partial t}(t,r) &= \sin(r)\frac{\partial x}{\partial r}(t,r) \qquad t \geq 0, r \in [0,2\pi]\\ x(0,r) &= x_0(r),\end{aligned}$$

i.e. $x(t,\cdot) = T(t)x_0$.

By considering the eigenfunctions $x^{(\lambda)}$ of the operator on the right hand side, which has point spectrum $\{\lambda \in \mathbb{C} : |\Re\lambda| < \frac{1}{p}\}$, as initial conditions we find solutions $x(t,r) = \exp(t\lambda)x^{(\lambda)}(r)$ with exponential growth rate λ. So the Lyapunov spectrum contains the interval $(-\frac{1}{p}, \frac{1}{p})$. The flow $T(t)$ has an explicit representation using characteristic curves (see [11]) which shows that on the *dense* subset $\{x_0 \in L^p : \pi \notin \text{ess supp } x_0\}$ of X the corresponding Lyaponov exponent is $-\frac{1}{p}$. This implies that there is no flag decomposition of X (otherwise all vectors with growth rates less than λ are contained in the non-dense subspace V_2), whence $\lambda = \kappa(\geq \frac{1}{p})$.

4 An Extension of Thieullen's Result to a Measurable Set-Up

An important class of examples covered neither by Ruelle's nor by Thieullen's result is given by stochastic partial differential equations of first order (cf. [11, 10]). In nice cases, the induced cocycle consists of isomorphisms and will, in general, not be uniformly measurable.

For a simple example for the second assertion, consider the following situation. Let X be a Hilbert space and A, B densely defined closed operators with common domain. Assume that A generates a strongly continuous semigroup $S(t)$ and that B generates a group $G(t)$ of bounded operators. Assume that S and G *commute*, an assumption which is quite strong.

Then the unique solution of the stochastic partial differential equation

$$dx = Ax\,dt + Bx\circ dw, \quad x(0) = x_0,$$

where w is a scalar Wiener process and $\circ$ denotes the Stratonovich differential, is given by

$$x(t,\omega) = G(w(t,\omega))S(t)x_0.$$

So the cocycle has the representation $T(t,\omega) = G(w(t,\omega))S(t)$. The assumption that $\omega \mapsto T(t,\omega)$ is uniformly measurable is equivalent to the uniform measurability of $t \mapsto G(t)$. A result of Hille and Phillips [5, p.280] shows that this is, in turn, equivalent to uniform continuity implying that B is bounded.

If we drop this assumption, the proof of Mañé and Thieullen of the multiplicative ergodic theorem doesn't work any more.

A (minor) effect is that X has to be separable to ensure the measurability of functions like $\omega \mapsto \|T(n,\omega)\|$ under strong measurability of $T(\cdot)$.

If the cocycle is only assumed to be strongly measurable, the proof of the fact that the spaces $E^\lambda(\omega)$ depend measurably on ω (where the set of completed subspaces $\mathcal{G}(X)$ is endowed with the Borel-σ-field of the Graßmann-manifold structure) breaks down.

To overcome this difficulty, one can work with a different notion of measurability of maps with values in the set $\mathcal{V}(X)$ of closed subspaces of X.

Definition $E: \Omega \to \mathcal{V}(X)$ is called *measurable* if for all $x \in X$ the real-valued function $\mathrm{dist}(x, E(\omega))$ is measurable.

Unfortunately, in order to show that $E^{\lambda}(\cdot)$ is measurable in this sense one has to assume that, if $E: \Omega \to \mathcal{V}(X)$ is as above, the subspaces $T(n, \vartheta^{-n}\omega)E(\vartheta^{-n}\omega)$ have to be closed.

For injective cocycles this is implied by

(H4) $T(\omega)$ has split image.

There is another effect of dropping the assumption that $T(\omega)$ is uniformly measurable which matters in the construction of the invariant complementary space $F^{\lambda}(\omega)$. The first step in the construction was to find *any* complementary subspace $G(\omega)$ such that $G(\cdot)$ is measurable and the norms of the projections associated to the splitting $X = E^{\lambda}(\omega) \oplus G(\omega)$ can be controlled. It is not clear how this can be done under our hypotheses for an arbitrary separable Banach space X.

If X were a Hilbert space, the obvious strategy is to take $G(\omega) = E^{\lambda}(\omega)^{\perp}$. This can be generalized. Since $E^{\lambda}(\omega)$ is finite-dimensional, we can pick a measurable basis $x_1(\omega), \ldots, x_d(\omega)$. Suppose

(H5) If $x \in X$, $\|x\| = 1$, there is a *unique* element $f \in X^*$ of norm one such that $f(x) = 1$ and the map $x \mapsto f$ is continuous.

This hypothesis is fulfilled for Banach spaces with a uniformly convex dual space X^* (c.f. Goldstein [4, p.24–31]), e.g. for the spaces $L^p(Y, \Sigma, \mu)$, (Y, Σ, μ) a measure space and $1 < p < \infty$. In this case f has the form $f(y) := |x(y)|^{p-2}x(y)$. Choose $x_1(\omega), \ldots, x_d(\omega)$ such that the associated f_i's are linearly independent. Then $G(\omega) := \bigcap_{i=1}^{d} \ker f_i(\omega)$ is complementary to $F^{\lambda}(\omega)$ and has the desired properties.

The following theorem is proved in [11, 10] for the Hilbert space case. The generalization using (H5) was pointed out to the author by Z. Brzezniak.

THEOREM 4 *Let $(\Omega, \mathcal{F}, P)$ be a probability space, $\vartheta: \Omega \to \Omega$ a P-ergodic invertible transformation and $T: \Omega \to \mathcal{L}(X)$ strongly measurable such that $T(\omega)$ is a.s. injective. Assume (H1), (H4) and (H5). Then the conclusions of Theorem 2 hold.*

We would like to stress the fact that it is not easy to check if this result gives information about exponential growth rates for a concrete problem since the quantities λ and κ are not easy to estimate (not to speak of exact calculations). Finite dimensional approximations do, in general, not help to decide e.g. if λ is positive. To see this effect consider on $X = l^2$ the random differential equation $\dot{x} = A(t, \omega)x$, where A is diagonal with entries $\xi_n(t, \omega)$, ξ_n a sequence of random processes which are jointly stationary and ergodic. The associated cocycle is again diagonal with entries $\exp \int_0^t \xi_n(s, \omega)\, ds$ so the Lyapunov exponents of any finite-dimensional subsystem are contained in $\{\mathbb{E}\xi_n(0, \cdot)) : n \in \mathbb{N}\}$ ($\mathbb{E}$ denotes the expected value). But if the sequences $\frac{1}{t} \int_0^t \xi_n(s, \omega)\, ds$ converge in a sufficiently non-uniform way, the top Lyapunov exponent will be bigger than $\sup\{\mathbb{E}\xi_n(0, \cdot) : n \in \mathbb{N}\}$ (cf. [6]).

References

[1] R. Abraham, J. E. Marsden, and T. Ratiu. *Manifolds, Tensor Analysis, and Applications.* Springer Verlag, Berlin, Heidelberg, New York, second edition, 1988.

[2] B. Beauzamy. *Introduction to Operator Theory and Invariant Subspaces.* North-Holland, Amsterdam, New York, Oxford, Tokyo, 1988.

[3] I. P. Cornfeld, S. V. Fomin, and Y. G. Sinai. *Ergodic Theory.* Springer Verlag, Berlin, Heidelberg, New York, 1982.

[4] J. A. Goldstein. *Semigroups of Linear Operators and Applications.* Oxford University Press, New York, 1985.

[5] E. Hille and R. S. Phillips. *Functional Analysis and Semi-Groups.* AMS, Providence, 1957.

[6] I. Lindemann. Private communication.

[7] R. Mañé. Lyapunov exponents and stable manifolds for compact transformations. In J. Palis, editor, *Geometric Dynamics*, number 1007 in Springer Lecture Notes, pages 522–577, 1983. (Proc. Rio de Janeiro, 1981).

[8] D. Ruelle. Private communication. 1988.

[9] D. Ruelle. Characteristic exponents and invariant manifolds in Hilbert space. *Ann. Math.*, 115:243–290, 1982.

[10] K.-U. Schaumlöffel. *Zufällige Evolutionsoperatoren für stochastische partielle Differentialgleichungen.* PhD thesis, Universität Bremen, 1990.

[11] K.-U. Schaumlöffel and F. Flandoli. A multiplicative ergodic theorem with application to a first order stochastic hyperbolic equation in a bounded domain. (To appear in *Stochastics*), 1990.

[12] R. Temam. *Infinite-Dimensional Dynamical Systems in Mechanics and Physics.* Springer Verlag, New York, Berlin, Heidelberg, London, Paris, Tokyo, 1988.

[13] P. Thieullen. Fibres dynamiques asymptotiquement compacts — exposants de Lyapunov. Entropie. Dimension. *Ann. Inst. Henri Poincaré, Anal. Non Linéaire*, 4(1):49–97, 1987.

STOCHASTIC FLOW AND LYAPUNOV EXPONENTS FOR ABSTRACT STOCHASTIC PDEs OF PARABOLIC TYPE

Franco Flandoli
Dipartimento di Matematica, Universita' di Torino, Torino, Italy

1. INTRODUCTION

In this paper we are concerned with certain properties of the stochastic flow, and their application to the analysis of Lyapunov exponents, associated to an abstract class of stochastic parabolic partial differential equations, covering a wide range of concrete parabolic problems.

The existence of the stochastic flow for concrete second order parabolic equations in bounded and unbounded domains has been treated in [RS], [T], [FS], [K]. Some abstract results can be obtained by the robust equation approach along the lines of [DIT], but only under rather restrictive conditions. No counterexample to the existence of the flow is known for parabolic problems (in contrast to the case of delay equations, cf. [M]), so that the question of general abstract affirmative results arises. As to the problem of Lyapunov exponents, some results are given in [FS] for second order parabolic equations in bounded domains, applying the infinite dimensional version of Oseledec Theorem due to [R]. A wide discussion of all these subjects can be found in [S].

A new method to prove the existence of stochastic flows for abstract parabolic equations has been proposed in [BF], based on the Hilbert-Schmidt property of the solution mapping. The abstract results of [BF] apply to several classes of concrete problems in bounded domains, like equations with Dirichlet or Neumann or periodic boundary conditions, equations of order greater than two, and systems of parabolic equations. Moreover, for all these examples, the abstract method allows us to prove the existence of the flow in various function spaces (under different compatibility conditions on the coefficients).

This paper is a continuation of [BF]. The final aim here is to study Lyapunov exponents by means of Oseledec type theorems. To this end the results of [BF] are not sufficient. A number of improvements concerning the continuity and uniformity in time of the flow are needed. These new properties are proved in section 2. Then in section 3 we apply these results to the analysis of Lyapunov exponents, along the lines of [FS]. Finally, some examples are sketched in section 4 (see [BF] for a more extensive discussion).

We remark that in our final result (Theorem 3.2), Lyapunov exponents are only given as upper limits in time, because the abstract approach of section 2 does not allow us to prove an integrability property of the flow which is usually employed to prove that upper and lower limits coincide (cf. [FS]). This property seems to be satisfied when the operators B^j in equation (2.2) are bounded, but the problem is open in general.

2. STOCHASTIC FLOW

Let $\Omega = \{f \in C([0,\infty);R^m), f(0)=0\}$ endowed with the topology of uniform convergence on compact sets. Let $\mathcal{F}$ be the Borel σ-field in Ω, P the Wiener measure, $w(t) = (w^1(t),...,w^m(t))$ the canonical Wiener process on $(\Omega,\mathcal{F},P)$, and $\mathcal{F}_t$ the σ-field generated by $\{w(s);\ 0 \leq s \leq t\}$.

Let H be a separable Hilbert space (norm $|.|_0$, inner product $(.,.)_0$), $A:D(A)\subset H\to H$ a strictly negative selfadjoint operator in H (i.e. $A=A^*$, $(Ax,x)_0\leq\nu|x|_0^2$, $\forall x\in D(A)$, for some $\nu<0$), e^{tA} ($t\geq0$) the analytic semigroup generated by A, $V_\alpha = D((-A)^{\alpha/2})$ (norm $|x|_\alpha := |(-A)^{\alpha/2}x|_0$ and corresponding inner product $(.,.)_\alpha$).

We consider the abstract stochastic equation in H (the assumptions on B^j will be specified later)

$$(2.1)\qquad du(t) = Au(t)dt + \sum_{j=1}^{m} B^j u(t)dw^j(t),\quad t\geq0,\qquad u(0)=u_0,$$

interpreted in the sequel always in the mild form

$$(2.2)\qquad u(t) = e^{tA}u_0 + \sum_{j=1}^{m}\int_0^t e^{(t-s)A}B^j u(s)dw^j(s),\quad t\geq0.$$

Definition 2.1. Given $b>a\geq0$ and $\alpha\in R$, we denote by $\mathcal{L}_\alpha(a,b)$ the space of all adapted processes $f\in L^2(\Omega\times[a,b];V_{\alpha+1})$ such that $f\in L^2(\Omega;C([a,b];V_\alpha))$. $\mathcal{L}_\alpha(a,b)$ will be endowed with the natural topology. Moreover, we say that $f\in\mathcal{L}_\alpha(a,\infty)$ if $f\in\mathcal{L}_\alpha(a,b)$ for all $b>a$.

Definition 2.2. Given $\alpha\in R$ assume that B^j are bounded linear operators from $V_{\alpha+1}$ to V_α. We say that problem (2.2) is well posed in V_α if for every $u_0\in L^2(\Omega;\mathcal{F}_0;V_\alpha)$ there exists a unique solution of (2.2) in $\mathcal{L}_\alpha(0,\infty)$ and the mappings $u_0\to u|_{[0,T]}$ are continuous from $L^2(\Omega;\mathcal{F}_0,V_\alpha)$ to $\mathcal{L}_\alpha(0,T)$, for all $T>0$.

Various conditions of well posedness, depending on the value of α, can be found in [BF], [D], [FS], [KR], [P] and others works.

Lemma 2.1. Let $\alpha<\beta$ be fixed real numbers. If (2.2) is well posed in V_α and V_β, then it is well posed in V_γ for all $\gamma\in[\alpha,\beta]$. Moreover, for every $u_0\in L^2(\Omega;\mathcal{F}_0;V_\alpha)$ and $\varepsilon>0$, the solution u belongs to $\mathcal{L}_\beta(\varepsilon,\infty)$, and the mappings $u_0\to u|_{[\varepsilon,T]}$ are continuous from $L^2(\Omega;\mathcal{F}_0;V_\alpha)$ into $\mathcal{L}_\beta(\varepsilon,T)$ for all $T>\varepsilon$.

Using this lemma we can prove the existence of the stochastic flow associated to equation (2.2) under a certain Hilbert-Schmidt assumption. This result improves a similar one of [BF].

Theorem 2.2. Assume that for a given $\alpha\in R$ equation (2.2) is well posed in both V_α and $V_{\alpha-\delta}$, where $\delta>0$ is such that $(-A)^{-\delta/2}$ is a Hilbert-Schmidt operator. Then there exists the stochastic flow associated to equation (2.2) in V_α, in the following sense. There exists a full measure set $\Omega_0\subset\Omega$ such that for all $\omega\in\Omega_0$ there is a family $\{\varphi_t(\omega),\ t\geq0\}$ of bounded linear operators in V_α with the following properties:

(2.6) $\varphi_0(\omega) = I$, $\varphi_t(\omega)\in L_2(V_\alpha)$, for all $t>0$ and $\omega\in\Omega_0$ (where $L_2(V_\alpha)$ denotes the space of all Hilbert-Schmidt operators in V_α);

(2.7) $t\to\varphi_t(\omega)$ is continuous from $(0,\infty)$ to $L_2(V_\alpha)$, for all $\omega\in\Omega_0$;

(2.8) $\omega\to\varphi_t(\omega)u_0$ is strongly measurable, for all $u_0\in V_\alpha$ and $t\geq0$;

(2.9) for every fixed $u_0 \in V_\alpha$ there exists a full measure set $\Omega(u_0) \in \Omega_0$ such that

$$u(t,\omega;u_0) = \varphi_t(\omega)u_0 \quad \text{for all } t\geq 0 \text{ and } \omega\in\Omega(u_0),$$

where $u(t,\omega;u_0)$ is the solution of (2.2) corresponding to u_0;

(2.10) for all $T>\varepsilon>0$ the function $c_{\varepsilon,T}(\omega) = \sup\{|\varphi_t(\omega)|^2_{L_2(V_\alpha)},\ \varepsilon\leq t\leq T\}$ is a random variable, and

$$E\,sup\{|\varphi_t(\omega)|^2_{L_2(V_\alpha)},\ \varepsilon\leq t\leq T\} < \infty;$$

(2.11) the family $\{\varphi_t(\omega),\ t\geq 0\}$ is unique in the following sense: if $\{\varphi_t'(\omega),\ t\geq 0\}$ is another family of bounded linear operators in V_α defined for ω in a full measure set Ω_0', satisfying at least property (2.9), then there is a full measure set $\Omega_0''\subset\Omega_0\cap\Omega_0'$ such that $\varphi_t(\omega) = \varphi_t'(\omega)$ for all $t\geq 0$ and $\omega\in\Omega_0''$;

(2.12) the following cocycle property for $\varphi_t(\omega)$, with respect to the canonical shift Θ_t on Ω defined as $(\Theta_t\,\omega)(s) = \omega(t+s)$ $(t,s>0)$ holds in the following sense: for every fixed $s\geq 0$ there exists a full measure set $\Omega_0(s)\subset\Omega_0$ such that $\Theta_s\Omega_0(s)\subset\Omega_0$ and

$$\varphi_t(\Theta_s\omega)\varphi_s(\omega) = \varphi_{t+s}(\omega) \quad \text{for all } t\geq 0 \text{ and } \omega\in\Omega_0(s).$$

Proof. Step 1 (construction of $\varphi_t(\omega)$).

Let $\{e_k\}$ be a complete orthonormal system in V_α, and let $D\subset V_\alpha$ be the countable dense set of all finite linear combinations of the form $\sum_{k=1}^{n} a_k\, e_k$, with $n\in N$ and $a_k\in Q$. For all $u_0\in D$ denote by $u(t,\omega;u_0)$ a continuous version of the solution of (2.2) corresponding to u_0, defined for all $t\geq 0$ and $\omega\in\Omega$ with values in V_α. Since D is countable, and the solution of (2.2) depends linearly on u_0, we can find a full measure set $\Omega_0'\subset\Omega$ such that

(2.13) $$u(t,\omega;\sum_{k=1}^{n} a_k\, e_k) = \sum_{k=1}^{n} a_k\, u(t,\omega;e_k) \quad \text{for all } \omega\in\Omega_0',\ t\geq 0,\ n\in N \text{ and } a_1,\dots,a_n\in Q.$$

Define also the random variables

$$\alpha_n(\omega) = \sum_{k=1}^{\infty} sup\{|u(t,\omega,e_k)|^2_\alpha,\ 1/n\leq t\leq n\},\quad n\in N,\ \omega\in\Omega.$$

From the assumptions of the theorem, and Lemma 2.1, for every $n\in N$ there exists a constant c_n such that

(2.14) $$E\alpha_n \leq c_n \sum_{k=1}^{\infty} |e_k|^2_{\alpha-\delta} \leq c_n \sum_{k=1}^{\infty} |(-A)^{-\delta/2} e_k|^2_\alpha < \infty.$$

This implies that all the random variables $\alpha_n(\omega)$ are a.s. finite, so that we can choose a full measure set

$\Omega_0'' \subset \Omega$ such that

(2.15) $\alpha_n(\omega) < \infty$ for all $\omega \in \Omega_0''$ and $n \in N$.

Let $\Omega_0 = \Omega_0' \cap \Omega_0''$. For $\omega \in \Omega_0$ and $t>0$, define the mapping $\varphi_t(\omega)$ from $\mathcal{D}$ to V_α as

(2.16) $\varphi_t(\omega)u_0 := u(t,\omega;u_0), \quad u_0 \in \mathcal{D}$.

If $u_0 = \sum_{j=1}^{m} a_k e_k$ and $t \in [1/n,n]$, then

(2.17) $$|\varphi_t(\omega)u_0|_\alpha^2 = |\sum_{j=1}^{m} a_k \varphi_t(\omega)e_k|_\alpha^2 \leq \sum_{j=1}^{m} a_k^2 \sum_{j=1}^{m} |\varphi_t(\omega)e_k|_\alpha^2 \leq \alpha_n(\omega)|u_0|_\alpha^2.$$

Since $\varphi_t(\omega)$ is linear on $\mathcal{D}$ (considered as a vector space on Q) and bounded by (2.17), one can extend $\varphi_t(\omega)$ in a unique way ($\mathcal{D}$ is dense in V_α) to a bounded linear operator in V_α, still denoted by $\varphi_t(\omega)$. Setting $\varphi_0(\omega) = I$ for all $\omega \in \Omega_0$, we have defined a family of bounded linear operators $\{\varphi_t(\omega), t \geq 0\}$ on a full measure set Ω_0.

Step 2 (proof of (2.6)-(2.11)).

Fix $\omega \in \Omega_0$, $t_0>0$, and $n \in N$ such that $t_0 \in (1/n,n)$. Since $\sum_{k=1}^{\infty} |\varphi_t(\omega)e_k|_\alpha^2 \leq \alpha_n(\omega) < \infty$, property (2.6) holds true. As to (2.7), given $\varepsilon>0$, by (2.15) there exists $M=M(\omega,n) \in N$ such that

$$\sum_{k=M+1}^{\infty} \sup\{|u(t,\omega;e_k)|_\alpha^2, 1/n \leq t \leq n\} < \varepsilon/2.$$

Hence, for $t \in [1/n,n]$,

$$|\varphi_t(\omega) - \varphi_{t_0}(\omega)|_{L_2(V_\alpha)}^2 \leq \varepsilon/2 + \sum_{k=1}^{M} |\varphi_t(\omega)e_k - \varphi_{t_0}(\omega)e_k|_\alpha^2$$

so that (2.7) follows from the continuity in t of $\varphi_t(\omega)e_k = u(t,\omega;e_k)$, $k=1,\dots,M$.

Let us now prove (2.9), which yields also (2.8) as an obvious consequence. Fix $u_0 \in V_\alpha$ and let $\{u_n\}$ be a sequence in $\mathcal{D}$ converging to u_0. Then, by the well-posedness of (2.2) in V_α, we can find a full measure set $\Omega(u_0)$ and a subsequence of $\{u_n\}$, still denoted by $\{u_n\}$ for simplicity, such that $u(t,\omega;u_n) \to u(t,\omega;u_0)$ uniformly in t over the compact sets, for all $\omega \in \Omega(u_0)$. Thus (2.9) follows as $n \to \infty$ from (2.16).

The inequality in (2.10) is an easy consequence of (2.14). As to the measurability of $c_{\varepsilon,T}$, first note that, for fixed $t>0$, $|\varphi_t(\omega)|_{L_2(V_\alpha)}$ is a random variable by (2.8) and the definition of Hilbert-Schmidt norm; second, by (2.7), the supremum in the definition of $c_{\varepsilon,T}$ can be taken only over the rational numbers in $[\varepsilon,T]$, proving the measurability of $c_{\varepsilon,T}$.

Finally, let us prove the uniqueness property (2.11). If $\{\varphi_t'(\omega), t \geq 0, \omega \in \Omega_0'\}$ is a second family with property (2.9), we can find a full measure set $\Omega_0'' \subset \Omega \cap \Omega_0'$ such that $\varphi_t(\omega)u_0 = \varphi_t'(\omega)u_0$ for all $t \geq 0$, $u_0 \in \mathcal{D}$, and $\omega \in \Omega_0''$. This identity yields (2.11) because $\varphi_t(\omega)$ and $\varphi_t'(\omega)$ are bounded operators (for fixed $t \geq 0$ and $\omega \in \Omega_0''$), and $\mathcal{D}$ is dense in V_α.

Step 3 (proof of the cocycle property (2.12)).

Let us fix $s\geq 0$. For all $u_0\in L^2(\Omega,\mathcal{F}_s,P)$ denote by $u(t,\omega;s;u_0)$ the solution in $\mathcal{L}_\alpha(s,\infty)$ of the equation

$$u(t) = e^{(t-s)A}u_0 + \sum_{j=1}^{m}\int_s^t e^{(t-\sigma)A}B^j u(\sigma)dw^j(\sigma), \quad t\geq s. \tag{2.18}$$

We will prove that:

(2.19) given $x\in V_\alpha$ there exists full measure set $\Omega(s,x)\subset\Omega_0$ such that $\Theta_s\Omega(s,x)\subset\Omega_0$ and

$$\varphi_{t-s}(\Theta_s\,\omega)x = u(t,\omega;s;x) \quad \text{for all } t\geq s \text{ and } \omega\in\Omega(s,x);$$

(2.20) given $u_0\in L^2(\Omega,\mathcal{F}_s,P)$ there exists a full measure set $\Omega(s,u_0)\subset\Omega_0$ such that $\Theta_s\Omega(s,u_0)\subset\Omega_0$ and

$$\varphi_{t-s}(\Theta_s\omega)u_0(\omega) = u(t,\omega;s;u_0) \quad \text{for all } t\geq s \text{ and } \omega\in\Omega(s,u_0);$$

(2.21) given $x\in V_\alpha$, there exists a full measure set $\Omega'(s,x)\subset\Omega_0$ such that

$$u(t,\omega;s;u_{s,x}) = u(t,\omega;s;x) \quad \text{for all } t\geq s \text{ and } \omega\in\Omega(s,x),$$

where $u_{s,x}(\omega) := u(t,\omega;0;x)$.

From these facts we can deduce (2.12) as follows. Let $\mathcal{D}\subset V_\alpha$ as in step 1. Let $\Omega_0(s)$ be the intersection of all $\Omega'(s,x)$ and $\Omega'(s,u_{s,x})$ when x varies in $\mathcal{D}$. If $\omega\in\Omega_0(s)$, $x\in\mathcal{D}$, and $t\geq s$, we have

$$\varphi_{t-s}(\Theta_s\omega)\varphi_s(\omega)x = \varphi_{t-s}(\Theta_s\omega)u(s,\omega;0;x) \tag{2.22}$$

$$= u(t,\omega;s;u_{s,x}) = u(t,\omega;0;x) = \varphi_t(\omega)x$$

(we have used (2.16) in the first and last identities). Since $\mathcal{D}$ is dense in V_α, the operator identity (2.13) follows from (2.22) (by change of time).

Let us prove (2.19). Given $x\in V_\alpha$, let $z(t,\omega)$ and $v(t,\omega)$ be the processes defined as

$$\begin{aligned} &z(t,\omega) = \varphi_t(\omega)x \\ &v(t,\omega) = z(t-s,\Theta_s\omega) = \varphi_{t-s}(\Theta_s\,\omega)x \end{aligned} \tag{2.23}$$

for $t\geq s$ and $\omega\in\Omega_0'$, where $\Omega_0'=\Omega_0\cap\Theta_s^{-1}\Omega_0$. Then, by (2.2) and (2.9), there exists a full measure set $\Omega_0''\subset\Omega_0'$ such that

$$z(t,\omega) = e^{tA}x + \sum_{j=1}^{m} I^j(t,\omega),$$

(2.24)

$$v(t,\omega) = e^{(t-s)A}x + \sum_{j=1}^{m} I^j(t-s,\theta_s\omega)$$

for all $\omega\in\Omega_0''$, where $I^j(t,\omega)$ is a continuous version of $\int_0^t e^{(t-\sigma)A}B^j z(\sigma)dw^j(\sigma)$. Let $L^j(t,\omega)$ be a continuous version of $\int_s^t e^{(t-r)A}B^j z(r-s,\theta_s.)dw^j(r)$. Using the definition of Ito integral (and the sample continuity of I^j and L^j) we can find a full measure set $\Omega_0'''\subset\Omega_0''$ such that

$$I^j(t-s,\theta_s\omega) = L^j(t,\omega) \quad \text{for all } t\geq s \text{ and } \omega\in\Omega_0'''.$$

Thus, recalling (2.23) and (2.24), on Ω_0''' we have

$$v(t) = e^{(t-s)A}x + \sum_{j=1}^{m} L^j(t) = e^{(t-s)A}x + \sum_{j=1}^{m}\int_s^t e^{(t-r)A}B^j v(r)dw^j(r), \quad \text{for all } t\geq s.$$

This equation has the unique solution $u=u(t,\omega;s;x)$ in $\mathcal{L}_\alpha(s,\infty)$; hence $v=u$ in $\mathcal{L}_\alpha(s,\infty)$, yielding (2.19) by definition of v.

Let us now prove (2.20). Fix $u_0\in L^2(\Omega,\mathcal{F}_s,P)$. We note that the expression $\varphi_t(\theta_s\omega)u_0(\omega)$ is well defined a.s.; its measurability will be clear from the following approximation argument.

Let $\{u_n\}$ be a sequence converging to u_0 in $L^2(\Omega,\mathcal{F}_s,P)$ and a.s., with $u_n = \sum_{k=1}^{n} x_{n,k}\, 1_{A_{n,k}}$ where $x_{n,k}\in V_\alpha$ and $A_{n,k}\in\mathcal{F}_s$. By (2.19), there exists a full measure set $\Omega_0'\subset\Omega_0$ such that

$$\varphi_{t-s}(\theta_s\omega)x_{n,k} = u(t,\omega;s;x_{n,k}) \quad \text{for all } n,k\in N,\ t\geq s,\ \omega\in\Omega_0'.$$

Moreover, by uniqueness in $\mathcal{L}_\alpha(s,\infty)$ of the solution of (2.18), there exists a full measure set $\Omega_0''\subset\Omega_0'$ such that

$$u(t,\omega;s;x_{n,k}1_{A_{n,k}}) = 1_{A_{n,k}}(\omega)u(t,\omega;s;x_{n,k}) \quad \text{for all } n,k\in N,\ t\geq s,\ \omega\in\Omega_0''$$

(this is proved by multiplying (2.18) by $1_{A_{n,k}}$). From the last two identities and the superposition property for (2.18), we can find a full measure set $\Omega_0'''\subset\Omega_0''$ such that

(2.25) $$\varphi_{t-s}(\theta_s\omega)u_n = u(t,\omega;s;u_n) \quad \text{for all } n\in N,\ t\geq s,\ \omega\in\Omega_0'''.$$

Since $u(t,\omega;s;u_n)$ converges to $u(t,\omega;s;u_0)$ in $\mathcal{L}_\alpha(s,T)$ for all $T>s$, there exists a subsequence of $\{u_n\}$, still denoted by $\{u_n\}$ for simplicity, and there exists a full measure set $\Omega_0''''\subset\Omega_0'''$, such that

(2.26) $u(t,\omega;s;u_n) \to u(t,\omega;s;u_0)$ uniformly in t over the compact sets, for all $\omega\in\Omega_0''''$.

From the a.s. convergence of $\{u_n\}$ to u_0, along with (2.25) and (2.26), we obtain (2.20).

Finally, (2.21) follows by uniqueness in $\mathcal{L}_\alpha(s,\infty)$ of the solution of equation (2.18) with $u_0=u_{s,x}$, using a standard computation. This completes the proof of the theorem.

3. LYAPUNOV EXPONENTS

Discrete-time Lyapunov exponents can be easily studied using Theorem 2.2 and Ruelle's extension of Oseledec theorem [R]:

Theorem 3.1. Let $t_0>0$ be fixed. Under the assumptions of Theorem 2.2 there exist:

(3.1.i) a full measure set $\Omega'=\Omega'(t_0)\subset\Omega_0$ (Ω_0 given by Theorem 2.2),

(3.1.ii) a decreasing sequence of real numbers $\{\lambda_k\,;\,k\in I\}$, where either $I=N$ and $lim_{n\to\infty}\lambda_n=-\infty$, or $I=\{1,...,s-1\}$ in which case we set $\lambda_s=-\infty$,

(3.1.iii) a family $\{L_k(\omega);\ k\in I,\ \omega\in\Omega'\}$ of subspaces of V_α, with $L_1(\omega)\supset L_2(\omega)\supset\ldots$ for all $\omega\in\Omega$, such that

$$(3.2)\qquad lim_{n\to\infty}\frac{1}{nt_0}\log|\varphi_{nt_0}(\omega)u_0|_\alpha=\lambda_k\qquad \text{iff } u_0\in L_k(\omega)-L_{k+1}(\omega),$$

for all $\omega\in\Omega$ and $k\in I$ (where $L_{s+1}(\omega):=\varnothing$ for all $\omega\in\Omega$ if $I=\{1,...,s\}$).

Proof. It is sufficient to apply Corollary 2.2 of [R] to the cocycle $T^n(\omega):=T(\Theta_{n-1}\omega)\text{o}\ldots\text{o}T(\omega)$, where $T(\omega):=\varphi_{t_0}(\omega)$. The assumptions of [R] are satisfied; in particular:
i) T is strongly measurable (cf. (2.8)),
ii) $T(\omega)$ is a.s. compact (in fact Hilbert-Schmidt, cf. (2.6)),
iii) $log^+|T(.)|\in L^1(\Omega,\mathcal{F},P)$ (obvious from (2.10)),
iv) Θ_1 is ergodic.
See also [FS] for a more detailed presentation of this argument.

Theorem 3.2. Under the hypotheses of Theorem 2.2, take any $t_0>0$, and let $\Omega'=\Omega'(t_0)$, $\{\lambda_k\,;\,k\in I\}$, $\{L_k(\omega);\ k\in I,\ \omega\in\Omega\}$ the corresponding full measure set, Lyapunov exponents and subspaces defined in Theorem 3.1. Then there exists a full measure set $\Omega''\subset\Omega'$ such that

$$(3.3)\qquad lim\ sup_{t\to\infty}\frac{1}{t}\log|\varphi_t(\omega)u_0|_\alpha=\lambda_k\qquad \text{iff } u_0\in L_k(\omega)-L_{k+1}(\omega),$$

for all $\omega\in\Omega'$ and $k\in I$.

Proof. The proof is classical, but we give it for completeness.

Let $t_0>0$ be fixed. For all $t>0$, denote by n_t the natural number such that $(n_t+1)t_0\le t<(n_t+2)t_0$, and set $\delta_t=t-n_tt_0$ (thus $t_0\le\delta_t\le 2t_0$). If $\omega\in\Omega$ and $k\in I$ we have

$$(3.4)\qquad |\varphi_t(\omega)u_0|_\alpha=|\varphi_{\delta_t}(\Theta_{n_tt_0}\omega)\varphi_{n_tt_0}(\omega)u_0|_\alpha\ \le$$

$$\leq \sup\{|\varphi_\delta(\Theta_{n_t t_0}\omega)|_{L_2(V_\alpha)}\ ;\ t_0\leq\delta_t\leq 2t_0\}\ |\varphi_{n_t t_0}(\omega)u_0|_\alpha\ =\lambda(\Theta_{n_t t_0}\omega)|\varphi_{n_t t_0}(\omega)u_0|_\alpha$$

where $\lambda(\omega) = \sup\{|\varphi_\delta(\omega)|_{L_2(V_\alpha)}\ ;\ t_0\leq\delta_t\leq 2t_0\}$. From (2.10), $log^+\lambda(\omega)\in L^1(\Omega,\mathcal{F},P)$, so that, by the Ergodic Theorem,

$$(3.5)\quad lim_{t\to\infty}\ \frac{1}{t}\ log^+\lambda(\Theta_{n_t t_0}\omega) = lim_{n\to\infty}\ \frac{1}{nt_0}\ log^+\lambda(\Theta_{nt_0}\omega) = 0 \quad \text{a.s.}$$

Therefore, from (3.4) it follows

$$(3.6)\quad lim\ sup_{t\to\infty}\ \frac{1}{t}\ \log|\varphi_t(\omega)u_0|_\alpha \leq lim\ sup_{t\to\infty}\ \frac{1}{t}\ log^+\lambda(\Theta_{n_t t_0}\omega)$$
$$+\ lim\ sup_{t\to\infty}\ \frac{1}{t}\ \log|\varphi_{n_t t_0}(\omega)u_0|_\alpha\ = \lambda_k \qquad \text{iff } u_0\in L_k(\omega)\text{-}L_{k+1}(\omega),$$

for all ω in a full measure subset Ω'' of Ω' (taking into account (3.5)). Property (3.3) follows now from (3.6), recalling (3.2).

4. EXAMPLE

Let $D\subset R^d$ be a bounded open domain with C^∞ boundary ∂D. Consider the following stochastic parabolic equation in D:

$$(4.1)\qquad du(t,x) = A_0u(t,x)dt + \sum_{j=1}^{m} B^ju(t,x)dw^j(t) \quad, t\in[0,T],\ x\in D,$$

with the initial condition

$$(4.2)\qquad u(0,x) = u_0(x) \quad, x\in D,$$

and either Dirichlet boundary condition

$$(4.3.i)\quad u(t,x) = 0 \quad, t\in[0,T],\ x\in\partial D\ ,$$

or Neumann boundary condition

$$(4.3.ii)\quad \frac{\partial u(t,x)}{\partial\nu_A} = 0 \quad, t\in[0,T],\ x\in\partial D$$

(in a hyper-rectangle D it is possible to treat also the case of periodic boundary conditions, as in [BF]). Here A_o and B^j are the differential operators

$$A_ou(x) = \sum_{i,j=1}^{d} \frac{\partial}{\partial x_j}(a_{ij}(x)\frac{\partial u(x)}{\partial x_i})\quad,\qquad B^ju(x) = \sum_{i=1}^{d} b_{ij}(x)\frac{\partial u(x)}{\partial x_i}$$

with regular coefficients $a_{ij}(x)$ and $b_{ij}(x)$ satisfying the usual joint ellipticity condition

$$\sum_{i,j=1}^{d} (a_{ij}(x) - \sum_{k=1}^{m} b_{ik}(x)b_{jk}(x))\xi_i \xi_j \geq \rho \sum_{i=1}^{d} \xi_i^2 \qquad \forall(\xi_1,\ldots,\xi_d)\in R^d,\ \forall x\in D$$

for some constant $\rho>0$, while $\frac{\partial u(x)}{\partial \nu_A}$ denotes the conormal derivative

$$\frac{\partial u(x)}{\partial \nu_A} = \sum_{i,j=1}^{d} a_{ij}(x)\frac{\partial u(x)}{\partial x_i}\nu_j(x)$$

where $\nu = (\nu_1,\ldots,\nu_d)$ is the outward normal to ∂D. Let us assume for simplicity that the matrix $\{a_{ij}(x)\}$ is symmetric, and that the coefficients $b_{ij}(x)$ have compact support contained in D(in the case of periodic boundary conditions it is sufficient to assume that the coefficients $b_{ij}(x)$ are periodic).

Let $H = L^2(D)$, $D(A) = \{u\in H^2(D)$: $u = 0$ on $\partial D\}$ in case (4.3.i), while $D(A) = \{u\in H^2(D)$: $\frac{\partial u}{\partial \nu_A} = 0$ on $\partial D\}$ in case (4.3.ii), $Af=A_0 f$, $\forall f\in D(A)$. Thanks to the assumptions on $b_{ij}(x)$, one can show as in [BF] that both problems (4.1-2-3i) and (4.1-2-3ii) are well posed in all the spaces V_α, for every value of α. Restricting ourselves to non-negative values of α for sake of simplicity, we note that V_α coincides with a subspace of $H^\alpha(D)$, determined by certain boundary conditions, and the graph topology in V_α is equivalent to the standard topology of $H^\alpha(D)$. Recall also that $(-A)^{-\delta/2}$ is Hilbert-Schmidt for $\delta > d/2$. Therefore, Theorems 2.2 and 3.2 can be applied. We have the following result:

(4.4) for every $\alpha\geq 0$, there exist the stochastic flows in V_α (in the sense of Theorem 2.2) for both the Dirichlet and the Neumann boundary value problems (4.1-2-3i) and (4.1-2-3ii).

Moreover, applying for instance Theorem 3.2 with $\alpha\in N$, we have:

(4.5) for every $n\in N$, there exist Lyapunov exponents $\{\lambda_k^{(n)}$; $k\in I^{(n)}\}$ and subspaces $\{L_k^{(n)}(\omega)$; $k\in I^{(n)}$, $\omega\in\Omega\}$ as in (3.1.ii-iii) such that, for all infinitly differentiable functions u_0 with compact support in D (therefore $u_0\in V_\alpha$ for any α) there exists a full measure set $\Omega(u_0)\subset\Omega$ such that

$$\lim\sup_{t\to\infty} \frac{1}{t} \log|u(t,\omega;u_0)|_{H^n(D)} = \lambda_k^{(n)} \qquad \text{iff } u_0\in L_k^{(n)}(\omega)\text{-}L_{k+1}^{(n)}(\omega),$$

for all $n\in N$, $\omega\in\Omega(u_0)$, and $k\in I^{(n)}$.

Similar results holds true in the case of periodic boundary conditions, as well as for higher order parabolic equations and for parabolic systems; preliminary results in this direction can be found in [BF].

Let us mention that the result (4.5) arises the problem to see if the Lyapunov exponents $\lambda_k^{(n)}$, or simply the top Lyapunov exponent $\lambda_1^{(n)}$, depend on n. From the properties of injection of the spaces , one can infer that

(4.6) $\lambda_1^{(0)} \leq \lambda_1^{(1)} \leq \ldots \leq \lambda_1^{(n)} \leq \ldots$

but it would be of interest to understand whether strict inequalities hold in (4.6).

REFERENCES

[BF] Z. Brzezniak, F. Flandoli, Regularity of solutions and random evolution operator for stochastic parabolic equations, Proc. 3rd Int. Conf. on Stochastic Partial Differential Equations, Trento 1990, to appear.

[D] G. Da Prato, Some results on linear stochastic evolution equations in Hilbert spaces by the semigroup method, *Stoch. Anal. Appl.* 1 (1983), 57-88.

[DIT] G. Da Prato, M. Iannelli, L. Tubaro, Some results on linear stochastic differential equations in Hilbert spaces, *Stochastics* 6 (1982), 105-116.

[FS] F.Flandoli, K.-U. Schaumlöffel, Stochastic parabolic equations in bounded domains: random evolution operator and Lyapunov exponents, *Stochastics* 29 (1990), 461-485.

[KR] N.V. Krylov, B.L. Rozovskii, Stochastic evolution equations, *J. Sov. Math.* 16 (1981), 1233-1277.

[K] H. Kunita, *Stochastic Flows and Stochastic Differential Equations*, to appear.

[M] S.-A. E. Mohammed, *Stochastic Functional Differential Equations*, Pitman, 1984.

[P] E. Pardoux, Equations du filtrage non linéaire, de la prédiction et du lissage, *Stochastics* 6 (1982), 193-231.

[RS] B.L. Rozovskii, A. Shimizu, Smoothness of solutions of stochastic evolution equations and the existence of a filtering transition density, *Nagoya Math. J.* 84 (1981), 195-208.

[R] D. Ruelle, Characteristic exponents and invariant manifolds in Hilbert space, *Ann. Math.* 115 (1982), 243-290.

[S] K.-U. Schaumlöffel, *Zufällige Evolutionsoperatoren für Stochastische Partielle Differentialgleichungen*, PhD thesis, Universität Bremen, 1990.

[T] L. Tubaro, Some results on stochastic partial differential equations by the stochastic characteristic method, *Stoch. Anal. Appl.* 6 (1988), 217-230.

THE LYAPUNOV EXPONENT FOR PRODUCTS OF INFINITE-DIMENSIONAL RANDOM MATRICES

R.W.R. Darling
Mathematics Department, University of South Florida
Tampa, Florida 33620-5700

SUMMARY. Consider a non-negative random matrix indexed by $\mathbb{Z}^d \times \mathbb{Z}^d$, with independent rows, and such that the distribution is invariant under translation down the diagonal. Multiply together independent random matrices with this same law, and define the Lyapunov exponent λ as the exponential growth rate of the sum of the entries in the zero row. For some examples derived from Oriented Percolation, there is positive probability that λ equals the log of the expected value of the sum of entries in the zero row of the original random matrix. The proofs, which are not new, use random walk arguments. Some unsolved problems are described.

0. INTRODUCTION

Let A_1 denote a random $\mathbb{Z}^d \times \mathbb{Z}^d$ matrix with non-negative real entries. Suppose that $(A_1(x, x+y), y \in \mathbb{Z}^d)$ has the same law for every $x \in \mathbb{Z}^d$, and the rows $(A_1(x,.), x \in \mathbb{Z}^d)$ are independent. Moreover let $A_1, A_2, \ldots$ be independent, identically distributed random matrices. Such random matrices, with entries in $\{0,1\}$, arise in the theory of oriented percolation in $\mathbb{Z}^d \times \mathbb{Z}_+$ (see Durrett [1988], Grimmett [1989]): $A_n(x,z) = 1$ if and only if there is an open bond from site $(x,n-1)$ to site (z,n), and $= 0$ otherwise; more details are given in the examples below.

Let $\mathcal{F}_n$ denote the σ-field $\sigma(A_1,\ldots,A_n)$, let

(0.1)
$$R_n \equiv \sum_y (A_1 \ldots A_n)(0,y)$$

where $A_1 \ldots A_n$ is the usual matrix product, and

(0.2)
$$\theta \equiv \sum_y \mathbb{E}[A_1(0,y)] = \mathbb{E}[R_1]$$

where we assume $1 \leq \theta < \infty$ (to avoid triviality). In the oriented percolation context, R_n denotes the number of open paths in the percolation structure from site $(0,0) \in \mathbb{Z}^d \times \mathbb{Z}_+$ to all sites whose second coordinate is n, and θ is the expected number of open bonds emanating from any site. From the matrix perspective, R_n can be written as $\|\delta_0 A_1 \ldots A_n\|_1$, where δ_0 is the row vector indexed by $\mathbb{Z}^d$ with 1 in the zero position and 0 elsewhere, and $\|v\|_1$ denotes the sum of the entries of a vector v indexed by $\mathbb{Z}^d$. The starting point for the study of Lyapunov exponents for the random matrices $A_1, A_2, \ldots$ is the following pair of simple results.

LEMMA 0.1. Let $Y_n \equiv \theta^{-n} R_n$; then $(Y_n, \mathcal{F}_n, n \geq 1)$ is a non-negative martingale, with $\mathbb{E}[Y_n] = 1$ for all n; consequently $Y_n \to Y$ a.s. for some Y with $\mathbb{E}|Y| < \infty$.

COROLLARY 0.2. The Lyapunov Exponent $\lambda \equiv \lim_{n\to\infty} \frac{1}{n} \log R_n$ exists a.s. $\in [-\infty, \log\theta]$, and $= \log\theta$ on $(Y > 0)$.

Remarks. (1) Showing $P(Y > 0) > 0$. It is possible that $\lambda = -\infty$ a.s. even when $\theta > 1$. For example in Oriented Bond Percolation in $\mathbb{Z}^1 \times \mathbb{Z}_+$, $P(\bigcup_n (R_n = 0)) = 1$ for $\theta \leq 1.25$; see the discussion in Durrett [1984]. Most of the remainder of this paper is devoted to finding conditions under which $P(Y > 0) > 0$.

(2) Problem of the "Gap". Obviously $\lambda = -\infty$ on $\bigcup_n (R_n = 0)$. Is it true that $(Y = 0) = \bigcup_n (R_n = 0)$ a.s., implying that λ is a.s. equal to either $-\infty$ or $\log\theta$ (i.e. no gap)? Or is $P(-\infty < \lambda < \log\theta) > 0$? See Section 4 for further discussion.

Proof. Breaking up the product $A_1 \ldots A_n$ in the obvious way, we see that

$$\mathbb{E}[R_n | \mathcal{F}_{n-1}] = \sum_z (A_1 \ldots A_{n-1})(0,z) \sum_y \mathbb{E}[A_n(z,y)] = \theta R_{n-1}$$

from which the Lemma easily follows, using the Martingale Convergence Theorem. For any positive integer k,

$$\lim_{n\to\infty} \log \theta^{-n} R_n \text{ exists and } \in (-\log k, \log k) \text{ on } (1/k < Y < k)$$

$$\Rightarrow$$

$$\lim_{n\to\infty} \frac{1}{n} \log R_n = \log\theta \text{ on } (1/k < Y < k)$$

and then take the union over $k \geq 1$ to prove the Corollary. □

1. THREE PERCOLATION EXAMPLES

Consider the following oriented percolation structure in $\mathbb{Z}^d \times \mathbb{Z}_+$. Let N be a non-empty finite subset of $\mathbb{Z}^d$, with cardinality $\ell+1$; for example N could be 0 together with some set of neighbours. A site (x,n) has an oriented bond, which may be open or closed, to each of the sites (y,n+1) such that $y-x \in N$. Such a bond is said to go from level n to level n+1, where the "level" of a site means the $\mathbb{Z}_+$-coordinate. The three kinds of probability model treated here are all "stochastic growth models" in the sense of Durrett and Schonmann [1987]; the first two are generalizations of classical models, and the third is introduced here. In all three, the probability distribution for bonds leaving (x,n) is the same (modulo translation) for all $(x,n) \in \mathbb{Z}^d \times \mathbb{Z}_+$, and bonds emerging from different sites are independent.

A. Oriented Bond Percolation. Here all bonds are independent, and each of them has a probability p of being open, where p is fixed $\in (0,1)$. In the notation of the previous section, this situation can be modelled by choosing the law of A_1 as follows:

(1.1)

$$(A_1(x,y): y-x \in N) \text{ are i.i.d. Bernoulli}(p) \text{ random variables,}$$
$$A_1(x,y) = 0 \text{ if } y-x \in N^c.$$

B. Oriented Site Percolation. Either all of the bonds emerging from site (x,n) are open, which happens with probability p, or all are closed, which happens with probability $1-p$. For this, let A_1 have independent rows, and such that row $A_1(x,.)$ has the following probability law:

(1.2)

$$A_1(x,.) = \begin{cases} \text{identically zero, with probability } (1-p) \\ 1_{N+x}(.) \text{ with probability } p \end{cases}$$

Here $1_{N+x}(.)$ denotes the row vector with 1 in column y whenever $y-x \in N$, and 0 elsewhere.

C. **Two-Parameter Growth Model.** This is a typical example from a large class of possible models. Suppose the finite subset N of $\mathbb{Z}^d$ contains 0, and let $\beta > 0$ and $\delta > 0$ with $\ell\beta + \delta < 1$, where $\ell \equiv |N|-1 \geq 1$. For each y such that $y-x \in N$ and $y \neq x$, the pair of bonds $(x,n)\to(x,n+1)$ and $(x,n)\to(y,n+1)$ is open, while other bonds leaving (x,n) are closed, with probability β; with probability $(1-\delta-\ell\beta)$, only the bond $(x,n)\to(x,n+1)$ is open; with probability δ, all bonds leaving (x,n) are closed. This could be a model for bacterial growth, in which an infected site either stays infected and infects one neighbour selected at random (each one selected with probability β), or stays infected without infecting a neighbour (probability $1-\delta-\ell\beta$), or becomes healthy (probability δ). For this, let A_1 have independent rows, and such that row $A_1(x,.)$ has the following probability law:

(1.3)

$$A_1(x,.) = \begin{cases} \text{identically zero, with probability } \delta \\ 1_{\{x,y\}}(.) \text{ with probability } \beta \text{ (for each } y \neq x, y-x \in N) \\ 1_{\{x\}}(.), \text{ with probability } 1-\delta-\ell\beta \end{cases}$$

Here $1_{\{x,y\}}$ denotes the row vector with 1 in columns x and y and 0 elsewhere; $1_{\{x\}}$ denotes the row vector with 1 in column x and 0 elsewhere.

2. RESULTS ON THE NUMBER OF OPEN PATHS

An **open path** from (0,0) to (y,n) means a sequence of n open bonds, starting at (0,0) and ending at (y,n). The focus of this paper is the random variable R_n, denoting the **number of open paths** from (0,0) to sites at level n. Of course $R_n \geq |\xi_n^{(0)}|$, where $\xi_n^{(0)}$ denotes the set of sites at level n which are connected to (0,0) by open paths. The inequality is usually strict, because many open paths from (0,0) may terminate at the same site at level n. A well-known quantity associated with models A and B (at least for certain choices of the set N) is the **critical probability** $p_{c,N}$ defined by

(2.1)

$$p_{c,N} \equiv \inf\{p: \Pr(\xi_n^{(0)} \neq 0 \text{ for all } n \geq 1) > 0\}$$

The **moment ratio** M_n means the quantity

(2.2)

$$M_n \equiv \frac{\mathbb{E}[R_n]}{\sqrt{\mathbb{E}[(R_n)^2]}}.$$

Another parameter associated with models such as A and B above is

(2.3)

$$p_{m,N} \equiv \inf\{p: \inf_n M_n > 0\}$$

which could be called the critical parameter for the moment ratio; it is easy to see that $M_n \equiv 1$ when $p = 1$ in both models, so $p_{m,N}$ is well defined. Cox and Durrett [1983] noted that $(\ell+1)^{-1} \le p_{c,N} \le p_{m,N} \le 1$. Since $\mathbb{E}[R_n] = \theta^n$, the proof of the following Lemma (compare Lemma 0.1 and Corollary 0.2) is straightforward.

LEMMA 2.1. *The martingale $(Y_n, \mathcal{F}_n)$ is bounded in L^2 if and only if $\inf_n M_n > 0$. If $p > p_{m,N}$, then $E[Y] = 1$, and so $P(\lambda = \log\theta) > 0$.*

Thus our remaining efforts are devoted to calculating $p_{m,N}$. Before stating the first theorem, another parameter needs to be introduced. Associated with the set N, there is a number $q_N \in (1/(\ell + 1),1]$ defined as follows. Consider the temporally and spatially homogeneous Markov chain on $\mathbb{Z}^d$ whose transition probability is given by

(2.4)

$$p_{x,y} \equiv 1/(\ell + 1) \text{ if } y - x \in N, \text{ and } = 0 \text{ otherwise.}$$

Let (X_n) and (X'_n) be two independent random walks in $\mathbb{Z}^d$ with this transition probability, both started at 0, and define

(2.5)

$$q_N \equiv \Pr(X_n = X'_n \text{ for some } n \ge 1) = \frac{\mathbb{E}[H]}{1 + \mathbb{E}[H]} \ (= 1 \text{ if } \mathbb{E}[H] = \infty)$$

where H denotes the number of times these two random walks meet after time 0. One may compute q_N as follows. Suppose that $\varphi(t)$, $t \in [-\pi,\pi]^d$, is the characteristic function for the one-step transition probability started at 0, namely

$$\varphi(t) = \sum_{y \in N} e^{it \cdot y}/(\ell + 1)$$

Using standard Fourier analysis techniques, as presented for example in Durrett [1991], p. 170, and the fact that $X_1 - X'_1$ has characteristic function $\varphi(t)\varphi(-t)$, it is easy to verify that

(2.6)

$$\mathbb{E}[H] = (2\pi)^{-d} \int_{-\pi}^{\pi} \cdots \int_{-\pi}^{\pi} \mathrm{Re}(1 - \varphi(t)\varphi(-t))^{-1}\, dt$$

and this is finite if and only if $\dim(\mathrm{span}(N)) \ge 3$ (see Spitzer [1976]), in which case $q_N < 1$.

A result equivalent to the following one appears in Cox and Durrett [1983], who attribute the proof to unpublished work of H. Kesten.

THEOREM 2.2.
For Oriented Site Percolation, $p_{m,N} = q_N$, while for Oriented Bond Percolation,

(2.7)

$$p_{m,N} = \frac{q_N}{1 + \ell(1 - q_N)}$$

Remarks and Applications.
(i) It is immediate from the proof that for $d = 1$ and $d = 2$, $p_{m,N} = 1$ for all N.
(ii) First note that q_N, and hence $p_{m,N}$ (unlike $p_{c,N}$) is numerically computable, and $p_{m,N}$ provides an explicit upper bound on $p_{c,N}$. For example, suppose N consists of the d (≥ 3) unit basis vectors in $\mathbb{R}^d$, and so $\ell = d-1$. Then by (2.6),
(2.8)

$$\mathbb{E}[H] = (2\pi)^{-3} \int_{-\pi}^{\pi}\int_{-\pi}^{\pi}\int_{-\pi}^{\pi} \frac{9}{6-2\cos(t_2-t_1)-2\cos(t_3-t_2)-2\cos(t_1-t_3)} \, dt_1\, dt_2\, dt_3$$

(iii) These results lead to various estimates on the behaviour of (R_n) for large n. For example, for both models, if $0 < a < 1$,
(2.9)

$$\frac{\mathbb{E}[R_n ; R_n \geq a(\ell+1)^n]}{\mathbb{E}[R_n]} \leq \frac{p^n}{M_\infty^2 a}$$

(iv) For Oriented Bond Percolation,
(2.10)

$$M_n \downarrow M_\infty \equiv \sqrt{\frac{1+\ell(1-q_N)-q_N p^{-1}}{(1-q_N)(\ell+p^{-1})}} \quad \text{for } p > p_{m,N}$$

For Oriented Site Percolation,
(2.11)

$$M_n \downarrow M_\infty \equiv \sqrt{\frac{p-q_N}{1-q_N}} \quad \text{for } p > p_{m,N}$$

In both cases, $\inf_n M_n = 0$ when $p = p_{m,N}$.

In order to state a corresponding result for the Two-Parameter Growth Model, define another transition probability on $\mathbb{Z}^d$ by
(2.12)

$$p_{x,y} \equiv \begin{cases} 0 \text{ if } y \in N^c \\ (1-\delta)/\theta \text{ if } y = x \\ \beta/\theta \text{ if } y \in N \text{ and } y \neq x \end{cases}$$

where $\theta \equiv 1 - \delta + \ell\beta$. Also define q_N as above, i.e. the probability that two independent realizations of this Markov chain ever meet, if they both start at 0; q_N is explicitly computable from the formulas (2.5) and (2.6), using the new transition probability (2.12), but this time it depends on β and δ as well as on N. As noted above, $q_N < 1$ if and only if $\dim(\text{Span}(N)) \geq 3$.

Introduce the following sets, by analogy with (2.1) - (2.3):
(2.13)

$$\Lambda \equiv \{(\beta,\delta)\colon 0 < \beta < 1/\ell,\ 0 < \delta < (1-\ell\beta)\}$$

$$\Lambda_{c,N} \equiv \{(\beta,\delta) \in \Lambda\colon \Pr(\xi_n^{\{0\}} \neq 0 \text{ for all } n \geq 1) > 0\}$$

$$\Lambda_{m,N} \equiv \{(\beta,\delta) \in \Lambda\colon \inf_n M_n > 0\}$$

THEOREM 2.3. *In any dimension*
(2.14)

$$\Lambda_{m,N} \subsetneq \Lambda_{c,N}.$$

If dim(Span(N)) ≥ 3,
(2.15)

$$\Lambda_{m,N} \equiv \{(\beta,\delta) \in \Lambda: f(\beta,\delta) < \frac{1}{q_N}\}$$

where

$$f(\beta,\delta) \equiv \frac{1 - 5\delta - 2\delta^2 + \ell\beta(3 - \beta + \ell\beta)}{(1 - \delta + \ell\beta)^2}$$

Remarks. (i) The complicated expression for $f(\beta,\delta)$ is approximately $(1.2 - 1/4\ell)$ when $\delta \approx 0$ and $\ell\beta \approx 1$. Since $\ell \geq \dim(\mathrm{Span}(N))$, this result is useful only if $d' \equiv \dim(\mathrm{Span}(N))$ is high enough so that $q_N < 4d'/(5d' - 1)$.
(ii)
(2.16)

$$M_n \downarrow M_\infty \equiv \sqrt{\frac{1 - q_N f(\beta,\delta)}{(1 - q_N) f(\beta,\delta)}} \text{ for } (\beta,\delta) \in \Lambda_{m,N}$$

and $\inf_n M_n = 0$ when $(\beta,\delta) \in (\Lambda_{m,N})^c$.

3. PROOFS

Define a set of paths in $\mathbb{Z}^d$ as follows:
(3.1)

$$\pi_n \equiv \{(0 = y_0, y_1, \ldots, y_n): y_{k+1} - y_k \in N,\ 0 \leq k \leq n-1\}.$$

A path γ in π_n is called open if $S_\gamma = 1$, where
(3.2)

$$S_\gamma \equiv A_1(0,y_1)A_2(y_1,y_2)\ldots A_n(y_{n-1},y_n), \text{ for } \gamma = (0,y_1,\ldots,y_n)$$

Thus the the **number of open oriented paths** from (0,0) to level n, denoted R_n, is the sum of the zero row of the matrix $A_1 \ldots A_n$, i.e.
(3.3)

$$R_n \equiv \sum_{y \in \mathbb{Z}^d} (A_1 \ldots A_n)(0,y) = \sum_{\gamma \in \pi_n} S_\gamma$$

From the definition of θ in (0.2), it follows that θ stands for $(\ell+1)p$ in models A and B, and $1-\delta+\ell\beta$ in model C (as mentioned earlier). As mentioned above, the following proof is essentially due to Kesten, and appears in Cox and Durrett [1983]. Note also the similarity to the methods used by Eckmann and Wayne [1989].

Proof of Theorem 2.2. Since $Y_n = \theta^{-n} R_n$ and $\mathbb{E}[R_n] = \theta^n$, it follows from (2.2) that
(3.4)

$$M_n = \mathbb{E}[(Y_n)^2]^{-1/2}$$

It will suffice to prove that, for Oriented Bond (resp. Site) Percolation, if $p > q_N/[\ell(1-q_N) + 1]$ (resp. $p > q_N$) then $\mathbb{E}[(Y_n)^2]$ is increasing in n, and
(3.5)

$$\lim_{n\to\infty} \mathbb{E}[(Y_n)^2] = \frac{(1-q_N)(\ell+p^{-1})}{1+\ell(1-q_N)-q_N p^{-1}} \quad (\text{resp. } \frac{1-q_N}{p-q_N})$$

Model A (Bond Percolation). Let Γ and Γ' denote random paths in π_n selected uniformly and independently; thus $\Pr(\Gamma = \gamma, \Gamma' = \gamma') = 1/(\ell+1)^{2n}$ for every γ and γ'. Define $b(\gamma,\gamma')$ to be the number of common bonds in paths γ and γ'. By regarding Γ and Γ' as sample paths of $(X_0, X_1,\ldots, X_n)$ and $(X'_0, X'_1,\ldots, X'_n)$ respectively (see (2.5)), we see that $b(\Gamma,\Gamma')$ has the same law as

$$J(n) \equiv |\{k: 1 \le k \le n,\ X_{k-1} = X'_{k-1} \text{ and } X_k = X'_k\}|$$
$$\le J(\infty) \equiv |\{k: 1 \le k < \infty,\ X_{k-1} = X'_{k-1} \text{ and } X_k = X'_k\}|.$$

Also define

$$H(n) \equiv |\{k: 0 \le k \le n-1,\ X_k = X'_k\}|,$$
$$H(\infty) \equiv |\{k: 0 \le k < \infty,\ X_k = X'_k\}|.$$

Note that $H(\infty) = H + 1$, for the H appearing in (2.5). By renewal theory, $H(\infty)$ is a geometric random variable, and

$$\Pr(H = m) = (1-q_N)q_N^{\,m-1},\ m = 1,2,\ldots .$$

Moreover

$$\Pr(J(\infty) = j \mid H(\infty) = h) = \binom{h}{j}\left(\frac{1}{\ell+1}\right)^j\left(\frac{\ell}{\ell+1}\right)^{h-j},\ j = 0,1,\ldots,h$$

Now

$$\mathbb{E}[(Y_n)^2] = ((\ell+1)p)^{-2n}\,\mathbb{E}\Big[\sum_{\gamma\in\pi_n} S_\gamma \sum_{\gamma'\in\pi_n} S_{\gamma'}\Big]$$

$$= ((\ell+1)p)^{-2n} \sum_{\gamma\in\pi_n}\sum_{\gamma'\in\pi_n} \Pr(S_\gamma = 1 = S_{\gamma'})$$

$$= ((\ell+1)p)^{-2n} \sum_{\gamma\in\pi_n}\sum_{\gamma'\in\pi_n} p^{2n} p^{-b(\gamma,\gamma')}$$

$$= \mathbb{E}[p^{-b(\Gamma,\Gamma')}]$$

$$= \mathbb{E}[p^{-J(n)}]$$

Consequently $\mathbb{E}[(Y_n)^2]$ is increasing in n, and by Monotone Convergence

$$\lim_n \mathbb{E}[(Y_n)^2] = \sup_n \mathbb{E}[(Y_n)^2] = \mathbb{E}[p^{-J(\infty)}]$$

$$= \sum_{h=1}^{\infty}\sum_{j=0}^{h} p^{-j}\Pr(J(\infty) = j \mid H(\infty) = h)(1-q_N)q_N^{\,h-1}$$

$$= \sum_{h=1}^{\infty}\sum_{j=0}^{h} p^{-j}\binom{h}{j}\left(\frac{1}{\ell+1}\right)^j\left(\frac{\ell}{\ell+1}\right)^{h-j}(1-q_N)q_N^{\,h-1}$$

$$= \frac{1-q_N}{q_N} \sum_{h=1}^{\infty} q_N{}^h \left[\frac{\ell+p^{-1}}{\ell+1}\right]^h$$

$$= \frac{(1-q_N)(\ell+p^{-1})}{1+\ell(1-q_N)-q_N p^{-1}}$$

provided $p > q_N/[\ell(1 - q_N) + 1]$, which completes the proof of (3.5).

Model B. This is easier than model A, for here $b(\Gamma,\Gamma')$ has the same law as $H(n)$, and so

$$\lim_n \mathbb{E}[(Y_n)^2] = \sup_n \mathbb{E}[(Y_n)^2] = \mathbb{E}[p^{-H(\infty)}]$$

$$= \frac{1-q_N}{q_N} \sum_{h=1}^{\infty} (q_N/p)^h$$

$$= \frac{1-q_N}{p-q_N}$$

provided $p > q_N$. □

Proof of Theorem 2.3. As before, we shall study the second moments of the martingale (Y_n). Given paths $\gamma \equiv (0=y_0,y_1,\dots,y_n)$ and $\gamma' \equiv (0=y_0',y_1',\dots,y_n')$ in π_n (refer to (3.2)), define the following quantities:

$u(\gamma) \equiv |\{j\colon y_j = y_{j-1}\}|$, the number of "up-moves",
$t(\gamma) \equiv |\{j\colon 0 \neq y_j - y_{j-1} \in \mathbb{N}\}|$, the number of "transverse moves",
$\tilde{u}(\gamma,\gamma') \equiv |\{j\colon y_j = y_j' = y_{j-1} = y_{j-1}'\}|$, the number of shared up-moves,
$\tilde{t}(\gamma,\gamma') \equiv |\{j\colon y_{j-1} = y_{j-1}' \neq y_j = y_j'\}|$, the number of shared transverse moves,
$\tilde{v}(\gamma,\gamma') \equiv |\{j\colon y_{j-1} = y_{j-1}'$, and either $y_j = y_{j-1}$ or $y_j' = y_{j-1}'$, but not both$\}|$, the number of simultaneous moves with the same starting point, of which one is up and the other is transverse.

Evidently $\Pr(S_\gamma = 1) = (1-\delta)^{u(\gamma)}\beta^{t(\gamma)}$, and it can also be shown that

$$\Pr(S_\gamma = 1 = S_{\gamma'}) = (1-\delta)^{u(\gamma)+u(\gamma')-\tilde{u}(\gamma,\gamma')-\tilde{v}(\gamma,\gamma')}\beta^{t(\gamma)+t(\gamma')-\tilde{t}(\gamma,\gamma')}$$
$$= \Pr(S_\gamma = 1)\Pr(S_{\gamma'} = 1)(1-\delta)^{-\tilde{u}(\gamma,\gamma')-\tilde{v}(\gamma,\gamma')}\beta^{-\tilde{t}(\gamma,\gamma')}$$

This formula is tricky, and is verified as follows. In the following table, the symbol "↑↗" means that path γ makes an up-move, and path γ' makes a transverse move; other symbols are similarly defined. The numbers in the table indicate the number of steps at which the various combinations of events occur. The unknowns b and c will eventually cancel out in the probability calculations.

Symbol	↑↑	↑↗	↗↗	↗↑
Same starting point	$\tilde{u}$	b	$\tilde{t}$	$\tilde{v} - b$
Different starting points	$c - \tilde{u}$	$u(\gamma) - c - b$	$t(\gamma') - u(\gamma) - \tilde{t} + c$	$u(\gamma') - c - \tilde{v} + b$
Total	c	$u(\gamma) - c$	$t(\gamma') - u(\gamma) + c$	$u(\gamma') - c$

The first eight entries in this table appear as exponents of $(1-\delta)$, β, β, β, $(1-\delta)^2$, $\beta(1-\delta)$, β^2, and $\beta(1-\delta)$ respectively, in the formula for the probability on the previous lines. Since $u(\gamma)+t(\gamma) = n = u(\gamma')+t(\gamma')$, the desired expression may be obtained.

Let (X_n) and (X_n') be two independent realizations of the Markov chain in $\mathbb{Z}^d$ with the transition probability $(p_{x,y})$ defined in (2.12), both started at 0. These Markov chains are designed so that

$$\Pr((X_0,\dots,X_n) = \gamma) = \theta^{-n}\Pr(S_\gamma = 1), \text{ and}$$

$$\theta^{-2n}\Pr(S_\gamma = 1 = S_{\gamma'}) = \Pr((X_0,\dots,X_n) = \gamma,\ (X_0',\dots,X_n') = \gamma')\beta^{-\tilde{t}}(1-\delta)^{-\tilde{u}-\tilde{v}}$$

where $\tilde{t} = \tilde{t}(\gamma,\gamma')$, $\tilde{u} = \tilde{u}(\gamma,\gamma')$, and $\tilde{v} = \tilde{v}(\gamma,\gamma')$. Imitating the reasoning of the last proof, we see that

$$\sup_n \mathbb{E}[(Y_n)^2] = \sup_n \theta^{-2n} \sum_{\gamma\in\pi_n}\sum_{\gamma'\in\pi_n} \Pr(S_\gamma = 1 = S_{\gamma'})$$

$$= \mathbb{E}[\beta^{-T}(1-\delta)^{-U-V}]$$

where

$$U \equiv |\{j\ge 1: X_{j-1} = X'_{j-1} = X_j = X'_j\}|,$$

$$T \equiv |\{j\ge 1: X_{j-1} = X'_{j-1} \ne X_j = X'_j\}|,$$

$$V \equiv |\{j\ge 1: X_{j-1} = X'_{j-1}, \text{ and either } X_j = X_{j-1} \text{ or } X'_j = X'_{j-1}, \text{ but not both}\}|$$

$$H(\infty) \equiv |\{j\ge 0: X_j = X'_j\}| \ge T+U+V$$

Abbreviating q_N to q, note that $\Pr(H = m) = q^{m-1}(1-q)$, for $m = 1,2,\dots$, and

$$\Pr(U+V = r,\ T = t \mid H(\infty) = m) = \binom{m}{r,t,m-r-t}\rho^r\tau^t(1-\rho-\tau)^{m-r-t}$$

where

$$\rho \equiv \Pr(\{X_j = X_{j-1}\}\cup\{X'_j = X'_{j-1}\} \mid X_{j-1} = X'_{j-1})$$

$$= \frac{1-\delta}{\theta}\left(2 - \frac{1-\delta}{\theta}\right)$$

$$\tau \equiv \Pr(X_j = X'_j \ne x \mid X_{j-1} = X'_{j-1} = x) = \ell(\beta/\theta)^2$$

Therefore

$$\sup_n \mathbb{E}[(Y_n)^2] = \sum_{m=1}^{\infty}(1-q)q^{m-1}\ \mathbb{E}[\beta^{-T}(1-\delta)^{-U-V} \mid H(\infty) = m]$$

$$= \sum_{m=1}^{\infty}(1-q)q^{m-1} \sum_{r+t\le m}(1-\delta)^{-r}\beta^{-t}\binom{m}{r,t,m-r-t}\rho^r\tau^t(1-\rho-\tau)^{m-r-t}$$

$$= \frac{1-q}{q}\sum_{m=1}^{\infty}[q(\frac{\rho}{1-\delta} + \frac{\tau}{\beta} + 1 - \rho - \tau)]^m = \frac{(1-q)g(\beta,\delta)}{1-qg(\beta,\delta)}$$

provided

$$g(\beta,\delta) \equiv (\frac{\rho}{1-\delta} + \frac{\tau}{\beta} + 1 - \rho - \tau) < \frac{1}{q}$$

Calculation shows that $g(\beta,\delta) = f(\beta,\delta)$. □

4. SUGGESTIONS FOR FURTHER WORK

(a) For the class of matrices described in the Introduction, find a general method for showing that $P(Y > 0) > 0$, which does not require the messy calculations given here for Model C, for example.

(b) Under further conditions, if necessary, determine whether there is "no gap" in Corollary 0.2, Remark (2); i.e. the only possible values of the Lyapunov exponent are $-\infty$ and $\log\theta$, a.s..

(c) It is clear that $\inf\{p: P(\lambda = \log\theta) > 0\} = \inf\{p: P(Y > 0) > 0\}$ in models A and B, but a priori we can only say that

$$p_{c,N} \le \inf\{p: P(\lambda > -\infty) > 0\} \le \inf\{p: P(Y > 0) > 0\}.$$

Are these three equal, and if not can the last two be evaluated? For example, in Lemma 2.1, we are using the inequality

$$\sup_n \mathbb{E}[Y_n;\ Y_n \ge T] \le (1/T)\sup_n \mathbb{E}[(Y_n)^2]$$

(c.f. (2.11)) to show that the martingale $\{Y_n\}$ is uniformly integrable when it is bounded in L^2. A better estimate for $\inf\{p: P(Y > 0) > 0\}$ may be obtainable by estimating $\mathbb{E}[Y_n;\ Y_n \ge T]$ in some other way. A simple calculation shows that

$$\mathbb{E}[Y_n;\ Y_n \ge T] \le p^{-n}\Pr(Y_n \ge T).$$

This suggests that large deviation estimates for Y_n may be useful here.

REFERENCES

COX, J.T. and DURRETT, R. (1983). Oriented percolation in dimensions d ≥ 4: bounds and asymptotic formulas. Math Proc. Camb. Phil. Soc. 93, 151-162.

DURRETT, R. (1984). Oriented percolation in two dimensions. Annals of Probability 12, 999-1040.

DURRETT, R. (1988). *Lecture Notes on Particle Systems and Percolation.* Wadsworth, Pacific Grove.

DURRETT, R. (1991). *Probability: Theory and Examples.* Wadsworth, Pacific Grove.

DURRETT, R. and SCHONMANN, R.H. (1987). Stochastic growth models. *Percolation theory and the Ergodic Theory of Interacting Particle Systems*, ed. H. Kesten, Springer, New York.

ECKMANN, J.-P., AND WAYNE, C.E. (1989). The largest Lyapunov exponent for random matrices and directed polymers in a random environment. Commun. Math. Phys. 121, 147-175.

GRIMMETT, G. (1989). *Percolation*. Springer, New York.

SPITZER, F. (1976). *Principles of Random Walk.* Springer, New York.

Acknowledgements. The author thanks Arunava Mukherjea for many helpful discussions, and Richard Durrett for locating an error and for references to the literature.

Lyapunov exponents and complexity for interval maps

Gerhard Keller, Universität Erlangen, FRG

1 Introduction

The amount of irregularity in trajectories of dynamical systems can be quantified in various ways. From a geometrical point of view, *Lyapunov exponents* measure the dependence of the future behaviour on small changes in the system's initial conditions. A statistician might use *entropy* as a measure of his uncertainty in predicting the future of the system from its past. *Algorithmic complexity* measures the amount of information needed to reproduce a finite section of the trajectory with a fixed precision on a universal computer.

In this note we examine the irregularity of trajectories produced by interval transformations. For very stable (contracting) and for very unstable (expanding) maps we see that the above three measures of irregularity essentially coincide. For some classes of maps in between, however, we find out that algorithmic complexity can describe aspects of irregularity to which entropy and Lyapunov exponents are unsensitive. The dynamics of these maps are so subtle that up to now rigorous results are available only for unimodal interval maps with negative Schwarzian derivative.

In the sequel the following notations are used constantly: A transformation $T:[0,1]\to[0,1]$ is called *piecewise monotone*, if there exists a partition $\mathcal{Z}$ of $[0,1]$ into finitely many intervals such that $T_Z = T_{|Z} : Z \to [0,1]$ is monotone and continuous. A piecewise monotone T is called *piecewise* C^r, if each T_Z is of class C^r and has bounded r-th derivative. The endpoints of the intervals $Z \in \mathcal{Z}$ are called the *critical points* of T.

For a piecewise monotone T let $\mathcal{Z}_n = \mathcal{Z} \vee T^{-1}\mathcal{Z} \vee \ldots \vee T^{-(n-1)}\mathcal{Z}$ be the partition of $[0,1]$ into monotonicity and continuity intervals of T^n. $\mathcal{Z}_n[x]$ denotes that interval in $\mathcal{Z}_n$ which contains x, $Z_\infty[x] := \cap_{n>0} Z_n[x]$. $Z_\infty[x]$ is an interval containing x, often it is just the set $\{x\}$.

The Lebesgue measure on $[0,1]$ is denoted by m, and by a measure we always mean a Borel probability measure. In order to avoid frequent normalizing by $\frac{1}{\log 2}$ in Section 3, we take henceforth all logarithms to base 2.

2 Lyapunov- and information exponents

2.1 Positive Lyapunov exponents and entropy for invariant measures

For a piecewise C^1 Transformation we define the *Lyapunov exponent at* x as

$$\lambda(x) = \lim_{n\to\infty} \frac{1}{n} \log|(T^n)'(x)| = \lim_{n\to\infty} \frac{1}{n} \sum_{k=0}^{n-1} \log|T'(T^k x)| \qquad (1)$$

if this limit exists and denote the corresponding lim sup by $\bar{\lambda}(x)$.

The *information exponent at* x of the partition $\mathcal{Z}$ with respect to some (not necessarily T-invariant) measure μ on $[0,1]$ is

$$I_{\mu,\mathcal{Z}}(x) = \lim_{n\to\infty} -\frac{1}{n}\log\mu(Z_n[x]), \tag{2}$$

and $\bar{I}_{\mu,\mathcal{Z}}(x)$ denotes again the corresponding lim sup.

The question arises, under which conditions the limits $\lambda(x)$ and $I_{\mu,\mathcal{Z}}(x)$ exist and what their respective values are. A first answer to the convergence problem is given by Birkhoff's ergodic theorem for $\lambda(x)$ and by the Shannon-McMillan-Breiman theorem for $I_{\mu,\mathcal{Z}}(x)$:

Theorem 1 *If μ is an ergodic T-invariant measure, then*

$$\lambda(x) = \lambda_\mu \quad \textit{and} \quad I_{\mu,\mathcal{Z}}(x) = h_\mu(\mathcal{Z}) \qquad \textit{for } \mu\textit{-a.e. } x.$$

Here

$$\lambda_\mu := \int \log|T'|\,d\mu$$

and

$$h_\mu(\mathcal{Z}) := \lim_{n\to\infty}\frac{1}{n}H_\mu(\mathcal{Z}_n), \quad H_\mu(\mathcal{Z}_n) := -\sum_{Z\in\mathcal{Z}_n}\mu(Z)\cdot\log\mu(Z).$$

Observe that $h_\mu(\mathcal{Z}) = h_\mu := \sup\{h_\mu(\alpha) : \alpha \text{ a finite partition}\}$, if $\mathcal{Z}$ is a generator for (T,μ), i.e. if $Z_n[x] \searrow \{x\}$ for μ-a.e. x. h_μ is called the entropy of (T,μ). If $h_\mu(\mathcal{Z}) > 0$, then $\mu(Z_\infty[x]) = \lim_{n\to\infty}\mu(Z_n[x]) = 0$ for μ-a.e. x by Theorem 1 (and hence for all x!), and as $Z_\infty[x]$ is always an interval, it follows easily that $\mathcal{Z}$ generates.

The numbers λ_μ and $h_\mu(\mathcal{Z})$ are related by a variant of Rohlin's formula, which was proved for a class of continuously differentiable interval maps by Ledrappier [Le]:

Theorem 2 *Consider a continuously differentiable piecewise monotone map T on $[0,1]$ with Hölder-continuous derivative and nondegenerate critical points,* [1] *and an ergodic T-invariant measure μ. Then*

1. $h_\mu \le \max\{0,\lambda_\mu\}$
2. *If $h_\mu(\mathcal{Z}) > 0$, then μ is absolutely continuous with respect to m if and only if $h_\mu = \lambda_\mu$ (Rohlin's formula).*

The inequality $h_\mu \le \max\{0,\lambda_\mu\}$ and the "only if"-implication generalize to higher dimensions [Ru, Pe].

[1] T has nondegenerate critical points $c_1,\dots,c_k$, if $T'(c_i) = 0$ for all i and $T'(x) \ne 0$ otherwise, and if there are numbers $C > 0$ and k_i^- (k_i^+) such that $C^{-1} \le |T'(x)|/|x-c_i|^{k_i^{-(+)}} \le C$ in a left (right) neighbourhood of c_i.

2.2 Exponents and entropy for Lebesgue measure

Although Lebesgue measure m on $[0,1]$ is in general not invariant under T, one is often interested in Lyapunov exponents $\lambda(x)$ and also in information exponents $I_m(x)$ (with respect to m!) for m-a.e. x. Our goal is to obtain, at least for special classes of maps, results analogous to those of Section 2.1, where the underlying measure μ was invariant.

We begin with some simple observations on $\bar{\lambda}(x)$ and $\bar{I}_{m,\mathcal{Z}}(x)$: It follows from (1) that

$$\bar{\lambda}(Tx) = \bar{\lambda}(x) \quad \text{if } T'(x) \neq 0. \tag{3}$$

Similarly, if the critical points of T are nondegenerate, then

$$\bar{I}_{m,\mathcal{Z}}(Tx) = \bar{I}_{m,\mathcal{Z}}(x), \text{ if } Tx \text{ is not a critical point of } T. \tag{4}$$

Analogous statements hold for λ and $I_{m,\mathcal{Z}}$ if the limit for x or for Tx exists. In order to prove (4) we show that there are constants $C > 0$ and $k \in \mathbb{N}$ (possibly depending on x) such that

$$C^{-1} \cdot m(Z_{n+1}[x])^k \leq m(Z_n[Tx]) \leq C \cdot m(Z_{n+1}[x]). \tag{5}$$

Observe that $T(Z_{n+1}[x]) \subseteq Z_n[Tx]$ for all n and that

$$m(TZ_{n+1}[x]) = \int_{Z_{n+1}[x]} |T'| \, dm.$$

Hence, if $T(Z_{n+1}[x]) = Z_n[Tx]$ for all but finitely many n, then (5) follows. As Tx is not a critical point of T, this is certainly true, if $Z_\infty[x] = \{x\}$. But if $Z_\infty[x]$ is a nondegenerate interval, then $m(Z_n[Tx])/m(TZ_{n+1}[x]) \leq 1/m(TZ_\infty[x]) < \infty$ for all n, and (5), possibly with a different constant, follows again.

Suppose now that (T,m) is ergodic, i.e. all measurable sets A satisfying $T^{-1}A = A$ have Lebesgue measure 0 or 1. [2] Then $\bar{\lambda}(x) =: \lambda_T$ and $\bar{I}_{m,\mathcal{Z}}(x) =: I_T$ are constant m-a.e. We call these quantities the *Lyapunov* and the *information exponent* of T, respectively.

Theorem 3 *Suppose (T,m) is ergodic. If there exists an m-absolutely continuous T-invariant probability measure (a.c.i.p.m.) μ on $[0,1]$, then $\lambda_T = \lambda_\mu$, $I_T = h_\mu(\mathcal{Z})$, and*

$$\lambda(x) = \lambda_\mu \quad \text{and} \quad I_{m,\mathcal{Z}}(x) = h_\mu(\mathcal{Z}) \quad \text{for } m\text{-a.e. } x. \tag{6}$$

Proof: As (T,m) is ergodic, also (T,μ) is ergodic. Hence $\lambda(x) = \lambda_\mu$ holds for a set of points x of positive Lebesgue measure in view of Theorem 1. By ergodicity of (T,m) it holds for m-a.e. x.

Also by Theorem 1, $I_{\mu,\mathcal{Z}}(x) = h_\mu(\mathcal{Z})$ for μ-a.e. x. As a routine application of the martingale theorem shows that $I_{m,\mathcal{Z}} = I_{\mu,\mathcal{Z}}$ μ-a.e. (cf. the proof of Theorem 3a in [Ke2]), the theorem follows in view of (4) from the ergodicity of (T,m). □

If T has no a.c.i.p.m. but a Bowen-Ruelle-Sinai measure (BRS measure), then still something can be said about Lyapunov-exponents provided $\inf |T'| > 0$.

[2] This happens e.g. for unimodal maps with negative Schwarzian derivative, if they have no attractive periodic orbit [BL, Ma]. These maps will be discussed in Section 2.3.

So let again (T,m) be ergodic. A measure ν on $[0,1]$ is a BRS measure, if for each $f \in C([0,1])$

$$\lim_{n\to\infty} \frac{1}{n}\sum_{j=0}^{n-1} f(T^j x) = \int f\,d\nu \quad \text{for } m\text{-a.e. } x. \tag{7}$$

It T is C^1 (or piecewise C^1 without periodic critical points), then it is easily seen that $\lambda(x) = \lambda_\nu$ for m-a.e. x.

Remark 1 If, for example, T has a stable periodic orbit $z, Tz, \ldots, T^p z = z$ attracting m-a.e. trajectory, then the uniform distribution on this periodic orbit is a BRS measure. If T is continuously differentiable in a neighbourhood of this orbit and if $(T^p)'(z) \neq 0$, then obviously $\lambda(x) = \lambda(z)$ for m-a.e. x. The other extreme example for a BRS measure is the measure μ from Theorem 3, which is BRS by Birkhoff's ergodic theorem.

Observe that these considerations do not apply to a BRS measure ν giving positive mass to each neighbourhood of a critical point c of T, because the function $f(x) = \log|T'(x)|$ has a singularity at c and is therefore unbounded on the support of ν, such that (7) does not apply to f. In Section 2.3 we discuss examples of very simple transformations T which have a BRS measure ν with $\lambda_\nu > 0$ although $\bar{\lambda}(x) = 0$ for m-a.e. x.

2.3 S-unimodal maps

Let $T : [0,1] \to [0,1]$ be a unimodal map with negative Schwarzian derivative and critical point c, e.g. $T(x) = ax(1-x)$, $(0 < a \leq 4, c = \frac{1}{2})$. The Schwarzian derivative of T is defined as $ST = \frac{T'''}{T'} - \frac{3}{2}(\frac{T''}{T'})^2$. It plays a very useful role for interval maps because of the following two properties:

1. If $ST \leq 0$, then $ST^n \leq 0$ for all iterates T^n of T.
2. $ST \leq 0$ if and only if $\frac{1}{\sqrt{|T'|}}$ is convex on each monotonicity interval of T.

The usefulness of 1. for interval maps was apparently first noticed by Singer [Si]. Singer, Guckenheimer [Gu], and Misiurewicz [Mi] were the first to make essential use of 2. or some weaker consequences of it.

Using these observations and a technique from symbolic dynamics originally developed by Hofbauer, the following results on Lyapunov- and information exponents were proved in [Ke2]:

Theorem 4 *Let T be a S-unimodal map. There is a constant λ_T such that*

$$\bar{\lambda}(x) = \lambda_T \quad \text{and} \quad I_{m,\mathcal{Z}}(x) = \max\{\lambda_T, 0\} \quad \text{for } m\text{-a.e. } x.$$

Furthermore:

1. *If $\lambda_T > 0$, then (T,m) is ergodic and there exists an a.c.i.p.m. μ on $[0,1]$. In particular, Theorem 3 applies, and μ is BRS.*
2. *If $\lambda_T < 0$, then T has a unique attracting periodic orbit which attracts m-a.e. orbit. The uniform distribution on this attracting orbit is BRS.*

The reader will notice that this theorem does not characterize the case $\lambda_T = 0$. From Guckenheimer's work [Gu] on the topological classification of S-unimodal maps we know that at least two different kinds of dynamics can produce $\lambda_T = 0$: There are maps with a stable but indifferent periodic orbit, [3] and there are maps with a solenoidal attractor (attracting also m-a.e. orbit) for which, in view of Theorem 4, only $\lambda_T = 0$ is possible. In this case the restriction of T to the attractor is topologically conjugate to an irreducible rotation on a compact group. So it has a unique invariant probability measure (Haar measure), and this is easily seen to be BRS.

It was a bit surprising, when Johnson [Jo] found maps without a.c.i.p.m. which have neither an attracting periodic orbit nor a solenoidal attractor. Obviously these maps must have $\lambda_T = 0$, and in contrast to all other cases it was not clear, whether these maps have BRS measures. This question was answered in the negative by Hofbauer and Keller [HK2] with the following

Theorem 5 *There are S-unimodal maps without BRS measure. Such maps exist even whithin the family of quadratic maps $T_a(x) = ax(1-x)$.*

Remark 2 These maps can be constructed with various additional properties. We describe some of them:

Denote by δ_x the unit mass in the point x, and let for a probability measure μ on $[0,1]$

$$\bar{\omega}_T(\mu) := \left\{ \text{all weak accumulation points of } \left(\frac{1}{n} \sum_{k=0}^{n-1} \mu \circ T^{-k} \right)_{n>0} \right\}.$$

1. Fix $0 \le h_0 < h_1 < \frac{1+\sqrt{5}}{2}$. There are maps T without BRS measure such that
   (a) $\{h_\nu(T) : \nu \in \bar{\omega}_T(\delta_c), \nu \text{ ergodic}\} = [h_0, h_1]$,
   (b) $\bar{\omega}_T(\delta_x) = \bar{\omega}_T(m) = \bar{\omega}_T(\delta_c)$ for m-a.e. x
2. There are maps T without BRS measure such that
   (a) $\bar{\omega}_T(\delta_c) = \{\delta_z\}$, where z is the unique *unstable* fix point $> c$ of T.
   (b) $\bar{\omega}_T(\delta_x) = \bar{\omega}_T(m) = \{\delta_z\}$ for m-a.e. x.

3 Algorithmic complexity of trajectories

3.1 Algorithmic complexity for invariant measures

Brudno [Br] introduced the notion of complexity for trajectories of dynamical systems. His starting point is Kolmogorov's definition [Ko] of the complexity of a finite word over a finite alphabet: Let $\mathcal{L}$ be a finite set (alphabet), $\mathcal{L}^*$ the set of finite words over $\mathcal{L}$, and let A be a universal partially recursive function from $\{0,1\}^*$ to $\mathcal{L}^*$. [4] The complexity of $w \in \mathcal{L}^*$ is defined as

[3] Indifferent means that the orbit has Lyapunov exponent 0.

[4] A can be "implemented " by a universal Turing machine. For our purposes it is nice to think of a two-tape machine with one read-only tape containing the input $u \in \{0,1\}^*$ and one read-and-write working-tape from which the result $A(u) \in \mathcal{L}^*$ can be read off when the computation stops (provided it stops at all for the particular value u). In nonmathematical terms one may think of a Personal Computer with unlimited storage capacity.

$$K(w) = \min\{|u| : u \in \{0,1\}^*, A(u) = w\},$$

where $|u|$ denotes the length of the word u. If $A(u) \neq w$ for all $u \in \{0,1\}^*$, then $K(w) = \infty$. [5] An elementary, heuristically obvious property of complexity is

$$K(vw) \leq K(v) + K(w) + \text{const}, \quad K(w) \leq K(vw) + \text{const} \tag{8}$$

with a constant depending only on the particular universal function A. Although not explicitly stated, this is implicitly used in [Br] and can be proved formally along the lines of the proof of [Br, Lemma 1.3].

Kolmogorov remarked already that this definition of complexity does not take into account the difficulty of actually producing the word w from its shortest codeword. Recently this problem was addressed in a review paper by Grassberger [Gr], but the concepts invented to describe this aspect of complexity have not yet developed to a theory which applies to general dynamical systems.

Suppose now $T : [0,1] \to [0,1]$ is piecewise monotone as in Section 1, and let $\mathcal{Z} = \{Z_1, \ldots, Z_l\}$ be its partition into monotonicity intervals. With each $x \in [0,1]$ we associate an infinite symbol sequence $(x_0x_1x_2\ldots) \in \mathcal{L}^\infty$, called itinerary, where $\mathcal{L} = \{1, \ldots, l\}$, $l = \text{card}(\mathcal{Z})$, and $x_i = j$ if $T^ix \in Z_j$. Brudno [Br] defines the complexity of the trajectory of x (with respect to the partition $\mathcal{Z}$) as

$$K(x; \mathcal{Z}) = \limsup_{n\to\infty} \frac{1}{n} K(x_0 \ldots x_{n-1})$$

and proves [6]

Theorem 6 *If μ is an ergodic T-invariant measure, then $K(x; \mathcal{Z}) = h_\mu(\mathcal{Z})$ for μ-a.e. x.*

The proof relies on the Shannon-McMillan-Breiman theorem. For the "$\geq$"-direction one combines it with the fact that there are not more than 2^k words $w \in \mathcal{L}^*$ with $K(w) \leq k$. For the reverse estimate we refer to the proof of Lemma 1, where a slightly more general inequality is derived.

In conclusion, if the reference measure μ is invariant, complexity gives μ-a.s. no information not already contained in the Lyapunov- or information exponent.

3.2 Complexity for Lebesgue measure

As in Section 2.2 we consider next the case where T has an a.c.i.p.m. μ. As $K(x; \mathcal{Z}) = K(Tx; \mathcal{Z})$ (which is quite evident from a heuristic understanding of how universal computers work; for a proof see [Br, Lemma 1.3]), we have as in Theorem 3

[5] This definition of complexity depends of course on the particular algorithm A and should be denoted by K_A, but as A is assumed to be universal, there is, for each partially recursive function B, a constant C_B such that $K_A(w) \leq K_B(w) + C_B$, and the dependence of K on the particular universal algorithm is, at least for very long words w, neglectible.

[6] Brudno proves his results for general dynamical systems.

Theorem 7 *Suppose (T, m) is ergodic. If there exists an a.c.i.p.m. μ on $[0,1]$, then $K(x; \mathcal{Z}) = h_\mu(\mathcal{Z})$ for m-a.e. x.*

If there is no a.c.i.p.m., Brudno [Br, Theorem 3.2] gives still the following upper bound for $K(x; \mathcal{Z})$, which we state here for interval maps:

Theorem 8 *Suppose $T : [0,1] \to [0,1]$ is piecewise monotone. Then*

$$K(x; \mathcal{Z}) \leq \sup\{h_\nu(\mathcal{Z}) : \nu \in \tilde{\omega}_T(\delta_x)\} \quad \textit{for all } x.$$

The corresponding lower estimate is not true, see [Br, Example 4.3.2, which can be adapted also to piecewise monotone maps]. Even if μ is a BRS measure for T, it seems to be unknown whether $K(x; \mathcal{Z}) = h_\mu(\mathcal{Z})$ for m-a.e. x. I expect, however, that a refinement of the construction in [HK2] would lead to examples of quadratic maps with a BRS measure of positive entropy but $K(x; \mathcal{Z}) = 0$ for m-a.e. x.

This motivates the following modified notion of complexity of a trajectory: For $L \in \mathbf{N}$ let

$$\begin{aligned} k_L(x; \mathcal{Z}) &= K(x_0 \ldots x_{L-1}) \quad \text{and} \\ K_L(x; \mathcal{Z}) &= \limsup_{n\to\infty} \frac{1}{n} \sum_{j=0}^{n-1} k_L(T^j x; \mathcal{Z}). \end{aligned}$$

Because of (8,)

$$k_{L+M}(x; \mathcal{Z}) \leq k_L(x; \mathcal{Z}) + k_M(T^L x; \mathcal{Z}) + \text{const},$$

whence $K_{L+M}(x; \mathcal{Z}) \leq K_L(x; \mathcal{Z}) + K_M(x; \mathcal{Z}) + \text{const}$, such that the limit

$$\tilde{K}(x; \mathcal{Z}) = \lim_{L\to\infty} \frac{1}{L} K_L(x; \mathcal{Z})$$

exists. Similarly, if μ is T-invariant, then

$$\tilde{K}_\mu(\mathcal{Z}) = \lim_{L\to\infty} \frac{1}{L} \int k_L(x; \mathcal{Z})\, d\mu(x)$$

exists. $\tilde{K}(x; \mathcal{Z})$ and $K(x; \mathcal{Z})$ are related by

Lemma 1 *$K(x; \mathcal{Z}) \leq \tilde{K}(x; \mathcal{Z})$, and if μ is an ergodic T-invariant measure, then*

$$K(x; \mathcal{Z}) = h_\mu(\mathcal{Z}) = \tilde{K}_\mu(\mathcal{Z}) = \tilde{K}(x; \mathcal{Z}) \quad \textit{for } \mu\textit{-a.e. } x.$$

Proof: To simplify the notation we skip $\mathcal{Z}$ in expressions like $k_L(x; \mathcal{Z})$ etc. For a word w of length nr, (8) implies

$$K(w) \leq \sum_{j=0}^{n-1} K(w_{jr} \ldots w_{(j+1)r-1}) + \text{const} \cdot n.$$

As also $|k_{nr}(T^ix)-k_{nr}(x)|\leq \text{const}\cdot i$ by (8), it follows that

$$\begin{aligned}
\limsup_{n\to\infty}\frac{1}{nr}k_{nr}(x) &\leq \limsup_{n\to\infty}\frac{1}{r}\sum_{i=0}^{r-1}\frac{1}{nr}k_{nr}(T^ix)+\limsup_{n\to\infty}\frac{1}{r}\sum_{i=0}^{r-1}\frac{1}{nr}|k_{nr}(T^ix)-k_{nr}(x)| \\
&\leq \limsup_{n\to\infty}\frac{1}{r}\sum_{i=0}^{r-1}\frac{1}{nr}\left(\sum_{j=0}^{n-1}k_r(T^{jr+i}x)+\text{const}\cdot n\right) \\
&= \limsup_{n\to\infty}\frac{1}{nr}\sum_{i=0}^{nr-1}\frac{1}{r}k_r(T^ix)+\frac{\text{const}}{r} \\
&\leq \frac{1}{r}K_r(x)+\frac{\text{const}}{r}.
\end{aligned}$$

As $|k_{rn+j}(x)-k_{rn}(x)|\leq \text{const}\cdot r$ $(0<j<r)$ by (8), this yields $K(x)\leq\frac{1}{r}K_r(x)+\frac{\text{const}}{r}$, and letting $r\to\infty$, we find $K(x)\leq \tilde{K}(x)$.

Suppose now that μ is an ergodic T-invariant measure. Let $\epsilon>0$. Using the Shannon-McMillan-Breiman Theorem we find $r\in\mathbb{N}$ and $\mathcal{W}\subseteq\mathcal{L}^r$ such that

$$\mu\left(\bigcup_{w\in\mathcal{W}}[w]\right)\geq 1-\epsilon \quad\text{and}\quad \text{card}(\mathcal{W})\leq 2^{r(h_\mu(\mathcal{Z})+\epsilon)},$$

where $[w]=\{v\in\mathcal{L}^\infty : v_i=w_i (i=0,\dots,r-1)\}$. To each word from $\mathcal{W}$ we can associate a binary word of length $[r(h_\mu(\mathcal{Z})+\epsilon)]+1$, to those from $\mathcal{L}^r\setminus\mathcal{W}$ a binary word of length $r([\log\text{card}(\mathcal{Z})]+1)$. Prefacing all codewords for $\mathcal{W}$ by the string 01 and doubling all digits in the codewords for $\mathcal{L}^r\setminus\mathcal{W}$, we obtain a code $\mathcal{C}$ and can associate with each word $w\in\mathcal{L}^{nr}$ a word $\Phi_r(w)\in\mathcal{C}^n$. $\Phi_r:\mathcal{L}^{nr}\to\mathcal{C}^n$ is injective, and there is a decoding algorithm for Φ_r.

For itineraries $(x_1x_2x_3\dots)$ arising from trajectories of T we find

$$\begin{aligned}
\frac{1}{nr}k_{nr}(x) &\leq \frac{1}{nr}|\Phi_r(x_0\dots x_{nr-1})|+\frac{\gamma_r}{nr} \\
&\leq \frac{1}{n}\sum_{j=0}^{n-1}\left(h_\mu(\mathcal{Z})+\epsilon+\frac{3}{r}+2(\log\text{card}(\mathcal{Z})+1)\cdot\chi_{\{(x_{jr}\dots x_{(j+1)r-1})\notin\mathcal{W}\}}\right)+\frac{\gamma_r}{nr} \\
&\leq h_\mu(\mathcal{Z})+\epsilon+\frac{3}{r}+2(\log\text{card}(\mathcal{Z})+1)\cdot\frac{1}{n}\sum_{j=0}^{n-1}\chi_{\{(x_{jr}\dots x_{(j+1)r-1})\notin\mathcal{W}\}}+\frac{\gamma_r}{nr},
\end{aligned}$$

where γ_r denotes the length of a "binary implementation" of the decoding algorithm for Φ_r. Hence

$$\tilde{K}(x)=\lim_{n\to\infty}\frac{1}{nr}K_{nr}(x)\leq h_\mu(\mathcal{Z})+\epsilon+\frac{3}{r}+2(\log\text{card}(\mathcal{Z})+1)\cdot\epsilon$$

for μ-a.e. x, using the ergodic theorem. In the limit $r\to\infty$ and $\epsilon\to 0$ this yields $\tilde{K}(x)\leq h_\mu(\mathcal{Z})$ for μ-a.e. x.

Putting things together with Theorem 7 we find for μ-a.e. x:

$$h_\mu(\mathcal{Z})=K(x)\leq\tilde{K}(x)\leq h_\mu(\mathcal{Z}),$$

and as $K_L(x)=\int k_L\,d\mu$ for μ-a.e. x by ergodicity, also $\tilde{K}(x)=\tilde{K}_\mu$ for μ-a.e. x.

At this point I want to remark that the correct definition of complexity for general topological dynamical systems involves, just as for entropy, open covers [Br]. Here we will simplify the discussion by looking only at continuous piecewise monotone interval maps with the property:

$$\text{No endpoint of the intervals } Z_i \in \mathcal{Z}, \text{ except possibly 0 and 1, is periodic.} \tag{9}$$

For such maps each invariant measure μ assigns mass 0 to these endpoints and all its preimages. Hence, the functions $k_L(\,.\,;\mathcal{Z})$, which are constant on intervals $Z \in \mathcal{Z}_L$ are μ-a.e. continuous. [7] This fact will be used repeatedly in the proof of

Lemma 2 *Suppose* $T:[0,1]\to[0,1]$ *is as above. Then*

$$\breve{K}(x;\mathcal{Z}) = \sup\{\breve{K}_\mu(\mathcal{Z}) : \mu \in \tilde{\omega}_T(\delta_x)\} \quad \textit{for all } x,$$

and

$$\breve{K}_\mu(\mathcal{Z}) = h_\mu(\mathcal{Z}) \tag{10}$$

for all T-invariant measures μ.

Proof: As above we skip $\mathcal{Z}$ in expressions like $k_L(x;\mathcal{Z})$.

"$\leq$:" For each $n \in \mathbf{N}$ there is $\mu_n \in \tilde{\omega}_T(\delta_x)$ such that $K_{2^n}(x) = \int k_{2^n}\,d\mu_n$. For $n > r$ it follows from (8) that

$$2^{-n}K_{2^n}(x) = 2^{-n}\int k_{2^n}\,d\mu_n \leq 2^{-r}\int k_{2^r}\,d\mu_n + 2^{-r}\cdot \text{const.} \tag{11}$$

Pick a subsequence (μ_{n_i}) converging weakly to some $\mu \in \tilde{\omega}_T(\delta_x)$. By (11),

$$\breve{K}(x) = \lim_{n\to\infty} 2^{-n}K_{2^n}(x) \leq 2^{-r}\int k_{2^r}\,d\mu + 2^{-r}\cdot\text{const}$$

for all $r > 0$. In the limit $r \to \infty$ this yields

$$\breve{K}(x) \leq \breve{K}_\mu.$$

"$\geq$:" If on the other hand $\mu \in \tilde{\omega}_T(\delta_x)$, then there is an increasing sequence (n_i) of integers such that for $L > 0$ holds

$$\int k_L\,d\mu = \lim_{i\to\infty}\frac{1}{n_i}\sum_{j=0}^{n_i-1} k_L(T^j x) \leq K_L(x),$$

whence $\breve{K}_\mu \leq \breve{K}(x)$.

---

[7] Assumption (9) is no severe restriction. It can be overcome by doubling the periodic endpoints and their preimages and endowing the enlarged space again with the order topology, see [HK1] or [Kel, 3.14,3.15].

To prove identity (10) for arbitrary T-invariant measures, note first that $\mu \mapsto h_\mu(\mathcal{Z})$ and $\mu \mapsto \tilde{K}_\mu(\mathcal{Z})$ are affine, upper semicontinuous functionals on the space of T-invariant probability measures. (For the proof of the semicontinuity one uses (8), the subadditivity of the sequence $(H_\mu(\mathcal{Z}_n))_n$, and property (9).) Hence (10) carries over from ergodic invariant measures (Lemma 1) to arbitrary invariant measures in view of the ergodic decomposition theorem for affine functionals, see e.g. [Gray, Theorem 8.9.1]. □

We note an immediate consequence of the last two lemmas:

Theorem 9 *Let T be a continuous piecewise monotone interval map satisfying (9). If T has a BRS measure μ, then $\tilde{K}(x;\mathcal{Z}) = h_\mu(\mathcal{Z})$ for m-a.e. x.*

3.3 S-unimodal maps

In analogy to Theorem 4 we now have

Theorem 10 *Let T be a S-unimodal map. There is a constant $\tilde{K}_T(\mathcal{Z}) \geq \max\{\lambda_T, 0\}$ such that*

$$\tilde{K}(x;\mathcal{Z}) = \tilde{K}_T(\mathcal{Z}) \quad \textit{for } m\textit{-a.e. } x.$$

Furthermore:

1. *If $\lambda_T > 0$, then $\tilde{K}_T(\mathcal{Z}) = \lambda_T$.*
2. *If T has an attracting periodic orbit or a solenoidal attractor (this includes the case $\lambda_T < 0$ and some of the cases with $\lambda_T = 0$, cf. Section 2.3), then $\tilde{K}_T(\mathcal{Z}) = 0$.*

Proof: If T has an attracting periodic orbit, then the itinerary $(x_0x_1x_2\ldots)$ of m-a.e. x is eventually periodic (with the same periodic pattern displayed by the itinerary of the periodic attractor). Hence $\tilde{K}(x) = 0$ for m-a.e. x.

Otherwise (T, m) is ergodic [BL], and as $\tilde{K}(Tx) = \tilde{K}(x)$, [8] $\tilde{K}(x) = \text{const} =: \tilde{K}_T$ for m-a.e. x.

If $\lambda_T > 0$, then T has an ergodic a.c.i.p.m. μ (Theorem 4), and $\tilde{K}_T = \tilde{K}(x) = h_\mu(\mathcal{Z}) = \lambda_T$ for m-a.e. x by Lemma 1 and Theorem 4.

It remains to consider the case of a solenoidal attractor. It is known (cf. [BL] for references) that the attractor is uniquely ergodic and that the unique invariant measure ν on the attractor has entropy 0. Hence $\tilde{K}(x) = 0$ for m-a.e. x by Lemma 2. □

There are, however, maps with $\lambda_T = 0$ but $\tilde{K}_T > 0$:

Theorem 11 *Let $0 < h < \frac{1+\sqrt{5}}{2}$. There are S-unimodal maps T (with or without BRS measure) such that $\lambda_T = 0$ and $\tilde{K}_T(\mathcal{Z}) = h$. Such maps exist even whithin the family of quadratic maps $T_a(x) = ax(1-x)$.*

Proof: This follows from Remark 2 and Lemma 2. □

---

[8] The proof is similar to that of the corresponding statement for $K(x)$, see [Br, Lemma 1.3].

References

[BL] A.M. Blokh, M.Yu. Lyubich, *Measurable dynamics of S-unimodal maps of the interval,* Preprint (1990), Stony Brook.

[Br] A.A. Brudno, *Entropy and the complexity of the trajectories of a dynamical system.* Trans. Moscow Math, Soc. **44** (1982), 127-151.

[Gr] P. Grassberger, *Complexity and forecasting in dynamical systems,* in: Measures of Complexity, Proceedings Rome, 1987, Lectures Notes in Physics **314**.

[Gray] R.M. Gray, *Probability, Random Processes, and Ergodic Properties,* Springer-Verlag, 1988.

[Gu] J. Guckenheimer, *Sensitive dependence to initial conditions for one dimensional maps,* Commun. Math. Phys. **70** (1979), 133-160.

[HK1] F. Hofbauer, G. Keller, *Ergodic properties of invariant measures for piecewise monotonic transformations,* Math. Zeitschrift **180** (1982), 119-140.

[HK2] F. Hofbauer, G. Keller, *Quadratic maps without asymptotic measure,* Commun. Math. Phys. **127** (1990), 319-337.

[Jo] S. Johnson, *Singular measures without restrictive intervals,* Commun. Math. Phys. **110** (1987), 185-190.

[Ke1] G. Keller, *Markkov extensions, zeta functions, and Fredholm theory for piecewise invertible dynamical systems,* Trans. Amer. Math. Soc. **314** (1989), 433-497.

[Ke2] G. Keller, *Exponents, attractors, and Hopf decompositions for interval maps,* Preprint, 1988, (to appear in Ergod. Th. & Dynam. Sys.)

[Ko] A.N. Kolmogorov, *Three approaches to the definition of the concept of the "amount of information",* Selected Transl. Math. Statist. and Probab. **7** (1968), 293-302.

[Le] F. Ledrappier, *Some properties of absolutely continuous invariant measures on an interval.* Ergod. Th.& Dynam. Sys. **1** (1981), 77-93.

[Ma] M. Martens, *Interval Dynamics,* Thesis (1990), Delft.

[Mi] M. Misiurewicz, *Absolutely continuous invariant measures for certain maps of an interval,* Publ. Math. I.H.E.S. **53** (1981), 17-51.

[Pe] Ya. Pesin, *Characteristic Lyapunov exponents and smooth ergodic theory.* Russian Math. Surveys **32** (1977), 54-114.

[Ru] D. Ruelle, *An inequality for the entropy of differentiable maps.* Bol. Soc. Bras. Mat. **9** (1978), 83-87.

[Si] D. Singer, *Stable orbits and bifurcation of maps of the interval,* SIAM J. Appl. Math. **35** (1978), 260-267.

An inequality for the Ljapunov exponent of an ergodic invariant measure for a piecewise monotonic map of the interval

FRANZ HOFBAUER

Institut für Mathematik, Universität Wien,
Strudlhofgasse 4, A-1090 Wien, Austria.

Abstract. We consider a piecewise monotonic and piecewise continuous map T on the interval. Under a weak condition on the derivative of T, we show for an ergodic invariant probability measure μ that $h_\mu \leq \max\{0, \lambda_\mu\}$, where h_μ denotes the entropy and λ_μ the Ljapunov exponent of μ.

In [1] methods are developped to investigate the Ljapunov exponent, the entropy and the Hausdorff dimension of an ergodic invariant probability measure μ for a piecewise monotonic map on $[0,1]$ and to prove a relation between these quantities. In this note these methods are used to get further information about the Ljapunov exponent of μ.

We consider the same class of transformations as in [1]. A map T on the interval $[0,1]$ is called piecewise monotonic, if there exists a finite or countable set $\mathcal{Z}$ of open, pairwise disjoint subintervals of $[0,1]$ such that $T|Z$ is strictly monotone and continuous for all $Z \in \mathcal{Z}$. In order to emphasize the dependence on $\mathcal{Z}$, sometimes we shall say that T is piecewise monotonic with respect to $\mathcal{Z}$. It suffices to define T on $\bigcup_{Z\in\mathcal{Z}} Z$. We are interested in $E_\mathcal{Z} := \bigcap_{i=0}^{\infty} T^{-i}(\bigcup_{Z\in\mathcal{Z}} Z)$, the set of all points, for which all iterates of T are defined. We consider an ergodic T-invariant probability measure μ concentrated on $\bigcup_{Z\in\mathcal{Z}} Z$, which implies that μ is concentrated on $E_\mathcal{Z}$. Its Ljapunov exponent is denoted by λ_μ and its entropy by h_μ. The aim of this note is to show that $h_\mu \leq \lambda_\mu$ if $h_\mu > 0$. For differentiable maps on compact manifolds such an inequality is shown in [4]. We shall prove it for piecewise monotonic maps T with a weak condition on the derivative of T.

In order to give the exact statement of the result we define p-variation. For $p > 0$ and $g : [0,1] \to \mathbf{R}$ define

$$\operatorname{var}^p g = \sup\{\sum_{i=1}^{m} |g(x_{i-1}) - g(x_i)|^p : m \in \mathbf{N},\ 0 \leq x_0 < x_1 < \cdots < x_m \leq 1\}$$

We say that a map T, which is piecewise monotonic with respect to $\mathcal{Z}$, has a derivative of bounded p-variation, if there exists a function $g : [0,1] \to \mathbf{R}$ with $\operatorname{var}^p g < \infty$ and $g(x) = 0$ for $x \in [0,1] \setminus \bigcup_{Z\in\mathcal{Z}} Z$, such that $g(x)$ is the derivative of T at x for almost all $x \in \bigcup_{Z\in\mathcal{Z}} Z$ with respect to Lebesgue measure. A function of bounded p-variation has at most countably many discontinuities. Hence two derivatives of T of bounded p-variation differ in an at most countable set. We fix a derivative of T of bounded p-variation and denote it by T'. If μ is an ergodic T-invariant probability measure on $E_\mathcal{Z}$, we can now define the Ljapunov exponent λ_μ by $\int \log |T'|\, d\mu$. If $h_\mu > 0$ then μ has no atoms, since μ is ergodic, and λ_μ does not depend on the choice of T'. As T' is bounded we have $\lambda_\mu \in [-\infty, \infty)$.

THEOREM 1. *Let T be piecewise monotonic with respect to $\mathcal{Z}$ with a derivative of bounded p-variation for some $p > 0$. Set $W_{\mathcal{Z}} = \{x : x \in T(Z)$ for infinitely many $Z \in \mathcal{Z}\}$ and suppose that $W_{\mathcal{Z}}$ is at most countable. If μ is an ergodic T-invariant probability measure on $E_{\mathcal{Z}}$ with $h_\mu > 0$ and if λ_μ is finite, then $h_\mu \le \lambda_\mu$.*

We mention two families of maps, for which the assumption, that $W_{\mathcal{Z}}$ is at most countable, is satisfied. The first one consists of that piecewise monotonic maps, for which $\mathcal{Z}$ is finite (this is the class of transformations usually called piecewise monotonic). In this case we have $W_{\mathcal{Z}} = \emptyset$. The second family is the class of piecewise monotonic maps T with countable $\mathcal{Z}$, which are defined and continuous on all of $[0, 1]$ and for which the set L of all limit points of endpoints of the intervals in $\mathcal{Z}$ is at most countable. If $x \in W_{\mathcal{Z}}$, then x has inverse images in countably many $Z \in \mathcal{Z}$. These inverse images have a limit point y, which is then in L. As T is continuous, we get $x = T(y) \in T(L)$. Hence $W_{\mathcal{Z}}$ is a subset of $T(L)$ and therefore at most countable. One can generalize this second family allowing additionally that T has finitely many discontinuities, at which onesided limits exist. A similar arguement shows that $W_{\mathcal{Z}}$ is at most countable.

Under the same assumptions as in Theorem 1, but with $\lambda_\mu > 0$ instead of $h_\mu > 0$, it is shown in [1], that $\mathrm{HD}(\mu) = h_\mu/\lambda_\mu$, where $\mathrm{HD}(\mu) = \inf_{\mu(X)=1} \mathrm{HD}(X)$, and $\mathrm{HD}(X)$ denotes the Hausdorff dimension of the set X. As $\mathrm{HD}(\mu) \le 1$ this implies $h_\mu \le \lambda_\mu$. Hence Theorem 1 follows from [1], if one knows that $\lambda_\mu > 0$. The additional information, which Theorem 1 gives, is that λ_μ has to be strictly positive, if $h_\mu > 0$. This might be useful, since the case $\lambda_\mu = 0$ is often the difficult one.

The main tools for the proof of Theorem 1 are already prepared in [1]. The first one is Lemma 1 of [1], whose proof can be read independently of the rest of [1]. It implies

LEMMA 1. *Let T be a map, which is piecewise monotonic with respect to $\mathcal{Z}$ and which has a derivative T' of bounded p-variation for some $p > 0$. Let μ be an ergodic T-invariant probability measure on $E_{\mathcal{Z}}$ with $h_\mu > 0$ such that λ_μ is finite. We write f for $\log|T'|$. Then, for each $\varepsilon > 0$, there is a finite or countable family $\mathcal{Y}$ of open, pairwise disjoint intervals refining $\mathcal{Z}$, which means that every $Y \in \mathcal{Y}$ is contained in an $Z \in \mathcal{Z}$, such that the following properties are satisfied.*

$$\mu(\bigcup_{Y \in \mathcal{Y}} Y) = 1 \tag{1}$$

$$\sup_{x,y \in Y} |f(x) - f(y)| \le \varepsilon \text{ for all } Y \in \mathcal{Y} \tag{2}$$

$$-\sum_{Y \in \mathcal{Y}} \mu(Y) \log \mu(Y) < \infty \tag{3}$$

PROOF: We apply Lemma 1 of [1] with $\psi = |T'|$. The assumptions of this lemma are satisfied, as $T'(0) = T'(1) = 0$, as $|T'|$ is of bounded p-variation for some $p > 0$, and as $\lambda_\mu = \int f d\mu$ is finite. Lemma 1 of [1] says that there is a finite or countable family $\mathcal{Y}$ of open, pairwise disjoint intervals satisfying (1), (2) and (3). Furthermore, $\mathcal{Y}$ is constructed in such a way that $\bigcup_{Y \in \mathcal{Y}} Y$ does not contain zeroes of $\psi = |T'|$. As T' equals zero outside $\bigcup_{Z \in \mathcal{Z}} Z$, we get that $\mathcal{Y}$ refines $\mathcal{Z}$. ∎

The second result of [1] we need is proved there using a generalization of the technique in [2], which is carried out in section 4 of [1]. Again this section 4 does not depend on the rest of [1]. If $\mathcal{Z}$ is as in the definition of piecewise monotonicity of T, set $\mathcal{Z}_n = \bigvee_{i=0}^{n-1} T^{-i}\mathcal{Z} = \{\bigcap_{i=0}^{n-1} T^{-i}(Z_i) \neq \emptyset : Z_i \in \mathcal{Z}\}$, which is again a family of open, pairwise disjoint intervals, since T is monotone on each $Z \in \mathcal{Z}$. Furthermore $\mathcal{Z}_{n+1}$ refines $\mathcal{Z}_n$ and T^n is monotone on each $Z \in \mathcal{Z}_n$. For x in $E_{\mathcal{Z}}$, which equals $\bigcap_{n=1}^{\infty} \bigcup_{Z\in\mathcal{Z}_n} Z$, and for $n \geq 1$ there is a unique $Z \in \mathcal{Z}_n$ with $x \in Z$. We call it $Z_n(x)$. The sets $Z_n(x)$ are open intervals satisfying $Z_{n+1}(x) \subset Z_n(x)$. Now we can state

Lemma 2. *Suppose that T is piecewise monotonic with respect to $\mathcal{Z}$, that T has a bounded derivative (almost everywhere with respect to Lebesgue measure), and that $W_{\mathcal{Z}}$ is at most countable. If μ is an ergodic T-invariant probability measure on $E_{\mathcal{Z}}$ with $h_\mu > 0$ and $-\sum_{Z\in\mathcal{Z}} \mu(Z)\log\mu(Z) < \infty$, then there exists a nontrivial interval J such that for μ-almost all $x \in E_{\mathcal{Z}}$ there are infinitely many n with $J \subset T^n(Z_n(x))$.*

Proof: The assumptions of Proposition 2 of [1] are listed above. In the proof of this proposition it is shown that there is a nontrivial interval J (called E there) which is independent of x and which occurs infinitely often in the sequence $(T^n(Z_{n+1}(x)))_{n\geq 0}$ for μ-almost all x. Since $Z_{n+1}(x) \subset Z_n(x)$ the desired result follows. ∎

Now we can give the proof of Theorem 1.

Proof: Suppose that $h_\mu > \lambda_\mu$. Choose numbers α and β such that $h_\mu > \alpha > \beta > \lambda_\mu$ and set $\varepsilon = \frac{1}{2}(\beta - \lambda_\mu) > 0$. We write f for $\log|T'|$. By Lemma 1 there is a finite or countable family $\mathcal{Y}$ of open, pairwise disjoint intervals refining $\mathcal{Z}$, such that (1), (2) and (3) are satisfied. As above for $\mathcal{Z}$ set $\mathcal{Y}_n = \bigvee_{i=0}^{n-1} T^{-i}\mathcal{Y}$. Again $\mathcal{Y}_{n+1}$ refines $\mathcal{Y}_n$. Set $E_{\mathcal{Y}} = \bigcap_{n=0}^{\infty} T^{-n}(\bigcup_{Y\in\mathcal{Y}} Y) = \bigcap_{n=1}^{\infty}\bigcup_{Y\in\mathcal{Y}_n} Y$. It follows from (1) that $\mu(E_{\mathcal{Y}}) = 1$. For $x \in E_{\mathcal{Y}}$ and $n \geq 1$ let $Y_n(x)$ be the unique element of $\mathcal{Y}_n$ containing x. The sets $Y_n(x)$ are open intervals satisfying $Y_{n+1}(x) \subset Y_n(x)$. We show that $\mathcal{Y}$ is a generator for μ which implies that $h_\mu = h_\mu(T, \mathcal{Y})$ (cf. [3]). To this end, for $x \in E_{\mathcal{Y}}$ set $A = \bigcap_{j=0}^{\infty} Y_j(x)$ which is a (may be trivial) interval. Suppose that $\mu(A) > 0$. Then $T^k(A) \cap A \neq \emptyset$ for some $k \geq 1$, since otherwise $\mu(\bigcup_{j=1}^{\infty} T^j(A)) < \mu(T^{-1}(\bigcup_{j=1}^{\infty} T^j(A)))$. For each $n \geq 1$ the interval $T^k(A)$ is contained in some element of $\mathcal{Y}_n$, which has to be $Y_n(x)$, since the elements of $\mathcal{Y}_n$ are pairwise disjoint and $T^k(A) \cap A \neq \emptyset$. Hence $T^k(A) \subset A$. As μ is ergodic it is concentrated on the invariant set $\bigcup_{j=0}^{k-1} T^j(A)$. But T maps the interval $T^j(A)$ monotonically to the interval $T^{j+1}(A)$ for $j \geq 0$ and hence $h_\mu = 0$, a contradiction. We have shown that $\mu(A) = 0$. Let C be the union of the sets A which are nontrivial intervals. Since there are at most countably many such A, we have shown that $\mu(C) = 0$. In the set $E_{\mathcal{Y}}\setminus C$, which has μ-measure one, the $\mathcal{Y}_n$'s separate points showing that $\mathcal{Y}$ is a generator for μ. From (3) and the Shannon-McMillan-Breiman-Theorem (cf. [3]) we get now that $\lim_{n\to\infty} -\frac{1}{n}\log\mu(Y_n(x)) = h_\mu(T, \mathcal{Y}) = h_\mu$ for μ-almost all $x \in E_{\mathcal{Y}}$, as μ is ergodic. By the choice of α there is a set $M_1 \subset E_{\mathcal{Y}}$ with $\mu(M_1) = 1$ such that

(4) $\quad$ for $x \in M_1$ there is a k with $\mu(Y_n(x)) \leq e^{-\alpha n}$ for $n \geq k$

Next set $S_n f(x) = \sum_{i=0}^{n-1} f(T^i(x))$. Since $\lambda_\mu = \int f\,d\mu$ is finite and μ is ergodic, the ergodic theorem implies the existence of a set $M_2 \subset E_{\mathcal{Y}}$ with $\mu(M_2) = 1$ such that $\lim_{n\to\infty} \frac{1}{n} S_n f(x) = \lambda_\mu$ for $x \in M_2$. For $x \in E_{\mathcal{Y}}$ and $n \geq 1$ we get from (2) that $|S_n f(x) - S_n f(y)| \leq n\varepsilon$ for all $y \in Y_n(x)$. By the choice of β and ε we have for $x \in M_2$

that $\frac{1}{n}S_nf(x) \leq \beta - \varepsilon$ for all large n. Therefore $\frac{1}{n}S_nf(y) \leq \beta$ for all $y \in Y_n(x)$ and all large n. We have shown that

(5) $$\text{for } x \in M_2 \text{ there is a } k \text{ with } \sup_{y \in Y_n(x)} S_nf(y) \leq n\beta \text{ for } n \geq k$$

As $\mathcal{Y}$ refines $\mathcal{Z}$, T is also piecewise monotonic with respect to $\mathcal{Y}$ and μ is concentrated on $E_{\mathcal{Y}}$ by (1). Since T is strictly monotone on each interval in $\mathcal{Z}$, we get $W_{\mathcal{Y}} \subset W_{\mathcal{Z}}$ and hence $W_{\mathcal{Y}}$ is at most countable. Furthermore, T' is bounded, since $\operatorname{var}^p T' < \infty$. By (3) and $h_\mu > 0$ we can apply Lemma 2 for $\mathcal{Y}$ instead of $\mathcal{Z}$. There is a nontrivial interval J and a set $M_3 \subset E_{\mathcal{Y}}$ with $\mu(M_3) = 1$ such that

(6) $$\text{for } x \in M_3 \text{ there are infinitely many } n \text{ with } J \subset T^n(Y_n(x))$$

Now fix an arbitrary $m \in \mathbb{N}$. Set $M = M_1 \cap M_2 \cap M_3$. Then $\mu(M) = 1$. By (4), (5) and (6), for each $x \in M$ there is an $n(x) \geq m$ such that $\mu(Y_{n(x)}(x)) \leq e^{-\alpha n(x)}$, such that $\sup_{y \in Y_{n(x)}(x)} S_{n(x)}f(y) \leq n(x)\beta$ and such that $J \subset T^{n(x)}(Y_{n(x)}(x))$. We denote the length of an interval I by $|I|$. Since $S_nf(y) = \log|(T^n)'(y)|$, the last two estimates imply

$$|Y_{n(x)}(x)| \geq |T^{n(x)}(Y_{n(x)}(x))|(\sup_{y \in Y_{n(x)}(x)} |(T^{n(x)})'(y)|)^{-1} \geq |J|e^{-n(x)\beta}$$

Together with $\mu(Y_{n(x)}(x)) \leq e^{-\alpha n(x)}$ this gives

(7) $$\mu(Y_{n(x)}(x)) \leq \frac{1}{|J|}e^{-n(x)(\alpha-\beta)}|Y_{n(x)}(x)| \leq \frac{1}{|J|}e^{-m(\alpha-\beta)}|Y_{n(x)}(x)|$$

since $\alpha > \beta$ and $n(x) \geq m$.

Set $\tilde{\mathcal{U}} = \{Y_{n(x)}(x) : x \in M\}$ which covers M. As two elements of $\bigcup_{j \geq 1} \mathcal{Y}_j$ are either disjoint or one contains the other, there is a subset $\mathcal{U}$ of $\tilde{\mathcal{U}}$, whose elements are pairwise disjoint and which still covers M. From this and (7) we get

$$1 = \mu(M) \leq \sum_{U \in \mathcal{U}} \mu(U) \leq \frac{1}{|J|}e^{-m(\alpha-\beta)} \sum_{U \in \mathcal{U}} |U| \leq \frac{1}{|J|}e^{-m(\alpha-\beta)}$$

As m was arbitrary and $\alpha > \beta$, we have arrived at a contradiction proving $h_\mu \leq \lambda_\mu$. ∎

REMARK: In Theorem 1 we have used the assumption $\lambda_\mu > -\infty$. This means that $f = \log|T'| \in L^1_\mu$, which is essential for the proof of (1) and (3) of Lemma 1. The other assumptions of Lemma 1 give no connection between T' and μ and hence no connection between $\mathcal{Y}$, determined by T' through (2), and μ (cf. the proof of Lemma 1 in [1]). Therefore I conjecture that (1) and (3) need not hold, if $f \notin L^1_\mu$.

For a piecewise monotonic map with finite $\mathcal{Z}$, Theorem 1 can be improved. A function $g : [0,1] \to \mathbb{R}$ is said to be regular, if $\lim_{y \downarrow x} g(y)$ and $\lim_{y \uparrow x} g(y)$ exist for all $x \in [0,1]$. A derivative of a piecewise monotonic map to be regular is defined in the same way as a derivative of bounded p-variation.

THEOREM 2. *Suppose that T is a piecewise monotonic map with finite $\mathcal{Z}$ which has a regular derivative. If μ is an ergodic invariant probability measure with $h_\mu > 0$ then $h_\mu \le \lambda_\mu$.*

PROOF: Choose α and β as in the proof of Theorem 1. For $x \in [0,1]$ define $h(x) = \max\{\gamma, |T'(x)|\}$, where $\gamma > 0$ is chosen such that $\int \log h d\mu < \beta$. It is easy to see that h is regular. Set $f = \log h$. Then also f is regular, since $\inf h \ge \gamma > 0$. Set $\varepsilon = \frac{1}{2}(\beta - \int f d\mu)$. There exists a finite family $\tilde{\mathcal{Y}}$ of open, pairwise disjoint intervals, covering $[0,1]$ up to a finite set, such that (2) holds for $\tilde{\mathcal{Y}}$. Set $\mathcal{Y} = \{Y \cap Z \ne \emptyset : Y \in \tilde{\mathcal{Y}},\ Z \in \mathcal{Z}\}$, which refines $\mathcal{Z}$. As μ has no atoms (μ is ergodic and $h_\mu > 0$) and as $\mu(\bigcup_{Z \in \mathcal{Z}} Z) = 1$, we get (1). Clearly (2) holds also for $\mathcal{Y}$ and (3) is trivial, since $\mathcal{Y}$ is finite. The rest of the proof is the same as for Theorem 1 using that T' is bounded, that f is μ-integrable and that $S_n f(y) \ge \log |(T^n)'(y)|$. ∎

REFERENCES

[1] F. Hofbauer, P. Raith: The Hausdorff dimension of an ergodic invariant measure for a piecewise monotonic map of the interval. Can. Math. Bull. (1991)

[2] G. Keller: Lifting measures to Markov extensions. Mh. Math. 108 (1989), 183-200.

[3] W. Parry: Topics in Ergodic theory. Cambridge Tracts in Math. 75. Cambridge: Cambridge Univ. Press 1981.

[4] D. Ruelle: An inequality for the entropy of differentiable maps. Bol. Soc. Bras. Mat. 9 (1978), 83-87.

Généralisation du théorème de Pesin pour l'α-entropie.

P. THIEULLEN [1]
Université de Paris Sud, Mathématiques
91405 Orsay Cedex, France

Abstract: Let M be a compact d-dimensional manifold, $t \in]0,1[$ and $\phi.M \to M$ a $\mathcal{C}^{1,t}$ diffeomorphism preserving a Lebesgue measure m. If we define $h_m(\alpha,x,\phi)$ the α-entropy of ϕ and $\mu_1(x) \geq \mu_2(x) \geq ... \geq \mu_d(x)$ the Lyapunov exponents of the tangent map$T\phi$, then for m-almost every x:

$$h_m(\alpha,x,\phi) = \sum_{i=1}^{d} [\mu_i(x)+\alpha]^+ \qquad (\forall\, 0 \leq \alpha \leq -\mu_d(x)),$$
$$h_m(\alpha,x,\phi) = \alpha d \qquad (\forall\, \alpha \geq -\mu_d(x)).$$

Resume: Soient M une variété compacte de dimension finie d, $t \in]0,1[$ et $\phi:M \to M$ un $\mathcal{C}^{1,t}$ difféomorphisme préservant une mesure de Lebesgue m. Si $h_m(\alpha,x,\phi)$ désigne l'α-entropie de ϕ et $\mu_1(x) \geq \mu_2(x) \geq ... \geq \mu_d(x)$, les exposants de Lyapunov de l'application tangente$T\phi$, alors pour m-presque tout x:

$$h_m(\alpha,x,\phi) = \sum_{i=1}^{d} [\mu_i(x)+\alpha]^+ \qquad (\forall\, 0 \leq \alpha \leq -\mu_d(x)),$$
$$h_m(\alpha,x,\phi) = \alpha d \qquad (\forall\, \alpha \geq -\mu_d(x)).$$

Remerciement: Je voudrais remercier le rapporteur pour l'attention qu'il a apportée à cet article, pour ses commentaires précis et les améliorations détaillées qu'il m'a demandé d'y inclure.

I. Notations et énoncé du résultat principal

Dans tout ce qui suit, M désigne une variété compacte riemannienne de dimension d, m une mesure de Lebesgue (cf. [Di], p. 158) de masse totale égale à 1 définie sur les boréliens $\mathcal{B}$ de M, $\phi:M \to M$ un $\mathcal{C}^{1,t}$ difféomorphisme préservant la mesure m, c'est à dire un difféomorphisme $\mathcal{C}^1$ de différentielle $T\phi$ Hölder d'exposant $t \in]0,1[$ vérifiant la propriété:

$$m(\phi^{-1}(B)) = m(B) \quad (\forall\, B \in \mathcal{B}).$$

M sera muni de la distance géodésique d(x,y) et on appellera $B(x,\varepsilon)$ la boule de centre x et de rayon ε pour cette distance.

---

[1] Université de Paris Sud
Mathématiques, bât.425
91405 Orsay Cedex

I.1 Rappel de la théorie d'Oseledec ([Os],[Ru],[Le])

En chaque point $x \in M$, l'image par l'application tangente $T_x\phi^n$ de la boule unité $B_x(0,1)$ de T_xM est un ellipsoïde dont les axes principaux ont pour longeur $\chi^n{}_1(x) \geq \chi^n{}_2(x) \geq \ldots \geq \chi^n{}_d(x)$. Le théorème de Kingman ([Ki], théorème I.4) appliqué à la suite sous additive de fonctions

$$f_n(x) = \log \| \Lambda^k T_x\phi^n \| = \sum_{i=1}^{k} \log \chi_i^n(x),$$

$$f_{m+n}(x) \leq f_m(x) + f_n \circ \phi^m(x),$$

(où Λ^k désigne le produit tensoriel k fois, [Sc]) donne l'existence pour presque tout x de la limite

$$\lim_{n \to \infty} \frac{1}{n} \log \chi_i^n(x) = \mu_i(x) \quad (\forall\, i = 1,2,\ldots,d),$$

où $\mu_1(x) \geq \mu_2(x) \geq \ldots \geq \mu_d(x)$ est la suite des exposants de Lyapunov de ϕ au point x. Comme ϕ et ϕ^{-1} sont de classe $\mathcal{C}^1$ sur M compacte, les exposants $\mu_i(x)$ sont des réels finis:

$$-\sup_{x \in M} \log \| T_x\phi^{-1} \| \leq \mu_d(x) \leq \mu_1(x) \leq \sup_{x \in M} \log \| T_x\phi \|,$$

(on notera $\| T_x\phi \| = \sup_{v \neq 0} \dfrac{\| T_x\phi.v \|}{\| v \|}$, $T_x\phi^n = T_x(\phi^n)$ et $\| T\phi^n \| = \sup_{x \in M} \| T_x\phi^n \|$ $(\forall\, n \in \mathbb{Z})$).

Soient $r = r(x)$ le nombre d'exposants de Lyapunov distincts, $(\lambda_1(x) > \ldots > \lambda_r(x))$ la suite de ces exposants distincts et $d_i(x)$ la multiplicité de $\lambda_i(x)$ dans la suite $(\mu_1(x), \ldots, \mu_d(x))$. On définit alors la filtration croissante instable

$$E_x^s = \{ v \in T_xM : \limsup_{n \to +\infty} \frac{1}{n} \log \|T_x\phi^{-n}.v\| \leq -\lambda_s(x) \} \qquad (\forall\, s = 1, \ldots, r).$$

ainsi que la filtration décroissante stable

$$F_x^s = \{ v \in T_xM : \limsup_{n \to +\infty} \frac{1}{n} \log \|T_x\phi^{n}.v\| \leq \lambda_s(x) \} \qquad (\forall\, s = 1, \ldots, r).$$

Le théorème d'Oseledec peut s'énoncer de la manière suivante:

Théorème I.1.1 [Os] Pour presque tout x de M, pour tout s=1,...,r

i) $T_xM = E_x^s \oplus F_x^{s+1}$, $T_x\phi(E_x^s) = E_{\phi(x)}^s$, $T_x\phi(F_x^s) = F_{\phi(x)}^s$, $\lambda_s \circ \phi(x) = \lambda_s(x)$,

ii) $\forall\, v \in F_x^s \setminus F_x^{s+1}$ $\quad \lim_{n \to +\infty} \frac{1}{n} \log \| T_x\phi^n.v \| = \lambda_s(x)$,

iii) $\lim_{n \to \pm\infty} \frac{1}{n} \log \max (\| P_{\phi^n(x)}^s \|, \| Q_{\phi^n(x)}^s \|) = 0$,

où $F_x^{r+1} = \{ 0 \}$ et (P_x^s, Q_x^s) désigne les deux projecteurs sur E_x^s (resp. sur F_x^{s+1}) parallèlement à F_x^{s+1} (resp. à E_x^s).

La troisième propriété montre que l'angle entre les sous espaces $(E_{\phi^n(x)}^s, F_{\phi^n(x)}^s)$ le long de l'orbite ne tend pas exponentiellement vite vers zéro. On remarque de plus qu'on peut définir une décomposition en somme directe de T_xM invariante par $T_x\phi$ en posant $G_x^s = E_x^s \cap F_x^s$, alors

iv) $T_xM = G_x^1 \oplus \ldots \oplus G_x^r$, $T_x\phi(G_x^s) = G_{\phi(x)}^s$,

v) $\forall\, v \in G_x^s$ $\quad \lim_{n \to \pm\infty} \frac{1}{n} \log \| T_x\phi^n.v \| = \lambda_s(x)$,

vi) $\lim_{n \to \pm\infty} \frac{1}{n} \log \| \pi_{\phi^n(x)}^s \| = 0$,

où $(\pi_x^1, \ldots, \pi_x^r)$ désigne les projecteurs associés à la décomposition: $T_xM = G_x^1 \oplus \ldots \oplus G_x^r$.

Si on appelle m l'élément de volume canonique associée à la métrique riemannienne (cf. [Sp], p. 9-

17), par définition de m, il existe une fonction $h:M\to\mathbb{R}$ borélienne strictement positive telle que

$$m(B) = \int_B h \, d\underline{m} \qquad (\forall B \in \mathcal{B}).$$

Comme $\underline{m}$ est de masse totale finie, h^{-1} est intégrable par rapport à m et si on appelle maintenant $J_x(\phi)$ le jacobien de ϕ au point x, alors

$$J_x(\phi^n) = \chi_1^n(x) \dots \chi_d^n(x) = \frac{h(x)}{h\circ\phi^n(x)} \quad m \text{ p.p.}$$

ce qui montre, en appliquant le théorème de Birkhoff à $h(x)/h\circ\phi(x)$,

$$\sum_{i=1}^{d} \mu_i(x) = \sum_{s=1}^{r} d_s(x)\lambda_s(x) = 0.$$

I.2 Définition de l'α-entropie et résultat principal

Pour tout $n\in\mathbb{N}$ et $\alpha \geq 0$ on définit une nouvelle distance:

$$d_n^\alpha(x,y) = \max_{0\leq k\leq n} \{d(\phi^k(x), \phi^k(y))e^{k\alpha}\}.$$

Une boule de centre x et de rayon η pour cette distance d_n^α est de la forme

$$B_n^\alpha(x,\eta) = B_n^{\alpha,\phi}(x,\eta) = \bigcap_{k=0}^{n} \phi^{-k}(B(\phi^k(x),\eta e^{-k\alpha})),$$

(pour chaque $\alpha > 0$ fixé, la distance entre deux orbites de deux points quelconques de cette boule décroit exponentiellement vite vers zéro).

On appelle α-entropie locale de la mesure invariante m, les réels positifs suivant:

$$\overline{h}_m(\alpha,x,\phi) = \lim_{\eta\to 0} \limsup_{n\to+\infty} -\frac{1}{n} \log m\,[B_n^\alpha(x,\eta)],$$

$$\underline{h}_m(\alpha,x,\phi) = \lim_{\eta\to 0} \liminf_{n\to+\infty} -\frac{1}{n} \log m\,[B_n^\alpha(x,\eta)].$$

Cette définition d'α-entropie généralise la notion usuelle d'entropie $h_m(\phi)$ correspondant au cas $\alpha = 0$. Brin et Katok ont en effet démontré l'équivalence suivante:

Théorème I.2.1 [Br-Ka] Pour presque tout x de M,

i) $\overline{h}_m(0,x,\phi) = \underline{h}_m(0,x,\phi) \overset{\text{déf}}{=} h_m(x,\phi)$,

ii) $\int h_m(x,\phi)\, dm(x) = h_m(\phi)$.

Par ailleurs, Pesin [Pe] a démontré dans le cas précisément où m était une mesure de Lebesgue:

Théorème I.2.2 [Pe] Pour tout difféomorphisme ϕ de classe $\mathcal{C}^2$ préservant une mesure de Lebesgue m, $h_m(\phi) = \int \sum_{i=1}^{d} \mu_i^+(x)\, dm(x)$.

Le but de cette article est de généraliser cette égalité (ou plutôt l'égalité locale) à l'α-entropie.

Théorème I.2.3 Pour presque tout $x\in M$

i) $\overline{h}_m(\alpha,x,\phi) = \underline{h}_m(\alpha,x,\phi) = \alpha d \qquad (\forall\, \alpha \geq -\mu_d(x))$,

ii) $\overline{h}_m(\alpha,x,\phi) = \underline{h}_m(\alpha,x,\phi) = \sum_{i=1}^{d} [\mu_i(x)+\alpha]^+ \quad (\forall\, 0 \leq \alpha \leq -\mu_d(x))$.

Ces inégalités s'appuient beaucoup sur le fait que la dimension fractale de la mesure m est précisément d (ou plus exactement égale à la dimension de Lyapunov ([Ec-Ru], chapitre 4-3)):

$$\lim_{\eta \to 0} \frac{\log m(B(x,\eta))}{\log \eta} = d \qquad m \text{ p.p.}$$

Il semble probable que le théorème I.2.2 reste encore vrai lorsque la dimension fractale et la dimension de Lyapunov coïncident, même si ϕ n'est plus supposée inversible.

II. Démonstrations

II.1 Cas $\alpha > -\mu_d(x)$

Dans ce cas là, seule la mesure de la dernière boule $\phi^{-n}(B(\phi^n(x),\eta e^{-n\alpha}))$ compte dans le calcul de $m(B_n^\alpha(x,\eta))$. Auparavant nous avons besoin d'établir le lemme suivant:

Lemme II.1.1 Pour tout $\varepsilon > 0$, il existe une fonction $K_\varepsilon: M \to \mathbb{R}_+$ mesurable telle que, pour presque tout x, pour tout $\eta > 0$ et pour tout $n \geq 0$

$$(K_\varepsilon(x))^{-1}\eta^d \leq m(B(x,\eta)) \leq K_\varepsilon(x)\,\eta^d,$$
$$e^{-n\varepsilon} K_\varepsilon(x) \leq K_\varepsilon(\phi^n(x)) \leq e^{+n\varepsilon} K_\varepsilon(x).$$

Démonstration Grâce à la compacité de M on peut trouver une constantye $C > 0$ telle que

$$C^{-1}\eta^d \leq \underline{m}(B(x,\eta)) \leq C\,\eta^d \qquad (\forall\, x \in M,\ \forall\, \eta > 0).$$

Comme la dérivée de Radon-Nikodym de m par rapport à $\underline{m}$ est h (cf. [LY], lemma 4.1.2),

$$\lim_{\eta \to 0} \frac{m(B(x,\eta))}{\underline{m}(B(x,\eta))} = h(x) \qquad m \text{ p.p.}$$

d'où l'existence de deux fonctions positives finies

$$S(x) = \sup_{\eta > 0} \ m(B(x,\eta))\,\eta^{-d} < +\infty \quad m \text{ p.p.}$$
$$I(x) = \inf_{\eta > 0} \ m(B(x,\eta))\,\eta^{-d} > 0 \quad m \text{ p.p.}$$

Comme

$$\phi(B(x,\eta \,\| T\phi^{-1} \|)) \supseteq B(\phi(x),\eta) \supseteq \phi(B(x,\eta \,\| T\phi \|^{-1})),$$
$$S\circ\phi(x) \leq \| T\phi^{-1} \|^d S(x) \quad \text{et} \quad I\circ\phi(x) \geq \| T\phi \|^{-d} I(x).$$

D'apres le lemme III.8 de Mañé $[Ma]_2$ on en déduit

$$\lim_{n \to +\infty} \frac{1}{n} \log S\circ\phi^n(x) = \lim_{n \to +\infty} \frac{1}{n} \log I\circ\phi^n(x) \quad m \ \text{ p.p.}$$

Le lemme II.1.1 est démontré en prenant par exemple

$$K_\varepsilon(x) = \sup_{n \geq 0} \ \{\max(S\circ\phi^n(x), I^{-1}\circ\phi^n(x))e^{-n\varepsilon}\}$$

CQFD

Démonstration du théorème I.2.3 (cas $\alpha > -\mu_d(x)$) On remarquera d'abord que ($\alpha \to \underline{h}_m(\alpha,x,\phi)$) est une fonction croissante en α et que pour tout $x \in M$, $\alpha \geq 0$ et $\eta > 0$

$$B_n^\alpha(x\ \eta) \subset \phi^{-n}(B(\phi^n(x),\eta e^{-n\alpha}))$$

et donc en utilisant l'invariance de m par ϕ et le lemme II.1.1 on obtient bien $\underline{h}_m(\alpha,x,\phi) \geq \alpha d$.

Réciproquement, appelons $M_\alpha = \{x \in M : -\mu_d(x) < \alpha\}$ et montrons que:

$$\int_{M_\alpha} \overline{h}_m(\alpha,x,\phi)\, dm(x) \leq \alpha d\, m(M_\alpha).$$

On peut supposer d'abord que $m(M_\alpha) > 0$ et appelons $A_N = \{ x \in M_\alpha : \| T_x\phi^{-N} \| < e^{N\alpha} \}$. Comme $\lim_{N\to\infty} \frac{1}{N} \log \| T_x\phi^{-N} \| = -\mu_d(x)$ m p.p., $\lim_{N\to\infty} m(A_N) = m(M_\alpha)$. Par continuité de $(x \to \| T_x\phi^{-N} \|)$, on peut trouver un voisinage d'ordre η_N de A_N sur lequel $\| T_x\phi^{-N} \| \le e^{N\alpha}$, en particulier

$$\forall x \in A_N \ \forall y \in B(x,\eta_N) \quad \| T_y\phi^{-N} \| \le e^{\alpha N}.$$

Par ailleurs, en choisissant $\nu \ge \alpha$ tel que $\| T_x\phi^{-1} \| \le e^\nu$ pour tout $x \in M$,

$$\forall x \in A_N^c \ \forall y \in B(x,\eta_N) \ \| T_y\phi^{-N} \| \le e^{N\nu},$$

$$\forall x \in M \ \forall \eta < \eta_N \quad \phi^{-N}(B(\phi^N(x),\eta)) \subseteq B(x, \eta e^{\beta\circ\phi^N(x)})$$

où $\beta(x) = N(\alpha \mathbb{1}_{A_N}(x) + \nu \mathbb{1}_{A_N^c}(x)) \ge N\alpha$.

Plus généralement pour tout $0 \le k \le n$

$$\phi^{-kN}(B(\phi^{nN}(x), \eta \exp - \sum_{i=1}^{n} \beta\circ\phi^{iN})) \subseteq B(\phi^{(n-k)N}(x), \eta \exp - \sum_{i=1}^{n-k} \beta\circ\phi^{iN}),$$

$$\phi^{-nN}(B(\phi^{nN}(x), \eta \exp - \sum_{i=0}^{n-1} \beta\circ\phi^{iN})) \subseteq B_n^{N\alpha,\phi^N}(x,\eta).$$

En appliquant le théorème de Birkhoff à β et de nouveau le lemme II.1.1 on obtient

$$\overline{h}_m(N\alpha,x,\phi^N) \le d \lim_{n\to\infty} \frac{1}{n} \sum_{i=0}^{n-1} \beta\circ\phi^{iN},$$

$$\overline{h}_m(N\alpha,x,\phi^N) \le dN\,[\, \alpha\, m(A_N \mid \mathcal{I}_N) + \nu\, m(A_N^c \mid \mathcal{I}_N)\,]$$

où $m(.\mid.)$ désigne la mesure conditionnelle par rapport à la tribu $\mathcal{I}_N$ des ensembles ϕ^N–invariants.

Enfin $\overline{h}_m(\alpha,x,\phi) = \frac{1}{N}\, \overline{h}_m(N\alpha,x,\phi^N)$ et par intégration sur M_α:

$$\int_{M_\alpha} \overline{h}_m(\alpha,x,\phi)\, dm(x) \le d\alpha\, m(A_N) + d\nu\, m(M_\alpha \setminus A_N).$$

On termine la démonstration en faisant tendre N vers $+\infty$. *CQFD*

II.2 Nombre de recouvrement local

Pour terminer la démonstration du théorème I.2.3, il suffit d'établir la proposition suivante. On appellera par la suite:

$$\Lambda(\alpha,x) = \sum_{i=1}^{d} [\alpha + \mu_i(x)]^+ \qquad (\forall\, 0 \le \alpha \le -\mu_d(x)),$$

$$\Lambda(\alpha,x) = \alpha d \qquad (\forall\, \alpha \ge -\mu_d(x)).$$

Proposition II.2.1 Pour m-presque tout x et pour tout $0 \le \alpha \le \beta$

i) $\overline{h}_m(\beta,x,\phi) - \overline{h}_m(\alpha,x,\phi) \le \Lambda(\beta,x) - \Lambda(\alpha,x)$,

ii) $\underline{h}_m(\beta,x,\phi) - \underline{h}_m(\alpha,x,\phi) \le \Lambda(\beta,x) - \Lambda(\alpha,x)$.

Démonstration du théorème I.2.3 (cas $0 \le \alpha \le -\mu_d(x)$). Les théorèmes I.2.1 et I.2.2 entrainent $\underline{h}_m(0,x,\phi) = \overline{h}_m(0,x,\phi) = \Lambda(0,x)$ m-p.p.. La première inégalité de II.2.1 donne donc $\overline{h}_m(\beta,x,\phi) \le \Lambda(\beta,x)$ $(\forall\, \beta \ge 0)$. La deuxième inégalité montre d'abord que $(\alpha \to \underline{h}_m(\alpha,x,\phi))$ est continue sur $\mathbb{R}^+$, et grâce à $\underline{h}_m(-\mu_d(x),x,\phi) = -d\mu_d(x) = \Lambda(-\mu_d(x),x)$, elle montre que alors $\underline{h}_m(\alpha,x,\phi) \ge \Lambda(\alpha,x)$ pour tout $0 \le \alpha \le -\mu_d(x)$. *CQFD*

La démonstration de la proposition II.2.1 se décompose en deux parties. On appelle nombre de recouvrement local $R_n^{\alpha,\beta}(x,\eta)$ le nombre minimum de boules $B_n^{\beta}(y,\eta/2)$ nécéssaires pour recouvrir $B_n^{\alpha}(x,2\eta)$:

$$R_n^{\alpha,\beta}(x,\eta) = \inf \{N : \exists\, y_1, \ldots , y_N \in M \mid B_n^{\alpha}(x,2\eta) \subseteq \bigcup_{i=1}^{N} B_n^{\beta}(y_i,\eta/2) \}.$$

La première partie est très générale.

Lemme II.2.3 Soient $(M,\mathcal{B})$ un espace de Borel standard (cf. [Pa], chapitre, V.2.2), m une mesure de probabilité, d_1, d_2 deux distances vérifiant

i) $((x,y) \in M^2 \to d_i(x,y))$ est une application mesurable,

ii) $m(\{ x \in M : m(B_i(x,\eta) = 0 \}) = 0 \quad (\forall\, \eta > 0)$,

où $B_1(x,\eta)$ et $B_2(x,\eta)$ désignent les boules associées et $R(x,\eta)$ le nombre minimum de boules $B_2(y,\eta/2)$ nécessaires pour recouvrir $B_1(x,2\eta)$. Alors pour tout $\lambda \in\,]0,1[$, $\eta > 0$

$$m(\{ x \in M : m(B_2(x,\eta)) \leq \lambda\, m(B_1(x,\eta)) / R(x,\eta) \}) \leq \lambda.$$

Démonstration Nous commençons par montrer une inégalité maximale plus générale. Pour toute fonction f intégrable positive on appellera

$$f_\eta(x) = \frac{1}{m(B_1(x,\eta))} \int_{B_1(x,2\eta)} f(y)\, m(dy).$$

Alors pour tout $\lambda \in\,]0,1[$

$$(*) \qquad m(\{ x \in M : f_\eta(x) \leq \lambda\, f(x) \}) \leq \lambda.$$

On peut commencer par supposer que le support de m est M et que $\inf_{x \in M} f(x) > 0$ (en prenant par exemple $f^{(\varepsilon)} = \varepsilon + f(x)$), alors

$$m(\{ x \in M : f_\eta(x) \leq \lambda\, f(x) \}) \leq \lambda \int \frac{f(x)}{f_\eta(x)}\, m(dx)$$

$$= \lambda \int \frac{f(x)\, m(B_1(x,\eta)}{\int_{B_1(x,2\eta)} f(y)\, m(dy)}\, m(dx)$$

(en utilisant l'identité $\mathbb{1}_{B_1(x,\eta)}(z) = \mathbb{1}_{B_1(z,\eta)}(x)$ et le théorème de Fubini, on obtient:)

$$= \lambda \int \Big[\int_{B_1(z,\eta)} \frac{f(x)}{\int_{B_1(x,2\eta)} f(y)\, m(dy)}\, m(dx) \Big]\, m(dz)$$

ce qui montre l'inégalité maximale (*) en remarquant

$$(\forall\, x \in B_1(z,\eta)) \qquad B_1(x,2\eta) \supseteq B_1(z,\eta).$$

On applique maintenant cette inégalité à $f(x) = \dfrac{1}{m(B_2(x,\eta))}$ et on remarque que

$$\int_{B_1(x,2\eta)} \frac{m(dy)}{m(B_2(y,\eta))} \leq R(x,\eta).$$

On peut en effet recouvrir $B_1(x,2\eta)$ par $R(x,\eta)$ boules $B_2(y_i,\eta/2)$ et de nouveau utiliser

$$(\forall\, y \in B_2(y_i,\eta/2)) \quad B_2(y,\eta) \supseteq B_2(y_i,\eta/2).$$

CQFD

Corollaire II.2.4 Pour presque tout x et pour tout $0 \leq \alpha \leq \beta$, $\eta > 0$,

$$\limsup_{n \to +\infty} \; -\frac{1}{n} \log \Big[\frac{m(B_n^{\beta}(x,\eta))}{m(B_n^{\alpha}(x,\eta))} \Big] \leq \limsup_{n \to +\infty} \frac{1}{n} \log R_n^{\alpha,\beta}(x,\eta).$$

Démonstration Fixons $\zeta > 0$ et appelons

$$A_n = \{ x \in M : m(B_n^{\beta}(x,\eta)) \leq e^{-n\zeta} \frac{m(B_n^{\alpha}(x,\eta))}{R_n^{\alpha,\beta}(x,\eta)} \}.$$

D'après le lemme précédent II.2.2, $m(A_n) \leq e^{-n\zeta}$, et d'apres Borel-Cantelli, presque tout x n'appartient pas à A_n à partir d'un certain rang n, d'où

$$\limsup_{n \to +\infty} \frac{1}{n} \log \frac{m(B_n^{\alpha}(x,\eta))}{m(B_n^{\beta}(x,\eta))\, R_n^{\alpha,\beta}(x,\eta)} \leq \zeta$$

pour tout $\zeta > 0$. *CQFD*

II.3 Métrique de Lyapunov et cônes invariants

La deuxième partie de la démonstration de II.2.1 consiste à majorer $R_n^{\alpha,\beta}(x,\eta)$ en fonction des exposants de Lyapunov. Comme dans tout ce qui suit nous allons travailler autour de l'orbite d'un point $x \in M$ fixé (générique pour m), nous commençons par remplacer ϕ par son relevé dans l'espace tangent au moyen de l'application exponentielle.

Notations II.3.1 Par compacité de M, on peut trouver des constantes $\varepsilon_0 > 0$, $C \geq 1$ telles que

i) $\exp_x$ soit un $\mathcal{C}^\infty$ difféomorphisme de $\{ v \in T_xM : \| v \| < \varepsilon_0 C \}$ sur $B(x,\varepsilon_0 C)$,

ii) $\max_{x \in M} \left(\| T_x\phi \| , \| T_x\phi^{-1} \| , \mathrm{lip}(\exp_x) , \mathrm{lip}(\exp_x^{-1}) \right) \leq C$,

iii) $\| D_v f_x - D_w f_x \| \leq C \| v - w \|^t \quad (\forall\ \| v \|, \|w \| < \varepsilon_0)$,

où $f_x = \exp_{\phi(x)}^{-1} \circ \phi \circ \exp_x$ sur $\{ v \in T_xM : \| v \| < \varepsilon_0 \}$, $D_v f_x$ est la différentielle de f_x au point $v \in T_xM$ (en particulier $D_0 f_x = T_x\phi$).

Pour simplifier on écrira par la suite $T_x = T_x\phi$. On construit maintenant sur chaque espace tangent une nouvelle métrique (dite de Lyapunov) $\| . \|_\varepsilon$ dépendant mesurablement de x et d'un paramètre ε qu'on fera tendre 0 à la fin de la démonstration, de manière que sur chaque sous espace G_x^s, T_x se comporte comme une homothétie de rapport $e^{\lambda_s(x)}$. On notera dans la suite $(\varepsilon : M \to]0,\varepsilon_0[)$ une fonction mesurable invariante par ϕ ($\varepsilon \circ \phi = \varepsilon$ m-p.p.).

Définition II.3.2 Pour presque tout x, pour tout $v = v_1 + \ldots + v_r \in T_xM$, où $v_s \in G_x^s$, on pose

$$\| v \|_\varepsilon = \max(\| v_1 \|_\varepsilon, \ldots, \| v_r \|_\varepsilon),$$

$$\| v_s \|_\varepsilon = \sqrt{r(x)} \sum_{n \in \mathbb{Z}} \exp(-n\, \lambda_s(x) - |n|\, \varepsilon(x))\ \| T_x^n v_s \|.$$

Rappelons les propriétés de la métrique de Lyapunov (cf. [Ka-St], théorème 2.2, p. 10):

Proposition II.3.3 Pour toute fonction ϕ-invariante $(\varepsilon : M \to]0,\varepsilon_0[)$, on peut construire une fonction mesurable $(\rho_\varepsilon : M \to]0,\varepsilon[)$ qui vérifie les propriétés suivantes:

i) $e^{-\varepsilon(x)} \leq \rho_\varepsilon \circ \phi(x) / \rho_\varepsilon(x) \leq e^{\varepsilon(x)}$,

ii) $\forall\ v \in T_xM \qquad \| v \| \leq \| v \|_\varepsilon \leq \| v \| (\rho_\varepsilon(x))^{-1}$,

iii) $\forall\ v \in G_x^s \qquad e^{\lambda_s - \varepsilon} \| v \|_\varepsilon \leq \| T_x v_s \|_\varepsilon \leq \| v \|_\varepsilon\, e^{\lambda_s + \varepsilon}$,

iv) $\forall\ v \in T_xM \qquad \| v \|_\varepsilon , \| w \|_\varepsilon \leq \rho_\varepsilon(x) \Rightarrow \| D_v f_x - D_w f_x \|_\varepsilon < \varepsilon(x)$,

v) $\| P^s_x \|_\varepsilon = \| Q^s_x \|_\varepsilon = 1.$

La construction fait intervenir les fonctions $\sum_{n\in\mathbb{Z}} \exp(-n\,\lambda_s(x) - |n|\frac{\varepsilon}{2}(x))\ \| T^n_x \mid G^s_x \|$, $\sum_{n\in\mathbb{Z}} \exp(-|n|\frac{\varepsilon}{2}(x))\ \max(\| P^s_{\phi^n(x)} \|, \| Q^s_{\phi^n(x)} \|)$, le lemme III.8 de Mañé $[\mathrm{Ma}]_2$ et pour la première fois le fait que $T\phi$ est Hölder d'exposant $t \in\ ]0,1[$.

On notera dans la suite $B_{x,\varepsilon}(v,\eta) = \{ w\in T_xM : \| v-w \|_\varepsilon < \eta \}$ une boule de centre v et de rayon η pour cette nouvelle métrique dans chaque espace tangent. Une propriété importante des systèmes hyperboliques est l'invariance des cônes par la transformation. On supposera de plus dans la suite que pour presque tout x de M:

$$2\varepsilon(x) < \min_{1\le s\le r} \left(e^{\lambda_s(x) - \varepsilon(x)} - e^{\lambda_{s+1}(x) + \varepsilon(x)} \right).$$

Proposition II.3.4 Pour presque tout $x\in M$, pour tout $0 \le s \le r$, v,w dans $B_{x,\varepsilon}(0,\rho_\varepsilon(x))$

i) si $\| P^s_x (v-w) \|_\varepsilon \ge \| Q^s_x (v-w) \|_\varepsilon$ alors
$\| P^s_{\phi(x)} (f_x(v)-f_x(w)) \|_\varepsilon \ge \| Q^s_{\phi(x)} (f_x(v)-f_x(w)) \|_\varepsilon$ et
$\| v-w \|_\varepsilon \le (e^{\lambda_s-\varepsilon} - \varepsilon)^{-1} \| f_x(v)-f_x(w) \|_\varepsilon$

ii) si $\| P^s_{\phi(x)} (f_x(v)-f_x(w)) \|_\varepsilon \le \| Q^s_{\phi(x)} (f_x(v)-f_x(w)) \|_\varepsilon$ alors
$\| P^s_x (v-w) \|_\varepsilon \le \| Q^s_x (v-w) \|_\varepsilon$ et
$\| f_x(v)-f_x(w) \|_\varepsilon \le (e^{\lambda_{s+1} + \varepsilon} + \varepsilon) \| v-w \|_\varepsilon.$

Démonstration La propriété iv) de la proposition II.3.3 donne

$$\| f_x(v)-f_x(w) - T_x(v-w) \|_\varepsilon \le \varepsilon \| v-w \|_\varepsilon.$$

On obtient le résultat en projettant cette inégalité sur E^s_x et F^{s+1}_x et en utilisant

$$\forall\ v\in E^s_x \qquad \| T_x v \|_\varepsilon \ge e^{\lambda_s - \varepsilon} \| v \|_\varepsilon,$$
$$\forall\ v\in F^{s+1}_x \qquad \| T_x v \|_\varepsilon \le e^{\lambda_{s+1}+\varepsilon} \| v \|_\varepsilon.$$

CQFD

On notera enfin $\mathcal{D}_{x,\varepsilon,n} = \{ v\in T_xM : \forall\ 0 \le k \le n \quad \| f^k_x(v) \|_\varepsilon < \rho_\varepsilon\circ\phi^k(x) \}$, le domaine de validité des propositions II.3.3 et II.3.4 pour la suite des transformations $(f^k_x)^n_{k=0}$ où pour tout $v\in \mathcal{D}_{x,\varepsilon,n}$

$$f^k_x = f_{\phi^{k-1}(x)} \circ \dots \circ f_{\phi(x)} \circ f_x$$
$$B^\alpha_{x,\varepsilon,n}(v,\eta) = \{ w\in \mathcal{D}_{x,\varepsilon,n} : \| f^k_x(w) - f^k_x(v) \| < \eta\, e^{-k\alpha}\ (\forall\ 0 \le k \le n) \}.$$

Corollaire II.3.5 Pour presque tout $x\in M$, pour tout $\eta\in]0,\rho_\varepsilon[$ et β tels que

$$e^{\lambda_{s+1} + \varepsilon} + \varepsilon \le e^{-\beta} \le e^{\lambda_s - \varepsilon} - \varepsilon.$$

Si $(v,w)\in \mathcal{D}_{x,\varepsilon,n}$, $\| Q^s_x(v-w) \|_\varepsilon < \eta$ et $\| P^s_{\phi^n(x)}(f^n_x(v) - f^n_x(w)) \|_\varepsilon < \eta e^{-n\beta}$ alors

$$\| f^k_x(v) - f^k_x(w) \|_\varepsilon < \eta e^{-k\beta} \quad (\forall\ 0 \le k \le n).$$

Démonstration Deux cas apparaîssent suivant que $f^k_x(v) - f^k_x(w)$ appartienne au cône instable ou stable:

Ou bien $\| P^s_{\phi^k(x)}(f^k_x(v) - f^k_x(w)) \|_\varepsilon \ge \| Q^s_{\phi^k(x)}(f^k_x(v) - f^k_x(w)) \|_\varepsilon$ alors

$$\| f^k_x(v) - f^k_x(w) \|_\varepsilon \le [\, e^{\lambda_s - \varepsilon} - \varepsilon]^{-(n-k)} \| f^n_x(v) - f^n_x(w) \|_\varepsilon.$$

Ou bien $\| P^s_{\phi^k(x)}(f^k_x(v) - f^k_x(w)) \|_\varepsilon \le \| Q^s_{\phi^k(x)}(f^k_x(v) - f^k_x(w)) \|_\varepsilon$ alors

$$\| f_x^k(v) - f_x^k(w) \|_\varepsilon \le [\, e^{\lambda_{s+1}+\varepsilon} + \varepsilon\,]^k \, \| v - w \|_\varepsilon . \qquad \textit{CQFD}$$

II.4 Démonstration de la proposition II.2.1 (cas $0 \le \alpha \le -\mu_d(x)$)

Appelons $R^{\alpha,\beta}_{x,\varepsilon,n}(v,\eta)$ le nombre minimum de boules $B^{\beta}_{x,\varepsilon,n}(v_i,\eta/2C)$ nécessaires pour recouvrir $B^{\alpha}_{x,\varepsilon,n}(v,2C\,\eta/\rho_\varepsilon(x))$. Alors une première étape de la démonstration de II.2.5 consiste à établir:

Lemme II.4.1 Pour presque tout $x\in M$, pour tout $0 \le \alpha \le \beta$

$$\limsup_{\eta\to 0}\ \limsup_{n\to+\infty}\ -\frac{1}{n}\log\frac{m(B^{\beta}_{n}(x,\eta))}{m(B^{\alpha}_{n}(x,\eta))} \le \limsup_{\varepsilon\to 0}\ \limsup_{\eta\to 0}\ \limsup_{n\to+\infty}\ \frac{1}{n}\log R^{(\alpha-\varepsilon)^+,\beta}_{x,\varepsilon,n}(0,\eta).$$

Démonstration Supposons d'abord $\alpha > 0$ et choisissons ε tel que $2\varepsilon(x) < \alpha$ m-p.p. Alors l'inégalité demandée découle du corollaire II.2.4 et de l'inégalité

$$R^{\alpha,\beta}_{n}(x,\eta) \le R^{\alpha-\varepsilon,\beta}_{x,\varepsilon,n}(0,\eta) \qquad (\forall\, \eta < (\rho_\varepsilon(x))^2/2C).$$

En effet

$$(\exp_x)^{-1}(B^{\alpha}_{n}(x,2\eta)) \subseteq B^{\alpha-\varepsilon}_{x,\varepsilon,n}(0,2C\eta\rho_\varepsilon(x)^{-1}),$$

et

$$(\exp_x)\,(B^{\beta}_{x,\varepsilon,n}(0,\eta/2C)) \subseteq B^{\beta}_{n}(x,\eta/2).$$

Dans le cas où $\alpha = 0$, comme ρ_ε est une fonction régulière le long des orbites (inégalités i) de la proposition II.3.3), le lemme 2 de Mañé s'applique $[\mathrm{Ma}]_1$. Pour tout $\eta < (\rho_\varepsilon(x))^2/2C$, $\delta\in\,]0,1[$ il existe un borélien A vérifiant $m(A) > 1-\delta$ et une partition $\mathcal{P}_n$ d'entropie finie telle que pour tout $x\in A$, $n \ge 0$,

$$\mathcal{P}_n(x,\eta) := \bigcap_{k=0}^{n} \phi^{-k}(\mathcal{P}_n\circ\phi^k(x)) \subseteq \bigcap_{k=0}^{n} \phi^{-k}(B(\phi^k(x),2\eta(\rho\circ\phi^k(x))^2/\rho_\varepsilon(x)).$$

Ce qui montre

$$(\exp_x)^{-1}\,(\mathcal{P}_n(x,\eta)) \subseteq B^{0}_{x,\varepsilon,n}(0,2C\eta/\rho_\varepsilon(x)) \quad (\forall\, x\in A,\ \forall\, n \ge 0).$$

Le nombre minimum de boules $B^{\beta}_{n}(y_i,\eta/2)$ nécessaires pour recouvrir $\mathcal{P}_n(x,\eta)$ est donc au plus $R^{0,\beta}_{x,\varepsilon,n}(0,\eta)$ pour tout $x\in A$ et $n \ge 0$. De la même manière que pour le corollaire II.2.4 on peut démontrer

$$\limsup_{n\to+\infty}\ -\frac{1}{n}\log\frac{m(B^{\beta}_{n}(x,\eta))}{m(\mathcal{P}_n(x,\eta))} \le \limsup_{n\to+\infty}\ \frac{1}{n}\log R^{0,\beta}_{x,\varepsilon,n}(0,\eta) \quad (\forall\, x\in A).$$

Par ailleurs, en utilisant le théorème de Brin et Katok [Br-Ka]

$$\underline{h}_m(0,x,\phi) = \overline{h}_m(0,x,\phi) = \lim_{\eta\to 0}\ \lim_{n\to+\infty}\ -\frac{1}{n}\log m(\mathcal{P}_n(x,\eta)),$$

on obtient pour presque tout $x\in M$

$$\limsup_{n\to+\infty}\ -\frac{1}{n}\log m(B^{\beta}_{n}(x,\eta)) \le \underline{h}_m(0,x,\phi) + \limsup_{n\to+\infty}\ \frac{1}{n}\log R^{0,\beta}_{x,\varepsilon,n}(0,\eta). \qquad \textit{CQFD}$$

Nous aurons aussi besoin du lemme suivant élémentaire (cf. [Li-Tz], proposition I.c.3):

Lemme II.4.2 Pour tout espace vectoriel normé de dimmension d, il existe une constante $\Gamma(d)$ ne dépendant que de la dimension d telle que, en appelant $r(\eta)$ le nombre minimum de boules de rayon η nécessaires pour recouvrir la boule unité, on a

$$r(\eta) \le \Gamma(d)\max(1,\eta^{-d}) \qquad (\forall\, \eta > 0).$$

Démonstration de la proposition II.2.1 Il ne reste plus maintenant qu'à majorer $R^{\alpha,\beta}_{x,\varepsilon,n}(0,\eta)$ pour tout $0 < \alpha < \beta$, $\eta\in\,]0,\rho_\varepsilon[$, $n \ge 0$ et ε suffisament petit dépendant de α et β. Plus généralement on cherchera à majorer $R^{\alpha,\beta}_{x,\varepsilon,n}(w,\eta)$ pour $w\in \mathcal{D}_{x,\varepsilon,n}$, ce qui permettra de supposer $\beta-\alpha$ petit: si $\alpha < \beta < \gamma$ on utilisera l'inégalité suivante:

$$R^{\alpha,\gamma}_{x,\varepsilon,n}(0,\eta) \le R^{\alpha,\beta}_{x,\varepsilon,n}(0,\eta) \sup_{w\in \mathcal{D}_{x,\varepsilon,n}} R^{\beta,\gamma}_{x,\varepsilon,n}(w,\eta).$$

Premier cas: $-\mu_d(x) = -\lambda_r(x) \le \alpha < \beta$. Fixons $w\in \mathcal{D}_{x,\varepsilon,n}$. On commence par recouvrir $B_{\phi^n(x),\varepsilon}(f^n_x(w), 2C\eta e^{-n\alpha}/\rho_\varepsilon(x))$ par des boules $B_{\phi^n(x),\varepsilon}(w_i, \eta e^{-n\beta}/4C)$, $i = 1,2,\ldots,N$ avec

$$N \le \Gamma(d) \left(\frac{\eta e^{-n\beta}/4C}{2C\eta e^{-n\alpha}/\rho_\varepsilon(x)}\right)^{-d}.$$

Quitte à doubler le rayon de ces boules, on peut trouver $(v_1,\ldots,v_N)\in \mathcal{D}_{x,\varepsilon,n}$ tels que

$$f^n_x(B^\alpha_{x,\varepsilon,n}(w, 2C\eta/\rho_\varepsilon(x)) \subseteq \bigcup_{i=1}^{N} B_{\phi^n(x),\varepsilon}(f^n_x(v_i), \eta e^{-n\beta}/2C).$$

Mais si $v\in B^\alpha_{x,\varepsilon,n}(x, 2C\eta/\rho_\varepsilon(x))$

$$\| f^n_x(v) - f^n_x(v_i) \|_\varepsilon < \eta e^{-n\beta}/2C,$$

comme

$$\| f^{n-k}_x(v) - f^{n-k}_x(v_i) \|_\varepsilon \le [e^{\lambda_r - \varepsilon} - \varepsilon]^{-k} \eta e^{-n\beta}/2C.$$

En supposant maintenant ε suffisament petit: $e^{\lambda_r - \varepsilon} - \varepsilon > \varepsilon^{-\beta}$,on obtient donc

$$v\in B^\beta_{x,\varepsilon,n}(v_i, \eta/2C) \ , \ R^{\alpha,\beta}_{x,\varepsilon,n}(w,\eta) \le \Gamma(d)\, \rho_\varepsilon^{-d}(x) \frac{e^{-n(\alpha-\beta)d}}{(8C)^{-d}},$$

$$\limsup_{n\to+\infty} \frac{1}{n} \log \sup_w R^{\alpha,\beta}_{x,\varepsilon,n}(w,\eta) \le (\alpha-\beta)d = \Lambda(\beta,x) - \Lambda(\alpha,x).$$

Deuxième cas. $-\lambda_s \le \alpha \le \beta \le -\lambda_{s+1}$, $\gamma\in]\alpha,\beta[$ et $w\in \mathcal{D}_{x,\varepsilon,n}$. On recouvre de nouveau $P^s_{\phi^n(x)}[B_{\phi^n(x),\varepsilon}(f^n_x(w), 2C\eta e^{-n\alpha}/\rho_\varepsilon(x))]$ par des boules $B_{\phi^n(x),\varepsilon}(b_j, \eta e^{-n\beta}/4C)$, $b_j\in E^s_{\phi^n(x)}$, $j = 1,\ldots,N$ avec

$$N \le \Gamma(\delta_s) \left(\frac{2C\eta e^{-n\alpha}/\rho_\varepsilon(x)}{\eta e^{-n\beta}/4C}\right)^{\delta_s}$$

où $\delta_s = d_1+\ldots+d_s = \dim(E^s_x)$.

Puis on recouvre $Q^s_x[B_{x,\varepsilon}(w, 2C\eta/\rho_\varepsilon(x))]$ par des boules $B_{x,\varepsilon}(a_i, \eta e^{-n(\beta-\gamma)}/4C)$, $a_i\in F^{s+1}_x$, $i = 1,\ldots,M$ avec

$$M \le \Gamma(d-\delta_s) \left(\frac{2C\eta/\rho_\varepsilon(x)}{\eta e^{-n(\beta-\gamma)}/4C}\right)^{d-\delta_s}.$$

On choisit pour chaque (i,j) un vecteur $v_{i,j} \in \mathcal{D}_{x,\varepsilon,n}$ tel que:

$$\| Q^s_x(v_{i,j} - a_i) \|_\varepsilon < \eta e^{-n(\beta-\gamma)}/4C \quad \text{et} \quad \| P^s_{\phi^n(x)}(f^n_x(v_{i,j}) - b_j) \|_\varepsilon < \eta e^{-n\beta}/4C.$$

Si maintenant ε est tel que

$$e^{\lambda_{s+1} + \varepsilon} + \varepsilon \le e^{-\gamma} \le e^{\lambda_s - \varepsilon} - \varepsilon,$$

en appliquant le principe d'invariance des cônes (corollaire II.3.5), pour tout $v\in B^\alpha_{x,\varepsilon,n}(w, 2C\eta/\rho_\varepsilon(x))$ on peut trouver deux indices i,j tels que

$$\| Q^s_x(v - v_{i,j}) \|_\varepsilon < \eta e^{-n(\beta-\gamma)}/2C \quad \text{et} \quad \| P^s_{\phi^n(x)}(f^n_x(v) - f^n_x(v_{i,j})) \|_\varepsilon < \eta e^{-n(\beta-\gamma)} e^{-n\gamma}/2C$$

et donc

$$\| f^k_x(v) - f^k_x(v_{i,j}) \|_\varepsilon \le \eta e^{-n(\beta-\gamma)} e^{-k\gamma}/2C \le \eta e^{-k\beta}/2C,$$

ce qui montre:

$$R^{\alpha,\beta}_{x,\varepsilon,n}(w,\eta) \le MN = \Gamma(\delta_s)\Gamma(d-\delta_s) \left(\frac{8C}{\rho_\varepsilon(x)}\right)^d e^{n(\beta-\alpha)\delta_s + n(\beta-\gamma)(d-\delta_s)},$$

$$\limsup_{n\to+\infty} \frac{1}{n} \log \sup_w R^{\alpha,\beta}_{x,\varepsilon,n}(w,\eta) \le \Lambda(\beta,x) - \Lambda(\alpha,x) + (\beta-\gamma)(d-\delta_s).$$

Il reste à faire tendre η vers 0, ε vers 0 et γ vers β. *CQFD*

Références

[Br-Ka] M. Brin & A. Katok. On local entropy. *Geometric Dynamics. Lecture Notes in Mathematics,* **1007**(1983), 30-38.

[Di] J. Dieudonné. *Eléments d'analyse. Tome III. Chapitre XVI et XVII.* Cahiers Scientifiques. (1970). Gauthier-Villars Editeurs.

[Ec-Ru] I. Eckmann & D. Ruelle. Ergodic theory and strange attractors. *Reviews of Modern Physics, Vol.* **57**(1985), 617-656.

[Ka-St] A. Katok & J.M.Strelcyn. Invariants manifolds, Entropy and Billiards; Smooth Maps with Singularities. *Lecture Notes in Mathematics, Vol.***1222**(1980). Springer-Verlag.

[Ki] J.F.C. Kingman. Subadditive processes. *Springer Lecture Notes in Math.* **539**(1976).

[Le] F. Ledrappier. Quelques propriétés des exposants caractéristiques. *Springer Lecture Notes in Math.* **1097** (1984).

[Li-Tz] J. Lindenstrauss & L. Tzafriri. *Classical Banach Spaces. Volume I.* Ergebnisse der Mathematik und ihrer Grenzgegiete, **97**(1979). Springer-Verlag.

[LY] F. Ledrappier & L.S. Young. The metric entropy of diffeomorphisms. *Ann. of Math.* **122**(1985), 509-574.

$[Ma]_1$ R. Mañé. A proof of Pesin formula. *Ergodic Theory and Dynamical Systems,* **1**(1981), *p.*95-102.

$[Ma]_2$ R. Mañé. Lyapunov exponents and stable manifolds for compact transformations. *Geometric Dynamic, Lecture Notes in Mathematics, Vol.***1007**(1983), *p.*522-577.

[Os] V.I. Oseledets. A multiplicative ergodic theorem. Lyapunov Characteristic numbers for dynamical systems. *Trudy Moskov. Math. Obsc.* **19**(1968), [=*Trans. Moscow Math. Soc.* **19**(1968), *p.* 197-221].

[Pa] K.R. Parthasarathy. *Probability measures on metric spaces.* Probability and Mathematical Statistics. **3**(1967). Academic Press.

[Pe] Ya. Pesin. Characteristic Lyapunov exponents and smooth ergodic theory. *Russian Mathematical Surveys,* **32:4**(1977), 55-114. From *Uspekhi Mat. Nauk* **32:4**(1977), 55-112.

[Ru] D. Ruelle. Characteristic exponents and invariant manifolds in Hilbert spaces. *Annals of Mathematics, Vol.* **115**(1982), *p.* 243-290.

[Sc] L. Schwartz. *Les tenseurs.* Actualités Scientifiques et Industrielles, **1376**(1975). Hermann.

[Sp] M. Spivak. *A Comprehensive Inroduction to Differential Geometry. Volume one.* (1970). Brandeis University.

SYSTEMS OF CLASSICAL INTERACTING PARTICLES WITH NONVANISHING LYAPUNOV EXPONENTS

MACIEJ P. WOJTKOWSKI

Department of Mathematics
University of Arizona
Tucson AZ 85721

ABSTRACT. We present a unified approach to the only two systems of many interacting particles for which nonvanishing of (some) Lyapunov exponents was established in all of the phase space

§0. Introduction.

In this paper we will discuss two classes of Hamiltonian systems for which nonvanishing of at least some Lyapunov exponents was rigorously established in all of the phase space. The first system is the gas of hard balls interacting by elastic collisions alone. They have equal mass and size and may be two dimensional (hard disks) or higher dimensional but the one dimensional gas of hard rods has no exponential instabilities (the metric entropy of the system is easily seen to be zero). The particles are assumed to live on a torus or in a perpendicular box (the case of the toral vessel is technically easier).

The other class which we will refer to as the system of falling balls is the system of point particles moving on a vertical line which also interact by elastic collisions and are subjected to a potential external field which forces the particles to fall down. To prevent the particles from falling into an abyss we introduce the hard floor and assume that the bottom particle bounces back upon collision with it. The masses of the particles are in general different (the system of equal masses is completely integrable since the elastic collision of equal masses in one dimension amounts to exchanging of the momenta).

In the gas of hard balls the shape of the particles plays a crucial role in making the trajectories diverge whereas the system of falling balls shows that pure change of momenta may also be sufficient.

So far these two models are the only systems with arbitrary finite number of degrees of freedom for which the nonvanishing of Lyapunov exponents was established in all of the phase space. This is by no means surprising since a typical Hamiltonian system is likely to have some of the phase space filled with quasiperiodic motions (invariant tori) for

Supported in part by the Sloan Foundation and the Arizona Center for Mathematical Sciences, sponsored by AFOSR Contract FQ8671-900589.

which Lyapunov exponents are automatically zero. Only very special interactions make the system free from quasiperiodic motions. Let us mention here an excellent survey by Strelcyn on the problem of coexistence of the two types of motion in Hamiltonian systems [St].

The hyperbolicity in a Hamiltonian system is associated with a special kind of monotonicity in the linearized equations. This monotonicity is defined in terms of a special cone which we call a sector. A sector in a linear symplectic space is defined by an ordered pair of transversal Lagrangian subspaces which can be thought of as its sides. Recently Bougerol [B] discovered that this kind of monotonicity plays a role in the theory of Kalman–Bucy filters.

In Section 1 we formulate all the necessary symplectic linear algebra. In Section 2 we first find the expansion coefficients for linear symplectic maps which are monotone in a more and less restrictive sense and then we describe how these different monotonicity properties affect the Lyapunov exponents of matrix cocycles with values in monotone symplectic matrices. In Section 3 we move the whole discussion from a linear setting to a manifold. In particular we define monotonicity of a symplectomorphism and a Hamiltonian system and its consequences for the nonvanishing of Lyapunov exponents. Finally in Section 4 we discuss the two models described above.

The gas of hard balls was extensively studied by Sinai [S1],[S2] , Chernov and Sinai [C-S] and Krámli, Simányi and Szász[K-S-S] (these papers contain extensive references). We cover here only the soft part of their work. The present approach was first applied to the gas of hard balls in [W2]. We relied there on the Sinai's reduction of the system of hard balls to the billiard system in a multidimensional domain. Here we show how it can be treated without this reduction. The system of falling balls was introduced and studied in [W3], [W4] and [C-W].

We stay away here from the question of ergodicity which is a much more difficult problem and was carefully studied in the above cited papers on the gas of hard balls. The main ideas in this direction belong to Sinai. The formulation of his method for general Hamiltonian systems and its applicability to the system of falling balls is studied in a recent paper by Liverani and the author [L-W].

While this paper was written the author enjoyed the hospitality and support of the Forschungsinstitut für Mathematik at ETH, Zürich.

§1. Some linear algebra.

Let W be a linear symplectic space of dimension $2n$ with the symplectic form ω. For example we call $W = \mathbb{R}^n \times \mathbb{R}^n$ the standard linear symplectic space if

$$\omega(w_1, w_2) = < \xi^1, \eta^2 > - < \xi^2, \eta^1 >,$$

$$\text{where } w_i = (\xi^i, \eta^i),\ i = 1,2, \quad \text{and } < \xi, \eta > = \xi_1\eta_1 + \cdots + \xi_n\eta_n.$$

The symplectic group $Sp(n, \mathbb{R})$ is the group of linear maps of W ($2n \times 2n$ matrices if $W = \mathbb{R}^n \times \mathbb{R}^n$) preserving the symplectic form i.e., $L \in Sp(n, \mathbb{R})$ if

$$\omega(Lw_1, Lw_2) = \omega(w_1, w_2)$$

for every $w_1, w_2 \in W$.

By definition a Lagrangian subspace of a linear symplectic space W is an n-dimensional subspace on which the restriction of ω is zero (equivalently it is a maximal subspace on which ω vanishes).

Definition 1. *Given two transversal Lagrangian subspaces V_1 and V_2 we define the sector between V_1 and V_2 by*

$$\mathcal{C} = \mathcal{C}(V_1, V_2) = \{w \in W | \omega(v_1, v_2) \geq 0 \text{ for } w = v_1 + v_2, v_i \in V_i, i = 1, 2\}$$

Equivalently we define first the quadratic form associated with an ordered pair of transversal Lagrangian subspaces

$$\mathcal{Q}(w) = \omega(v_1, v_2)$$

where $w = v_1 + v_2, v_i \in V_i, i = 1, 2$, is the unique decomposition. We have now

$$\mathcal{C} = \{w \in W | \mathcal{Q}(w) \geq 0\}.$$

In the case of the standard symplectic space, $V_1 = \mathbb{R}^n \times \{0\}$ and $V_2 = \{0\} \times \mathbb{R}^n$ we get

$$\mathcal{Q}((\xi, \eta)) = < \xi, \eta >$$

and

$$\mathcal{C} = \{(\xi, \eta) \in \mathbb{R}^n \times \mathbb{R}^n | < \xi, \eta > \geq 0\}.$$

Since any two pairs of transversal Lagrangian subspaces are symplectically equivalent we have the option of considering only this case without any loss of generality. But the coordinate free formulation may have some advantages.

It is natural to ask if a sector determines uniquely its sides. It is not a vacuous question since there are many Lagrangian subspaces in the boundary of a sector. The answer is positive.

Proposition 1. *For two pairs of transversal Lagrangian subspaces V_1, V_2 and V_1', V_2' if*

$$\mathcal{C}(V_1, V_2) = \mathcal{C}(V_1', V_2')$$

then

$$V_1 = V_1' \text{ and } V_2 = V_2'.$$

Moreover V_1 and V_2 are the only isolated Lagrangian subspaces contained in the boundary of the sector $\mathcal{C}(V_1, V_2)$.

Proof. We will obtain both statements simultaneously by describing explicitly all Lagrangian subspaces V in the boundary of the sector $\mathcal{C}(V_1, V_2)$.

Let $P_1 : V \to V_1$ and $P_2 : V \to V_2$ be the natural projections, i.e., for every $w \in V$

$$P_1 w + P_2 w = w.$$

Since V lies in the boundary of the sector we have for all $w \in V$

$$\mathcal{Q}(w) = \omega(P_1 w, P_2 w) \equiv 0.$$

On the other hand since V is a lagrangian subspace we have for any $w_1, w_2 \in V$

$$0 \equiv \omega(w_1, w_2) = \omega(P_1 w_1, P_2 w_2) + \omega(P_2 w_1, P_1 w_2),$$

so that the bilinear form is symmetric

$$\omega(P_1 w_1, P_2 w_2) = \omega(P_1 w_2, P_2 w_1).$$

Hence it must vanish on V which shows that the images of P_1 and P_2 (in V_1 and V_2 respectively) are skeworthogonal subspaces. Taking into account that V_1 and V_2 are transversal Lagrangian subspaces we conclude that the sum of the dimensions of the images cannot exceed n.

More precisely let $Ker P_i \subset V$ be the kernel of the projection P_i, $Im P_i \subset V_i$ be its image and the dimension of $Ker P_i$ be equal to k_i, $i = 1, 2$. The dimension of $Im P_i$ is equal to $n - k_i$ and we have $(n - k_1) + (n - k_2) \leq n$, i.e.,

$$k_1 + k_2 \geq n.$$

At the same time

$$Ker P_i \subset Im P_{3-i}, \quad i = 1, 2,$$

so that

$$k_i \leq n - k_{3-i}.$$

Thus we actually have the equality $k_1 + k_2 = n$ and we get

$$Ker P_i = Im P_{3-i}, \quad i = 1, 2,$$

which implies that

$$V = Im P_1 + Im P_2.$$

It is also clear that for any choice of a subspace $Y_1 \subset V_1$ if we take $Y_2 \subset V_2$ to be the intersection of V_2 with the skeworthogonal complement of Y_1 then $Y_1 + Y_2$ is a Lagrangian subspace in the boundary of the sector.

Having described all such subspaces we see that we can continuously vary them except for V_1 and V_2 which are hence the only isolated Lagrangian subspaces in the boundary of the sector. □

Based on the notion of the sector between two transversal Lagrangian subspaces (or the quadratic form $\mathcal{Q}$) we define two monotonicity properties of a linear symplectic map.

Definition 2. *Given the sector C between two transversal Lagrangian subspaces we call a linear symplectic map L monotone if*

$$LC \subset C$$

and strictly monotone if

$$LC \setminus \{0\} \subset int C.$$

There are several other ways to describe monotonicity and strict monotonicity of a linear symplectic map.

Theorem 1. *L is (strictly) monotone if and only if* $\mathcal{Q}(Lw) \geq \mathcal{Q}(w)$ *for every* $w \in W$ ($\mathcal{Q}(Lw) > \mathcal{Q}(w)$ *for every* $w \in W$, $w \neq 0$).

The fact that staying in a cone implies the increase of the quadratic form is a manifestation of a very special geometric structure of a sector and does not hold for cones defined by general quadratic forms. This theorem was first proved in [W2] in coordinate language. We repeat the proof here in a coordinate free fashion.

First for a pair of transversal Lagrangian subspaces V_1 and V_2 and a linear map $L : W \to W$ we can define the following 'block' operators:

$$A : V_1 \to V_1, B : V_2 \to V_1$$
$$C : V_1 \to V_2, D : V_2 \to V_2.$$

They are uniquely defined by the requirement that for any $v_1 \in V_1, v_2 \in V_2$

$$L(v_1 + v_2) = Av_1 + Bv_2 + Cv_1 + Dv_2.$$

We will need the following two Lemmas

Lemma 1. *If L is monotone with respect to the sector between* V_1 *and* V_2 *then* LV_1 *is transversal to* V_2 *and* LV_2 *is transversal to* V_1.

Proof. Suppose that to the contrary there is $0 \neq \bar{v}_1 \in V_1$ such that $L\bar{v}_1 \in V_2$. We choose $\bar{v}_2 \in V_2$ so that

$$\mathcal{Q}(\bar{v}_1 + \bar{v}_2) = \omega(\bar{v}_1, \bar{v}_2) > 0.$$

We have also

$$\omega(\bar{v}_1, \bar{v}_2) = \omega(L\bar{v}_1, B\bar{v}_2 + D\bar{v}_2) = \omega(L\bar{v}_1, B\bar{v}_2).$$

Let $v_\epsilon = \bar{v}_1 + \epsilon\bar{v}_2$. We have that for $\epsilon > 0$ $v_\epsilon$belongs to int$C$. Hence also $\mathcal{Q}(Lv_\epsilon) \geq 0$ for $\epsilon > 0$. On the other hand

$$\mathcal{Q}(Lv_\epsilon) = \epsilon^2\omega(B\bar{v}_2, D\bar{v}_2) - \epsilon\omega(L\bar{v}_1, B\bar{v}_2)$$

which is negative for sufficiently small positive ϵ.

This contradiction proves the Lemma. □

It follows from Lemma 1 that the operators $A : V_1 \to V_1$ and $D : V_2 \to V_2$ are invertible.

Lemma 2. *If L is (strictly) monotone with respect to the sector between* V_1 *and* V_2 *then* L^{-1} *is (strictly) monotone with respect to the sector between* V_2 *and* V_1.

Proof. We have

$$W = C(V_1, V_2) \cup \text{int}C(V_2, V_1).$$

Hence if

$$LC(V_1, V_2) \subset C(V_1, V_2)$$

then

$$C(V_1, V_2) \subset L^{-1}C(V_1, V_2)$$

and finally

$$L^{-1}\text{int}C(V_2,V_1) \subset \text{int}C(V_2,V_1).$$

The last property is easily seen to be equivalent to the monotonicity of L^{-1}. □

Proof of Theorem 1. Using the definitions and the symplecticity of L we have for $w = v_1 + v_2, v_i \in V_i, i = 1,2$,

$$\begin{aligned} \mathcal{Q}(w) &= \omega(v_1,v_2) = \omega(Lv_1,Lv_2) \\ &= \omega(Av_1 + Cv_1, Bv_2 + Dv_2) = \omega(Av_1,Dv_2) + \omega(Cv_1,Bv_2), \end{aligned} \tag{1}$$

where we have used the vanishing of ω on V_1 and V_2.

We have also

$$\begin{aligned} \mathcal{Q}(Lw) &= \omega(Av_1 + Bv_2, Cv_1 + Dv_2) \\ &= \omega\left(Av_1 + Bv_2, CA^{-1}(Av_1 + Bv_2)\right) + \omega(Av_1,Dv_2) + \omega(Bv_2,Dv_2) \\ &\quad - \omega\left(Av_1, CA^{-1}Bv_2\right) - \omega\left(Bv_2, CA^{-1}Bv_2\right). \end{aligned} \tag{2}$$

Note that $\omega\left(v_1, A^{-1}Bv_2\right) = 0$ for all $v_1 \in V_1$ and $v_2 \in V_2$. Hence by symplecticity of L we get also for all $v_1 \in V_1$ and $v_2 \in V_2$

$$\omega\left(Lv_1, LA^{-1}Bv_2\right) = \omega\left(Av_1 + Cv_1, Bv_2 + CA^{-1}Bv_2\right) = 0$$

or equivalently

$$\omega\left(Av_1, CA^{-1}Bv_2\right) = \omega(Bv_2, Cv_1). \tag{3}$$

Further

$$\begin{aligned} \omega\left(A^{-1}Bv_2, v_2\right) &= \omega\left(LA^{-1}Bv_2, Lv_2\right) = \omega\left(Bv_2 + CA^{-1}Bv_2, Bv_2 + Dv_2\right) \\ &= \omega(Bv_2,Dv_2) + \omega\left(CA^{-1}Bv_2, Bv_2\right). \end{aligned} \tag{4}$$

Putting together (1) , (2) , (3) and (4) we obtain

$$\mathcal{Q}(Lw) - \mathcal{Q}(w) = \omega\left(Av_1 + Bv_2, CA^{-1}(Av_1 + Bv_2)\right) + \omega\left(A^{-1}Bv_2, v_2\right).$$

It is thus enough to show that the following quadratic forms are positive semidefinite,

$$\mathcal{P}(v_1) = \omega(Av_1, Cv_1), \; v_1 \in V_1,$$

$$\mathcal{R}(v_2) = \omega\left(A^{-1}Bv_2, v_2\right), \; v_2 \in V_2.$$

$\mathcal{P}$ is positive semidefinite because $LV_1 \subset \mathcal{C}$ and $\mathcal{P}(v_1) = \mathcal{Q}(Lv_1)$.

To show that $\mathcal{R}$ is positive semidefinite note that

$$L\left(A^{-1}Bv_2 - v_2\right) = CA^{-1}Bv_2 - Dv_2 \in V_2.$$

Hence by Lemma 2

$$-\omega\left(A^{-1}Bv_2, v_2\right) \leq 0.$$

In the strictly monotone case we see immediately that $\mathcal{P}$ must be positive definite. So if $\mathcal{Q}(Lw) - \mathcal{Q}(w) = 0$ then $v_1 + A^{-1}Bv_2 = 0$ and also $Av_1 + Bv_2 = 0$.

Now we get $Lw = Cv_1 + Dv_2 \in V_2$ and by Lemma 2 $w \notin \mathcal{C}$ or $w = 0$. The first case would give $\mathcal{Q}(w) < 0 = \mathcal{Q}(Lw)$ so that w must be zero.

(A posteriori we can conclude that in the strictly monotone case $\mathcal{R}$ must be positive definite.) □

As a byproduct of the proof we get the following useful observation

Proposition 2. *A monotone map L is strictly monotone if and only if*

$$LV_i \setminus \{0\} \subset intC, \ i = 1, 2.$$

Comparing the above proof with the one in [W2] we must conclude that the coordinate free calculations are not necessarily simpler or more illuminating than those in the standard linear symplectic space with the standard pair of transversal Lagrangian subspaces. We abandon now for some time the coordinate free environment.

Let

$$L = \begin{pmatrix} A & B \\ C & D \end{pmatrix}$$

be a symplectic map of the standard symplectic space $\mathbb{R}^n \times \mathbb{R}^n$. A, B, C, D are now just $n \times n$ matrices.

First of all let us describe those symplectic matrices which are monotone in the weakest form, namely they preserve the quadratic form $\mathcal{Q}$. We will call such matrices $\mathcal{Q}$-isometries.

Proposition 3. *If L is a linear symplectic map and*

$$LC = C$$

then

$$L = \begin{pmatrix} A & 0 \\ 0 & A^{*-1} \end{pmatrix}.$$

In particular it preserves the quadratic form $\mathcal{Q}$

$$\mathcal{Q} \circ L = \mathcal{Q}.$$

Proof. If $LC = C$ then L maps also the boundary of the sector C onto itself. It follows from Proposition 1 that both sides of the sector stay put under L. Hence $B = C = 0$. By symplecticity $D = A^{*-1}$. □

By Lemma 1 given a monotone L we can always factor out the following $\mathcal{Q}$-isometries on the left and on the right.

$$L = \begin{pmatrix} A & B \\ C & D \end{pmatrix} = \begin{pmatrix} A & 0 \\ 0 & A^{*-1} \end{pmatrix} \begin{pmatrix} I & \cdot \\ \cdot & \cdot \end{pmatrix} = \begin{pmatrix} \cdot & \cdot \\ \cdot & I \end{pmatrix} \begin{pmatrix} D^{*-1} & 0 \\ 0 & D \end{pmatrix}.$$

Symplecticity of L forces further unique factorizations

$$L = \begin{pmatrix} A & 0 \\ 0 & A^{*-1} \end{pmatrix} \begin{pmatrix} I & 0 \\ P & I \end{pmatrix} \begin{pmatrix} I & R \\ 0 & I \end{pmatrix} = \begin{pmatrix} I & K \\ 0 & I \end{pmatrix} \begin{pmatrix} I & 0 \\ S & I \end{pmatrix} \begin{pmatrix} D^{*-1} & 0 \\ 0 & D \end{pmatrix}$$

with symmetric P, R and K, S. Moreover monotonicity forces $P \geq 0$, $R \geq 0$ and $K \geq 0$, $S \geq 0$ (it was essentially shown in the proof of Theorem 1). Strict monotonicity means that $P > 0$, $R > 0$ and $K > 0$, $S > 0$.

It is useful to introduce an intermediate notion between monotonicity and strict monotonicity.

Definition 3. *A monotone map L is called exactly monotone if $C^*B \neq 0$. (Note that $C^*B = PR = D^{-1}SKD$.)*

We will use the following simple criterion of exact monotonicity.

Proposition 4. *If for a monotone L*

$$LV_2 \setminus \{0\} \subset intC$$

and

$$LV_1 \setminus \{0\} \cap intC \neq \emptyset$$

then L is exactly monotone.

Proof. The fact that

$$LV_2 \setminus \{0\} \subset \mathrm{int}C$$

is equivalent to

$$\omega(Bv_2, Dv_2) > 0, \quad \text{for every} \quad v_2 \in V_2, v_2 \neq 0.$$

This implies that B is invertible so that the exact monotonicity reduces to $C \neq 0$. This is guaranteed by the second condition. □

The meaning and role of exact monotonicity will become clear in the next section. For now let us note that for a monotone L the matrix C^*B is a product of two symmetric positive semidefinite matrices and hence it has real nonnegative eigenvalues. Further if all the eigenvalues of C^*B are zero then the matrix has to be zero.

§2. Expansion properties of linear monotone maps and Lyapunov exponents of matrix cocycles with values in monotone maps.

We are now going to compute the expansion coefficients for strictly and exactly monotone symplectic matrices. It is not surprising that we will look at the expansion of Q.

For a monotone matrix

$$L = \begin{pmatrix} A & B \\ C & D \end{pmatrix}$$

we introduce for $w \in \mathrm{int}C$ the coefficient of expansion at w

$$\beta(w, L) = \sqrt{\frac{Q(Lw)}{Q(w)}}$$

and its infimum

$$\sigma(L) = \inf_{w \in \mathrm{int}C} \beta(w, L).$$

$\sigma(L)$ does not change if L is multiplied on the left or on the right by Q-isometries.

As mentioned before C^*B has only real nonnegative eigenvalues. Let us denote them by $0 \leq u_1 \leq \cdots \leq u_n$. It is proven in [L-W] that

Proposition 5.

$$\sigma(L) = \min_{1\le i\le n} \left(\sqrt{1+u_i} + \sqrt{u_i}\right) = \min_{1\le i\le n} \exp \sinh^{-1} \sqrt{u_i}$$

and if L is strictly monotone the infimum is attained at some $w \in intC$, i.e.,

$$\sigma(L) = \beta(w, L).$$

For monotone but not strictly monotone linear maps the minimal coefficient of expansion of $\mathcal{Q}$ on vectors is equal to 1 but the expansion of $\mathcal{Q}$ on Lagrangian subspaces may still be bigger than one. It is indeed so for exactly monotone maps. More precisely we consider the restriction of $\mathcal{Q}$ to a Lagrangian subspace and its image under L. Then L increases the n-dimensional volume defined by these restrictions. It turns out that the minimal coefficient of expansion of this volume is bigger than one if and only if L is exactly monotone.

Let $U : \mathbb{R}^n \to \mathbb{R}^n$ be a linear map and

$$gU = \{(\xi, \eta) \in \mathbb{R}^n \times \mathbb{R}^n | \eta = U\xi\}$$

be its graph. gU is a Lagrangian subspace if and only if U is symmetric and further for a symmetric U $gU \subset C$ if and only if $U \ge 0$.

For a monotone L and $gU \subset C$ LgU is again a Lagrangian subspace and the graph of a linear map which we denote by LU i.e., $gLU = LgU$.

L acts on Lagrangian subspaces by the following Möbius transformation

$$LU = (C + DU)(A + BU)^{-1}.$$

If $U > 0$ then also $LU > 0$. For such U we equip gU and gLU with the scalar products obtained by restricting the form $\mathcal{Q}$ to these subspaces. Now we define $\alpha(U, L)$ to be the coefficient of volume expansion by L acting from gU to gLU as measured with respect to these scalar products.

Let us introduce the minimal coefficient of volume expansion

$$\varrho(L) = \inf_{U>0} \alpha(U, L).$$

Proposition 6. *For a monotone L*

$$\varrho(L) = \prod_{i=1}^{n} \left(\sqrt{1+u_i} + \sqrt{u_i}\right) = \exp \sum_{i=1}^{n} \sinh^{-1} \sqrt{u_i}$$

and if L is strictly monotone the infimum is attained at a unique $U > 0$, i.e.,

$$\varrho(L) = \alpha(U, L).$$

Proof. Let us denote the right hand side of the equality by $s(L)$. It was proven in [W2] that for a strictly monotone L

$$\alpha(U, L) \ge s(L)$$

and the equality is attained at a unique $U > 0$.

Both $\alpha(U, L)$ and $s(L)$ are continuous functions of L which gives us the inequality also for all monotone maps. It remains to show that for monotone but not strictly monotone L the infimum of $\alpha(U, L)$ does not exceed $s(L)$. To that end for any $\epsilon > 0$ we choose a strictly monotone matrix L_ϵ so close to the identity that $s(L_\epsilon L) < s(L)+\epsilon$. Let $U_\epsilon > 0$ be such that

$$\alpha(U_\epsilon, L_\epsilon L) = s(L_\epsilon L).$$

But $\alpha(U, L_\epsilon L) > \alpha(U, L)$ for any $U > 0$. Hence

$$s(L) \leq \alpha(U_\epsilon, L) < \alpha(U_\epsilon, L_\epsilon L) = s(L_\epsilon L) < s(L) + \epsilon$$

which ends the proof. □

We see that the infimum of $\alpha(U, L)$ is strictly bigger than 1 if and only if L is exactly monotone.

Now that we have established the expansion properties for strictly and exactly monotone maps it will follow easily that matrix cocycles with values in such linear maps have nonzero Lyapunov exponents.

Let $T : X \to X$ be a measurable map preserving a probability measure ν on X and let $L : X \to Sp(n, \mathbb{R})$ be a measurable map such that

$$\int_X \ln^+ \|L(x)\| d\nu(x) < +\infty \quad \text{where} \quad \ln^+ a = \max(0, \ln a).$$

By the Multiplicative Ergodic Theorem of Oseledets we have that for ν-almost all $x \in X$ and all $w \neq 0$

$$\lim_{k \to +\infty} \frac{1}{k} \ln \|L^k(x) w\|$$

exists and assumes at most $2n$ values which we call Lyapunov exponents of the matrix cocycle. By $L^k(x)$ we denote the following product

$$L(T^{k-1}x) \ldots L(Tx) L(x).$$

In the symplectic case the Lyapunov exponents come in pairs of opposite numbers: $-\lambda_1 \leq \cdots \leq -\lambda_n \leq 0 \leq \lambda_n \leq \cdots \leq \lambda_1$ and if 0 is an exponent it has even multiplicity. The exponents are in general functions of x but mere ergodicity of T is sufficient to make them constant almost everywhere.

Theorem 2. *If $L : X \to Sp(n, \mathbb{R})$ has values in monotone matrices then*

$$\int_X \lambda_1(x) d\nu(x) \geq \int_X \ln \sigma(L(x)) d\nu(x)$$

and

$$\int_X (\lambda_1(x) + \cdots + \lambda_n(x)) d\nu(x) \geq \int_X \ln \varrho(L(x)) d\nu(x).$$

The idea of the proof. These inequalities follow easily from the fact that the form Q and the volume element it defines on a Lagrangian subspace in the sector are not bigger than the standard scalar product and the respective volume element.

In the first case using Proposition 5 we have for any $w \in \mathrm{int}C$

$$\|L^k(x)w\| \geq \sqrt{Q(L^k(x)w)} = \beta\left(L^{k-1}(x)w, L\left(T^{k-1}x\right)\right)\ldots\beta(w, L(x))\sqrt{Q(w)} \geq$$
$$\geq \sigma\left(L\left(T^{k-1}x\right)\right)\ldots\sigma(L(x))\sqrt{Q(w)}$$

and we obtain the inequality by taking logarithms of both sides and invoking the Birkhoff Ergodic Theorem.

In the second case we look at the growth of volumes on Lagrangian subspaces in the sector. Using Proposition 6 we get

$$|\det\left(L^k(x)|_{gU}\right)| \geq \alpha\left(U, L^k(x)\right) \text{const} =$$
$$= \alpha\left(L^{k-1}(x)U, L\left(T^{k-1}x\right)\right)\ldots\alpha(U, L(x)) \text{const} \geq \varrho\left(L\left(T^{k-1}x\right)\right)\ldots\varrho(L(x)) \text{const}$$

where the constant is the ratio of the two volume elements in gU and naturally depends on U but does not depend on k. Proceeding as in the first case we obtain the desired inequality. The details can be found in [W1], [W2] . □

§3. Monotonicity of a symplectomorphism and a Hamiltonian system.

We now want to translate the notion of monotonicity from a linear symplectic space and a linear symplectic map to a symplectic manifold and a symplectomorphism or a Hamiltonian system. This is done in a very natural way.

First of all given a symplectic manifold (M^{2n}, ω) we choose two transversal bundles of Lagrangian subspaces $V_1(x), V_2(x) \subset T_xM, x \in M$. Having in view the applications we allow the bundles to be discontinuous and even defined only almost everywhere. There are cases where piecewise continuity of the bundles may be important ([B-G],[K], [L-W]) but in this paper it is enough to assume measurability alone (hence there are no global obstructions to the existence of such bundles).

Let $\Phi : M \to M$ be a symplectomorphism i.e., a diffeomorphism preserving the symplectic form.

Definition 4. *We call $\Phi : M \to M$ monotone with respect to the bundle of sectors $C(V_1(x), V_2(x)), x \in M$, if for almost all $x \in M$*

$$D_x\Phi C(x) \subset C(\Phi x).$$

Given a monotone symplectomorphism Φ we call $x \in M$ (or the orbit of x) eventually strictly (exactly) monotone if there is $n(x) \geq 1$ such that $D_x\Phi^{n(x)}$ is strictly (exactly) monotone with respect to the sectors

$$C(V_1(x), V_2(x)) \text{ and } C\left(V_1\left(\Phi^{n(x)}x\right), V_2\left(\Phi^{n(x)}x\right)\right).$$

In the case of continuous time i.e., a Hamiltonian system we have to modify the above definition because the Hamiltonian vector field is automatically preserved by the flow. So we require that the Hamiltonian vector field which we denote by ∇H, where $H : M \to \mathbb{R}$ is the Hamiltonian, does not vanish and belongs to one, say the first one, of the Lagrangian subspaces:

$$\nabla H(x) \in V_1(x), x \in M.$$

This condition allows to define a sector in the factor space

$$T_x\{H = const\}/_{\nabla H(x)}$$

i.e., the factor of the skeworthogonal complement of $\nabla H(x)$ by the one dimensional subspace generated by $\nabla H(x)$. Indeed if only $\nabla H(x) \in V_1(x)$ then the restrictions of $V_1(x)$ and $V_2(x)$ to $T_x\{H = const\}$ factor onto two transversal Lagrangian subspaces in the factor symplectic space.

Now we define a Hamiltonian system to be monotone if for almost every $x \in M$ and for every $t \geq 0$ the projection of $D\varphi^t$ to the described above factor space is monotone (with respect to the appropriate sectors). By $\varphi^t : M \to M, t \in \mathbb{R}$ we denoted the flow defined by the Hamiltonian vector field.

Further we define $x \in M$ (or the trajectory of x) to be eventually strictly (exactly) monotone if there is $t(x) \geq 0$ such that the projection of $D\varphi^{t(x)}$ to the factor space is strictly (exactly) monotone.

One can naturally define the Lyapunov exponents both for a symplectomorphism and a Hamiltonian system. In the definition of Section 2 we take for the matrix valued function the derivative of the symplectomorphism or the time one map of the Hamiltonian flow. For a monotone symplectomorphism Theorem 2 implies that on the set of eventually strictly (exactly) monotone points all (some of) the Lyapunov exponents are different from zero. Similarly for a monotone Hamiltonian system all but two (some of) the Lyapunov exponents are different from zero on the set of eventually strictly (exactly) monotone points. (The two zero exponents are always there - there is no expansion in the direction of the Hamiltonian vector field and another exponent is forced to be zero by the symplecticity.)

What one may find surprising in this criterion is that we do not assume any uniform properties of the sectors, they can shrink to one Lagrangian subspace at some places and be very wide elsewhere. The explanation lies in the ergodic theory: a typical orbit spends most of its time in the large set where one has uniform estimates of the sectors.

In a smooth (or at least piecewise smooth) system if we can find one strictly monotone orbit then by continuity all nearby orbits are strictly monotone so that we get all Lyapunov exponents different from zero on a subset of positive measure.

It is worth noting that in the case of a Hamiltonian system both monotonicity and eventual strict or exact monotonicity can be expressed (and checked) at the level of the whole phase space without taking explicitly the quotients. Here is how we do it. We write down the quadratic form Q defined by the sectors and then differentiate it carefully. To guarantee monotonicity the derivative should be nonnegative at least for all vectors tangent to a given level of the Hamiltonian $(T_x\{H = const\})$. Having checked that we look at vectors in $V_1(x)$ and $V_2(x) \cap T_x\{H = const\}$ and convince ourselves that the only vectors that will never in the future get strictly inside the sector are parallel to $\nabla H(x)$. This by Proposition 2 guarantees strict monotonicity. Similarly by Proposition 4 our orbit is eventually exactly monotone if all of V_2 enters the interior of the sector and some of the vectors from V_1 do.

§4. Applications.

The gas of hard balls.

The Hamiltonian is

$$H = \sum_{i=1}^{N} \frac{p_i^2}{2}$$

where $p_i \in \mathbb{R}^d$ is the momentum of i-th particle, $i = 1, \dots, N, d \geq 2$ and we take the mass of a particle to be equal to one. The Hamiltonian differential equations are then linear

$$\dot{q} = p$$
$$\dot{p} = 0,$$

where $q = (q_1, \dots, q_N) \in \mathbb{R}^d \times \dots \times \mathbb{R}^d = \mathbb{R}^{dN}$ are positions of the particles and $p = (p_1, \dots p_N) \in \mathbb{R}^{dN}$ the momenta.

Our phase space is then a linear symplectic space. All its tangent spaces are naturally identified with it but for the sake of clarity we will distinguish between a point in the phase space and its tangent space by placing δ in front of q and p. We choose the constant Lagrangian subspaces in the $(\delta q, \delta p)$-space $\mathbb{R}^{dN} \times \mathbb{R}^{dN}$

$$V_1 \equiv \mathbb{R}^{dN} \times \{0\} = \{dp_1 = \dots = dp_N = 0\}$$

and

$$V_2 \equiv \{0\} \times \mathbb{R}^{dN} = \{dq_1 = \dots = dq_N = 0\}.$$

The quadratic form $\mathcal{Q}$ is then equal to

$$\mathcal{Q}((\delta q, \delta p)) = < \delta q, \delta p > = \sum_{i=1}^{N} < \delta q_i, \delta p_i > .$$

We see that $\nabla H \in V_1$ and

$$\frac{d}{dt}\mathcal{Q}((\delta q, \delta p)) = < \dot{\delta q}, \delta p > + < \delta q, \dot{\delta p} > = < \delta p, \delta p > \geq 0$$

since the linearized equations are

$$\dot{\delta q} = \delta p$$
$$\dot{\delta p} = 0.$$

Hence we get monotonicity but so far we did not describe the system completely - there are also collisions between particles. Mathematically speaking the collisions are described by a symplectomorphism defined on the boundary of the phase space. More precisely not all positions of the balls are allowed since they cannot overlap.

The configurations with at least two balls touching each other form then the boundary of the phase space. The intersection of this boundary with a given total energy level, say $\{H = \frac{1}{2}\}$, has a canonical symplectic structure (its tangent can be identified with the factor space by the line spanned by the velocity vector). Collisions are then described by a symplectomorphism of this symplectic manifold (a gross simplification – this boundary has many discontinuities and so we only get piecewise smoothness but it is still possible to talk about Lyapunov exponents). So we have here a combination of a Hamiltonian system and a symplectomorphism. Such systems were discussed abstractly in [W3] and they are called there flows with collisions, the boundary being called the collision manifold.

It is not hard to see that the traces of the Lagrangian subspaces in the tangent space of the collision manifold are also transversal Lagrangian subspaces with respect to the canonical symplectic structure there. Moreover the respective form $\mathcal{Q}$ is just the restriction of the previous one. To be able to apply Theorem 2 we have also to check monotonicity of the collision map with respect to these restricted sectors.

We begin by describing analytically an elastic collision of two balls. Let $q_1^{\pm}, q_2^{\pm}$ be the positions of the centers of the colliding balls and $p_1^{\pm}, p_2^{\pm}$ their momenta (velocities) respectively immediately before (superscript $^-$) and immediately after (superscript $^+$) the collision. Let us assume for simplicity that the distance of the two balls is 1,

$$e = q_2 - q_1, \|e\| = 1,$$

i.e., the radius of the balls is $\frac{1}{2}$.

In an elastic collision the balls exchange the components of their velocities in the direction of the line connecting their centers, the direction of e, while the orthogonal components of their velocities do not change. We get then the collision map

$$p_1^+ = p_1^- - < p_1^-, e > e + < p_2^-, e > e$$
$$p_2^+ = p_2^- - < p_2^-, e > e + < p_1^-, e > e$$

and the other coordinates are not changed.

The collision manifold is given locally by

$$\|q_2 - q_1\| = 1$$

and its tangent space by

$$< \delta q_2 - \delta q_1, e >= 0.$$

Differentiating the collision map we get

$$\delta p_1^+ = \delta p_1^- + < \delta p_2^- - \delta p_1^-, e > e + < p_2^- - p_1^-, \delta q_2 - \delta q_1 > e + < p_2^- - p_1^-, e > (\delta q_2 - \delta q_1)$$
$$\delta p_2^+ = \delta p_2^- - < \delta p_2^- - \delta p_1^-, e > e - < p_2^- - p_1^-, \delta q_2 - \delta q_1 > e - < p_2^- - p_1^-, e > (\delta q_2 - \delta q_1)$$

Restricting the derivative to the tangent space we get

$$\mathcal{Q}\left((\delta q^+, \delta p^+)\right) - \mathcal{Q}\left((\delta q^-, \delta p^-)\right) =< p_1^- - p_2^-, e > \left(\delta q_2^- - \delta q_1^-\right)^2.$$

Thus monotonicity is verified also for the collisions. (The inequality $< p_1^- - p_2^-, e >> 0$ holds automatically for all nondegenerate collisions. It simply means that the balls are indeed colliding and not flying away from each other.)

It is very important to realize that when approaching collision before we can apply the derivative of the collision map or the formula for the increase of the form $\mathcal{Q}$ to a vector $(\delta q, \delta p)$ we have to project it onto the tangent space of the collision manifold, $< \delta q_2 - \delta q_1, e >= 0$, along the velocity vector $(p, 0)$

$$(\delta q, \delta p) \mapsto (\delta q - \lambda p, \delta p), \quad \lambda = \frac{< \delta q_2 - \delta q_1, e >}{< p_2 - p_1, e >}.$$

This projection reflects the fact that nearby trajectories do not arrive at the collision at the same time. It does not change the value of the form $\mathcal{Q}$ but it does change the tangent vector.

Now we want to look for orbits which are eventually exactly or strictly monotone. First of all we note that the dynamics between collisions puts V_2 immediately strictly inside the sector so the whole issue is what happens to V_1 (which is preserved between collisions). Using Proposition 4 we see that exactness occurs after any nondegenerate ($< p_1^- - p_2^-, e > \neq 0$) collision. Indeed there are always vectors in V_1 on which the form $\mathcal{Q}$ gets increased in the collision which means that the vector enters the interior of the sector. Hence we are guaranteed that there is at least one nonzero Lyapunov exponent almost everywhere in the phase space. One can try to get more quantitative statement using the inequalities in Theorem 2, this was done in [W2].

Eventual strict ergodicity is much more subtle. First of all we have to distinguish now between the system of particles in a box and on a torus. In the case of a box we have collisions with the walls which change the total momentum. In the case of a torus the total momentum is preserved. We will discuss briefly only this case. The extra first integrals of the motion force $2d$ Lyapunov exponents to be zero. To look at the other Lyapunov exponents we have to perform the Hamiltonian reduction (fixing the center of mass). This reduction fits nicely with our formalism because all the Hamiltonian vector fields generated by the first integrals (d components of the total momentum) lie in V_1. As a result in order to establish eventual strict monotonicity of an orbit in the reduced system it is sufficient to look at vectors in V_1 which fail to enter C and establish that they must belong to the $d+1$-dimensional subspace spanned by $\nabla H = (p, 0)$ and the d generators of simultaneous translations of all the balls $\{\delta q_1 = \cdots = \delta q_N, \delta p = 0\}$).

The formula for the increase of the form $\mathcal{Q}$ shows that if a vector in V_1 does not enter C as a result of a nondegenerate collision of the first two particles then

$$\delta q_2^- - \delta q_1^- = 0.$$

If we take into account the necessary projection preceding the application of the derivative of the collision map it means that $\delta q_2 - \delta q_1$ is parallel to $p_2 - p_1$ (then the projection makes the two components equal). It is then clear that there are many special orbits and vectors in V_1 which do not enter C for a long time (ever).

The problem now is to establish how big the collection of these orbits really is. Krámli, Simányi and Szász [K-S-S] were able to show that such orbits form a set of codimension 2 in the case of three balls and has measure zero in the case of four balls.

In the system with arbitrary many balls Chernov and Sinai [C-S] give an example of an eventually strictly monotone orbit thus showing that the set of such orbits has nonempty interior (positive measure). We will give another example to demonstrate how our methods work. We assume that N balls move in a large torus $\mathbb{T}^d = \mathbb{R}^d/(a\mathbb{Z})^d$ where $a > N$. At time zero the balls have their centers on a line l with the direction very close but not equal to $e_1 = (1,0,\dots,0) \in \mathbb{R}^d$. The balls are indexed according to the order of their centers on the line l. The first ball has the momentum p parallel to l and it is moving towards the second ball which is at rest. All the other balls are also at rest. After the first collision the first ball will be at rest and the second will be moving with momentum p towards the third ball. After $N-1$ collisions the N-th ball will be moving with momentum p and all the other balls will be at rest with their centers still on l. If the direction of l is sufficiently close to e_1 then the N-th ball will go around the torus and emerge on the other side colliding with the first ball. But at their collision since the direction of l is different from e_1 the line through their centers is not parallel to p. Hence the first ball will start moving towards the second with the momentum which is not parallel to p. We have the same situation in the following $N-2$ collisions: when the k-th ball collides with the next its momentum is not parallel to p, $k = 1,\dots,N-2$.

We claim that after these $2N-2$ collisions we have strict monotonicity. To prove this we will describe all vectors in V_1 which stay in V_1 after all these collisions. Let

$$(\delta q_1^-,\dots,\delta q_N^-,0,\dots,0) \in V_1$$

be such a vector at time zero and let

$$(\delta q_1^+,\dots,\delta q_N^+,0,\dots,0) \in V_1$$

be its image after N collisions. Since there is no increase of Q in the first collision $\delta q_2^- - \delta q_1^-$ must be parallel to p so that

$$\delta q_2^- = \delta q_1^- + \lambda_1 p.$$

Since the second ball is at rest we get also

$$\delta q_1^+ = \delta q_2^-.$$

Similarly using the first $N-1$ collisions we obtain

$$\delta q_k^+ = \delta q_{k+1}^- = \delta q_k^- + \lambda_k p, \quad k = 1,\dots,N-1.$$

In the $N+1$ collision $\delta q_2^+ - \delta q_1^+ = \lambda_2 p$ must be parallel to the momentum of the first particle which is not parallel to p. Hence $\lambda_2 = 0$. Similarly using the following collisions we get that $\delta q_k^+ - \delta q_{k-1}^+ = \lambda_k p$ must be parallel to the momentum of the moving particle which is not parallel to p. Hence by necessity $\lambda_k = 0$ for $k = 2,\dots,N-1$. It follows that

$$\delta q_1^- = \delta q_2^- - \lambda_1 p$$
$$\delta q_k^- = \delta q_2^-$$

for $k = 2, \dots, N$ which proves our claim.

As a result all Lyapunov exponents (that can be different from zero) are guaranteed to be different from zero at least in a neighborhood of this orbit (with positive Lebesgue measure). The ergodicity of the system forces the Lyapunov exponents to be constant but the scheme for the proof of ergodicity formulated by Chernov and Sinai in [C-S] requires the knowledge that the set of orbits which are not eventually strictly monotone not only has measure zero but that also its complement is connected (it has codimension 2). It was so far established only for three balls by Krámli, Simányi and Szász [K-S-S])

The system of falling balls.

The Hamiltonian of the system is

$$H = \sum_{i=1}^{N} \left(\frac{p_i^2}{2m_i} + m_i U(q_i) \right)$$

where q_i are the positions and $p_i = m_i v_i$ the momenta of the particles, $q_i, p_i \in \mathbb{R}, i = 1, \dots, N$, and $U(q)$ is the potential of the external field .

The differential equations then are

$$\dot{q}_i = \frac{p_i}{m_i}$$
$$\dot{p}_i = -m_i U'(q_i),$$

$i = 1, \dots, N$.

We choose the following Lagrangian subspaces

$$V_1 = \{dp_1 = \dots = dp_N = 0\} \quad \text{and} \quad V_2 = \{dh_1 = \dots = dh_N = 0\}.$$

where $h_i = \frac{p_i^2}{2m_i} + m_i U(q_i), i = 1, \dots, N$, are individual energies of the particles.

We have

$$dh_i = \frac{p_i dp_i}{m_i} + m_i U'(q_i)\, dq_i,$$

$i = 1, \dots, N$, so that V_1 and V_2 are indeed transversal if only $U' \neq 0$, i.e., the external field is actually present.

With this choice of sectors we have $\nabla H \in V_2$.

The form $\mathcal{Q}$ is equal to

$$\mathcal{Q}((\delta q, \delta p)) = \sum_{i-1}^{N} \left(\delta q_i \delta p_i + \frac{p_i}{m_i^2 U'(q_i)} (\delta p_i)^2 \right).$$

The linearized differential equations are

$$\dot{\delta q_i} = \frac{\delta p_i}{m_i}$$
$$\dot{\delta p_i} = -m_i U''(q_i)\, \delta q_i,$$

$i = 1, \ldots, N$, which gives us

$$\frac{d}{dt}\mathcal{Q}((\delta q, \delta p)) = \sum_{i-1}^{N} -\frac{U''(q_i)}{m_i^2 (U'(q_i))^2} \left(\frac{p_i \delta p_i}{m_i} + m_i U'(q_i)\delta q_i\right)^2$$
$$= \sum_{i-1}^{N} -\frac{U''(q_i)}{m_i^2 (U'(q_i))^2} (dh_i((\delta q, \delta p)))^2 .$$

So we have monotonicity provided

$$U''(q) \leq 0.$$

We have yet to take into account collisions between particles and the collision of the first particle with the floor.

Let us consider the collision of the first particle with the second. The collision manifold is given by $\{q_1 = q_2\}$ (point particles), its tangent space is $\{\delta q_1 = \delta q_2\}$. The collision map is linear

$$p_1^+ = \gamma p_1^- + (1+\gamma) p_2^-$$
$$p_2^+ = (1-\gamma) p_1^- - \gamma p_2^-,$$

where $\gamma = \frac{m_1 - m_2}{m_1 + m_2}$, the superscript $^-$ refers to momenta before the collision, the superscript $^+$ to momenta after the collision, and the other coordinates stay unchanged.

In particular in a collision of two particles the top particle increases its momentum while the bottom particle decreases its momentum by the same amount Δp

$$p_1^+ = p_1^- - \Delta p$$
$$p_2^+ = p_2^- + \Delta p,$$

where

$$\Delta p = \frac{2m_1 m_2}{m_1 + m_2}\left(v_1^- - v_2^-\right) > 0.$$

$\left(v_1^- - v_2^-\right) > 0$ means only that the particles indeed collide (and do not fly away from each other).

We can now compute

$$\mathcal{Q}((\delta q^+, \delta p^+)) - \mathcal{Q}((\delta q^-, \delta p^-)) = \frac{\gamma}{U'(q_1)}\Delta p \left(\frac{\delta p_1^-}{m_1 U'(q_1)} - \frac{\delta p_2^-}{m_2 U'(q_2)}\right)^2 .$$

Monotonicity is guaranteed by the assumption that $m_1 \geq m_2 \geq \cdots \geq m_N$ and $U' > 0$ which means that the particles are falling (accelerated down). (We could do it the other way around $m_1 \leq m_2 \leq \cdots \leq m_N$ and $U' < 0$ but then we would have to close our system by a hard ceiling rather than hard floor and we would get back to essentially the same system.)

In the case of the collision of the first particle with the floor we have the following collision map

$$p_1^+ = -p_1^-,$$

and the rest of coordinates stay unchanged. We readily obtain

$$\mathcal{Q}\left(\left(\delta q^{+},\delta p^{+}\right)\right)-\mathcal{Q}\left(\left(\delta q^{-},\delta p^{-}\right)\right)=-\frac{2p_{1}^{-}}{m_{1}^{2}U'(0)}\left(\delta p_{1}\right)^{2}$$

which is nonnegative without further restrictions.

Inspecting the system for eventual strict monotonicity we see that between collisions V_1 enters the sector immediately if only we assume $U'' < 0$ while V_2 is preserved.

In a collision of i-th and $i+1$ particles a vector from V_2 enters the sector unless $(\delta p_i, \delta p_{i+1})$ is parallel to $(m_i U'(q_i), m_{i+1} U'(q_{i+1}))$ or (m_i, m_{i+1}) (remember that at the collision $q_i = q_{i+1}$). This relation survives both the dynamics between collisions and any projection on the collision manifold. Moreover if a vector from V_2 does not enter the sector in a collision then it is not changed by this collisions. The collision with the floor fits nicely into this analysis. Let us note that in our system all the collisions that can happen will happen. As a result we obtain that the only vector in V_2 which will never enter the sector is such that $(\delta p_1, \ldots, \delta p_N)$ is parallel to $(m_1 U'(q_1), \ldots, m_N U'(q_N))$ which taking into account that

$$dh_1 = \cdots = dh_N = 0$$

means that the vector is parallel to ∇H. So we get eventual strict monotonicity of almost all orbits.

We have also to ensure that our system is closed (that the particles will not fly away to infinity). This is done by assuming that the total energy is smaller then the escape energy of the top particle.

Let us summarize the conditions which guarantee eventual strict monotonicity in all of the phase space.

If $U'(q) > 0$ and $U''(q) < 0$ for all $q, 0 \le q \le q_{max}$ where q_{max} is the highest point the top mass can reach under the energy constraint $H = const$, i.e.,

$$H=\sum_{i=1}^{n-1} m_i U(0) + m_n U(q_{max}),$$

and if the masses are decreasing $m_1 > \ldots > m_n$ then the system has all but one Lyapunov exponents different from zero.

The case of constant acceleration or some masses being equal is much more subtle and the difficulties encountered are similar to those in the study of the gas of hard balls. Just as there eventual exact monotonicity in all of the phase space and eventual strict monotonicity on a subset of positive measure are easy to establish but eventual strict monotonicity in all of the phase space seems to be a very elusive goal. The details can be found in [W3] and [W4].

It was shown in [C-W] that the condition on the masses is essential. A linear periodic orbit is constructed there which turns out to be linearly stable if the masses above are bigger than the masses below. It basically excludes eventual strict monotonicity in all of the phase space.

References

[B] P. Bougerol, *Filtre de Kalman Bucy et Exposant de Lyapounov*, this volume.

[B-G] K. Burns, M. Gerber, *Continuous invariant cone families and ergodicity of flows in dimension three*, Erg.Th.Dyn.Syst. **9** (1989), 19-25.

[C-W] J.Cheng, M.P.Wojtkowski, *Linear stability of a periodic orbit in the system of falling balls*, The Geometry of Hamiltonian Systems, Proceedings of a Workshop Held June 5-16,1989 MSRI Publications, Springer Verlag 1991 (ed. Tudor Ratiu).

[C-S] N.I.Chernov, Ya.G.Sinai, *Ergodic properties of some systems of 2-dimensional discs and 3-dimensional spheres*, Russ.Math.Surveys **42** (1987), 181-207.

[K] A.Katok, *Invariant cone families and stochastic properties of smooth dynamical systems*, preprint (1988).

[K-S-S] A.Krámli, N.Simányi D.Szász, *Three billiard balls on the ν-dimensional torus is a K-flow*, preprint (1988).

[L-W] C.Liverani, M.P.Wojtkowski, *Ergodicity of Hamiltonian systems*, (in preparation).

[S] Ya.G.Sinai, *Dynamical systems with elastic reflections*, Russ.Math.Surveys **25** (1970), 137–189.

[St] J.-M. Strelcyn, *The "coexistence problem" for conservative dynamical systems: a review*, preprint (1989) (to appear in Colloquium Mathematicum).

[W1] M.P.Wojtkowski, *Invariant families of cones and Lyapunov exponents*, Erg.Th.Dyn.Syst. **5** (1985), 145–161.

[W2] M.P.Wojtkowski, *Measure theoretic entropy of the system of hard spheres*, Erg.Th.Dyn.Syst. **8** (1988), 133–153.

[W3] M.P.Wojtkowski, *A system of one dimensional balls with gravity*, Comm.Math.Phys. **126** (1990), 507-533.

[W4] M.P.Wojtkowski, *The system of one dimensional balls in an external field. II*, Comm.Math.Phys. **127** (1990), 425-432.

Lyapunov exponents from time series

Joachim Holzfuss and Ulrich Parlitz
Institut für Angewandte Physik
TH Darmstadt, Schloßgartenstr. 7, 6100 Darmstadt
Federal Republic of Germany

Introduction

Nonlinear dynamical systems are intrinsically linked with chaos. Even simple systems, such as a driven nonlinear oscillator, can respond with irregular oscillations though the driving force is a purely periodic sine wave. To characterize such behaviour, linear methods of data analysis such as Fourier spectra and correlations, must have limited abilities.

Aperiodic behaviour of physical systems, which has been encountered so often, is now being analyzed to find characterizations of the underlying process in terms of nonlinear dynamics. Phase space analysis has been developed to analyze data with methods from nonlinear dynamical systems theory that classify behaviour using dynamical invariants. Fractal dimensions and Lyapunov exponents are the most prominent candidates, they express complexity and predictability of a process and are a measure for chaos [7, 17, 18].

Calculation of these invariants from a time series of a single observable is possible with the use of time shifted coordinates and the construction of diffeomorphic equivalent trajectories in an embedding phase space [21, 24]. So far, most of the methods used for determining Lyapunov exponents use a linear approach with least squares estimates of the linear time evolution in tangent space. We present a numerical method that is based on the interpolation of the local flow using radial basis functions. It is able to extract the spectrum of Lyapunov exponents and it is very robust against influences of noise and changes of method inherent parameters. Especially the "spurious exponents" arising from the embedding procedure can be distinguished from the "real exponents" describing the dynamics.

Lyapunov exponents

Lyapunov exponents describe the overall expanding and contracting behaviour of phase space volumes and separation rates of slightly different initial states.

The flow map

$$\begin{aligned} \Phi^t :\ & \mathbb{R}^n \to \mathbb{R}^n \\ & \mathbf{x} \mapsto \Phi^t(\mathbf{x}) \end{aligned} \tag{1}$$

describing the dynamical system acts on the n–dimensional state space $M = \mathbb{R}^n$ and is generated by a vector field v

$$\dot{\mathbf{x}} = v(\mathbf{x}), \quad \mathbf{x} \in \mathbb{R}^n, \quad t \in \mathbb{R}. \tag{2}$$

To gather information about the time evolution of infinitesimally small perturbed initial states, the linearized flow map

$$\begin{aligned} D_{\mathbf{x}}\Phi^t : \; T_{\mathbf{x}}M &\to T_{\Phi^t(\mathbf{x})}M \\ \mathbf{u} &\mapsto D_{\mathbf{x}}\Phi^t\mathbf{u} \end{aligned} \tag{3}$$

has to be considered.

The linearized flow map $D_{\mathbf{x}}\Phi^t$ is given by an invertible $n \times n$ matrix describing the time evolution of a vector $\mathbf{u}$ in tangent space. For ergodic systems the Lyapunov exponents are defined as the logarithms of the eigenvalues μ_i $(1 \le i \le m)$ of the positive and symmetric limit matrix

$$\Lambda_{\mathbf{x}} = \lim_{t\to\infty} \left[D_{\mathbf{x}}\Phi^{t*} \, D_{\mathbf{x}}\Phi^t\right]^{\frac{1}{2t}} \tag{4}$$

as given by the theorem of Oseledec [20] (* denotes transposition). The Lyapunov exponents are the logarithmic growth rates

$$\lambda_i = \lim_{t\to\infty} \frac{1}{t} \ln \left\|D_{\mathbf{x}}\Phi^t\mathbf{e}_i\right\|, \quad (1 \le i \le m), \tag{5}$$

where $\{\mathbf{e}_i : 1 \le i \le m\}$ are basis vectors that span the eigenspaces of $\Lambda_{\mathbf{x}}$. When starting the numerical computation with an arbitrary set of basis vectors, renormalization has to be applied after some evolution time Δt [2, 12], because almost all vectors tend to fall into the most growing direction for $t \to \infty$. Equation (5) can be written as a product of local linearized flow maps $D_{\mathbf{x}^j}\Phi^{\Delta t}$, $0 \le j \le N-1$, $\Delta t = t/(N-1)$ along the orbit points $\mathbf{x}^j = \Phi^{j\Delta t}(\mathbf{x})$

$$\lambda_i = \lim_{t\to\infty} \frac{1}{t} \ln \left\| \prod_{j=0}^{N-1} D_{\mathbf{x}^j}\Phi^{\Delta t} \, \mathbf{e}_i \right\| \tag{6}$$

and, using the QR–decomposition for renormalization [16]

$$\lambda_i = \lim_{k\to\infty} \frac{1}{k\Delta t} \sum_{j=0}^{N-1} \ln r_{ii}^j \quad . \tag{7}$$

r_{ii}^j are local expansion rates. They are the diagonal elements of the upper triangular matrices R^j occurring upon the repeated QR–decomposition along the orbit [8, 13].

In the case of experiments usually a measured time series of a single observable is known instead of the governing equations of motion. To construct trajectories in phase space in this case, the embedding method of time shifted samples is used [21, 24, 25]. Let

$$\begin{aligned} p : \; \mathbb{R}^m &\to \mathbb{R} \\ \Psi^t(\mathbf{z}_o) &\mapsto p\left(\Psi^t(\mathbf{z}_o)\right) \end{aligned} \tag{8}$$

be an observable of the system described by some unknown flow map Ψ^t acting on an m–dimensional state space $\mathbb{R}^m$. It has been shown for compact manifolds of dimension m, that the set

$$\left\{p\left(\Psi^t(\mathbf{z}_o)\right), p\left(\Psi^{t+T}(\mathbf{z}_o)\right), \ldots, p\left(\Psi^{t+2mT}(\mathbf{z}_o)\right) \mid T \in \mathbb{R}^+,\ t \to \infty\right\} \tag{9}$$

is diffeomorphic to the positive limit set of $\Psi^t(\mathbf{z}_o)$ under generic conditions [24, 25]. T is called the time shift constant and $n = 2m + 1$ is the embedding dimension. Because the Lyapunov exponents are a dynamical invariant, we can use the flow

$$\begin{aligned} &\Phi^t : \mathbb{R}^n \to \mathbb{R}^n \\ &\mathbf{x}_o \mapsto \Phi^t(\mathbf{x}_o) = \left(p\left(\Psi^t(\mathbf{z}_o)\right), p\left(\Psi^{t+T}(\mathbf{z}_o)\right), \ldots, p\left(\Psi^{t+2mT}(\mathbf{z}_o)\right)\right) \end{aligned} \tag{10}$$

in the embedding space $\mathbb{R}^n$ for further calculations. If the dimension m of Ψ^t is unknown, n has to be increased until an embedding is achieved.

Approximation of the linearized flow map

For the computation of Lyapunov exponents the linearizations $D_{\mathbf{x}^j}\Phi^{\Delta t}$ of the unknown flow map $\Phi^{\Delta t}$ at successive orbit points $\mathbf{x}^j$ in embedding space have to be determined. Earlier work [8, 13, 22, 23, 26] tried to approximate $D_{\mathbf{x}^j}\Phi^{\Delta t}$ from the time evolution of difference vectors, making a linear ansatz. That is imposing that the difference vectors in state space can be considered as tangent vectors. In some cases this can lead to stable results, but the method has to be used with caution, because nonlocal and nonlinear effects can introduce problems (see [13] for details). Other methods have been published [3,4].

We propose a method based on the local approximation of flow maps by radial basis functions. Given M neighbouring orbit points $\mathbf{x}^k \in U(\mathbf{x}) \subset \mathbb{R}^n, 1 \le k \le M$ around a given point $\mathbf{x} = \mathbf{x}^j$ in state space. The unknown flow map maps the state points to their time evolved images (fig. 1)

$$\begin{aligned} \Phi^{\Delta t} :\ & \mathbb{R}^n \to \mathbb{R}^n \\ & \mathbf{x}^k \mapsto \Phi^{\Delta t}(\mathbf{x}^k) \quad . \end{aligned} \tag{11}$$

Now we are looking for an interpolating function f that satisfies

$$\begin{aligned} &f : \mathbb{R}^n \to \mathbb{R}^n \\ &f(\mathbf{x}^k) = \Phi^{\Delta t}(\mathbf{x}^k) \end{aligned} \tag{12}$$

For each component we do the following ansatz:

$$\begin{aligned} &f_i : \mathbb{R}^n \to \mathbb{R} \\ &f_i(\mathbf{x}) = \sum_{k=1}^{M} c_{ik}\sqrt{r^2 + \|\mathbf{x} - \mathbf{x}^k\|^2}, \quad c_{ik} \in \mathbb{R}, \quad r \in \mathbb{R}^+ \quad . \end{aligned} \tag{13}$$

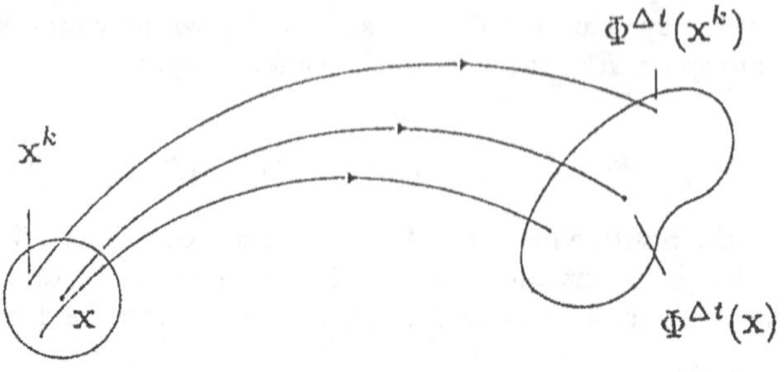

Fig. 1: Mapping of the state points by the local flow.

The c_{ik} are coefficients, r is a stiffness parameter and $\|\cdot\|$ the Euclidean norm. The radial basis functions f are smooth except for $r = 0$ [5, 6, 9]. r is chosen to be in the same range as the norm values. Calculating the r.h.s. of eq. (13) for all given values of $\mathrm{x}^k \in U(\mathrm{x})$ and $f_i(\mathrm{x}^k)$ leaves an M-dimensional system of linear equations for each component. It is solved numerically for the coefficients c_{ik} e.g. with a singular value decomposition. The Jacobian of the interpolating function $D_{\mathrm{x}}f : \mathbb{R}^n \to \mathbb{R}^n$ is an approximation of the linearized flow map $D_{\mathrm{x}}\Phi^{\Delta t}$. It is determined by differentiation:

$$D_{\mathrm{x}}f = \left(\left.\frac{\partial f_i}{\partial x_j}\right|_{\mathrm{x}}\right) \qquad \text{with}$$
$$\frac{\partial f_i}{\partial x_j} = \sum_{k=1}^{M} c_{ik} \frac{x_j - x_j^k}{\sqrt{r^2 + \|\mathrm{x} - \mathrm{x}^k\|^2}} \quad , \tag{14}$$

where x_j^k and x_j denote the j-th component of the state points x^k around x. From the Jacobians at sequential points $\mathrm{x} = \mathrm{x}^j = \Phi^{j\Delta t}(\mathrm{x}_o)$ on the attractor the Lyapunov exponents are computed the usual way.

Numerical results

The method has been tested with various changes of method inherent parameters. Time series of two different dynamical systems have been considered. The $x(t)$ variable of the driven Duffing oscillator

$$\ddot{x} + D\dot{x} + x + x^3 = F\cos\omega t \qquad D = 0.2,\ F = 40,\ \omega = 2\pi/T_o = 1$$

as an example for a continuous system has been taken as time series. Calculating the Lyapunov exponents from the variational equations (see [2]) yields $\lambda_1 = 1.0$, $\lambda_2 = 0.0$, $\lambda_3 = -2.8$ bits/T_o.

The time series has been sampled with a sampling rate of $f_s = 1/t_s = 2\pi$ points per period of the driving force. The time shift constant for the construction of the trajectory in the embedding phase space is $T = 1t_s$, the evolution time $\Delta t = 1t_s$. The value for the time shift is close to an "optimal" value. It has been obtained with a higher dimensional dynamical analogon to the mutual information function [10, 11, 14, 19]. The number of data points is $N = 20000$ and the number of neighbouring points for the interpolation is about $M = 30$.

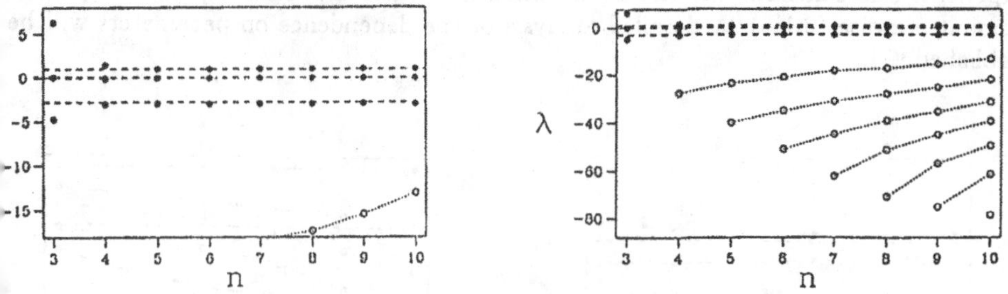

Fig. 2: Exponents from the time series of a Duffing oscillator as a function of the embedding dimension.
a) blowup b) all exponents

Figs. 2 show the calculated exponents as a function of the embedding dimension. The upper three exponents converge very good to the correct values of the Lyapunov exponents. All other exponents are more negative than the lowest real exponent (fig. 2a). Each additional exponent is more negative than the previous most negative one, when increasing the embedding dimension (fig. 2b). The choice of the parameters T and Δt is not very critical. The method is robust in a range of "educated guesses". It affects however the values of the spurious exponents. Lowering Δt e.g. pushes them to very low values.

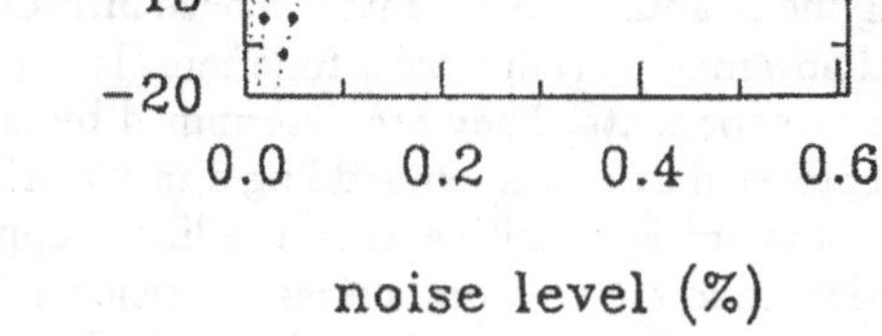

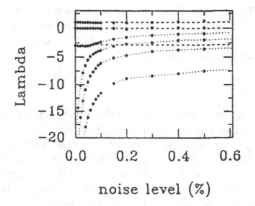

Fig. 3: Dependence of the calculated exponents on Gaussian noise. The embedding dimension is $n = 6$. Same data as in fig. 2.

To test the dependence on the influence of noise, different amounts of Gaussian noise have been added to the data. The result is shown in fig. 3. Shown are the calculated exponents as a function of the standard deviation of the noise relative to the maximal extent. With increasing noise level the spurious exponents approach the lowest real exponent. The calculation of the nonnegative exponents is possible up to a very large amount of noise (0.1%). A detailed analysis of the dependence on parameters will be published.

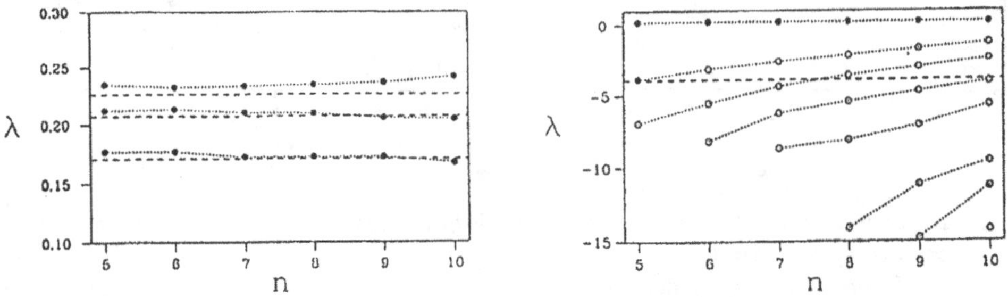

Fig. 4: Exponents for a higher dimensional Hénon type map as a function of the embedding dimension. a) blowup b) all exponents

To test the method with a more complex time series, a data set from a higher dimensional version of the Hénon map is analyzed [1].

$$
\begin{aligned}
(x_1)_{n+1} &= a - (x_{D-1})_n^2 - b(x_D)_n \\
(x_i)_{n+1} &= (x_{i-1})_n \qquad i = 2, ..., D
\end{aligned}
$$

with $D = 3$, $a = 1.76$, $b = 0.1$. The Lyapunov exponents are 0.226, 0.207, 0.171 and −3.923 bits/iteration, i.e. the attractor has three positive exponents. 20000 data points have been used for the calculation. Fig. 4 shows, that the calculation resolves all three positive exponents with a good accuracy.

Conclusion

A method for calculating the Lyapunov exponents has been introduced. It is based on the interpolation of the local flow map by radial basis functions. It allows the computation of positive, zero and negative exponents. They are determined by looking for convergence of the calculated exponents with increased embedding dimension. The choice of method inherent parameter values is not as sensitive as in the linear approach [13]. Especially the value for the evolution time Δt between successive renormalizations can be much larger as in the linear case, where it has to be very small. This may be of significant interest, if the method of interpolation of the local flow is used for nonlinear prediction [15]. The so-called "spurious" exponents that occur due to the embedding, are even smaller than the most negative exponent. Their values change, when the evolution time or the noise level is changed, whereas the correct exponents stay constant. The method is robust against noise up to a quite high level.

The authors wish to acknowledge interesting discussions with the nonlinear dynamics group at IAP, TH Darmstadt and the use of the computing facilities. The work has been supported in part by the Deutsche Forschungsgemeinschaft via the SFB 185 – Nichtlineare Dynamik.

References

1. G. Baier & M. Klein, Discrete Steps up the Dynamic Hierarchy, Phys. Lett. A **151**, 6 (1990), 281 – 284.

2. G. Benettin & L. Galgani, Lyapunov characteristic exponents for smooth dynamical systems and for Hamiltonian systems; a method for computing all of them. Part I: Theory, Part II: Numerical Application, Meccanica **15** (1980), 9 – 30.

3. P. Bryant, R. Brown & H. Abarbanel, Lyapunov Exponents from Observed Time Series, Phys. Rev. Lett. **65**, 13 (1990), 1523 – 1526.

4. K. Briggs, An improved method for estimating Liapunov exponents of chaotic time series, Phys. Lett. A **151**, 1,2 (1990), 27 – 32.

5. D.S. Broomhead & D. Lowe, Multivariable functional interpolation and adaptive networks, Complex systems 2 (1988), 321 – 355.

6. M. Casdagli, Nonlinear Prediction of chaotic time series, Physica D 35 (1989), 335 – 356.

7. J.P. Eckmann & D. Ruelle, Ergodic theory of chaos and strange attractors, Rev. Mod. Phys. **57** (1985), 617 – 656.

8. J.-P. Eckmann, S.O. Kamphorst, D. Ruelle & S. Ciliberto, Lyapunov exponents from a time series, Phys. Rev. A **34**, 6 (1986), 4971 – 4979.

9. R. Franke, Scattered Data Interpolation: Tests of some Methods, Math. of comp., **38**, 157 (1982), 181 – 200.

10. A.M. Frazer & H.L. Swinney, Independent coordinates for strange attractors from mutual information, Phys. Rev. A **33** (1986), 1134.

11. A.M. Frazer, Information and Entropy in Strange Attractors, IEEE Trans. Information Th., vol. 35, 2 (1989), 245 – 262.

12. K. Geist, U. Parlitz & W. Lauterborn, Comparison of Different Methods for Computing Lyapunov Exponents, Prog. Theor. Phys. **83**, 5 (1990), 875 – 893.

13. J. Holzfuss & W. Lauterborn, Liapunov exponents from a time series of acoustic chaos, Phys. Rev. A **39**, 4 (1989), 2146 – 2152.

14. J. Holzfuss & W. Lauterborn, Nonlinear Dynamics of Acoustic Cavitation Noise, in: Frontiers of Nonlinear Acoustics, Proceedings of the 12th ISNA, Eds.: M. F. Hamilton & D. T. Blackstock, Elsevier, London (1990), 464 – 469.

15. J. Holzfuss & U. Parlitz, to be published.

16. A.S. Householder, Unitary triangularization of a nonsymmetric matrix, J. Assoc. Comput. Mach. **5** (1958), 339 – 342.

17. W. Lauterborn & J. Holzfuss, Evidence for a low-dimensional strange attractor in acoustic turbulence, Phys. Lett. A **115**, 8 (1986), 369 –372.

18. W. Lauterborn & J. Holzfuss, Acoustic Chaos, Int. Journal of Bifurcation and Chaos, **1**, 1 (1991), 13 – 26.

19. W. Liebert & H.G. Schuster, Proper choice of the time delay for the analysis of chaotic time series, Phys. Lett. A **142**, (1989), 107 – 111.

20. V.I. Oseledec, A multiplicative ergodic theorem. Lyapunov characteristic numbers for dynamical systems, Trans. Moscow Math. Soc. **19** (1968), 197 – 231.

21. N.H. Packard, J.P. Crutchfield, J.D. Farmer & R.S. Shaw, Geometry from a time series, Phys. Rev. Lett. **45** (1980), 712 – 716.

22. M. Sano & Y. Sawada, Measurement of the Lyapunov spectrum from a chaotic time series, Phys. Rev. Lett. **55**,10, (1985), 1082 – 1085.

23. R. Stoop & P.F. Meier, Evaluation of Lyapunov exponents and scaling functions, J. Opt. Soc. Am. B **5**, (1988), 1037 – 1045.

24. F. Takens, Detecting strange attractors in turbulence, in: Dynamical systems and turbulence, eds. D.A. Rand and L.-S. Young, Lecture notes in mathematics, Vol. 898, Springer, Berlin (1981), 366 – 381.

25. F. Takens, Invariants related to dimension and entropy, in: Atas do 13° colóqkio brasileiro de matemática, Rio de Janeiro (1983), 353 – 359.

26. J.A. Vastano & E.J. Kostelich, Comparison of algorithms for determining Lyapunov exponents from experimental data, in: Dimensions and entropies in chaotic systems. Ed. G. Mayer-Kress, Springer, Berlin (1986), 100 – 107.

LYAPUNOV EXPONENTS IN STOCHASTIC STRUCTURAL DYNAMICS

S. T. Ariaratnam
Solid Mechanics Division
Faculty of Engineering, University of Waterloo
Waterloo, Ontario, Canada, N2L 3G1
Wei-Chau Xie
CANDU Operations, Atomic Energy of Canada Ltd.
Mississauga, Ontario, Canada, L5K 1B2

Abstract

The role of Lyapunov exponents in problems of stochastic structural dynamics is illustrated through three examples. Emphasis is placed on the explicit evaluation of the largest Lyapunov exponent for each example by numerical and asymptotic methods.

1. Introduction

Lyapunov exponents play an important role in modern theories of nonlinear structural dynamics. In this paper, their application to the dynamics of engineering structures is illustrated through three problems in structural engineering. The first concerns localization of stress wave propagation through randomly disordered periodic structures. The second deals with the problem of the sensitivity to stochastic disturbance of pitch-fork bifurcation typically encountered in the buckling of slender columns and plates under stochastically fluctuating axial loads. The third example is on the stochastic stability of coupled linear elastic systems. For each example, the largest Lyapunov exponent is evaluated and its significance is discussed.

2. Localization in disordered structures

Many engineering structures are constructed by assembling several identical units end-to-end to form a large spatially periodic structure, some examples being long space antennae, periodic truss structures for space platforms and long pipelines continuous over several supports. Such periodic structures behave like band-pass filters when transmitting stress waves. If damping is neglected, the stress wave propagates without attenuation of amplitude in the frequency pass-bands but suffers attenuation in the frequency stop-bands (Brillouin, 1946). However, due to unavoidable defects in manufacture and assembly, no real structure can be perfectly periodic, but is rather disordered. The disorder can be in the geometry of the structure or in the material properties. In disordered structures, amplitudes of waves with frequencies in the pass-bands will also be attenuated. This means that energy imparted at some location in the structure cannot propagate indefinitely into the structure but dies rapidly away from the source. This is the phenomenon of localization in which the steady-state response of the structure decays exponentially away from the source of disturbance. It is therefore important to study the localization behaviour of randomly disordered periodic structures and to evaluate the localization factor, which is the spatial rate of decay of amplitude of a stress wave propagating in the structure. As an application the extent of damage that is spread in a structure due to impact at some location can be estimated. The localization behaviour may also be used to serve as a "damping" mechanism in periodic structures, especially for structures in outer space where atmospheric damping is small.

In the following, the localization factor is related to the largest Lyapunov exponent associated with a product of random matrices and a formula for evaluating this Lyapunov exponent is established.

Consider an element numbered n in a linear periodic structure (Figure 1). The element is modelled by the dynamic transfer matrix, $\mathbf{T}$, which relates the state amplitude vector $\mathbf{x}_{n-1}$ on the left-side of the element to that on the right-side, $\mathbf{x}_n$, by the linear transformation $\mathbf{x}_n = \mathbf{T}\mathbf{x}_{n-1}$. In linear elastic structures the state vector $\mathbf{x}$ usually involves generalized displacements and forces. The transfer matrix $\mathbf{T}$ is a square matrix of even dimension and is a function of the frequency ω of the disturbance propagating in the structure. For a perfectly periodic structure, the transfer matrix is the same for each element and therefore the state vector, $\mathbf{x}_n$, after n elements is related to that at the beginning, $\mathbf{x}_0$, by $\mathbf{x}_n = \mathbf{T}^n\mathbf{x}_0$. Depending on the nature of the eigenvalues of T, the waves propagating in a periodic structure are described as travelling waves and attenuating waves, which occur in alternating frequency bands known as pass-bands and stop-bands. If the eigenvalues of T are complex and of the form $e^{\pm ik}$ ($k \in \mathbf{R}$), the corresponding wave frequency is in the pass-band and the waves travel in the form $e^{\pm ikn}$, where k is a real wave number, the positive and negative signs indicating left and right travelling waves, respectively. On the other hand, if the eigenvalues of T are of the form $e^{\pm\alpha}$ or $e^{\pm\alpha+i\pi}$ ($\alpha \in \mathbf{R}$), the corresponding frequency is in the stop-band and the wave amplitudes after travelling n elements are attenuated by the factor $e^{\pm\alpha n}$, in which the real exponent α implies attenuating waves.

When the periodic structure is disordered randomly due to variability in geometry, material properties and manufacturing conditions, the transfer matrix for each element is not the same but is a function of the parameters of disorder. In this case the state vector after n elements is related to that at the beginning by $\mathbf{x}_n = \mathbf{T}_n\mathbf{T}_{n-1}\cdots\mathbf{T}_1\mathbf{x}_0$, where $\mathbf{T}_1, \mathbf{T}_2, \cdots, \mathbf{T}_n$ are random matrices. Let $\mathbf{T}$ be the average transfer matrix and $\mathbf{X}$ denote the matrix whose columns are the eigenvectors of $\mathbf{T}$. Then the wave-transfer matrix $\mathbf{W}_n$ is related to $\mathbf{T}_n$ by the transformation $\mathbf{W}_n = \mathbf{X}^{-1}\mathbf{T}_n\mathbf{X}$. The matrix $\mathbf{W}_n$ relates the vector amplitudes of the left– and right–travelling waves, $\mathbf{L}_{n-1}$ and $\mathbf{R}_{n-1}$, respectively at the beginning of element n to those at the end of the element by the transformation

$$\begin{Bmatrix} \mathbf{L}_n \\ \mathbf{R}_n \end{Bmatrix} = \mathbf{W}_n \begin{Bmatrix} \mathbf{L}_{n-1} \\ \mathbf{R}_{n-1} \end{Bmatrix}, \tag{2.1}$$

or $\mathbf{y}_n = \mathbf{W}_n\mathbf{y}_{n-1}$, where $\mathbf{y}_n = \mathbf{X}^{-1}\mathbf{x}_n = \{\mathbf{L}_n^T, \mathbf{R}_n^T\}$. The matrix $\mathbf{W}_n$ is, in general, complex. Hence,

$$\mathbf{y}_n = \mathbf{W}_n \mathbf{W}_{n-1} \cdots \mathbf{W}_1 \mathbf{y}_0 = \mathbf{A}_n \mathbf{y}_0, \tag{2.2}$$

where $\mathbf{A}_n = \mathbf{W}_n \mathbf{W}_{n-1} \cdots \mathbf{W}_1$. We define the Euclidean norm $\|\mathbf{y}_n\|$ of vector $\mathbf{y}_n$ by $\|\mathbf{y}_n\|^2 = (\bar{y}_n, y_n)$ where (,) denotes the scalar product and the over bar the complex conjugate. Then $\|\mathbf{y}_n\|^2 = \mathbf{y}_0^* \mathbf{A}_n^* \mathbf{A}_n \mathbf{y}_0$, where * denotes the operation of transposition and taking of complex conjugate. Defining a Lyapunov exponent by

$$\lambda(\mathbf{y}_0) = \lim_{n \to \infty} \frac{1}{n} \log \|\mathbf{y}_n(\mathbf{y}_0)\| , \tag{2.3}$$

it is easily shown using Rayleigh's principle that the largest Lyapunov exponent λ_{max} is given by

$$\lambda_{max} = \lim_{n \to \infty} \frac{1}{n} \log \sigma_{max}, \tag{2.4}$$

where σ_{max} is the largest singular value of the matrix $\mathbf{A}_n$, i.e. σ_{max} is equal to the square root of the largest eigenvalue of the matrix $\mathbf{A}_n^* \mathbf{A}_n$. To establish a formula for λ_{max}, it is necessary to make the assumption that the matrices $\mathbf{W}_n$ are independent and identically distributed. The asymptotic properties of the product of such random matrices were first studied by Furstenberg (1963). The following derivation is due to Khas'minskii (1967).

Let $\mathbf{s}_n = \mathbf{y}_n / \|\mathbf{y}_n\|$. Then $\mathbf{s}_n = \mathbf{W}_n \mathbf{y}_{n-1} / \|\mathbf{W}_n \mathbf{y}_{n-1}\| = \mathbf{W}_n \mathbf{s}_{n-1} / \|\mathbf{W}_n \mathbf{s}_n\|$, $\|\mathbf{s}_n\| = 1$. Hence $\mathbf{s}_0, \mathbf{s}_1, \cdots, \mathbf{s}_n, \cdots$ represents a Markov chain on the unit hypersphere $\|\mathbf{s}\| = 1$. Suppose that the chain is ergodic with invariant probability measure $\nu(d\mathbf{s})$. Letting $\rho_n = \log \|\mathbf{y}_n\|$, it is easily seen that

$$\rho_n = \rho_0 + \sum_{k=1}^{n} \log \|\mathbf{W}_k \mathbf{s}_{k-1}\| . \tag{2.5}$$

Since the matrices $\mathbf{W}_1, \mathbf{W}_2, \cdots, \mathbf{W}_n, \cdots$ are independent, the pair $(\mathbf{W}_n, \mathbf{s}_{n-1})$ also forms a Markov chain. Under sufficiently broad conditions (Khas'minskii, 1980) the chain satisfies the strong law of large numbers, and hence

$$\lim_{n \to \infty} \frac{\rho_n}{n} = \lim_{n \to \infty} \sum_{k=1}^{n} \frac{1}{n} \log \|\mathbf{W}_k \mathbf{s}_{k-1}\| = E[\log \|\mathbf{W}\mathbf{s}\|], \qquad \text{w.p.1},$$

so that

$$\lambda = \lim_{n \to \infty} \frac{1}{n} \log \|\mathbf{y}_n\| = \int \int \log \|\mathbf{W}\mathbf{s}\| P(d\mathbf{W}) \nu(d\mathbf{s}), \tag{2.6}$$

where $P(d\mathbf{W})$ is the common probability distribution of the matrices $\mathbf{W}_1, \mathbf{W}_2, \cdots$ The exponent λ given by the above formula has been shown to correspond to the largest Lyapunov exponent λ_{max} w.p.1 (Oseledec, 1968). In order to obtain the probability measure $\nu(d\mathbf{s})$, the invariance of probability mass on the unit hypersphere $\|\mathbf{s}\| = 1$ is employed. Referring to Figure 2, if the region R_1 is mapped to region R_2 by the transformation $\mathbf{W}$, then equating the probability masses on R_1 and R_2,

$$\int 1_{R_1}(\mathbf{s})\nu(d\mathbf{s}) = \int\int 1_{R_1}(\frac{\mathbf{Ws}}{||\mathbf{Ws}||})P(d\mathbf{W})\nu(d\mathbf{s}), \tag{2.7}$$

where $1_R(\cdot)$ is the indicator function which takes the value unity when the argument lies on R and zero if it lies elsewhere. It is usually difficult to solve this integral equation for $\nu(d\mathbf{s})$ analytically even in the two-dimensional case when the hypersphere becomes the unit circle S^1. Approximate methods or numerical simulation techniques have to be employed.

In view of the difficulty of solving for the probability measure $\nu(d\mathbf{s})$, the discussion from now on will be restricted to mono-coupled, one-dimensional structures whose transfer matrices are of dimension 2. The wave transfer matrix $\mathbf{W}$ is then of the form

$$\mathbf{W}_n = \begin{bmatrix} \frac{1}{t_n} & -\frac{r_n}{t_n} \\ -(\frac{r_n}{t_n})^* & \frac{1}{t_n^*} \end{bmatrix}, \tag{2.8}$$

where t_n is the transmission coefficient, which is the complex amplitude of a wave emerging from the right of the nth element when a wave of unit amplitude is incident at the left, and r_n is the reflection coefficient, which is the complex amplitude of the reflected wave when a wave of unit amplitude is incident on the nth element from the left. The star here denotes the complex conjugate. For a perfectly periodic structure, $t_n = e^{-ik}$ and $r_n = 0$ where k is the wave number. The matrix $\mathbf{A}_n = \prod_{j=1}^{n} \mathbf{W}_j$ is of the form

$$\mathbf{A}_n = \begin{bmatrix} \frac{1}{\tau_n} & -\frac{\rho_n}{\tau_n} \\ -(\frac{\rho_n}{\tau_n})^* & (\frac{1}{\tau_n})^* \end{bmatrix}, \tag{2.9}$$

where τ_n, ρ_n, $(|\rho_n|<1)$, are the transmission and reflection coefficients, respectively of the n-element disordered structure. The eigenvalues of $\mathbf{A}_n^*\mathbf{A}_n$ are then

$$\sigma_{1,2}^2 = \frac{1}{|\tau_n|^2}(1 \pm |\rho_n|)^2. \tag{2.10}$$

Hence, $\sigma_{\max} = ||\mathbf{A}_n|| = (1 + |\rho_n|)/|\tau_n|$, so that from equation (2.4), the largest Lyapunov exponent is

$$\lambda_{\max} = \lim_{n\to\infty} \frac{1}{n} \log \sigma_{\max} = -\lim_{n\to\infty} \frac{1}{n} \log |\tau_n|. \tag{2.11}$$

Therefore, $|\tau_n| \approx e^{-n\lambda_{\max}}$ as $n \to \infty$ Since $|\tau_n|^2$ is the ratio of transmitted energy to incident energy for the disordered structure, the amplitude of a wave travelling in the structure propagates according to $e^{-n\lambda_{max}}$ as $n \to \infty$ w.p.1. Hence the largest Lyapunov exponent is by definition the localization factor.

Let α be the vector parameter of disorder. Then the wave transfer matrix is a function of α, $\mathbf{W}(\alpha)$. In two-dimensions, a vector $\mathbf{s}$ on the unit circle is of the form $\mathbf{s} = \frac{1}{\sqrt{2}}(e^{i\phi}, e^{-i\phi})^T$ where, in view of $\mathbf{s}(\phi+\pi) = -\mathbf{s}(\phi)$, we may take $0 \leq \phi \leq \pi$; so that only the top-half of the unit circle is considered. Then the formula (2.7) for λ_{max} may be written

$$\lambda_{max} = \int\int \log\|\mathbf{W}(\alpha)\mathbf{s}(\phi)\| p(\alpha)\mu(\phi) d\alpha d\phi, \tag{2.12}$$

where $p(\alpha)$ is the probability density of α and $\nu(d\mathbf{s}) = \mu(\phi)d\phi$. If, as a first-order approximation, $\mu(\phi)$ is taken to be the uniform density $1/\pi$, the formula (2.12) yields a first perturbation approximation to λ_{max} for small disorder (Kissel, 1988, Ariaratnam, 1990).

In the following a numerical simulation scheme is developed to determine $\mu(\phi)$ from which λ_{max} can be calculated for disorder of any magnitude. Let the wave transfer matrix $\mathbf{W}$ in equation (2.8) be written in the form

$$\mathbf{W} = \begin{bmatrix} a_1+ia_2 & b_1+ib_2 \\ b_1-ib_2 & a_1-ia_2 \end{bmatrix}. \tag{2.13}$$

Taking $\mathbf{s}_n = \frac{1}{\sqrt{2}}\{e^{i\phi_n} \;\; e^{-i\phi_n})^T$, we get after transformation by $\mathbf{W}$,

$$\hat{\mathbf{s}}_{n+1} = \mathbf{W}\mathbf{s}_n = \rho_{n+1}\frac{1}{\sqrt{2}}\{e^{i\phi_{n+1}} \;\; e^{-i\phi_{n+1}}\}^T, \tag{2.14}$$

which, when normalized to unit length, becomes

$$\mathbf{s}_{n+1} = \frac{1}{\sqrt{2}}\{e^{i\phi_{n+1}} \;\; e^{-i\phi_{n+1}}\}, \quad \hat{\mathbf{s}}_{n+1} = \rho_{n+1}\mathbf{s}_{n+1}, \tag{2.15}$$

where

$$\rho_{n+1}^2 = [(a_1+b_1)\cos\phi_n - (a_2-b_2)\sin\phi_n]^2 + [(a_2+b_2)\cos\phi_n + (a_1-b_1)\sin\phi_n]^2, \tag{2.16}$$

$$\phi_{n+1} = \tan^{-1}\left[\frac{(a_2+b_2)\cos\phi_n + (a_1-b_1)\sin\phi_n}{(a_1+b_1)\cos\phi_n - (a_2-b_2)\sin\phi_n}\right]. \tag{2.17}$$

Then, starting with some initial value ϕ_0, the iteration formula (2.17) can be applied to obtain the probability density $\mu(\phi)$ which after several iterations will become independent of ϕ_0 due to the ergodicity of the Markov chain $\phi_0, \phi_1, \cdots, \phi_n, \cdots$. Equation (2.12) can then be employed to evaluated λ_{max} numerically.

As an example a long, continuous, elastic beam on several simple supports is considered (Figure 3). The length of each span is l, the mass per unit length is μ and the flexural rigidity is EI. The transfer matrix for a span relates the angles of rotation θ

and the bending moments M at adjacent supports (Yang and Lin, 1975, Lin, 1976):

$$\begin{Bmatrix} \phi_n \\ (\frac{lM}{EI})_n \end{Bmatrix} = \mathbf{T} \begin{Bmatrix} \phi_{n-1} \\ (\frac{lM}{EI})_{n-1} \end{Bmatrix}, \qquad \mathbf{T} = \begin{bmatrix} \cos k & \beta/\sqrt{\overline{\omega}} \\ -\sqrt{\overline{\omega}}\sin^2 k/\beta & \cos k \end{bmatrix}, \tag{2.18}$$

where

$$\cos k = \frac{\sinh\sqrt{\overline{\omega}}\cos\sqrt{\overline{\omega}} - \cosh\sqrt{\overline{\omega}}\sin\sqrt{\overline{\omega}}}{\sinh\sqrt{\overline{\omega}} - \sin\sqrt{\overline{\omega}}}, \qquad \beta = \frac{C_4}{S_3},$$

$$S_3 = \frac{\sinh\sqrt{\overline{\omega}} - \sin\sqrt{\overline{\omega}}}{2}, \qquad C_4 = \frac{\cosh\sqrt{\overline{\omega}}\cos\sqrt{\overline{\omega}} - 1}{2},$$

and the non-dimensional frequency $\overline{\omega}$ is defined by

$$\overline{\omega} = \omega\left(\frac{\mu l^4}{EI}\right)^{\frac{1}{2}}.$$

Suppose that the length of each span l is disordered and let $\eta = l/\langle l\rangle$, where $\langle l\rangle = l_0$. The wave transfer matrix is found to be

$$\mathbf{W}(\eta) = \begin{bmatrix} \frac{1}{t} & -\frac{r}{t} \\ -(\frac{r}{t})^* & (\frac{1}{t})^* \end{bmatrix}, \tag{2.19}$$

where

$$\frac{1}{t} = \cos(k\eta) + i\left[\frac{\beta(\eta)\sin k}{2\beta} + \frac{\beta\sin^2(k\eta)}{2\beta(\eta)\sin k}\right],$$

$$-\frac{r}{t} = i\left[\frac{\beta\sin^2(k\eta)}{2\beta(\eta)\sin k} - \frac{\beta(\eta)\sin k}{2\beta}\right].$$

Whenever the argument η appears, it is implied that the underlying trigonometric and hyperbolic functions should have $\sqrt{\overline{\omega}}$ replaced by $\sqrt{\overline{\omega}}\,\eta$.

The pass-bands are determined by the condition $|\mathrm{tr}(\mathbf{T})| < 0$, i.e.

$$\frac{\sinh\sqrt{\overline{\omega}}\cos\sqrt{\overline{\omega}} - \cosh\sqrt{\overline{\omega}}\sin\sqrt{\overline{\omega}}}{\sinh\sqrt{\overline{\omega}} - \sin\sqrt{\overline{\omega}}} < 1. \tag{2.20}$$

Suppose that η is normally distributed with $\langle\eta\rangle = 1$, $\sigma_\eta = \sigma$. One defines a new random variable $\alpha = \eta - 1$, so that $\langle\alpha\rangle = 0$, $\langle\alpha^2\rangle = \sigma^2$. Then the random variable α is $N(0,\sigma^2)$. Equation (2.17) is used to find the probability density $\mu(\phi)$, where the coefficients are given by

$$a_1(\alpha) = \cos k(\alpha+1), \qquad a_2(\alpha) = \frac{\beta(\alpha+1)\sin k}{2\beta} + \frac{\beta \sin^2 k(\alpha+1)}{2\beta(\alpha+1)\sin k},$$

$$b_1(\alpha) = 0, \qquad b_2(\alpha) = -\frac{\beta(\alpha+1)\sin k}{2\beta} + \frac{\beta \sin^2 k(\alpha+1)}{2\beta(\alpha+1)\sin k},$$

Taking $\sigma=0.1$ and the number of iterations $N_s=10^7$, the probability density $\mu(\phi)$ is obtained at 500 discrete points $\mu(\phi_n)$, $\phi_n=n\pi/500$, $n=0,1,\cdots,499$. A typical result for $\mu(\phi)$ is shown in Figures 4 for $\overline{\omega}=100$. Equation (2.12) is then applied to determine the largest Lyapunov exponent λ; the numerical results are plotted in Figure 5. The uniform distribution $\mu^0=1/\pi$ is also used in equation (2.12) to obtain λ, and the result obtained using a perturbation method (Kissel, 1988, Ariaratnam 1989) is also plotted in Figure 5 for comparison. It can be observed that when $\overline{\omega}$ is near the ends of pass-bands, the probability density $\mu(\phi)$ differs from the uniform distribution μ^0 significantly, which causes large discrepancies in the values of the largest Lyapunov exponent.

Kissel (1988) simulated λ using a different numerical scheme. The simulated results were compared to the analytical results obtained using a perturbation method with the assumption of a uniform density for $\mu(\phi)$. Kissel concluded that for a fairly small number of iterations (in this case 1001), the simulated results agreed well with the analytical results. However, it should be pointed out that his analytical result, is only a first order approximation, which first uses the uniform density μ^0 to replace the actual density $\mu(\phi)$ and then retains terms only up to the first order in the variance of the disorder parameter α. Therefore, the analytical result cannot be taken as the reference for verifying the accuracy of computer simulation; a much larger number of iterations has to be used to increase the accuracy of simulation. Again it can be seen clearly from Figure 5 that the error due to replacing $\mu(\phi)$ by the uniform distribution is very prominent when $\overline{\omega}$ is large and when $\overline{\omega}$ is close to the ends of pass-bands.

3. Stochastic perturbation of bifurcations

In problems concerning the dynamic stability of slender structures such as axially loaded columns and flat plates, one is led to the study of a nonlinear differential equation of the form

$$\ddot{q} + 2\beta\dot{q} - [\gamma + \sigma\xi(t)]\, q + \alpha q^3 = 0, \quad \alpha, \beta > 0, \tag{3.1}$$

where q represents a generalized displacement, β the damping coefficient and α, γ, σ are constants. The parameter γ corresponds to the mean axial load on the structure and σ is the intensity of the load fluctuation $\xi(t)$, which may be regarded as a zero mean, ergodic random process. In the absence of the load fluctuation, the structure undergoes a supercritical pitchfork bifurcation from the trivial equilibrium configuration to a deflected non-trivial configuration as the loading parameter γ is increased from negative to positive values. A question of interest is to determine the shift in the bifurcation point as a result of the load fluctuation. For this purpose it is necessary to evaluate the variation of the largest Lyapunov exponent of the linearized system

$$\ddot{q} + 2\beta\dot{q} - [\gamma + \sigma\xi(t)]\,q = 0, \tag{3.2}$$

for small values of γ near the bifurcation value of zero. Assuming that $\xi(t)$ may be approximated by a "unit" white noise process, equation (3.2) may be written in the Itô form

$$dq_1 = q_2 dt, \quad dq_2 = -(\beta q_2 + \gamma q_1)dt - \sigma q_1 dW, \tag{3.3}$$

where $W(t)$ is the "unit" Wiener process. Introducing the scaling $q_1 = x_1$, $q_2 = \sigma^{1/2} x_2$, $\beta = \sigma^{a_2}\hat{\beta}$, $\gamma = \sigma^{a_3}\hat{\gamma}$, where $\hat{\beta}$, $\hat{\gamma} = O(1)$, these equations become

$$dx_1 = \sigma^{a_1} x_2 dt, \;\; dx_2 = -\sigma^{a_2}\hat{\beta} x_2 dt - \sigma^{a_3 - a_1}\hat{\gamma} x_1 dt - \sigma^{1-a_1} x_1 dW. \tag{3.4}$$

In order that the right hand sides of both of equations (3.4) may have comparable influence, a_1, a_2, a_3 must be chosen so that $a_1 = a_2 = a_3 - a_1 = 2(1 - a_1)$, implying that $a_1 = a_2 = 2/3$, $a_3 = 4/3$. Hence, writing $\epsilon = \sigma^{1/3}$, equations (3.4) take the form

$$dx_1 = \epsilon x_2 dt, \quad dx_2 = -\epsilon(\hat{\beta} x_2 + \hat{\gamma} x_1)dt - \epsilon^{1/2} x_1 dW. \tag{3.5}$$

Following Khas'minskii (1967), polar coordinates (a,ϕ) are now introduced via the relations $y_1 = a\cos\phi$, $y_2 = a\sin\phi$. If one defines a p–th norm of the vector $\mathbf{y} = (y_1, y_2)$ by $P = a^p$, a new pair of Itô equations for P and ϕ are obtained by applying Itô's lemma:

$$dP = \epsilon p P f(\phi)\, dt - \epsilon^{1/2} p P \sin\phi\cos\phi\, dW, \quad d\phi = -\epsilon F(\phi)\, dt - \epsilon^{1/2}\cos^2\phi\, dW, \tag{3.6}$$

where

$$f(\phi) = (1 + \hat{\gamma})\sin\phi\cos\phi - 2\hat{\beta}\sin^2\phi + \tfrac{1}{2}[\cos^4\phi + (p-1)\sin^2\phi\cos^2\phi]\,,$$

$$F(\phi) = 1 - (1 + \hat{\gamma})\cos^2\phi + 2\hat{\beta}\sin\phi\cos\phi + \sin\phi\cos^3\phi\,.$$

Following Wedig (1988, 1989), a linear stochastic transformation $S = T(\phi)P$, $-\frac{1}{2}\pi \le \phi \le \frac{1}{2}\pi$, is now applied which leads to the Itô equation for S:

$$dS = \epsilon\Big\{\tfrac{1}{2}\cos^4\phi\, T_{\phi\phi} + [p\sin\phi\cos^3\phi - F(\phi)]\, T_\phi + p f(\phi) T\Big\} P\, dt -$$

$$- \epsilon^{1/2}\cos\phi[\cos\phi\, T_\phi + p\sin\phi\, T]\, P\, dW\,. \tag{3.8}$$

For a bounded and non-singular transformation $T(\phi)$, both processes P and S are expected to have the same growth behaviour in t. The function $T(\phi)$ is now chosen so that the drift term in equation (3.8) is independent of the phase process ϕ, so that equation (3.8) is of the form

$$dS = \epsilon\Lambda S\, dt + \epsilon^{1/2} S g(\phi)\, dW\,. \tag{3.9}$$

The transformation T is then governed by the eigenvalue problem

$$\tfrac{1}{2}\cos^4\phi T_{\phi\phi} - [1 - (1 + \hat{\gamma})\cos^2\phi + 2\hat{\beta}\sin\phi\cos\phi + (1-p)\sin\phi\cos^3\phi]\, T_\phi +$$

$$+ \tfrac{1}{2}p\left[(p-1)\sin^2\phi\cos^2\phi+\cos^4\phi+2(1+\hat{\gamma})\sin\phi\cos\phi-4\hat{\beta}\sin^4\phi\right]T = \Lambda T. \tag{3.10}$$

The Lyapunov exponent λ of system (3.5) is related to Λ by the relation (Kozin and Sugimoto, 1977, Molchanov, 1978, Arnold, 1984) $\lambda = \lim_{p\to 0}(\epsilon\Lambda)/p$. It now remains to solve the eigenvalue problem (3.10) for Λ. Since the coefficients in equation (3.10) are π–periodic, the following series expansion is assumed for $T(\phi)$:

$$T(\phi) = \sum_{n=0}^{\infty}(u_n\cos 2n\phi + v_n\sin 2n\phi). \tag{3.11}$$

Substituting this expansion in equation (3.10) and equating coefficients of like trigonometric terms leads to a system of infinitely many homogeneous linear equations for the unknowns coefficients u_n, v_n. The existence of non-trivial solution requires that the determinant of the coefficient matrix be equal to zero, from which the eigenvalue Λ can be obtained in principle. In practice, only a finite number of terms is considered to obtain an approximate value for the eigenvalue Λ. Taking five terms u_0, u_1, v_1, u_2, v_2 into consideration in equation (3.11), the vanishing of the determinant of the 5X5 coefficient matrix gives

$$\begin{vmatrix} a-\Lambda & (p+2)(4\hat{\beta}-1)/8 & (p+2)(1+\hat{\gamma})/4 & -(p+2)(p+4)/32 & 0 \\ p(\hat{\beta}+1/4) & a-(\Lambda+3/4) & -1+\hat{\gamma} & (p+4)(4\hat{\beta}-3)/8 & (p+4)(1+\hat{\gamma})/4 \\ p(1+\hat{\gamma})/2 & 1-\hat{\gamma} & a-(\Lambda+3/4) & -(p+4)(1+\hat{\gamma})/4 & (p+4)(4\hat{\beta}-3)/8 \\ -p(p-2)/16 & (p-2)(4\hat{\beta}+3)/8 & -(p-2)(1+\hat{\gamma})/4 & a-(\Lambda+3) & -2(1-\hat{\gamma}) \\ 0 & (p-2)(1+\hat{\gamma})/16 & (p-2)(4\hat{\beta}+3)/8 & 2(1-\hat{\gamma}) & a-(\Lambda+3) \end{vmatrix} = 0,$$

where $a = p[\frac{1}{16}(p+2)-\hat{\beta}]$. Solving for Λ and using the Molchanov condition yields, after some lengthy calculation,

$$\lambda = -\beta + 0.28308\,\sigma^{\frac{2}{3}}\left[1+1.04564\hat{\gamma}+0.56352\hat{\gamma}^2+O(\hat{\gamma}^3)\right], \tag{3.12}$$

where $\hat{\gamma}=\gamma/\sigma^{4/3}$. In the case of the nilpotent system for which $\beta=\gamma=0$, equation (3.12) gives $\lambda=0.28308\,\sigma^{2/3}$, which is consistent with the exact result $0.28931\sigma^{2/3}$ obtained by Ariaratnam and Xie (1990). The validity of the approximate result (3.12) is compared with that obtained by a digital simulation procedure due to Wedig (1989). The two results, shown in Figure 6, are found to be in good agreement for small values of γ. The shift in the bifurcation value of the loading parameter γ may be obtained by setting $\lambda=0$ in equation (3.12).

. Lyapunov exponents of coupled linear systems

The systems considered are described by stochastic differential equations of the form

$$\ddot{q}_i + 2\sum_{j=1}^{n}\beta_{ij}\dot{q}_j + \omega_i^2 q_i + \omega_i \xi(t)\sum_{j=1}^{n} k_{ij}q_j = 0, \qquad i = 1,2,\cdots,n, \tag{4.1}$$

where the q_i are generalized coordinates, β_{ij} are damping constants, ω_i are natural frequencies, and k_{ij} are constants. The excitation is represented by $\xi(t)$, which is taken to be an ergodic stochastic process with zero mean value and a sufficiently small correlation time. Equations (4.1) describe exactly the parametrically excited motion of certain non-gyroscopic, discrete, linear elastic systems with n degrees-of-freedom about the equilibrium configuration $q_i(t)=0$. They also describe approximately the motion of certain continuous elastic structures whose equations of motion have been discretized by some suitable technique such as Rayleigh-Ritz, Galerkin, finite differences or finite elements. It will be seen later that small cross-damping terms such as β_{ij}, $i \neq j$ have no effect on the solution in the first approximation.

In equation (4.1), the generalized coordinates $q_i(t)$ and velocities $\dot{q}_i(t)$ are transformed to polar coordinates by means of the relations

$$q_i = a_i\cos\Theta_i, \quad \dot{q}_i = -a_i\omega_i\sin\Theta_i, \quad \Theta_i = \omega_i t + \theta_i, \quad i = 1,2,\cdots,n. \tag{4.2}$$

Then, one obtains the equations of motion in terms of $a_i(t)$ and $\theta_i(t)$:

$$\begin{aligned} \dot{a}_i &= -\frac{2}{\omega_i}\sum_{j=1}^{n}\beta_{ij}a_j\omega_j\sin\Theta_j\sin\Theta_i + \xi(t)\sum_{j=1}^{n}k_{ij}a_j\cos\Theta_j\sin\Theta_i, \\ \dot{\theta}_i &= -\frac{2}{a_i\omega_i}\sum_{j=1}^{n}\beta_{ij}a_j\omega_j\sin\Theta_j\cos\Theta_i + \frac{1}{a_i}\xi(t)\sum_{j=1}^{n}k_{ij}a_j\cos\Theta_j\cos\Theta_i. \end{aligned} \tag{4.3}$$

It is supposed that the damping constants and the stochastic perturbation are small, i.e. $\beta_{ij}=O(\epsilon)$, $S(\omega)=O(\epsilon)$, $\Psi(\omega)=O(\epsilon)$, $0<|\epsilon|\ll 1$, where $S(\omega)$ and $\Psi(\omega)$ denote the cosine and sine power spectral densities of the stochastic process $\xi(t)$, defined by

$$S(\omega) + i\,\Psi(\omega) = 2\int_0^\infty E[\xi(t)\xi(t+\tau)]\,e^{i\omega\tau}d\tau\ ,$$

$E[\cdot]$ denoting the expectation operation. Under these conditions, the method of stochastic averaging (Stratonovich, 1963, Khas'minskii, 1966) may be applied to equations (4.3) to yield the following Itô equations for the averaged amplitudes a_i and phase angles ϕ_i, whose solutions provide a uniformly valid first approximation to the exact values (Ariaratnam and Srikantaiah, 1978):

$$da_i = m_{a_i}dt + \sum_{j=1}^{n}\sigma_{ij}dW_j, \quad d\theta_i = m_{\theta_i}dt + \sum_{j=1}^{n}\mu_{ij}dB_j, \tag{4.4}$$

where $W_j(t)$, $B_j(t)$, $j=1,2,\cdots,n$ are mutually independent unit Wiener processes.

The n degrees-of-freedom system given by equation (4.1) is difficult to study in its general form. Hence, the discussion from now on will be restricted to two degrees-of-freedom systems described by the equations:

$$\ddot{q}_1 + 2\beta_{11}\dot{q}_1 + 2\beta_{12}\dot{q}_2 + \omega_1^2 q_1 + \omega_1(k_{11}q_1 + k_{12}q_2)\xi(t) = 0\,,$$
$$\ddot{q}_2 + 2\beta_{21}\dot{q}_1 + 2\beta_{22}\dot{q}_2 + \omega_2^2 q_2 + \omega_2(k_{21}q_1 + k_{22}q_2)\xi(t) = 0\,, \tag{4.5}$$

in which, by a suitable scaling of co-ordinates, it is always possible to take $k_{12} = \pm k_{21} = k > 0$, without loss of generality. The product $|k_{12}k_{21}| = k^2$, however, remains invariant under scaling. The results derived for the two degrees-of-freedom system may be generalized to n degrees-of-freedom systems ($n > 2$) under certain conditions on the spectral density distribution of the parametric excitation.

For the two degrees-of-freedom system, the amplitude equations of (4.4) become

$$da_1 = m_1 dt + \sigma_{11}dW_1 + \sigma_{12}dW_2\;, da_2 = m_2 dt + \sigma_{21}dW_1 + \sigma_{22}dW_2\,, \tag{4.6}$$

where

$$m_1 = \left[-\beta_1 + \frac{3}{16}k_{11}^2 S(2\omega_1) + \frac{1}{8}k_{12}k_{21}S^-\right]a_1 + \frac{1}{16}k_{12}^2 S^+ \frac{a_2^2}{a_1}\,,$$

$$m_2 = \left[-\beta_2 + \frac{3}{16}k_{22}^2 S(2\omega_2) + \frac{1}{8}k_{12}k_{21}S^-\right]a_2 + \frac{1}{16}k_{21}^2 S^+ \frac{a_1^2}{a_2}\,,$$

$$[\sigma\sigma^T]_{11} = \frac{1}{8}k_{11}^2 S(2\omega_1)a_1^2 + \frac{1}{8}k_{12}^2 S^+ a_2^2\,,$$

$$[\sigma\sigma^T]_{22} = \frac{1}{8}k_{22}^2 S(2\omega_2)a_2^2 + \frac{1}{8}k_{21}^2 S^+ a_1^2\,,$$

$$[\sigma\sigma^T]_{12} = [\sigma\sigma^T]_{21} = \frac{1}{8}k_{12}k_{21}S^- a_1 a_2\,,$$

$$S^{\pm} = S(\omega_1 + \omega_2) \pm S(\omega_1 - \omega_2)\,, \qquad \beta_1 = \beta_{11}\,, \qquad \beta_2 = \beta_{22}\,.$$

It may be noted that the averaged amplitude vector (a_1, a_2) is a two-dimensional diffusion process and that the coefficients of the right hand side terms of equations (4.6) are homogeneous in a_1, a_2 of degree one. Hence, the procedure of Khas'minskii (1967) may be employed to derive an expression for the largest Lyapunov exponent of the amplitude process (Ariaratnam, 1977). To this end a further logarithmic polar transformation is applied:

$$\rho = \tfrac{1}{2}\log(a_1^2 + a_2^2), \quad \phi = \mathrm{Tan}^{-1}\left(\frac{a_2}{a_1}\right), \quad 0 \le \phi \le \tfrac{1}{2}\pi\,,$$

and, by the use of Itô's differential rule or otherwise, the following pair of Itô equations governing ρ, ϕ are obtained:

$$d\rho = Q(\phi)dt + \Sigma(\phi)dW\,, \quad d\phi = \Phi(\phi)dt + \Psi(\phi)dW\,, \tag{4.7}$$

where $W(t)$ is a Wiener process of unit intensity and

$$Q(\phi) = \lambda_1\cos^2\phi + \lambda_2\sin^2\phi \pm \frac{1}{8}k^2S^- + \Psi^2(\phi),$$

$$\Phi(\phi) = -\frac{1}{2}(\lambda_1-\lambda_2)\sin2\phi+\frac{1}{16}[2(\lambda_1+\lambda_2+\beta_1+\beta_2)-k^2S(\omega_1\pm\omega_2)]\sin4\phi+\frac{1}{8}k^2S^+\cot2\phi,$$

$$\Sigma^2(\phi) = (\lambda_1+\beta_1)\cos^4\phi + (\lambda_2+\beta_2)\sin^4\phi + \frac{1}{8}k^2S(\omega_1\pm\omega_2)\sin^2 2\phi, \tag{4.8}$$

$$\Psi^2(\phi) = \frac{1}{8}k^2S^+ + \frac{1}{8}[2(\lambda_1+\lambda_2+\beta_1+\beta_2)-k^2S(\omega_1\pm\omega_2)]\sin^2 2\phi,$$

in which the upper sign is taken when $k_{12}=k_{21}=k$, and the lower sign when $k_{12}=-k_{21}=k$. The constants λ_1, λ_2 are defined by

$$\lambda_1 = -\beta_1 + \frac{1}{8}k_{11}^2S(2\omega_1), \qquad \lambda_2 = -\beta_2 + \frac{1}{8}k_{22}^2S(2\omega_2), \tag{4.9}$$

which, as will be seen later, are the Lyapunov exponents of the two uncoupled single degree-of-freedom systems that result when the coupling coefficients k_{12}, k_{21} are set equal to zero.

From the second of equations (4.7), it is clear that the ϕ—process is itself a diffusion on the first quadrant of the unit circle. If $\Psi(\phi)$ vanishes in $[0,\frac{1}{2}\pi]$, the diffusion process is *singular*, otherwise it is *non-singular*. It can be shown that $\Psi(\phi)$ vanishes at $\phi=\phi_0=\frac{1}{4}\pi$ only when

(i) for $k_{12}k_{21}>0$, $k_{11} = k_{22} = 0$ and $S(\omega_1 - \omega_2) = 0$;

(ii) for $k_{12}k_{21}<0$, $k_{11} = k_{22} = 0$ and $S(\omega_1 + \omega_2) = 0$.

In the following, the evaluation of the largest Lyapunov exponent of the averaged system (4.7) will be examined for both non-singular and singular cases.

Non-singular case

It can be shown that the boundaries $\phi=0$, $\phi=\frac{1}{2}\pi$ are both *entrance points* in the sense of Feller (1952), and hence the stationary probability flux is zero. Thus, there is no accumulation of probability mass at the boundaries, and the ϕ—process is ergodic throughout the interval $0\leq\phi\leq\frac{1}{2}\pi$. The invariant density $\mu(\phi)$ which satisfies the Fokker-Planck equation is given by

$$\mu(\phi) = \frac{C}{\Psi^2(\phi)U(\phi)}, \tag{4.10}$$

where

$$U(\phi) = \exp\left[-2\int^{\phi}\{\Phi(t)\Psi^{-2}(t)\}dt\right]$$

$$= \frac{1}{\sin 2\phi} \exp\left[\frac{\lambda_2-\lambda_1}{2a} \int^{\cos 2\phi} \frac{dt}{1-bt^2/a}\right], \tag{4.10}$$

the constants a, b being given by

$$a = \frac{1}{8}[2(\lambda_1+\lambda_2+\beta_1+\beta_2)+k^2S(\omega_1\mp\omega_2)],$$

$$b = \frac{1}{8}[2(\lambda_1+\lambda_2+\beta_1+\beta_2)-k^2S(\omega_1\pm\omega_2)],$$

and C is the normalizing constant. Since the constant a is always positive, the form of the integral in equation (4.11) depends on the sign of the constant b.

For $b>0$, i.e. $\lambda_1+\lambda_2+\beta_1+\beta_2>\frac{1}{2}k^2S(\omega_1\pm\omega_2)$, the invariant density $\mu(\phi)$ is of the form

$$\mu(\phi) = \frac{C\sin 2\phi}{\Psi^2(\phi)} \exp\left[\frac{\lambda_1-\lambda_2}{2\sqrt{\Delta}} \tanh^{-1}\frac{b\cos 2\phi}{\sqrt{\Delta}}\right], \quad (b>0), \tag{4.12a}$$

where C is determined from the normalization condition and is found to be

$$C = \frac{1}{2}(\lambda_1-\lambda_2)\operatorname{csch}\left(\frac{\lambda_1-\lambda_2}{2\sqrt{\Delta}}\tanh^{-1}\frac{b}{\sqrt{\Delta}}\right), \quad (b>0). \tag{4.12a'}$$

The constant Δ is defined by $\Delta = ab$.

For $b<0$, the hyperbolic term in (4.12) is to be replaced appropriately by its trigonometric counterpart, while, for $b=0$, the right hand side of (4.12) is to be replaced by its limit as $b\to 0$. Stated explicitly these expressions are

$$\mu(\phi) = \frac{C\sin 2\phi}{\Psi^2(\phi)} \exp\left[-\frac{\lambda_1-\lambda_2}{2\sqrt{-\Delta}} \tan^{-1}\frac{b\cos 2\phi}{\sqrt{-\Delta}}\right], \quad (b<0), \tag{4.12b}$$

$$C = \frac{1}{2}(\lambda_1-\lambda_2)\operatorname{csch}\left[\frac{\lambda_1-\lambda_2}{2\sqrt{-\Delta}}\tan^{-1}\frac{b}{\sqrt{-\Delta}}\right], \quad (b<0), \tag{4.12b'}$$

and

$$\mu(\phi) = \frac{C\sin 2\phi}{\Psi^2(\phi)} \exp\left[\frac{(\lambda_1-\lambda_2)\cos 2\phi}{2a}\right], \quad (b=0), \tag{4.12c}$$

$$C = \frac{1}{2}(\lambda_1-\lambda_2)\operatorname{csch}\left(\frac{\lambda_1-\lambda_2}{2a}\right), \quad (b=0), \tag{4.12c'}$$

Employing Khas'minskii's (1967) formulation, the largest Lyapunov exponent of system (4.6) is given by

$$\lambda = E[Q(\phi)] = \int_0^{\frac{1}{2}\pi} Q(\phi)\mu(\phi)d\phi. \tag{4.13}$$

Substituting from equations (4.8) and (4.12) into equation (4.13) and performing the indicated integration yields the following expression for the Lyapunov exponent, in which the fact that $k^2 = |k_{12}k_{21}|$ has been taken into account.

(i) if $k_{11}^2 S(2\omega_1) + k_{22}^2 S(2\omega_2) > 4|k_{12}k_{21}|S(\omega_1 \pm \omega_2)$, i.e. $\Delta_0 > 0$,

$$\lambda = \frac{1}{2}\left[(\lambda_1+\lambda_2) + (\lambda_1-\lambda_2)\coth\left(\frac{\lambda_1-\lambda_2}{\Delta_0^{1/2}}\alpha\right)\right] + \frac{1}{8}k_{12}k_{21}S^-, \quad (b>0)\,, \tag{4.14a}$$

where $\alpha = \cosh^{-1}(K/2|k_{12}k_{21}|S^+)$;

(ii) if $k_{11}^2 S(2\omega_1) + k_{22}^2 S(2\omega_2) < 4|k_{12}k_{21}|S(\omega_1 \pm \omega_2)$, i.e. $\Delta_0 < 0$,

$$\lambda = \frac{1}{2}\left\{(\lambda_1+\lambda_2) + (\lambda_1-\lambda_2)\coth\left[\frac{\lambda_1-\lambda_2}{(-\Delta_0)^{1/2}}\alpha\right]\right\} + \frac{1}{8}k_{12}k_{21}S^-, \quad (b<0)\,, \tag{4.14b}$$

where $\alpha = \cos^{-1}(K/2|k_{12}k_{21}|S^+)$.

(iii) if $k_{11}^2 S(2\omega_1) + k_{22}^2 S(2\omega_2) = 4|k_{12}k_{21}|S(\omega_1 \pm \omega_2)$, i.e. $\Delta_0 = 0$,

$$\lambda = \frac{1}{2}\left\{(\lambda_1+\lambda_2) + (\lambda_1-\lambda_2)\coth\left[\frac{4(\lambda_1-\lambda_2)}{|k_{12}k_{21}|S^+}\right]\right\} + \frac{1}{8}k_{12}k_{21}S^-, \quad (b=0)\,. \tag{4.14c}$$

The constants K and Δ_0 are defined as

$$K = k_{11}^2 S(2\omega_1) + k_{22}^2 S(2\omega_2) - 2k_{12}k_{21}S^-, \quad \Delta_0 = \frac{1}{64}[K^2 - 4k_{12}^2 k_{21}^2 (S^+)^2]\,.$$

In conditions (i), (ii), (iii), the upper (plus) sign is to be taken when $k_{12}k_{21} > 0$, and the lower (minus) sign when $k_{12}k_{21} < 0$.

The Lyapunov exponent for a single degree-of-freedom system, which was first obtained by Stratonovich and Romanovskii (1958), may be deduced from equation (4.14). Thus, setting the coupling coefficients k_{12}, k_{21} to zero, it is evident that $\Delta_0 > 0$, $\alpha \to +\infty$, and equation (4.14a) then gives $\lambda = \lambda_1$ if $\lambda_1 > \lambda_2$ and $\lambda = \lambda_2$ if $\lambda_2 > \lambda_1$, confirming that the expression (4.14) is, in fact, the largest Lyapunov exponent of the system.

Singular case

It has been found that $\Psi(\phi)$ vanishes at $\phi_0 = \frac{1}{4}\pi$ when

(i) $k_{12}k_{21} > 0$, $\quad k_{11} = k_{22} = 0$, $\quad S(\omega_1 - \omega_2) = 0$,

(ii) $k_{12}k_{21} < 0$, $\quad k_{11} = k_{22} = 0$, $\quad S(\omega_1 + \omega_2) = 0$.

In both cases, $\phi_0 = \frac{1}{4}\pi$ is a singular point. To determine the nature of the singular point, the sign of the drift coefficient $\Phi(\phi)$ at this point has to be checked. From equation (4.8), one obtains (see, e.g. Mitchell and Kozin, 1974)

(i) if $\beta_1 > \beta_2$, then $\Phi(\frac{1}{4}\pi) > 0$; the singular point $\phi_0 = \frac{1}{4}\pi$ is therefore a *right* or *forward shunt*;

(ii) if $\beta_2>\beta_1$, then $\Phi(\frac{1}{4}\pi)<0$; the singular point $\phi_0=\frac{1}{4}\pi$ is therefore a *left* or *backward shunt*;

(iii) if $\beta_2=\beta_1=\beta$, then $\Phi(\frac{1}{4}\pi)=0$; the singular point $\phi_0=\frac{1}{4}\pi$ is therefore a *trap*.

In the following, these three cases will be discussed in some detail.

(i) $\beta_1>\beta_2$. For $\beta_1>\beta_2$, the singular point $\phi_0=\frac{1}{4}\pi$ is a right shunt. This means that even if an initial point ϕ is in the left half interval $(0,\frac{1}{4}\pi)$, it will eventually be shunted across to the right half interval $(\frac{1}{4}\pi,\frac{1}{2}\pi)$ and remain there forever. Hence, the probability density $\mu(\phi)$ is concentrated in the right half of the interval $0\leq\phi\leq\frac{1}{2}\pi$. The density $\mu(\phi)$ of the invariant measure is governed by the Fokker-Planck equation with solution of the form

$$\mu(\phi)=\begin{cases} 0, & 0\leq\phi\leq\frac{1}{4}\pi, \\ \dfrac{C}{\Psi^2(\phi)U(\phi)}, & \frac{1}{4}\pi<\phi\leq\frac{1}{2}\pi, \end{cases} \tag{4.15}$$

where

$$\begin{aligned} U(\phi) &= \exp\Big[-2\int^{\phi}\Phi(t)\Psi^{-2}(t)dt\Big] \\ &= \frac{1}{\sin 2\phi}\exp\Big[-\frac{4(\beta_1-\beta_2)}{k^2S(\omega_1\pm\omega_2)}\sec 2\phi\Big]. \end{aligned} \tag{4.16}$$

The constant C in equation (4.1) is determined by the normalization condition and is found to be

$$C=(\beta_1-\beta_2)\exp\Big[\frac{4(\beta_1-\beta_2)}{k^2S(\omega_1\pm\omega_2)}\Big].$$

Substituting from equations (4.8) and (4.15) in equation (4.13) and performing the indicated integration results in, after replacing k^2 by $|k_{12}k_{21}|$,

$$\lambda=-\beta_2+\frac{1}{8}|k_{12}k_{21}|S(\omega_1\pm\omega_2)\,,\quad \beta_1>\beta_2\,, \tag{4.17}$$

where the upper sign is taken when $k_{12}k_{21}>0$, and the lower sign when $k_{12}k_{21}<0$.

(ii) $\beta_2>\beta_1$. For $\beta_2>\beta_1$, the singular point $\phi=\frac{1}{4}\pi$ is a left shunt and the invariant probability density of the ϕ—process is now concentrated in the left half of the interval $0\leq\phi\leq\frac{1}{2}\pi$. The density $\mu(\phi)$ of the invariant measure is given by

$$\mu(\phi) = \begin{cases} \dfrac{C}{\Psi^2(\phi)U(\phi)}, & 0 \le \phi < \tfrac{1}{4}\pi, \\ 0, & \tfrac{1}{4}\pi \le \phi \le \tfrac{1}{2}\pi, \end{cases} \tag{4.18}$$

where $U(\phi)$ is given by equation (4.16). The constant C determined by the normalization condition is

$$C = (\beta_2 - \beta_1)\exp\left[\frac{4(\beta_2-\beta_1)}{k^2 S(\omega_1 \pm \omega_2)}\right].$$

Substituting from equation (4.8) and (4.18) in equation (4.13), one obtains the largest Lyapunov exponent as

$$\lambda = -\beta_1 + \frac{1}{8}|k_{12}k_{21}|S(\omega_1 \pm \omega_2), \qquad \beta_2 > \beta_1 . \tag{4.19}$$

(iii) $\beta_1 = \beta_2 = \beta$. For $\beta_1 = \beta_2 = \beta$, the singular point $\phi_0 = \frac{1}{4}\pi$ is a trap. This means that regardless of where the initial point ϕ is situated, it will eventually be attracted to the point $\phi_0 = \frac{1}{4}\pi$ and remain there forever. The density $\mu(\phi)$ of the invariant measure is the Dirac delta function concentrated at $\frac{1}{4}\pi$:

$$\mu(\phi) = \delta(\phi - \tfrac{1}{4}\pi), \qquad 0 \le \phi \le \tfrac{1}{2}\pi . \tag{4.20}$$

From equations (4.8), (4.13) and (4.18), the largest Lyapunov exponent is found to be:

$$\begin{aligned} \lambda &= \int_0^{\frac{1}{2}\pi} Q(\phi)\,\delta(\phi - \tfrac{1}{4}\pi)\,d\phi = Q(\tfrac{1}{4}\pi) \\ &= -\beta + \frac{1}{8}|k_{12}k_{21}|S(\omega_1 \pm \omega_2), \qquad \beta_1 = \beta_2 = \beta . \end{aligned} \tag{4.21}$$

It may be noted that the result (4.21) can also be obtained from either (4.17) or (4.19) by taking $\beta_1 = \beta_2 = \beta$.

Again, in equations (4.17), (4.19) and (4.21), the upper sign is to be taken when $k_{12}k_{21} > 0$ and the lower sign when $k_{12}k_{21} < 0$.

Generalization to multi-degrees-of-freedom systems

Consider now the n degrees-of-freedom system (4.1) Suppose the spectral density $S(\omega)$ of $\xi(t)$ has significant values only over a bandwidth $\Delta\omega_0$ around a frequency ω_0 and is zero outside this band, and $S(\omega) = O(\epsilon)$, $0 < |\epsilon| \ll 1$. Then the correlation time τ_c of the process $\xi(t)$ is $O(1/\Delta\omega_0)$ while the relaxation time τ_r of the response process $\mathbf{q}(t)$ is $O(1/\epsilon)$. Under this condition, the application of the stochastic averaging procedure is justified (Stratonovich, 1963).

From the results of the previous sections, the largest Lyapunov exponents for the n degrees-of-freedom system (4.1) may be deduced:

(i) $k_{ij}k_{ji} > 0$

If $\omega_0 = \omega_i + \omega_j$,

$$\lambda = -\min(\beta_i, \beta_j) + \frac{1}{8}k_{ij}k_{ji}S(\omega_i + \omega_j), \qquad (4.22)$$

where $\beta_i = \beta_{ii}$.

If $\omega_0 = |\omega_i - \omega_j|$,

$$\lambda = -\frac{1}{2}(\beta_i + \beta_j) - \frac{1}{8}k_{ij}k_{ji}S(\omega_i - \omega_j) + \frac{1}{2}(\beta_i - \beta_j)\coth\left[\frac{4(\beta_i - \beta_j)}{k_{ij}k_{ji}S(\omega_i - \omega_j)}\right]. \qquad (4.23)$$

(ii) $k_{ij}k_{ji} < 0$

If $\omega_0 = |\omega_i - \omega_j|$,

$$\lambda = -\min(\beta_i, \beta_j) - \frac{1}{8}k_{ij}k_{ji}S(\omega_i - \omega_j), \qquad (4.24)$$

If $\omega_0 = \omega_i + \omega_j$,

$$\lambda = -\frac{1}{2}(\beta_i + \beta_j) + \frac{1}{8}k_{ij}k_{ji}S(\omega_i + \omega_j) - \frac{1}{2}(\beta_i - \beta_j)\coth\left[\frac{4(\beta_i - \beta_j)}{k_{ij}k_{ji}S(\omega_i + \omega_j)}\right]. \qquad (4.25)$$

(iii) $\omega_0 = 2\omega_i$

$$\lambda = \lambda_i = -\beta_i + \frac{1}{8}k_{ii}^2 S(2\omega_i). \qquad (4.26)$$

This last result can also be obtained from equation (4.22) by taking $i = j$.

As an application, the flexural-torsional stability of a simply supported, uniform, narrow, rectangular, elastic beam subjected to a stochastically varying concentrated lateral load or end moments can be studied using the method developed here (Xie, 1990).

Acknowledgement

This research was supported by the Natural Sciences and Engineering Research Council of Canada through Grant No.A-1815.

References

Ariaratnam, S.T. (1977) Discussion to paper by Kozin and Sugimoto. *Stochastic Problems in Dynamics* (ed. B.L Clarkson; Pitman Press, London), 34.

Ariaratnam, S.T. (1990) Localization in randomly disordered periodic structures. *Proceedings of Workshop on Stochastic Structural Dynamics and Stability* (August 29, 1989, ed. G.I. Schuëller, Institute of Engineering Mechanics, University of Innsbruck, Austria), 41-55.

Ariaratnam, S.T. and Srikantaiah, T.K. (1978) Parametric instabilities in elastic structures under stochastic loading, *Journal of Structural Mechanics* 6(4), 349-365.

Ariaratnam, S.T. and Xie, Wei-Chau (1990) Lyapunov exponent and rotation number of a two-dimensional nilpotent stochastic system. *Dynamics and Stability of Systems* **5(1)**, 1-9.

Arnold, L. (1984) A formula connecting sample and moment stability of linear stochastic systems. *SIAM Journal on Applied Mathematics* **44**, 793-802.

Brillouin, L. (1946) *Wave Propagation in Periodic Structures: Electric Filters and Crystal Lattices* (McGraw-Hill, New York; reprinted 1953 by Dover Publications, New York).

Furstenberg, H. (1963) Noncommuting random products. *Transactions of the American Mathematical Society* **108**(3), 377-428.

Khas'minskii, R.Z. (1966) A limit theorem for the solutions of differential equations with random right-hand sides. *Theory of Probability and Its Applications* **11**, 390-406 (English translation).

Khas'minskii, R.Z. (1967) Necessary and sufficient conditions for the asymptotic stability of linear stochastic systems. *Theory of Probability and Its Applications* **12**, 144-147 (English translation).

Khas'minskii, R.Z. (1980) *Stochastic Stability of Differential Equations* (Sijthoff & Noordhoff) (English translation).

Kissel, G.I. (1988) *Localization in Disordered Periodic Structures* (PhD Thesis, Department of Aeronautics and Astronautics, Massachusetts Institute of Technology, USA).

Kozin, F. and Sugimoto, S. (1977) Relations between sample and moment stability for linear stochastic differential equations. *Proceedings of the Conference on Stochastic Differential Equations and Applications* (Park City, Utah, USA, 1976) (ed. J. David Mason; Academic Press, New York).

Lin, Y.K. (1976) Random vibration of periodic and almost periodic structures. *Mechanics Today* **3**, 93-124.

Molchanov, S.A. (1978) The structure of eigenfunctions of one-dimensional unordered structures. *Math USSR Izvestija* **12**, 69-101 (in Russian).

Oseledec, Y.I. (1968) A multiplicative ergodic theorem. Lyapunov characteristic numbers for dynamical systems. *Transactions of the Moscow Mathematical Society* **19**, 197-231 (English translation).

Wedig, W. (1988) Lyapunov exponents of stochastic systems and related bifurcation problems. *Stochastic Structural Dynamics - Progress in Theory and Applications* (eds S.T. Ariaratnam, G.I. Schuëller, I. Elishakoff; Elsevier, London), 315-327.

Wedig, W. (1989) Pitchfork and Hopf bifurcations in stochastic systems - effective methods to calculate Lyapunov exponents. *Effective Stochastic Analysis I* (eds P. Krée, W. Wedig; Springer-Verlag, Berlin) (to appear).

Yang, J.N. and Lin, Y.K. (1975) Frequency response functions of a disordered periodic beam. *Journal of Sound and Vibration* **38**, 317-340.

Xie, Wei-Chau (1990) *Lyapunov Exponents and Their Application in Structural Dynamics* (PhD Thesis, Department of Civil Engineering, University of Waterloo, Canada).

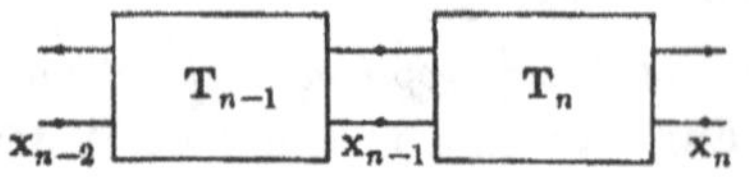

Figure 1 Transfer matrix

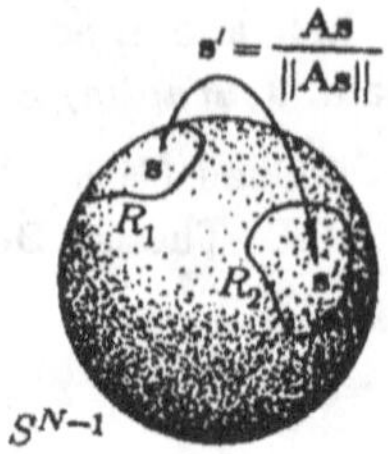

Figure 2 Markov chain on unit hypersphere S^{N-1}

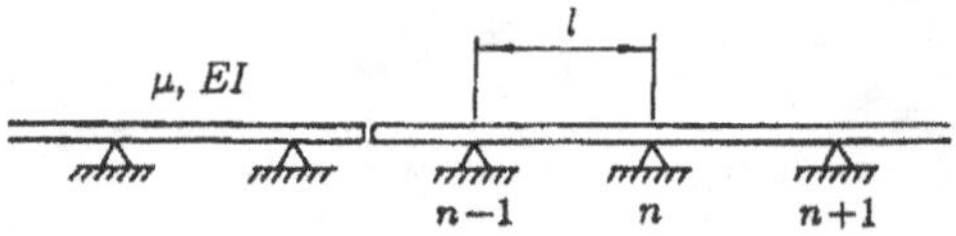

Figure 3 Long continuous beam on simple supports

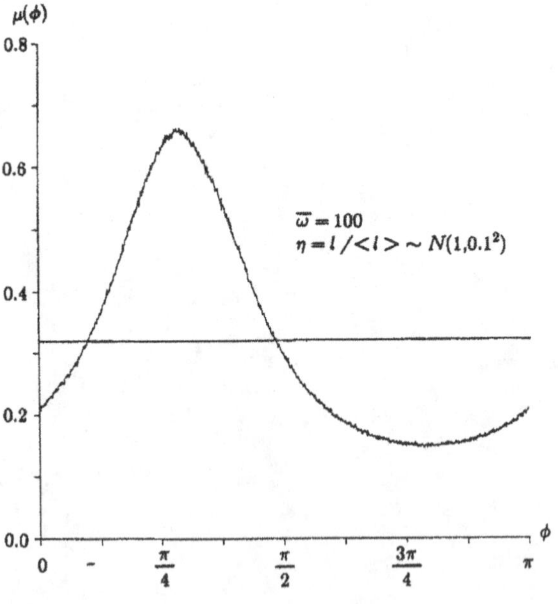

Figure 4 Invariant probability density $\mu(\phi)$

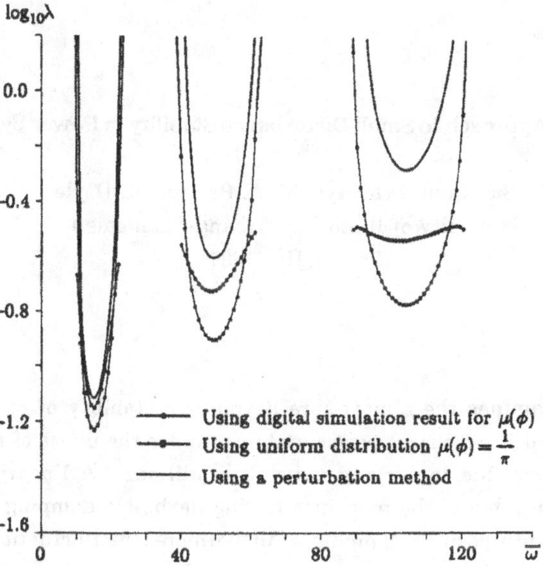

Figure 5 Localization factor for continuous beam

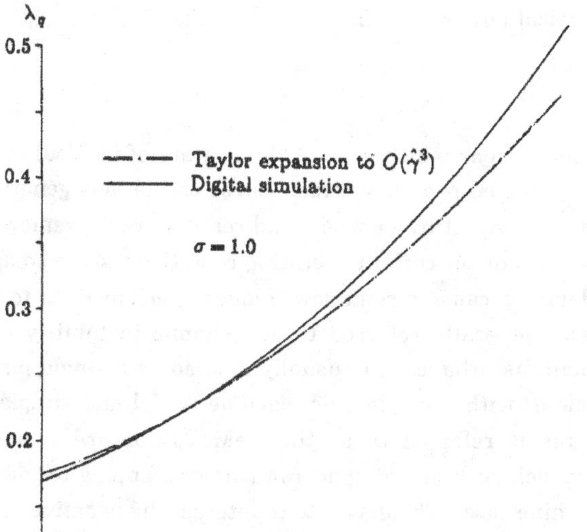

Figure 6 Largest Lyapunov exponent for

$$\ddot{q}-[\gamma+\sigma\xi(t)]\,q=0$$

Stochastic Approach to Small Disturbance Stability in Power Systems

N. Sri Namachchivaya, M. A. Pai and M. Doyle
University of Illinois at Urbana-Champaign
Urbana, IL 61801

Abstract

This paper examines the almost-sure asymptotic stability of coupled synchronous machines encountered in electrical power systems, under the effect of fluctuations in the interconnection system due to varying network conditions. A linearized multimachine model is assumed with one of the machines having negligible damping weakly coupled to the other machines with positive damping. Furthermore, the fluctuations are assumed to contain both harmonically varying and stochastically varying components. For small intensity excitations, the physical processes are approximated by a diffusive Markov process defined by a set of Itô equations. Results pertaining to the almost-sure asymptotic stability are derived using the maximal Lyapunov exponent obtained for the Itô equations. Assumptions made for the modeling and analysis are consistent with possible operating conditions in an electrical power system.

Introduction

Spontaneous oscillations which grow with time are of particular concern in some of the large scale interconnected power systems. These oscillations generally occur when the system has a longitudinal structure or when load centers are far removed from generation. It has been observed that under certain operating conditions, some small perturbation will result in negative damping causing some low frequency eigenvalues to cross into the R.H. plane. This problem, generally referred to as dynamic instability (or, more correctly, instability under small disturbances), is usually studied as a single machine connected to an infinite bus problem with a single axis machine model and simple voltage regulator. In the literature, this is referred to as the DeMello-Concordia model [1]. Concepts formulated for this model, such as net synchronizing or damping torque, cannot be carried over to the multi-machine case. The device to counteract the negative damping is known as the power system stabilizer which consists of lead-lag networks between the local frequency and input to voltage regulator. The multi-machine system is often studied with the simple classical model for analyzing location or coordination of the power system stabilizers. In this paper, the concept of almost sure stability is proposed as an application to the multi-machine case. The multi-machine system with the classical model is linearized and under the assumptions of randomly varying loads, the fluctuations in the parameters (synchronizing coefficients) of the multi-machine system are assumed to be a combination of both periodic functions and stochastic processes. These synchronizing coefficients are generally complex functions of continuously varying load conditions and stochasticity is the dominant feature of these parameters.

There are several probabilistic methods which have been presented in the literature for investigating the stability of stochastic dynamical systems. Most investigations have been concerned with the stability in moments and almost sure or sample stability. In practice, it is generally most desirable to examine the almost-sure sample stability and the results thus obtained hold true with probability one. It is clear that for first order Itô equations, the regions of stability for higher integer moments are included in the regions of stability of lower integer moments and, furthermore, all integer moment stability regions are included in the region of sample stability. Recently, an exact relation between conditions for sample stability and conditions of moment stability was established in Kozin and Sugimoto [2] and extended by Arnold [3]. It has been shown by Kozin and Sugimoto [2] that the sample stability region in some parameter space is a limit of the regions of the pth moment stability as $p \downarrow 0$. However, it is difficult to obtain sample stability in such a fashion, since the conditions for the pth moment stability are more difficult to obtain for arbitrary p in multidimensional multiplicative systems. For this reason, sample stability in this paper is obtained using a very well developed concept of Lyapunov characteristic exponents (see e.g., Khasminskii [4] and Arnold and Klieman [5]). Furthermore, it is well known that in regions where the system is stable with probability one, the second moments may grow exponentially. It is, therefore, important from the design point of view that the higher order moments decay in some parameter region. To this end, mean square stability conditions are also obtained in this work.

General results for almost sure and moment stability are obtained using the method of stochastic averaging. The mathematical model is somewhat analogous to that of Loparo and Blankenship [7], but with the difference that transfer conductances in the power system model are not neglected and parametric harmonic excitations are also included in this analysis.

System Model

Consider a multi-machine system of n synchronous machines. The classical model is represented for each ith machine as a voltage of constant magnitude E_i and varying phase angle δ_i behind the transient reactance x'_{di} of that machine. E_i and δ_i can be calculated from the load flow data. Let H_i (in secs) be the inertia constant of the machine on a common system base and P_{mi} is the mechanical input power in per unit. In the network, all loads are replaced by constant admittances based on the load flow data and all the passive nodes are eliminated excepting the nodes of the machines behind the transient reactances. This results in an admittance matrix Y with the elements $Y_{ij} = G_{ij} + j\,B_{ij}$ where

G_{ij} = transfer conductance between machines i and j ($i \neq j$)
= self conductance for i = j
B_{ij} = transfer susceptance between machines i and j ($i \neq j$)
= self susceptance for i = j

The nonlinear swing equations for each machine can be derived as [6]

$$\frac{H_i}{\pi f}\frac{d^2\delta_i}{dt^2} + D_i\frac{d\delta_i}{dt} = P_{mi} - E_i^{\,2}G_{ii} - \Big(\sum_{j\neq i}^{n} C_{ij}\sin\delta_{ij} + D_{ij}\cos\delta_{ij}\Big)\,,\quad i,j=1,2,\ldots,n \tag{1}$$

where D_i = damping of the ith machine, f is the frequency in Hz and

$$\delta_{ij} = \delta_i - \delta_j\,,\quad C_{ij} = E_i\,E_j\,B_{ij},\quad D_{ij} = E_i\,E_j\,G_{ij}$$

It can be shown that δ_i is also the rotor angle of the ith machine with respect to a synchronously rotating reference axis in electrical radians. More compactly, equation (1) can be written as

$$M_i\frac{d^2\delta_i}{dt^2} + D_i\frac{d\delta_i}{dt} = P_i - P_{ei} \tag{2}$$

where $M_i = H_i/\pi f$

$P_i = P_{mi} - E_i^2\,G_{ii}$

$$P_{ei} = \sum_{j\neq i}^{n}(C_{ij}\sin\delta_{ij} + D_{ij}\cos\delta_{ij})$$

The initial conditions for (2), $\delta_i(0) = \delta_i^o$, are calculated from the load flow data assuming constant voltage behind reactance and $\dot{\delta}_i(0) = 0$. Thus, δ_i^o is also the solution of real power equations

$$P_i - P_{ei} = 0\,,\quad i = 1,2,\ldots n$$

Thus, (2) can be written as

$$M_i\frac{d^2\delta_i}{dt^2} + D_i\frac{d\delta_i}{dt} = -\sum_{j\neq i}^{n}[C_{ij}(\sin\delta_{ij} - \sin\delta_{ij}^o) + D_{ij}(\cos\delta_{ij} - \cos\delta_{ij}^o)]$$

$$= F_i(\delta_1 \ldots \delta_n, \delta_1^o \ldots \delta_n^o) \tag{3}$$

Linearizing (3) around δ_i^o and introducing the incremental variable $\hat{\delta}_i = \delta_i - \delta_i^o$

$$M_i \frac{d^2}{dt^2} \hat{\delta}_i + D_i \frac{d}{dt} \hat{\delta}_i + \sum_{j=1}^{n} k_{ij} \hat{\delta}_i = 0 \tag{4}$$

where $$-k_{ij} = \left. \frac{\partial F_i}{\partial \delta_j} \right|_{\delta=\delta^o}$$

In power system analysis, $G_{ij}(i \neq j)$ represents the transfer conductance and is assumed small [6]. Moreover, the quantity $\delta_i^o - \delta_j^o = \delta_{ij}^o(i \neq j)$ is also small. Note that under the conditions $G_{ij} = 0$ $(i \neq j)$, $k_{ij} = k_{ji}$ $(i \neq j)$, i.e. the matrix $K = [k_{ij}]$ is symmetric. In general, K is not symmetric. However, since $C_{ij} = C_{ji}$ and $D_{ij} = D_{ji}$, K can be decomposed into a symmetric part and an asymmetric part where it is assumed that the non-symmetric components of k_{ij} $(i \neq j)$ are small for the reasons given above..

As indicated in the introduction, the elements k_{ij} are complex functions of the continuously but slowly varying network conditions, such as voltage magnitudes, phase angles, loads, etc. Hence, we can assume the elements of K to consist of an average part plus a fluctuating part. There are many situations in power systems in which the fuctuating part consists of cyclic loads [18] in addition to randomly varying loads [6]. It is worth noting that in Reference [7], only randomly varying loads are considered. Thus, we assume k_{ij} (i=1,...n, j=1,...n) as $k_{ij} = \overline{k}_{ij} + \Delta k_i$ where $\overline{k}_{ij}$ is the average part of the matrix K, and Δk_{ij} the incremental time varying part which may include contributions from the unsymmetric part also. If we wish to investigate the effects of time dependent fluctuations, we cast Eq. (4) in a decoupled form in matrix notation as follows:

$$M \frac{d^2 \hat{\delta}}{dt^2} + D \frac{d \hat{\delta}}{dt} + (\overline{K} + \Delta K) \hat{\delta} = 0 \tag{5}$$

where $M = \mathrm{Diag}\,(M_1 \ldots . M_n)$
$D = \mathrm{Diag}\,(D_1 \ldots . D_n)$
$\hat{\delta} = [\hat{\delta}_1 \ldots\ldots\ldots \hat{\delta}_n]^t$

$\overline{K} = \overline{K}_{sym} + \overline{K}_{asy}$, with $\overline{K}_{sym}$ symmetric and $\overline{K}_{asy}$ small.

Since $\overline{K}_{sym}$ is symmetric, Eq. (5) can be written in a decoupled normal form by the change of coordinates $x = Q\hat{\delta}$, as in Loparo and Blankenship [7], where Q = n x n orthogonal matrix. Assuming such a transformation, Eq. (5) becomes

$$\frac{d^2}{dt^2} x_i + 2\alpha_i \frac{dx_i}{dt} + \omega_i^2 x_i = \sum_{j=1}^{n} \Delta\hat{k}_{ij} x_j \tag{6}$$

The details of the transformation from Eq. (5) to Eq. (6) are omitted for the sake of brevity. Note that the fluctuating part of Δk_{ij} of Eq. (5) has been re-introduced in Eq. (6) in terms of the normal coordinates and now includes the symmetric terms from the matrix $\overline{K}$ which are small.

In physical terms, the elements of $\Delta\hat{K} = (\Delta\hat{k}_{ij})$ are due to the transfer conductance and susceptance terms which reflect the loads in the original network because of the constant impedance approximation [6]. The load variations in the network, which may be cyclic in nature [18], as reflected in Eq. (6), are assumed to be a combination of harmonically and stochastically varying parts in $\Delta\hat{k}_{ij}$. Thus, the final equations of motion can be written in the form

$$\ddot{x}_i + 2\alpha_i\dot{x}_i + \omega_i^2 x_i = \varepsilon^{1/2} A_{ij} x_j \xi(t) + \varepsilon\mu\, B_{ij} x_j \sin \upsilon t \tag{7}$$

where A_{ij} and B_{ij} are elements of constant matrices. In this equation, we shall assume, as in [6], that machine 1 has small damping compared to the rest of the machines. Introducing a detuning parameter λ to unfold the subharmonic resonance and new time τ such that $\tau = \upsilon t = \omega_o(1 - \varepsilon\lambda)t$, and putting $\alpha_i/\omega_o = \beta_i$, $\omega_i/\omega_o = \kappa_i$, $k_{ij} = A_{ij}/\omega_o{}^2$ and $p_{ij} = B_{ij}/\omega_o{}^2$, the equations can be rewritten as

$$\begin{aligned} \frac{du}{d\tau} &= C_o u + \kappa_1^{-1} [\varepsilon^{1/2} C_1 y\, \xi(t) + \varepsilon\mu\, C_2 y \sin\tau + \varepsilon\overline{C}_o u] \\ \frac{dv}{d\tau} &= D_o v + [\varepsilon^{1/2} D_1 y\, \xi(t) + \varepsilon\mu\, D_2 y \sin\tau + \varepsilon\overline{D}_o v] \end{aligned} \tag{8}$$

where

$$y = [u,v]^t,\ u_1 = x_1,\ u_2 = \kappa_1^{-1}\dot{x}_1 ;\ v_{2(i-1)-1} = x_i,\ v_{2(i-1)} = \dot{x}_i \quad i = 2,3,\ldots n$$

$$C_0 = \begin{bmatrix} 0 & \kappa_1 \\ -\kappa_1 & 0 \end{bmatrix},\ C_1 = \begin{bmatrix} 0 & 0 & 0 & 0 \ldots.. \\ k_{11} & 0 & k_{12} & 0 \ldots.. \end{bmatrix}_{2x2n}$$

$$\overline{C}_0 = \begin{bmatrix} 0 & 0 \\ -2\lambda\kappa_1^2 & -2\beta_1 \end{bmatrix}, \quad C_2 = \begin{bmatrix} 0 & 0 & 0 & 0 \dots \\ p_{11} & 0 & p_{12} & 0 \dots \end{bmatrix}_{2x2n}$$

$$D_0 = \mathrm{diag}\left\{\begin{bmatrix} 0 & 1 \\ -\kappa_j^2 & -2\beta_j \end{bmatrix}\right\}, \quad \overline{D}_0 = \mathrm{diag}\left\{\begin{bmatrix} 0 & 1 \\ -2\lambda\kappa_j & -4\lambda\beta_j \end{bmatrix}\right\} \quad j = 2,3,\dots n$$

$$D_1 = \begin{bmatrix} 0 & 0 & 0 & 0 & \dots\dots & 0 & 0 \\ k_{21} & 0 & k_{22} & 0 & & k_{2n} & 0 \\ 0 & 0 & 0 & 0 & \dots\dots & 0 & 0 \\ \vdots & \vdots & \vdots & \vdots & & \vdots & \vdots \\ k_{n1} & 0 & k_{n2} & 0 & \dots\dots & k_{nn} & 0 \end{bmatrix}$$

$$D_2 = \begin{bmatrix} 0 & 0 & 0 & 0 & \dots\dots & 0 & 0 \\ p_{21} & 0 & p_{22} & 0 & & p_{n2} & 0 \\ 0 & 0 & 0 & 0 & \dots\dots & 0 & 0 \\ \vdots & \vdots & \vdots & \vdots & & \vdots & \vdots \\ p_{n1} & 0 & p_{n2} & 0 & \dots\dots & p_{nn} & 0 \end{bmatrix}$$

In order to obtain a set of equations in "standard form", the above equations are transformed from u and v to the new variables a, φ and z by means of the following transformations:

$$u_1 = a \sin\Phi\,, \quad u_2 = a\cos\Phi\,, \quad \Phi = \kappa_1\tau + \varphi$$

$$v_{2j-1} = \left(\cos \kappa_{j+1}\tau + \frac{\beta_{j+1}}{\kappa_{j+1}} \sin \kappa_{j+1}\tau\right) z_{2j-1} + \sin \kappa_{j+1}\tau \frac{z_{2j}}{k_{j+1}}$$

$$v_{2j} = \left(-\kappa_{j+1} \sin \kappa_{j+1}\tau\right) z_{2j-1} + \left(\cos \kappa_{j+1}\tau - \frac{\beta_{j+1}}{\kappa_{j+1}} \sin \kappa_{j+1}\tau\right) z_2$$

This procedure yields a set of equations in the form

$$\frac{d}{d\tau}\begin{Bmatrix} a \\ \varphi \end{Bmatrix} = \frac{\varepsilon^{1/2}}{\kappa_1}\left[TC_1\begin{Bmatrix} u(a,\varphi,t) \\ v(z,t) \end{Bmatrix}\xi(t)\right] + \frac{\varepsilon}{\kappa_1}\left[\mu TC_2\begin{Bmatrix} u(a,\varphi,t) \\ v(z,t) \end{Bmatrix}\sin\tau\right] \tag{9}$$

$$+ \left[T\bar{C}_0 \begin{Bmatrix} a \sin \Phi \\ a \cos \Phi \end{Bmatrix} \right]$$

$$\frac{dz}{d\tau} = E_o Z + \varepsilon^{1/2} \left[SD_1 \begin{Bmatrix} u(a,\varphi,t) \\ v(z,t) \end{Bmatrix} \xi(t) \right] + \varepsilon \left[\mu SD_2 \begin{Bmatrix} u(a,\varphi,t) \\ v(z,t) \end{Bmatrix} \sin\tau + S\bar{D}_o\, v(z,t) \right] \tag{10}$$

where

$$T = \begin{bmatrix} \sin \Phi & \cos \Phi \\ \frac{\cos \Phi}{a} & -\frac{\sin \Phi}{a} \end{bmatrix}$$

$$S = \text{diag} \left\{ \begin{bmatrix} \cos\kappa_j - \frac{\beta_j}{\kappa_j} \sin \kappa_j\tau & -\frac{\sin \kappa_j\tau}{\kappa_j} \\ \kappa_j \sin \kappa_j\tau & \cos\kappa_j\tau + \frac{\beta_j}{\kappa_j} \sin \kappa_j \tau \end{bmatrix} \quad j = 2,3,\ldots n \right\}$$

The above equations contain both the rapidly oscillating flutter mode and decaying stable modes along with rapidly varying stochastic components. The asymptotic behavior of such a system is studied using the modified method of averaging due to Papanicoloau and Kohler [8]. This procedure, as in [7,9] yields a set of Itô equations for the amplitude a(t) and phase $\varphi(t)$ as

$$\begin{aligned} da &= m_a(a,\varphi)\, dt + \sigma_{aj}\, dw_j \\ d\varphi &= m_\varphi(a,\varphi)\, dt + \sigma_{\varphi j}\, dw_j \end{aligned} \tag{11}$$

where

$$m_a = \{ \frac{3}{16\kappa_1^2}\, k_{11}^2\, S_{\xi\xi}(2\kappa_1) - \sum_{j=2}^{n} \frac{k_{1j}k_{j1}}{8\kappa_1\kappa_j} \left\{ \hat{S}_{\xi\xi}(\Omega_{1j}^-) - \hat{S}_{\xi\xi}(\Omega_{1j}^+) \right\}$$

$$- \frac{\beta_1}{\kappa_1} + \mu \frac{p_{11}}{4\kappa_1} \cos 2\varphi \}\, a$$

$$m_\varphi = \left\{ - \frac{k_{11}^2}{8\kappa_1^2}\, \psi_{\xi\xi}(2\kappa_1) - \sum_{j=2}^{n} \frac{k_{1j}k_{j1}}{8\kappa_1\kappa_j} \{ \hat{\psi}_{\xi\xi}(\Omega_{1j}^+) - \hat{\psi}_{\xi\xi}(\Omega_{1j}^-) \} \right\}$$

$$+ \lambda\kappa_1 - \mu \frac{p_{11}}{4\kappa_1} \sin 2\varphi$$

$$[\sigma\sigma^T]_{aa} = \frac{a^2 k_{11}^2}{8\kappa_1^2} S_{\xi\xi}(2\kappa_1), \quad [\sigma\sigma^T]_{\varphi\varphi} = \frac{k_{11}^2}{8\kappa_1^2} [2S_{\xi\xi}(0) + S_{\xi\xi}(2\kappa_1)]$$

and

$$S_{\xi\xi}(\kappa) = 2\int_0^\infty R_{\xi\xi}(\tau) \cos\kappa\tau \, d\tau, \qquad \psi_{\xi\xi}(\kappa) = 2\int_0^\infty R_{\xi\xi}(\tau) \sin\kappa\tau \, d\tau$$

$$\widehat{S}_{\xi\xi}(\Omega_{ij}^{\pm}) = 2\int_0^\infty \{e^{-\beta_j\tau} R_{\xi\xi}(\tau) \cos(\Omega_{ij}^{\pm}\tau)\} \, d\tau, \quad R_{\xi\xi}(t) = E[\xi(t)\xi(t+\tau)]$$

$$\widehat{\psi}_{\xi\xi}(\Omega_{ij}^{\pm}) = 2\int_0^\infty \{e^{-\beta_j\tau} R_{\xi\xi}(\tau) \sin(\Omega_{ij}^{\pm}\tau)\} \, d\tau, \quad \Omega_{ij}^{\pm} = \kappa_i \pm \kappa$$

Almost-Sure Asymptotic Stability

In this section, a very well developed theory of almost sure stability using the concept of Lyapunov exponents will be utilized. Lyapunov exponents are a generalization of the characteristic exponent (Floquet exponents defined for periodic orbits) so that more general nonperiodic orbits can be characterized. These are indeed the average exponential rates of divergence or convergence of nearby orbits in phase space. A brief description of this theory is given in the Appendix.

Several investigators have invoked this theory in the study of stochastic differential equations. Kozin and Prodromou [11] and Nishioka [12], for example, applied the necessary and sufficient condition given by Eq. (A8) to some second-order Itô differential equations. In this paper, for the first time, Khasminskii's theory is applied to determine the top Lyapunov exponent for a multidegree-of-freedom system under both harmonic and stochastic excitations.

By making the change of variables, $u = \ln |a|$, $\psi = 2(\phi)$, the Itô stochastic differential equations for u and ψ are obtained using the Itô differential rule as

$$du = [-\rho + \mu^* \cos\psi] \, d\tau + \frac{k_{11}}{2\kappa_1}\left(\frac{S_{\xi\xi}(2\kappa_1)}{2}\right)^{1/2} dw_u \tag{12}$$

$$d\psi = [\Delta - 2\mu^* \sin\psi] \, d\tau + (2D)^{1/2} \, dw_\psi \tag{13}$$

where

$$\Delta = 2\kappa_1\lambda - \left(\frac{k_{11}}{2\kappa_1}\right)^2 \psi_{\xi\xi}(2\kappa_1) - \sum_{j=2}^{n} \frac{k_{1j}k_{j1}}{4\kappa_1 \kappa_j} [\hat{\psi}_{\xi\xi}(\Omega_{1j}^{+}) - \hat{\psi}_{\xi\xi}(\Omega_{1j}^{-})]$$

$$D = \frac{k_{11}^2}{4\kappa_1} [2S_{\xi\xi}(0) + S_{\xi\xi}(2\kappa_1)]$$

$$\rho = \frac{\beta_1}{\kappa_1} - \frac{1}{2}\left(\frac{k_{11}}{2\kappa_1}\right)^2 S_{\xi\xi}(2\kappa_1) + \sum_{j=2}^{n} \frac{k_{1j}k_{j1}}{8\kappa_1\kappa_j} [\hat{S}_{\xi\xi}(\Omega_{1j}^{-}) - \hat{S}_{\xi\xi}(\Omega_{1j}^{+})]$$

$$\mu^* = \frac{\mu p_{11}}{4\kappa_1}$$

It is clear that the stochastic process $\psi(t)$ generated on a circle by Eq. (13) is Markov and since the diffusion process is non-singular, it is ergodic on the entire circle [13]. The invariant measure of the process $\psi(t)$ is governed by

$$\frac{\partial^2 p}{\partial \psi^2} - \frac{\partial}{\partial \psi} [(\bar{\Delta} - 2\bar{\mu} \sin\psi)p] = 0 \tag{14}$$

where

$$\bar{\Delta} = \Delta/D \quad \text{and} \quad \bar{\mu} = \mu^*/D$$

The unique solution satisfying both the periodicity condition and normality condition

$$p(\phi) = p(\phi + 2\pi) \quad \text{and} \quad \int_0^{4\pi} p(\psi)\, d\psi = 1$$

can be written as [14]

$$p(\psi) = \frac{\exp[\bar{\Delta}(\psi+\pi) + 2\bar{\mu}\cos\psi]}{8\pi^2 I_{i\bar{\Delta}}(-2\bar{\mu})\, I_{-i\bar{\Delta}}(-2\bar{\mu})} \int_{\psi}^{\psi+2\pi} \exp[-(\bar{\Delta}x + 2\bar{\mu}\cos x)]\, dx \tag{15}$$

Now by integrating the equation for u from 0 to t and using the fact that ψ is ergodic in $[0,4\pi]$, yields

$$\Lambda = \lim_{t\to\infty} \frac{1}{t}(u(t) - u(0)) = \lim_{t\to\infty} \frac{1}{t} \ln \left\| \frac{a(t)}{a(o)} \right\|$$

$$= -\rho + \lim_{t\to\infty} \frac{1}{t} \int_0^t \mu^* \cos\psi\, d\tau = -\rho + \mu^* E[\cos\psi] \tag{16}$$

where Λ is the maximal Lyapunov exponent. Then, according to Khasminskii [13], if Λ exists, then for any arbitrary $x_o(\neq 0)$, the system is asymptotically stable when $\Lambda < 0$. The value of Λ is obtained as

$$\Lambda = -\rho + \mu^* \int_0^{4\pi} \cos\psi\, p(\psi)\, d\psi$$

The integral part of Λ is denoted by J, i.e.

$$J = \int_0^{4\pi} \cos\psi\, p(\psi)\, d\psi = \left(\frac{1}{N}\right) \int_0^{4\pi} \left\{ \int_\psi^{\psi+2\pi} \cos\psi \exp\{-[\bar{\Delta}(x-\psi) - 4\bar{\mu} \sin\frac{(x+\psi)}{2} \sin\frac{(x-\psi)}{2}]\}\, dx \right\} d\psi$$

where N is the normalization constant of the invariant measure and is given by

$$N = 8\pi^2 \exp[-\bar{\Delta}\pi] \cdot I_{i\bar{\Delta}}(-2\bar{\mu}) I_{-i\bar{\Delta}}(-2\bar{\mu})$$

Introducing the variable $\tau = x-\psi$, and changing the order of integration, we obtain, as in [13],

$$J = -\frac{4\pi}{N} \int_o^{2\pi} \sin\frac{\tau}{2}\, I_1\left(-4\bar{\mu} \sin\frac{\tau}{2}\right) \exp[-\bar{\Delta}\tau]\, d\tau$$

Letting $y = 1/2\,(\pi-\tau)$ for $0 < \tau < \pi$ and $y = 1/2\,(\tau-\pi)$ for $\pi < \tau < 2\pi$, J reduces to

$$J = -\frac{8\pi}{N} \exp(-\bar{\Delta}\pi) \left\{ \int_o^{\pi/2} \cos[(1-i2\bar{\Delta})y]\, I_1(-4\bar{\mu}\cos y)\, dy + \int_o^{\pi/2} \cos[(1+i2\bar{\Delta})y]\, I_1(-4\bar{\mu}\cos y)\, dy \right\}$$

Once again, making use of formulas given in Gradshteyn and Ryzhik [16], the expression for Λ simplifies to

$$\Lambda = -\rho - \frac{\mu^*}{2} \left\{ \frac{I_{1-i\bar{\Delta}}(-2\bar{\mu})}{I_{-i\bar{\Delta}}(-2\bar{\mu})} + \frac{I_{1+i\bar{\Delta}}(-2\bar{\mu})}{I_{i\bar{\Delta}}(-2\bar{\mu})} \right\} \quad (17)$$

In the above expression, κ_1 can be eliminated by using the fact that near subharmonic resonances $\kappa_1 = 1/2$. Then the condition for almost-sure asymptotic stability is given in terms of the original system parameters as

$$2k_{11}^2 S_{\xi\xi}(1) - \mu\, p_{11} \left\{ \frac{I_{1-\bar{\Delta}}(-2\bar{\mu})}{I_{-\bar{\Delta}}(-2\bar{\mu})} + \frac{I_{1+i\bar{\Delta}}(-2\bar{\mu})}{I_{i\bar{\Delta}}(-2\bar{\mu})} \right\}$$

$$< \; 8\beta_1 + \sum_{j=2}^{n} \frac{k_{1j}k_{j1}}{\kappa_j} \left\{ \hat{S}_{\xi\xi}(\Omega_{1j}^-) - \hat{S}_{\xi\xi}(\Omega_{1j}^+) \right\}$$

where

$$\bar{\Delta} = \frac{\lambda - k_{11}^2 \psi_{\xi\xi}(1) - \sum_{j=2}^{n} (k_{1j}k_{j1}/2\kappa_j) \{\hat{\psi}_{\xi\xi}(\Omega_{1j}^+) - \hat{\psi}_{\xi\xi}(\Omega_{1j}^-)\}}{(k_{11}^2/2)\,[2S_{\xi\xi}(0) + S_{\xi\xi}(1)]}$$

$$\bar{\mu} = \frac{\mu\, p_{11}}{k_{11}^2\,[2S_{\xi\xi}(0) + S_{\xi\xi}(1)]}$$

The above results can be further simplified for the special case where the detuning parameters and the sine spectrum (broad band noise) are taken to be identically zero, i.e., $\bar{\Delta} = 0$. For this case, the invariant measure $p(\psi)$ and stability conditions reduce to

$$p_{sp}(\psi) = \exp\,[2\bar{\mu} \cos \psi] \,/\, 4\pi\, I_o(-2\bar{\mu})$$

and

$$k_{11}^2 S_{\xi\xi}(1) - \mu\, p_{11} \left\{ \frac{I_1(-2\bar{\mu})}{I_o(-2\bar{\mu})} \right\} < 4\,\beta_1 + \sum_{j=2}^{n} \frac{k_{1j}k_{j1}}{2\kappa_j} \{\hat{S}_{\xi\xi}(\Omega_{1j}^-) - \hat{S}_{\xi\xi}(\Omega_{1j}^+)\}$$

respectively.

Moment Stability

It is well known that even though sample solutions may be stable with probability one, the mean square response of the system for the same parameter values may grow exponentially. However, the connection between sample and moment stabilities was established by Kozin and Sugimoto [2] for first and second order linear stochastic systems and extended by Arnold [3] using the Lyapunov exponent for the pth mean. According to this, the "sample stability criteria will include samples that are stable in some p^{th} moment, no matter how small p may be. This implies that for certain regions the sample solutions may decay very slowly (especially for small p), which certainly would not be useful from the view of applications to the design of stable systems [2]". It is, therefore, important to consider higher order moments to specify a certain rate of decay. To this end, mean square stability is considered in this section.

In order to obtain the moment stability, the Itô equations in a and ϕ are transformed to rectangular coordinates z_1 and z_2 by means of the transformation

$$z_1 = a \sin \phi, \quad z_2 = a \cos \phi$$

Once again making use of Itô's differential rule, yields

$$dz_1 = [-(\frac{p_{11}}{2} + \rho^*) z_1 + (\lambda/2 - a) z_2] dt + \sigma_{11}(z) dw_1 + \sigma_{12}(z) dw_2$$

$$dz_2 = [-(\lambda/2 - a) z_1 + (\frac{p_{11}}{2} - \rho^*) z_2] dt + \sigma_{21}(z) dw_1 + \sigma_{22}(z) dw_2$$

where

$$\rho^* = 2\beta_1 + \sum_{j=2}^{n} \frac{k_{1j} k_{j1}}{4\kappa_j} [\hat{S}_{\xi\xi}(\Omega_{1j}^-) - \hat{S}_{\xi\xi}(\Omega_{1j}^+)] + \frac{k_{11}^2}{2} [S_{\xi\xi}(0) - S_{\xi\xi}(1)]$$

$$a = k_{11}{}^2 \psi_{\xi\xi}(1) + \sum_{j=2}^{n} \frac{k_{1j} k_{j1}}{4\kappa_j} \left[\hat{\psi}_{\xi\xi}(\Omega_{1j}^+) - \hat{\psi}_{\xi\xi}(\Omega_{1j}^-)\right]$$

The differential equations governing the second moments can be found by applying the Itô differential rule to the quantities z_1^2, z_2^2 and $z_1 z_2$. This procedure yields three ordinary differential equations for the moments $E[z_1^2]$, $E[z_2^2]$ and $E[z_1 z_2]$. From these equations the condition for mean square stability can be obtained as

$$\rho > 2k_{11}^2 S_{\xi\xi}(1)$$

$$p_{11} < \{(\lambda - 2a)^2 + [\rho + k_{11}{}^2 (2S_{\xi\xi}(0) - S_{\xi\xi}(1))]^2\} \frac{\rho - 2 k_{11} S_{\xi\xi}(1)}{[\rho + k_{11}^2 (2S_{\xi\xi}(0) - S_{\xi\xi}(1))]}$$

where

$$\rho = 4\beta_1 + \sum_{j=2}^{n} \frac{k_{1j} k_{j1}}{2\kappa_j} [\hat{S}(\Omega_{1j}^-) - \hat{S}(\Omega_{1j}^+)]$$

Conclusions

In this paper we have proposed an approach for small disturbance stability of multi-machine systems with randomly varying load conditions. In the reduced model of the power system, we assume that these load variables contain a combination of harmonically varying and stochastically varying components.

Both the method of extended stochastic averaging and the concept of Lyapunov exponents have been utilized to study the almost-sure stability of power systems under both harmonic and stochastic parametric fluctuations. The extended method of averaging is used to reduce the multi-degree of freedom system to a set of Itô equations in amplitude and phase. Using the fact that the Itô equation for the logarithm of the amplitude contains only the phase angle, the method of Lyapunov exponents leads to an almost sure asymptotic stability condition which depends on the amplitude of the harmonic excitation, the spectral density of the stochastic excitation and the detuning parameter. It should be mentioned that even though the approximate Itô equations are stable with probability one, the original system may or may not be almost surely stable due to the weak convergence of the averaged equations. However, the method presented and illustrated here offers a new way of testing almost sure stability of dynamic systems under both harmonic and stochastic excitations.

Acknowledgments

This research was partially supported by the National Science Foundation (NSF) through Grant MSS-90-57437. The second author also wishes to acknowledge the support of NSF through Grant ECS-87-19055.

Appendix

Both in the study of linear multiplicative stochastic differential equations and in the investigation of the stability of the steady state solution of a nonlinear stochastic system, one encounters

$$\dot{x} = A(t)x\,, \quad x(0) = x_o \qquad x\,\varepsilon R^n \tag{A1}$$

where A(t) is a stationary stochastic process. Then, according to Oseledec (see e.g., Arnold and Klieman [5]), the exponential growth rate (i.e. Lyapunov exponent) of the corresponding solution $x(t;\,x_o)$ of equation (A1) is given by

$$\lambda\,(x_o) = \lim_{t \to \infty} \frac{1}{t} \log |x(t;x_o)| \quad \text{a.s.} \tag{A2}$$

Moreover, if A(t) is ergodic, the random variable $\lambda(x_o)$ can take on any finitely many non-random values

$$\lambda_{min} = \lambda_p < \lambda_{p-1} < .... < \lambda_1 = \lambda_{max} \tag{A3}$$

with possible multiplicities. The original results for the white noise multiplicative excitation were obtained by Khasminskii [4] by projecting the system in R^n onto a sphere S^{n-1}. Equation (A1) for multiplicative white noise excitation can be converted, using the Wong and Zakai [10] correction, into a set of Itô equations of the form

$$dx_i = \sum_{j=1}^{n} b_i^j x_j dt + \sum_{r=1}^{m} \sum_{j=1}^{n} \sigma_{ir}^j x_j \, dw_r \tag{A4}$$

where b_i^j are constants after the correction and w_r are mutually independent Wiener processes. By making the change of variables $s = x/||x||$ and $\rho = \log||x||$, where $||x||$ denotes the Euclidean norm $(\Sigma\, x_i{}^2)^{1/2}$, Khasminskii showed that the n-dimensional Markov process generated by Eq. (A4) is mapped onto the surface of an n-dimensional sphere. Using Itô's formula the stochastic differential equation for ρ can be written as

$$d\rho = Q(s)dt + \sum_{r=1}^{n} s^T\sigma(r)s \; d\xi_r(t) \tag{A5}$$

where $\sigma(r) = \sigma_{ij}^j$ (i=1,2,...,m), Q(s) is given by

$$Q(s) = s^T B s + \frac{1}{2} \operatorname{tr}A(s) - s^T A(s) s, \quad A(s) = \sum_{\substack{k=1 \\ t=1}}^{m} \sigma_{ir}^{k} \sigma_{jr}^{t} s_k s_t \tag{A6}$$

Substituting (A6) into (A5), integrating the resulting equation and dividing by t, one obtains

$$\frac{\rho(t) - \rho(0)}{t} = \frac{1}{t} \int_0^t (Q\, s(\tau))\, d\tau + \frac{1}{t} \int_0^t \sum_{r=1}^{n} [\sigma_r(\tau)s, s]\, d\xi_r(\tau) \tag{A7}$$

For the ergodic case, one has

$$\lim_{t \to \infty} \frac{\rho(t) - \rho(0)}{t} = E[Q(s)] \quad \text{w.p.1} \tag{A8}$$

where the expectation is over the invariant measure of the s-process. Equation (A8) yields a necessary and sufficient condition for almost sure sample asymptotic stability in terms of E[Q(s)]. If E[Q(s)] is negative, the samples are stable with probability one.

References

1. F. P. deMello and C. Concordia, 1969. IEEE Trans. Power Appar. Syst., pp. 316-329. Concepts of synchronous machine stability as affected by excitation control.

2. F. Kozin and S. Sugimoto, 1977. Proceedings of the Conference on Stochastic Differential Equations and Applications, ed: J. David Mason, Academic Press, New York. Relations between sample and moment stability for linear stochastic differential equations.

3. L. Arnold, 1984. SIAM J. of Applied Mathematics, Vol. 44, No. 4, pp. 793-802. A formula connecting sample and moment stability of linear stochastic systems.

4. R. Z. Khasminiskii, 1967. Theory of Probab. Appl., Vol. 12, pp. 144-147. Necessary and sufficient conditions for the asymptotic stability of linear stochastic systems.

5. L. Arnold and W. Kliemann, 1983. Probabilistic Analysis and Related Topics, Vol. 3, ed. A.T. Bharucha-Reid, Academic Press, New York, Qualitative theory of stochastic systems.

6. M. A. Pai, Energy Function Analysis for Power System Stability, Kulwer Academic Publisher, Boston, 1989.

7. K. A. Laparo and G. L. Blankenship, 1985. IEEE Transactions on Circuits and Systems, Vol. 32, No. 2, pp. 177-184. A probabilistic mechanism for small disturbance instabilities in electric power systems.

8. G. C. Papanicolaou and W. Kohler, 1976. Communications in Mathematical Physics, Vol. 46, pp. 217-232. Asymptotic analysis of deterministic and stochastic equations with rapidly varying components.

9. N. Sri Namachchivaya and Y. K. Lin, 1988. Probabilistic Engineering Mechanics, Vol. 3, No. 3, pp. 159-167. Application of stochastic averaging for nonlinear systems with high damping.

10. E. Wong and M. Zakai, 1965. International Journal of Engineering Sciences, Vol. 3, pp. 213-229. On the relationship between ordinary and stochastic differential equations.

11. F. Kozin and S. Prodromou, 1971. SIAM Journal of Applied Mathematics, Vol. 21, pp. 413-424. Necessary and sufficient conditions for almost-sure sample stability of linear Itô equation.

12. K. Nishioka, 1976. Kodai Math. Sem. Rep., 27, pp. 211-230. On the stability of two dimensional linear stochastic systems.

13. R. Z. Khasminiskii, 1980. Stochastic Stability of Differential Equations. Sijthoff and Noordhoff.

14. R. L. Stratonovich, 1967. Topics in the Theory of Random Noise, Vol. II, Gordon and Breach.

15. N. Sri Namachchivaya, 1989. J. of Sound and Vibration, Vol. 132, pp. 301-314. Mean square stability of a rotating shaft under combined harmonic and stochastic excitations.

16. I. S. Gradshteyn and I.M. Ryzhik, 1980. Table of Integrals Series, and Products. Academic Press.

17. R.R. Mitchell and F. Kozin, 1979. SIAM J. of Applied Mathematics, Vol. 27, No. 4, pp. 571-606. Sample stability of second order linear differential equation with wide band noise coefficients.

18. J. A. Pinello and J. E. Van Ness, 1971. IEEE Transactions on Power Apparatus and Systems, Vol. PAS-90, pp. 1856-1862. Dynamic response of a large power system to a cyclic load produced by nuclear reactor.

LYAPUNOV EXPONENTS AND INVARIANT MEASURES OF EQUILIBRIA AND LIMIT CYCLES

W. Wedig, University of Karlsruhe

Abstract:

Khasminskii's projection on circles, spheres or hyperspheres leads to the top Lyapunov exponents of dynamic systems. Provided there exists an invariant measure, the multiplicative ergodic theorem of Oseledec can be reduced to a finite integral on the projection angles. This technique is demonstrated by non-linear deterministic systems with self-exciting terms and by linear systems with parametric excitations by white noise. The paper emphasizes different numerical methods to solve Liouville or Fokker-Planck equations and to determine the invariant measures of dynamic systems.

1. Simulation of stochastic bifurcations

To motivate research in bifurcation and stability problems, we consider the mechanical system of a uniform beam under axial excitations $u(t)$. According to figure 1, the beam model possesses the length l, the bending stiffness EI, the axial stiffness EA, the mass μ per unit length and the external viscous damping with the coefficient β.

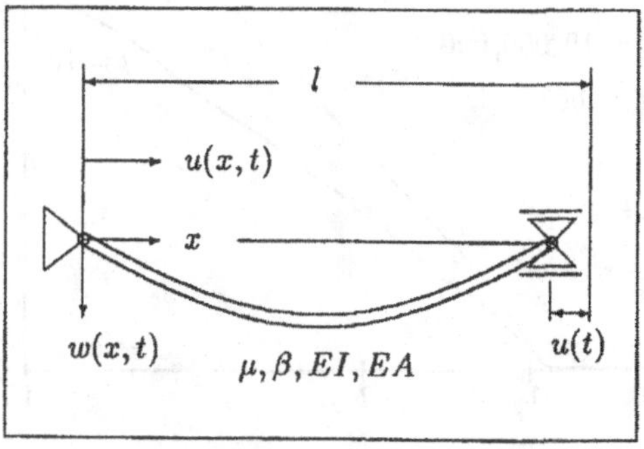

Figure 1: Beam under axial excitations

Following Weidenhammer (1969), the transverse vibration $w(x,t)$ of the beam is described by the following boundary problem.

$$EIw_{xxxx} + \beta w_t + \mu w_{tt} - \frac{1}{l}EA\,[u(t) + \frac{1}{2}\int_0^l w_{xx}^2 dx]\, w_{xx} = 0, \tag{1}$$

$$w(0,t) = w(l,t) = w_{xx}(0,t) = w_{xx}(l,t) = 0, \qquad 0 \le x \le l. \tag{2}$$

Herein, subscripts t and x denote partial derivations of $w(x,t)$ with respect to the time t and to the length coordinate x, respectively. The governing equation of motion (1) provides

Kirchhoff's assumption that the longitudinal waves $u(x,t)$ can approximately be replaced by the end displacement $u(t)$ of the beam. The homogeneous boundary conditions (2) are valid for the special case of simply hinged supports at both ends of the beam. Applying the first mode approximation $w(x,t) = T(t)\sin \pi x/l$, the boundary conditions (2) are satisfied and the partial differential equation (1) is reduced to an ordinary one of the following form.

$$w(x,t) = T(t)\sin \pi x/l, \qquad \omega_1^2 = \pi^4 EI/(\mu l^4), \tag{3}$$

$$\ddot{T}(t) + 2D\omega_1\dot{T}(t) + \omega_1^2\,[1 + \frac{Al}{I\pi^2}u(t) + \gamma T^2(t)]\,T(t) = 0. \tag{4}$$

Herein, dots denote time derivatives applied to the mode function $T(t)$, ω_1 is the first natural frequency of the free beam vibrations, D is a dimensionless damping measure and γ characterizes the cubic restoring of the beam. In the stochastic case, the given end displacement $u(t)$ can be modelled by stationary white noise $\dot{W}_t$ normalized by the intensity parameter σ. From this it follows that $T(t)$ is also a stochastic process. It is replaced by X_t where subscript t denotes the time dependency. Thus, the equation (4) reads as follows.

$$\ddot{X}_t + 2D\omega_1\dot{X}_t + \omega_1^2[1 + \sigma\dot{W}_t + \gamma X_t^2]X_t = 0, \tag{5}$$

$$E(\dot{W}_t) = 0, \quad E(\dot{W}_t\dot{W}_s) = \delta(t-s), \quad t \geq 0. \tag{6}$$

The bifurcation equation (5) possesses two different solutions: The trivial solution or equilibrium $X_t \equiv 0$ and the bifurcated solution $X_t^2 > 0$.

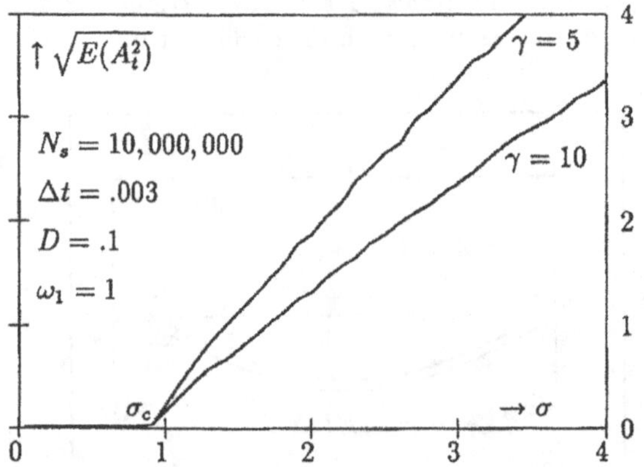

Figure 2: Simulated bifurcation diagram

Before going into more details, we perform numerical experiments to show basic effects of this bifurcation problem. For this purpose, we follow the concept of Khasminskii (1967) by introducing polar coordinates via the amplitude process A_t and the phase process Ψ_t.

$$A_t = \sqrt{X_t^2 + (\dot{X}_t/\omega_1)^2}, \qquad \Psi_t = \arctan \dot{X}_t/(\omega_1 X_t). \tag{7}$$

We achieve this transformation by calculating the increments of the phase Ψ_t and of the natural logarithm of the amplitude process A_t.

$$\begin{aligned} d\Psi_t &= -\omega_1(1 + 2D\sin\Psi_t\cos\Psi_t)dt - \omega_1\sigma\cos^2\Psi_t dW_t \\ &\quad -\omega_1\gamma A_t^2\cos^4\Psi_t dt - \omega_1^2\sigma^2\sin\Psi_t\cos^3\Psi_t dt, \qquad (8) \\ d\log A_t &= -2D\omega_1\sin^2\Psi_t dt - \omega_1\gamma A_t^2\sin\Psi_t\cos^3\Psi_t dt \\ &\quad +\tfrac{1}{2}\omega_1^2\sigma^2\cos^2\Psi_t\cos 2\Psi_t dt - \tfrac{1}{2}\omega_1\sigma\sin 2\Psi_t dW_t. \qquad (9) \end{aligned}$$

Herein, W_t is the normalized Wiener process. It is normally distributed with zero mean. It possesses uncorrelated increments with the mean square $E(dW_t^2) = dt$. The $(\omega_1\sigma)^2$ - terms in the equations (8) and (9) follow from the application of Ito's calculus. More details of this calculus are given e.g. by Arnold (1974).

Applying a forward Euler scheme (Wedig, 1988) to the stochastic differential equations (8) and (9), both processes Ψ_t and A_t are simulated for the system data $\omega_1 = 1s^{-1}$ and $D = .1$ in order to obtain the root mean square of the stationary amplitude in dependence on the intensity σ of the parametric white noise excitation. The time step selected was $\Delta t = .003s$ applied for $N_s = 10{,}000{,}000$ sample points. Figure 2 shows results obtained for the two different restoring parameters $\gamma = 5$ and $\gamma = 10$. According to the plotted results, there is a critical intensity parameter or bifurcation point σ_c where the trivial solution $A_t \equiv 0$ bifurcates into stationary non-trivial amplitudes $A_t^2 > 0$. In particular, we recognize in the figure above that the bifurcation point is not influenced by the γ - parameter of the cubic system restoring. For decreasing parameters $\gamma \to 0$, the bifurcation curve of the amplitude becomes vertical indicating an exponential growth in this linear case.

2. Multiplicative white noise systems

To investigate the stability of the equilibrium solution $A_t \equiv 0$, the phase equation (8) and the log - amplitude equation (9) are linearized with respect to $A_t << 1$. Normalized by $\omega_1 = 1$, both equations take the following forms.

$$\begin{aligned} d\Psi_t &= -[1 + (2D + \sigma^2\cos^2\Psi_t)\sin\Psi_t\cos\Psi_t]dt - \sigma\cos^2\Psi_t dW_t, \qquad (10) \\ d\log A_t &= -[2D\sin^2\Psi_t - \tfrac{1}{2}\sigma^2\cos^2\Psi_t\cos 2\Psi_t]dt - \tfrac{1}{2}\sigma\sin 2\Psi_t dW_t. \qquad (11) \end{aligned}$$

The linear amplitude equation (11) can easily be integrated. Inserted into the multiplicative ergodic theorem of Oseledec (1968), the top Lyapunov exponent λ is obtained as

$$\lambda = \lim_{t\to\infty}\frac{1}{t}\log\frac{A_t}{A_0} = -\lim_{t\to\infty}\frac{1}{t}\int_0^t [2D\sin^2\Psi_\tau - \tfrac{1}{2}\sigma^2\cos^2\Psi_\tau\cos 2\Psi_\tau]d\tau. \qquad (12)$$

Herein, A_0 denotes the initial value of the amplitude. Its asymptotic growth is completely determined by the phase equation (10). If there exists an invariant measure of the phase, the time average in (12) can be replaced by the corresponding ensemble average.

$$\lambda = -\int_{-\pi/2}^{\pi/2}[2D\sin^2\psi - \tfrac{1}{2}\sigma^2\cos^2\psi(\cos^2\psi - \sin^2\psi)]p(\psi)d\psi. \qquad (13)$$

Thus, the multiplicative ergodic theorem (12) is reduced to the finite integral (13) defined on the angle range $|\psi| \leq \pi/2$ and calculable by the density $p(\psi)$ of the invariant measure.

2.1. Invariant measures

The invariant measure of the phase process determines the top Lyapunov exponent of the trivial solution $A_t \equiv 0$. The phase represents a one-dimensional diffusion process rotating clockwise on the unit circle. Its density $p(\psi)$ is given by the Fokker-Planck equation associated with (10). In the stationary case, this diffusion equation is integrated with respect to the angle variable ψ leading to a linear first order differential equation for the density $p(\psi)$ of the invariant angle measure.

$$\tfrac{1}{2}\sigma^2 \cos^4 \psi\, p'(\psi) + [1 + (2D - \sigma^2 \cos^2 \psi) \sin \psi \cos \psi]\, p(\psi) = C, \qquad (14)$$

$$-\pi/2 \leq \psi \leq \pi/2, \qquad p(\pm\pi/2) = C, \qquad \int_{-\pi/2}^{+\pi/2} p(\psi) d\psi = 1. \qquad (15)$$

Herein, derivatives with respect to the angle ψ are abbreviated by primes. C is the constant of integration to be calculated by the normalization condition, noted in (15). The solution of (14) is doubly periodic. It possesses two singularities at the boundaries $\psi = \pm\pi/2$. Provided that $p'(\pm\pi/2)$ is finite, the integration constant C is equal to $p(\pm\pi/2)$.

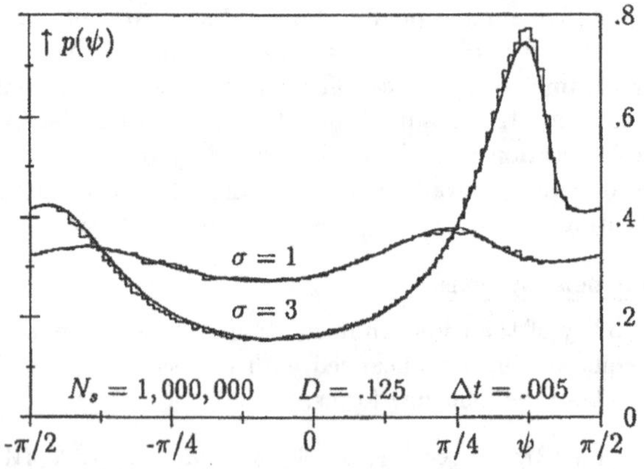

Figure 3: Density of the invariant measure

The stationary formulation (14) implies that the rotating phase is restricted to the finite angle range $|\psi| \leq \pi/2$. This *mod* π - restriction is physically possible also in a pathwise realization of the phase process Ψ_t. If Ψ_t exceeds the finite angle range during a time simulation, one can reset it by means of the periodicity condition $\psi(t+\pi) = \psi(t)$ without any influence on the Itô equations of the phase (10) and amplitude process (11). We realize all these properties of existence, stationarity and periodicity in a purely numerical way by applying a forward Euler scheme to the phase equation (10).

$$\Psi_{n+1} = \Psi_n - [1 + (2D + \sigma^2 \cos^2 \Psi_n) \sin \Psi_n \cos \Psi_n]\Delta t - \sigma \cos^2 \Psi_n \Delta W_n, \qquad (16)$$

$$\Delta W_n = \sqrt{\Delta t}\, R_n, \quad n = 0, 1, 2, \ldots \qquad E(R_n) = 0, \quad E(R_n^2) = 1. \qquad (17)$$

This simulation was performed with the step size $\Delta t = .005$ for $N_s = 1,000,000$ sample points. The Wiener increments ΔW_n were generated by a sequence of numbers R_n normally distributed and mutually independent (see e.g. Kloeden and Platen, 1989). Figure 3 shows obtained results in form of histograms. The estimated phase densities are valid for the damping value $D = .1$ and for the two excitation intensities $\sigma = 1$ and $\sigma = 3$. Note that the diffusion term in (16) is vanishing in the singular points $\psi = \pm\pi/2$. This is the reason why the density estimates converge rapidly in the two boundary regions.

2.2. Backward schemes

The histograms, obtained in figure 3, are underlaid by smooth lines calculated by a direct integration of the associated density equation (14). According to Wedig (1989), this integration is performed by means of a backward scheme which regularizes the numerical integration of the singular diffusion equation (14). If $\Delta\psi = \pi/N$ is a sufficiently small step size of the discrete angles $\psi_n = -\pi/2 + n\Delta\psi$ for $n = 0, 1, 2, ..N$, one can replace the differential quotient $p'(\psi)$ in (14) by the backward difference form $(p_n - p_{n-1})/\Delta\psi$ in order to obtain the following regular recurrence formula.

$$p_n = \frac{\Delta\psi C + \frac{1}{2}\sigma^2 \cos^4 \psi_n p_{n-1}}{\Delta\psi[1 + (2D - \sigma^2 \cos^2 \psi_n) \sin\psi_n \cos\psi_n] + \frac{1}{2}\sigma^2 \cos^4 \psi_n}, \tag{18}$$

$$p_0 = p(-\pi/2) = C, \quad p_N = p(\pi/2) = C, \qquad n = 1, 2, 3, ...N. \tag{19}$$

This recursion satisfies the periodicity condition $p_0 = p_N = p(\pm\pi/2)$. Hence, the formula (18) can be started with arbitrary initial values p_0 at the left side of the angle range and will allways end with the same density value p_N at the right hand side. Subsequently, the p_0 - value is determined by the normalization of $p(\psi)$. The smooth lines in figure 3 represent densities $p(\psi)$ calculated by applying the formula (18) for $N = 1,000$. Both histograms of Monte-Carlo simulations and smooth lines of the backward algorithm show a sufficiently good coincidence.

The applied backward algorithm can considerably be improved by corresponding higher order schemes. For their derivation, we expand the density $p(\psi)$ by a backward Taylor series.

$$p(\psi - \Delta\psi) = p(\psi) - \frac{1}{1!}\Delta\psi p'(\psi) + \frac{1}{2!}\Delta\psi^2 p''(\psi) - + ... , \tag{20}$$

$$p(\psi) = p(\psi_n) = p_n, \quad p(\psi - \Delta\psi) = p_{n-1}, \quad n = 1, 2, ...N. \tag{21}$$

By means of the diffusion equation (14), the first derivative $p'(\psi)$ can be expressed by the integration constant C and the density $p(\psi)$, itself. The same can be achieved for the second derivative $p''(\psi)$. Both are inserted into the Taylor expansion (20) leading to the following second order backward scheme.

$$p_n = \frac{g_n(\Delta\psi C + g_n p_{n-1}) + \frac{1}{2}\Delta\psi^2(f_n + g_n')C}{g_n(\Delta\psi f_n + g_n) + \frac{1}{2}\Delta\psi^2(g_n' f_n + f_n^2 - f_n' g_n)}. \tag{22}$$

Herein, g_n and f_n are abbreviations of the coefficients $g(\psi)$ and $f(\psi)$ of the diffusion equation (14) to be calculated at the discrete angles $\psi_n = -\pi/2 + n\Delta\psi$ for $n = 1, 2, 3, ...N$.

$$g(\psi)p'(\psi) = C - f(\psi)p(\psi), \qquad g'(\psi) = -2\sigma^2 \sin\psi \cos^3\psi, \tag{23}$$

$$g(\psi) = \tfrac{1}{2}\sigma^2 \cos^4\psi, \quad f(\psi) = 1 + (2D - \sigma^2\cos^2\psi)\sin\psi\cos\psi, \tag{24}$$

$$f'(\psi) = (2D - \sigma^2\cos^2\psi)(\cos^2\psi - \sin^2\psi) + 2\sigma^2\sin^2\psi\cos^2\psi. \tag{25}$$

Obviously, the second order scheme (23) coincides with the first order one (18) for $\Delta\psi^2 = 0$. In figure 4, we show evaluations of the second order backward scheme in order to give an impression of the improved convergence. For $\sigma = .5$ and $D = 2$, the applied step numbers are $N = 100$ and $N = 150$. For $\sigma = 25$ and $D = .4$, results are given for $N = 50$ and $N = 100$. In both cases, the highest step number is denoted by thick lines since the obtained curves are graphically not changed for further decreasing step sizes $\Delta\psi = \pi/N$.

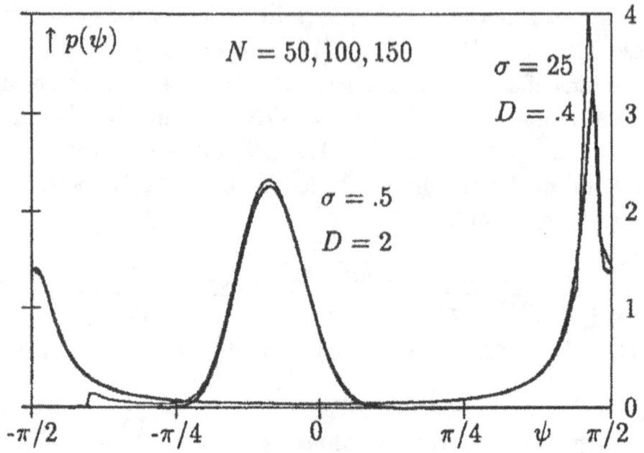

Figure 4: Second order backward scheme

Finally, it is worth to note that the applied backward algorithm can easily be extended to higher order schemes. For this purpose, we need the k th derivatives $g^{(k)}$ and $f^{(k)}$ of the coefficient functions $g(\psi)$ and $f(\psi)$, noted in (24).

$$g^{(k)} = 2^{k-2}\sigma^2[(2^k \cos 4\psi + 4\cos 2\psi)\cos k\pi/2 + +(2^k \sin 4\psi + 4\sin 2\psi)\cos(k+1)\pi/2], \qquad k = 1,2,3,\ldots, \tag{26}$$

$$f^{(k)} = 2^k D[\sin 2\psi \cos k\pi/2 - \cos 2\psi \cos(k+1)\pi/2] + \tfrac{1}{2}g^{(k+1)}. \tag{27}$$

Subsequently, the Fokker-Planck equation (23) is differentiated k times.

$$g\, p^{(k+1)} + f\, p^{(k)} + \sum_{l=0}^{k-1}\binom{k}{l}(g^{(k-l)}\, p^{(l+1)} + f^{(k-l)}\, p^{(l)}) = 0. \tag{28}$$

This result represents a recursion formula. Applied for $k = 1,2,3,\ldots$, it allows one to calculate all higher order derivatives $p^{(k+1)}(\psi)$ in dependence on the density $p(\psi)$. Insertion into the Taylor series (20) leads to $p_n = p(\psi_n)$ in dependence on $p_{n-1} = p(\psi_{n-1})$.

2.3. Lyapunov exponents

In the following, we apply the second order backward scheme (22) in order to determine the Lyapunov exponent $\lambda(\sigma, D)$ in dependence on the damping value D and on the excitation intensity σ. The step number selected was $N = 300$. The calculations are performed by starting the recursion (22) with the initial value $p_0 = 1$. The subsequent density values

p_n for $n = 1, 2, ...N$ are collected in order to calculate the normalization integral, noted in (15), and the ensemble average (13) of the Lyapunov exponent applying Simpson's rule of integration. Thus, the end of the recursion gives simultaneously the numerical values of both integrals related to the unknown initial value p_0. Since p_0 is determined by the normalization condition, the obtained average (13) has finally to be divided by the normalization integral (15) in order to obtain the correct Lyapunov exponent $\lambda(\sigma, D)$.

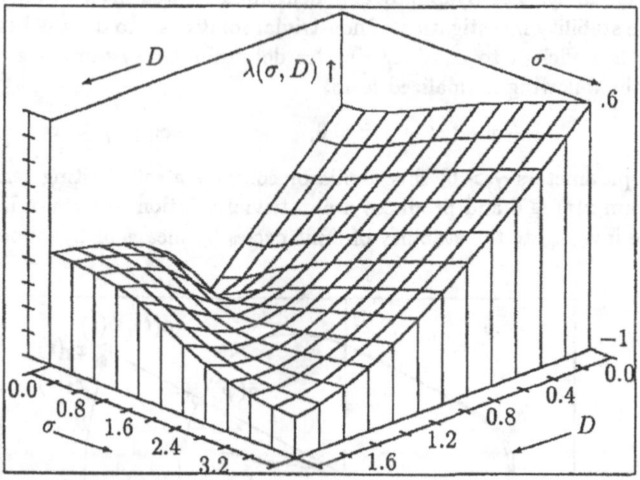

Figure 5: Distribution of the Lyapunov exponents

Figure 5 shows the calculated Lyapunov exponents over the parameter range $0 \leq D \leq 2$ and $0 \leq \sigma \leq 4$ in a three-dimensional picture. The vertical axis is scaled for the values $-1 \leq \lambda \leq .6$. The back corner point $\sigma = D = 0$ determines the zero level of the stability boundary. Below this level, the Lyapunov exponents are negative; i.e. the trivial solution $A_t \equiv 0$ or equilibrium of the system is asymptotically stable with probability one. For $D = 0$ and increasing excitation intensities $\sigma > 0$, the Lyapunov exponents become positive and the trivial solution is unstable. For increasing damping values $D > 0$, the Lyapunov exponents decrease up to a certain optimal line where they increase again. The Lyapunov exponent distribution of figure 5 can be compared with similar results of Monte-Carlo simulations obtained by Arnold and Kliemann (1981) for parametric excitations by coloured noise. Recently, extensions are given in Wei-Chau Xie's Ph.D. thesis (1990) supervised by Ariaratnam.

Finally, it is worth to note that the Lyapunov exponents, calculated above, coincide with the eigenvalues of the unperturbed system ($\sigma = 0$). For this special deterministic case, the applied backward schemes give the density of the invariant measure in the following form.

$$p_n = \frac{C}{1 + 2D \sin\psi_n \cos\psi_n} \qquad n = 0, 1, 2, 3, ...N. \qquad (29)$$

For $\sigma = 0$, it represents the exact solution of the diffusion equation (14) provided that $D < 1$. For $D \geq 1$, the stationary density $p(\psi)$ degenerates to the Dirac delta function $p(\psi) = \delta(\psi - \psi_0)$ concentrated at $\psi_0 = -\frac{1}{2}\arcsin(1/D)$. Naturally, this singular overdamped case has to be regularized by non-vanishing noise intensities $\sigma > 0$ in order to avoid to small step sizes $\Delta\psi$ or vice versa, numerical instabilities of the backward schemes.

3. Limit cycles of nonlinear systems

The performed analysis of Lyapunov exponents and related invariant measures can be extended to the stability investigation of non-trivial solutions. To discuss basic aspects of such problems, it is sufficient to consider simpler deterministic systems, e.g. the Van der Pol equation of the following normalized form.

$$\ddot{x} + x + (x^2 - \gamma)\dot{x} = 0, \qquad -\infty \leq \gamma \leq \infty. \tag{30}$$

For positive parameters $\gamma > 0$, this oscillator contains a self-exciting term. It destabilizes the equilibrium $x(t) \equiv 0$ and produces a non-trivial solution in form of limit cycles. In the following, we investigate the stability of limit cycles by means of Lyapunov exponents.

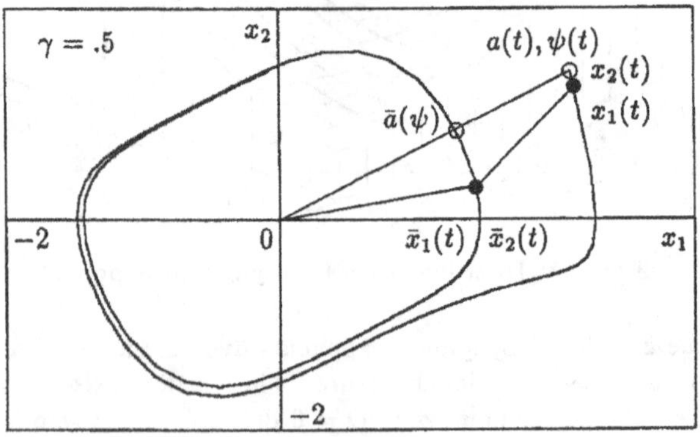

Figure 6: Simulation of the Van der Pol equation

3.1. Different stability concepts

The simulation of limit cycles can be performed in rectangular or polar coordinates, respectively. Applying the rectangular coordinates of the displacement x_1 and the velocity x_2, the oscillator equation (30) is transformed into the following first order system.

$$x_1 = x, \qquad \dot{x}_1 = x_2, \tag{31}$$

$$x_2 = \dot{x}, \qquad \dot{x}_2 = -x_1 - (x_1^2 - \gamma)x_2. \tag{32}$$

Subsequently, we introduce the amplitude a and the phase ψ. The transformed system equations read as follows.

$$x_1 = a\cos\psi, \qquad \dot{a} = a(\gamma - a^2\cos^2\psi)\sin^2\psi, \tag{33}$$

$$x_2 = a\sin\psi, \qquad \dot{\psi} = (\gamma - a^2\cos^2\psi)\sin\psi\cos\psi - 1. \tag{34}$$

In figure 6, we show simulation results obtained for the parameter $\gamma = .5$ by applying a forward Euler scheme. The unperturbed limit cycle or closed orbit is denoted by $\bar{x}_1(t), \bar{x}_2(t)$ or by $\bar{a}(\psi)$, respectively. The perturbed solutions are $x_1(t), x_2(t)$ or $a(t), \psi(t)$. They are simulated by starting with initial values outside the closed orbit $\bar{a}(\psi)$. Also in figure 6, we sketched two different stability concepts. The first one is the kinematic concept introduced by Lyapunov. It measures the distance between perturbed solutions $x_1(t), x_2(t)$ and unperturbed ones $\bar{x}_1(t), \bar{x}_2(t)$ at same times t. The Euclidean distance between both is marked by filled in circles. Since the associated perturbation equations are coupled with the unperturbed solutions, the kinematic concept of Lyapunov leads to a four-dimensional problem. Moreover, this concept includes phase differences at same times which are without any interest in many practical situations.

The second stability concept goes back to Poincaré. It measures the distance between the perturbed time solutions $a(t), \psi(t)$ and the unperturbed limit cycle $\bar{a}(\psi)$ at different times but at the same phase angle $\psi(t)$. In figure 6, this amplitude metric is represented by hollow circles. It can be shown that there exists an invariant measure which allows one to derive associated Lyapunov exponents without any knowledge of the perturbed solutions $a(t)$ and $\psi(t)$. The amplitude metric between both refers to the geometrical configuraion of the closed orbit $\bar{a}(\psi)$ which can be calculated by eliminating the time variable t. For this reason, we divide the amplitude equation (33) and the phase equation (34) to obtain a first order equation of the following form.

$$\frac{d\bar{a}}{d\psi} = \frac{\bar{a}(\gamma - \bar{a}^2\cos^2\psi)\sin^2\psi}{(\gamma - \bar{a}^2\cos^2\psi)\sin\psi\cos\psi - 1} = Q(\bar{a}, \psi). \tag{35}$$

Herein, $a(t)$ of equation(33) can be replaced by $\bar{a}(\psi)$ provided that the integration of (35) is started in any known initial point $\bar{a}(\psi_0)$ of the closed orbit. Thus, the stability investigation is now reduced to a three-dimensional problem. The equations (33) and (34) are applied to simulate the perturbed solutions $a(t)$ and $\psi(t)$. The first order equation (35) defines the unperturbed solution or closed orbit $\bar{a}(\psi)$ without any time dependency. Note that the right-hand side of equation (35) is abbreviated by $Q(\bar{a}, \psi)$ applied in the following.

3.2. Orbital invariant measures

According to Wedig (1990), a common time basis of both unperturbed limit cycle $\bar{a}(\psi)$ and perturbed solutions $a(t), \psi(t)$ can be reintroduced. For this purpose, we replace the time independent angle increment $d\psi$ of equation (35) by the time dependent angle increment, given in equation (34).

$$d\bar{a} = Q(\bar{a}, \psi)d\psi = Q(\bar{a}, \psi)[(\gamma - a^2\cos^2\psi)\sin\psi\cos\psi - 1]dt. \tag{36}$$

Note that the equation (36), now obtained, contains two different amplitudes: the unperturbed amplitude $\bar{a}(\psi(t))$ and the perturbed one $a(t)$. Taking the difference $\Delta = a - \bar{a}$ of both amplitudes and introducing it into (33), (34) and (36), we eliminate the perturbed amplitude $a(t)$ and derive the following three equations for $\bar{a}(t), \psi(t)$ and $\Delta(t)$.

$$\begin{aligned} \dot{\bar{a}} &= Q\{[\gamma - (\bar{a} + \Delta)^2\cos^2\psi]\sin\psi\cos\psi - 1\}, & (37)\\ \dot{\psi} &= [\gamma - (\bar{a} + \Delta)^2\cos^2\psi]\sin\psi\cos\psi - 1, & (38)\\ \dot{\Delta} &= \Delta[\gamma - (3\bar{a}^2 + 3\bar{a}\Delta + \Delta^2)\cos^2\psi]\sin^2\psi \\ &\quad + \Delta Q(\bar{a}, \psi)(2\bar{a} + \Delta)\cos^3\psi\sin\psi. & (39) \end{aligned}$$

Now, these equations can be simulated in a common time basis. The simulation is started with the initial values $\Delta(0) = \Delta_0$, $\psi(0) = \psi_0$ and $\bar{a}(0) = \bar{a}(\psi_0)$. Δ_0 and ψ_0 are arbitrary, meanwhile $\bar{a}(\psi_0)$ represents one limit cycle point to be known.

According to the multiplicative ergodic theorem, an orbital Lyapunov exponent λ_O is introduced by the amplitude metric defined by the deviation $\Delta(t)$ of both amplitudes.

$$\lambda_O = \lim_{t\to\infty} \frac{1}{t} \log \frac{\Delta(t)}{\Delta(0)}, \qquad \Delta(t) = a(t) - \bar{a}(\psi(t)), \tag{40}$$

$$\dot{\Delta} = \Delta[(\gamma - 3\bar{a}^2 \cos^2 \psi) \sin \psi + 2\bar{a}Q(\bar{a}, \psi) \cos^3 \psi] \sin \psi. \tag{41}$$

The asymptotic stability investigation of interest implies $\Delta(t) \to 0$. Hence, the nonlinear $\Delta(t)$ - equation (39) can be linearized to the form (41). This scalar equation is integrated with respect to the initial value $\Delta(0)$. Taking the natural logarithm of the solution $\Delta(t)$ and inserting it into the multiplicative ergodic theorem (40) yields

$$\lambda_O = \lim_{t\to\infty} \frac{1}{t} \int_0^t [(\gamma - 3\bar{a}^2 \cos^2 \psi) \sin \psi + 2\bar{a}Q(\bar{a}, \psi) \cos^3 \psi] \sin \psi d\tau, \tag{42}$$

$$\lambda_O = \int_{-\pi/2}^{\pi/2} [(\gamma - 3\bar{a}^2(\psi) \cos^2 \psi) \sin \psi + 2\bar{a}Q(\bar{a}, \psi) \cos^3 \psi] \sin \psi p(\psi) d\psi. \tag{43}$$

Provided that there exists a closed orbit $\bar{a}(\psi)$ and an invariant measure of the associated angle $\psi(t)$, the time average (42) can be reduced to the finite integral (43) where $p(\psi)$ is the density of the invariant angle measure. For $\gamma << 1$, a first approximation of $p(\psi)$ leads to the uniform distribution $p(\psi) = 1/\pi$, already applied by Wedig (1990).

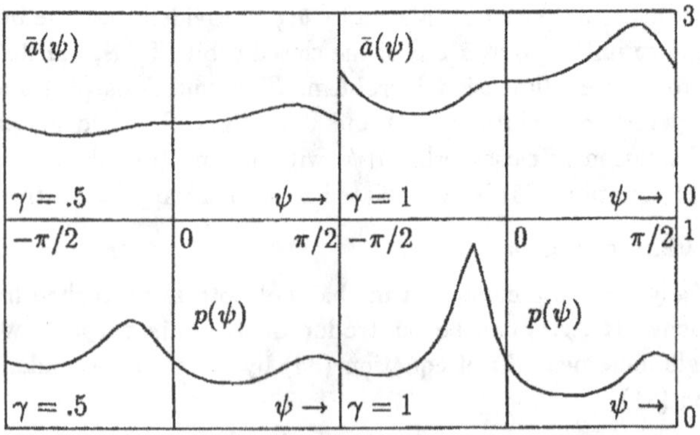

Figure 7: Amplitudes and angle densities

An exact evaluation of (43) is proposed by L. Arnold (private communication during the Oberwolfach Conference, 1990). In a similar manner, as already performed by Wedig (1989), one can investigate the Liouville equation associated with $\bar{a}(t)$ and $\psi(t)$.

$$\frac{\partial}{\partial \bar{a}}\{[\bar{a}(\gamma - \bar{a}^2 \cos \psi) \sin^2 \psi] p(\bar{a}, \psi)\} + \frac{\partial}{\partial \psi}\{[(\gamma - \bar{a}^2 \cos^2 \psi) \sin \psi \cos \psi - 1] p(\bar{a}, \psi)\} = 0, \tag{44}$$

For $\bar{a}(\psi)$: $$\frac{\partial}{\partial \psi}[(\gamma - \bar{a}^2(\psi) \cos^2 \psi) \sin \psi \cos \psi - 1] p(\psi) = 0, \qquad |\psi| \le \pi/2. \tag{45}$$

For the special case of periodic orbits $\bar{a}(\psi)$, the two-dimensional density $p(\bar{a},\psi)$ degenerates to a Dirac delta function which reduces (44) to the one-dimensional form (45). Provided that $\bar{a}(\psi)$ is known, (45) can easily be integrated leading to the periodic density $p(\psi)$.

$$p(\psi) = \frac{C}{1-[\gamma-\bar{a}^2(\psi)\cos^2\psi]\sin\psi\cos\psi}, \qquad \int_{-\pi/2}^{\pi/2} p(\psi)d\psi = 1. \tag{46}$$

Herein, C is an integration constant to be calculated by the normalization of $p(\psi)$ in the angle range $|\psi| \leq \pi/2$. Figure 7 shows corresponding evaluations of $\bar{a}(\psi)$ and $p(\psi)$ for the two parameters $\gamma = .5$ and $\gamma = 1$.

3.3. Orbital and kinematic Lyapunov exponents

In figure 8, one finds numerical results of the Lyapunov exponents λ_O. To calculate them, we apply a forward Euler scheme to equation (35) or to its modified version

$$\frac{d\bar{a}^2}{d\psi} = \frac{2\bar{a}^2(\psi)[\gamma-\bar{a}^2(\psi)\cos^2\psi]\sin^2\psi}{[\gamma-\bar{a}^2(\psi)\cos^2\psi]\sin\psi\cos\psi - 1}. \tag{47}$$

After some iteration steps, we find closed orbits or that initial value $\bar{a}^2(-\pi/2)$ which satisfies the periodicity condition $\bar{a}^2(-\pi/2) = \bar{a}^2(\pi/2)$. The numerical solution $\bar{a}^2(\psi)$ determines the density (46) of the invariant measure. Subsequently, we normalize $p(\psi)$ and calculate the integral (43) of the orbital Lyapunov exponent λ_O. For negative parameters $\gamma < 0$, a periodic solution $\bar{a}^2(\psi)$ can be simulated by starting with negative initial values $\bar{a}^2(\psi_0) < 0$ and applying negative angle increments in (47). Analytical evaluations are performable by means of Fourier expansions.

$$\bar{a}^2(\psi) = c_0 + \sum_{n=1}^{\infty}(c_n\cos 2n\psi + s_n\sin 2n\psi), \tag{48}$$

$$c_0 = 4\gamma, \qquad p(\psi) = 1/\pi, \qquad \lambda_O = -\gamma. \tag{49}$$

The insertion of the Fourier series (48) into the limit cycle equation (47) leads to a nonlinear system for the Fourier coefficients. For $\gamma << 1$, we obtain the first approximations, noted in (49). They are tangents of the corresponding numerical results, drawn in figure 8.

It is finally interesting to compare the orbital Lyapunov exponents λ_O with corresponding results of kinematic Lyapunov exponents λ_K. For this purpose, we introduce the coordinate perturbation $\Delta_1(t)$ and $\Delta_2(t)$ into the equations (31) and (32) and derive the following perturbation equations.

$$x_1 = \bar{x}_1 + \Delta_1, \qquad \dot{\Delta}_1 = \Delta_2, \tag{50}$$

$$x_2 = \bar{x}_2 + \Delta_2, \qquad \dot{\Delta}_2 = -(1+2\bar{x}_1\bar{x}_2)\Delta_1 - (\bar{x}_1^2-\gamma)\Delta_2. \tag{51}$$

To obtain the top Lyapunov exponent, we transform both equations (50) and (51) by means of the polar coordinates α and φ.

$$\Delta_1 = \alpha\cos\varphi, \qquad \dot{\alpha} = -\alpha[2\bar{x}_1\bar{x}_2\cos\varphi + (\bar{x}_1^2-\gamma)\sin\varphi]\sin\varphi, \tag{52}$$

$$\Delta_2 = \alpha\sin\varphi, \qquad \dot{\varphi} = -1 - [2\bar{x}_1\bar{x}_2\cos\varphi + (\bar{x}_1^2-\gamma)\sin\varphi]\cos\varphi. \tag{53}$$

Subsequently, the amplitude equation (52) is integrated. Taking the natural logarithm of the $\alpha(t)$ - solution and inserting it into the multiplicative ergodic theorem of Oseledec yields

$$\lambda_K = -\lim_{t\to\infty}\frac{1}{t}\int_0^t [2\bar{x}_1\bar{x}_2\cos\varphi + (\bar{x}_1^2-\gamma)\sin\varphi]\sin\varphi d\tau. \tag{54}$$

In figure 8, one finds numerical evaluations of (54) in the parameter range $0 \le \gamma \le 5$. Applying a forward Euler scheme to the equations (31), (32) and (53), the simulation is started with non-vanishing initial values $x_1(0), x_2(0)$ and $\varphi(0)$. After some iteration steps, stationary solutions $\bar{x}_1(t)$ and $\bar{x}_2(t)$ are approached. Inserting them into (54), the time average is evaluated by step approximations applying the scan rate $\Delta t = .01$. The numerical results obtained are $|\ \lambda_K\ | < .05$ for all parameters $\gamma = 0(.5)5$. For decreasing scan rates, the simulated $|\ \lambda_K\ |$ - values are diminished. This property indicates that the kinematic top Lyapunov exponent λ_K is vanishing for positive γ - parameters.

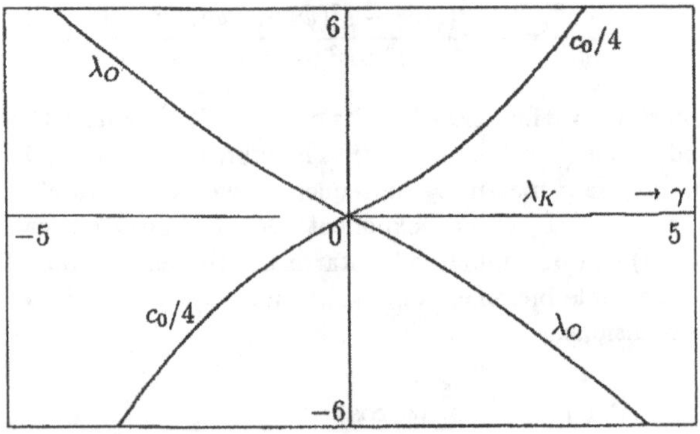

Figure 8: Orbital and kinematic Lyapunov exponents

4. Conclusions

The present paper is devoted to the stability analysis of dynamical systems and related bifurcation problems. As an example, we studied the mechanical system of a uniform column simply supported and axially excited by white noise. Applying a first mode approximation, the equation of motion is reduced to a nonlinear second order system with multiplicative white noise. The simulated bifurcation diagram shows that there exists a critical excitation intensity beyond which the equilibrium bifurcates into a non-trivial stationary solution.

The bifurcation point is calculated by means of Lyapunov exponents associated with the trivial solution. Applying Khasminskii's projection on unit circles, the angle process leads to a singular Fokker-Planck equation which is efficiently solved by means of backward differences. The paper introduces higher order backward schemes. They regularize the singularities of the one-dimensional diffusion equation and give the periodic density of the invariant angle measure without any iteration procedure. The invariant measure reduces the multiplicative ergodic theorem of Oseledec to a finite integral which is easily evaluable.

The investigation of Lyapunov exponents and associated invariant measures is extended to non-trivial solutions of deterministic limit cycle systems. For the example of the Van der Pol equation, two different stability concepts are discussed: the kinematic concept of Lyapunov and the orbital concept of Poincaré. Both lead to different Lyapunov exponents. In particular, it is numerically verified that the kinematic top Lyapunov exponent is vanishing and is therefore not applicable to investigate the asymptotic stability of limit cycle solutions.

References

1. Weidenhammer, F. 1969. Biegeschwingungen des Stabes unter axial pulsierender Zufallslast. VDI-Berichte Nr. 135: 101-107.

2. Khasminskii, R.Z. 1967. Necessary and sufficient conditions for asymptotic stability of linear stochastic systems. Theor. Prob. and Appls. 12: 144-147.

3. Arnold, L. 1974. Stochastic Differential Equations, Theory and Applications. New York, Wiley.

4. Wedig, W. 1988. Simulation and analysis of mechanical systems with parameter fluctuations. To appear in Proceedings of the Oberwolfach Conference on Random Partial Differential Equations (ed. by U. Hornung), Springer, Lecture Notes in Mathematics.

5. Oseledec, V.I. 1968. A multiplicative ergodic theorem, Lyapunov characteristic numbers for dynamical systems. Trans. Moscow Math. Soc. 19: 197-231.

6. Kloeden, P.E. and Platen, E. 1989. A survey of numerical methods for stochastic differential equations. Stochastic Hydrology and Hydraulics 3, 155-178.

7. Wedig, W., Vom Chaos zur Ordnung, Mitteilungen GAMM (R. Mennicken ed.), ISSN 0936-7195, 2 (1989) 3-31.

8. Arnold, L. & Kliemann, W. 1981. Qualitative theory of stochastic systems. In: Probabilistic Analysis and Related Topics (ed. by A.T. Bharucha-Reid). Vol. 3, New York: Academic Press.

9. Xie, Wei-Chau. 1990. Lyapunov exponents and their applications in structural dynamics. Ph D thesis, supervised by S. T. Ariaratnam and presented to the University of Waterloo, Canada.

10. Wedig, W. 1990. Lyapunov exponents and invariant measures for isotropic limit cycles. To appear in Proceedings of the Second International Conference on Stochastic Structural Dynamics, Boca Raton, Florida, May 1990, Florida Atlantic University, Center for Applied Stochastics Research.

Address: Prof. Dr.-Ing. W. Wedig, Institute for Technical Mechanics,
University of Karlsruhe, D-7500 Karlsruhe 1, FRG

Sample Stability of Multi-Degree-of-Freedom Systems

Christian G. Bucher
Institute of Engineering Mechanics, University of Innsbruck, Austria

ABSTRACT: An approximate computational procedure to determine the top Lyapunov exponent of linear multi-degree-of-freedom systems with random parametric excitation is presented. The concept is based on the assumption of Markov properties of the response or, equivalently, white noise properties of the excitation. The procedure is very efficient since it relies on well developed algorithms from linear algebra. Comparison with available exact results and numerical results from Monte-Carlo-Simulation show excellent agreement.

1 INTRODUCTION

In the analysis of mechanical systems the case of parametric excitation occurs quite frequently. A classical example in structural mechanics is the beam with pulsating axial force. Moreover, there is a considerable number of applications in mechanical and structural engineering where cases of random parametric excitations (e.g. wind, earthquake loading) become increasingly important (e.g. Shih and Lin 1982, Lin and Ariaratnam 1980, Bucher and Lin 1988a). The long term behavior of such systems may be significantly influenced by the temporal variation of system parameters (e.g. stiffness). For sufficiently high levels of this parametric excitation the response may become unstable leading to inadmissible growth of stresses in the system.

The development of mathematical tools for the analysis of stochastic stability (e.g. Arnold et al. 1985) is largely based on Markovian properties of the system response under consideration, i.e. (filtered) white noise is assumed to excite the system. For the case of non-white excitations only very few numerical results are available, most of them being sufficient conditions for stability (Infante 1968, Kozin 1986, Lin et al. 1986, Ariaratnam and Xie, 1988). Even for the white noise case those stability boundaries which can be computed in a straightforward way are sufficient ones in terms of e.g. second moment stability. As far as sample stability is concerned, the computational tools are not that well developed as yet. The exact results available for the simple oscillator (Mitchell and Kozin 1974) heavily rely on analytical solutions for the Fokker-Planck-equation. This strategy, however, appears not to be suitable for systems of higher order. First order perturbation results near a Hopf bifurcation point of the unperturbed (i.e. deterministic) system can be obtained relatively easy, e.g. by means of a stochastic averaging technique (e.g.Sri Namachchivaya and Lin 1988, Bucher 1990).

It is the purpose of this paper to present an extension and improvement of the above mentioned results on sample stability. The computational procedure is applied to SDOF systems for verification with available exact results and to MDOF systems to show the numerical efficiency.

2 METHOD OF ANALYSIS

2.1 Concepts

In the context of asymptotic stability in the Lyapunov sense the major task is to determine the behavior of a perturbed solution as time tends to infinity. In fact, the analysis of a system linearized around a particular solution whose stability is to be investigated is sufficient for most practical cases. Hence in the following attention will be focussed on the stability of the trivial solution of a linear system governed by the differential equation

$$\dot{x} = Ax + Bx\,\xi(t) \tag{1}$$

in which x is a state vector of order n, A and B are matrices of order nxn and $\xi(t)$ is a stationary wide band scalar random process with zero mean and autocorrelation function $R_{\xi\xi}(\tau)$. An overdot denotes differentiation with respect to time.

Since eq (1) contains random quantities the stability of the trivial solution $x \equiv 0$ can be described in probabilistic terms only. In general, the concept of sample or almost sure stability is utilized. This can be stated as follows

$$P\left[\lim_{t\to\infty} \|x_t\| = 0\right] = 1 \tag{2}$$

in which P[.] denotes probability and $\|.\|$ is a vector norm. x_t is the perturbed solution at time t. A different stability criterion arises from the definition of convergence in mean square leading to

$$\lim_{t\to\infty} E\left[\|x_t\|^2\right] = 0 \tag{3}$$

For most engineering applications the concept of sample stability is more appealing since it is related to directly observable events while mean square stability requires averaging over an ensemble which, in reality, cannot be observed. However, as shown in the following, mean square stability can be dealt with considerably easier in the case of systems driven by white noise.

In the following, the differential equation (1) is replaced by an equivalent system of Ito equations

$$dx = m dt + \sigma dW(t) \tag{4}$$

in which W(t) denotes a unit Wiener process and the drift and diffusion terms m, σ in eq(4) can be computed from

$$m = Ax + \frac{D_0}{2}B^2x = Fx; \quad \sigma = \sqrt{D_0}Bx \tag{5}$$

and the noise intensity D_0 is given by

$$D_0 = \int_{-\infty}^{\infty} R_{\xi\xi}(\tau)d\tau \tag{6}$$

Hence eq(4) can be written in the form

$$dx = Fx\, dt + \sqrt{D_0}Bx\, dW(t) \tag{7}$$

For the purpose of evaluating eqs.(2) and (3) the following norm $||x||$ is introduced:

$$||x|| = (x^TMx)^{1/2} \tag{8}$$

In the above equation M is an arbitrary symmetric, positive definite matrix. By applying Ito's differential rule the increment of the p-th norm $||x||^p$ can be written as

$$\begin{aligned} d||x||^p &= \frac{p}{2}||x||^{p-2}\cdot x^T(F^TM + MF + D_0B^TMB)x\, dt \\ &+ \frac{p}{2}(\frac{p}{2} - 1)||x||^{p-4}D_0[x^T(B^TM + MB)x]^2dt \\ &+ \frac{p}{2}||x||^{p-2}\cdot \sqrt{D_0}x^T(B^TM + MB)x\, dW(t) \end{aligned} \tag{9}$$

For the special case p = 2 eq(9) reduces to

$$\begin{aligned} d||x||^2 &= x^T(F^TM + MF + D_0B^TMB)x\, dt \\ &+ \sqrt{D_0}x^T(B^TM + MB)x\, dW(t) \end{aligned} \tag{10}$$

Additionally, the Ito-equation for the logarithm of the norm is given as follows

$$\begin{aligned} d\ln||x|| &= \frac{1}{2}||x||^{-2}\cdot x^T(F^TM + MF + D_0B^TMB)x\, dt \\ &+ \frac{1}{4}||x||^{-4}D_0[x^T(B^TM + MB)x]^2dt \\ &+ ||x||^{-2}\cdot \sqrt{D_0}x^T(B^TM + MB)x\, dW(t) \end{aligned} \tag{11}$$

Eqs.(9) - (11) form the basis for investigation of the stochastic stability of the system (7).

2.2 Sample stability and moment stability

As stated by Arnold et al. 1986 the stability of the p-th norm is determined by the sign of the following quantity

$$q(p) = \frac{d}{dt} E\,[\|x\|^p] \tag{12}$$

in which $q(p)<0$ indicates stability. Due to the fact that $dW(t)$ is uncorrelated with its own past, the expectation in eq(12) can be calculated from eq(9) yielding

$$q(p) = \frac{p}{2} E\Big[\|x\|^{p-2}\Big\{ x^T(F^TM + MF + D_0B^TMB)\,x + D_0\big(\tfrac{p}{2}-1\big)\Big[\frac{x^T(B^TM + MB)x}{\|x\|^4}\Big]^2\Big\}\Big] \tag{13}$$

and, specifically for p=2

$$q(2) = E\big[x^T(F^TM + MF + D_0B^TMB)x\big] \tag{14}$$

Since this equation contains a linear combination of second moments it becomes evident that the stability of the second norm is equivalent to the stability of the second moments.
Sample stability is determined by the sign of the top Lyapunov exponent λ defined by

$$\lambda = \lim_{p\to 0} \frac{q(p)}{p} = E\,[\,d \ln \|x\|\,]$$

$$= \frac{1}{2} E\Big[\frac{x^T(F^TM + MF + D_0B^TMB)x}{x^TMx}\Big] - \frac{D_0}{4} E\Big[\Big(\frac{x^T(B^TM+MB)x}{x^TMx}\Big)^2\Big] \tag{15}$$

As shown by Arnold et al. 1986 $q(p)<0$ for any $p>0$ is a sufficient condition for $\lambda<0$, i.e. for sample stability.
Rewriting eq.(15) in the form of

$$\lambda = \frac{1}{2} E_1 - \frac{D_0}{4} E_2 \tag{16}$$

it is seen that E_1 can be computed immediately if the positive definite matrix M is chosen to be a solution to the eigenvalue problem

$$F^TM + MF + D_0B^TMB = \mu M \tag{17}$$

Then eq(16) can be simplified to

$$\lambda = \frac{\mu}{2} - \frac{D_0}{4} E\Big[\Big(\frac{x^T(B^TM+MB)x}{x^TMx}\Big)^2\Big] \tag{18}$$

which again shows that $\mu<0$ (i.e. stability of the second moments) is a sufficient condition for sample stability. The remaining expectation in eq.(18) can easily be determined from sample functions based on discrete time Monte Carlo simulation. It should be noted that the eigenvalue of the mapping defined by eq(17) with maximum real part (deciding stability) is real and isolated provided $D_0\neq0$. Hence iteration techniques converging to the maximum real eigenvalue μ can be successfully applied.

By invoking the multiplicative ergodic theorem (see e.g. Arnold et al. 1986, Wedig 1990) the Lyapunov exponent can, in principle, be obtained directly as

$$\lambda = \lim_{n\to\infty} \frac{1}{n\Delta t} \sum_{i=1}^{n} \ln \frac{\|x(t_i)\|}{\|x(t_{i-1})\|} \tag{19}$$

However, this straightforward application of the multiplicative ergodic theorem in conjunction with Monte Carlo simulation methods requires a very large number of simulations. An explanation is given as follows: Based on the Ito-equation for $\ln\|x\|$ (eq.(11)) the randomly fluctuating part Fluct($\ln\|x\|$) of $\ln\|x\|$ during discrete time simulation can be estimated by means of

$$\text{Fluct}(\ln\|x\|) = \|x\|^{-2} \cdot \sqrt{D_0}\, x^T(B^TM + MB)x \frac{1}{2\sqrt{\Delta t}} u \tag{20}$$

in which u is a standard Gaussian variable. It becomes quite clear from the above equation that the fluctuations tend to become more and more pronounced as the time increment Δt tends to zero, i.e. the noise process is approaching white noise. The number of simulations must be increased in order to reduce the statistical uncertainties involved. Hence the application of eq.(18) which avoids such rapid fluctuations is clearly preferable.

2.3 An approximate solution

In the next step, an approximative assumption regarding the joint probability density function of the state vector components is made. This assumption is based on the asymptotic behavior of the state vector covariance matrix as time tends to infinity. In analogy to eq.(12), differential equations for the statistical moments can be obtained. For the second moments $E[x_ix_j]$, i.e. the covariance matrix C_x, these differential equations can be written as

$$\frac{d}{dt} C_x = FC_x + C_xF^T + D_0BC_xB^T \tag{21}$$

It is seen that the covariance matrix C_x will be asymptotically dominated by the component C solving the eigenvalue problem

$$\mu C = FC + CF^T + D_0BCB^T \tag{22}$$

where μ is the largest eigenvalue (equal to largest eigenvalue in eq(17)). Hence, as $t \to \infty$ the covariance matrix C_x will be proportional to C and decay exponentially provided $\mu < 0$. Hence, the transformation

$$x = Lu \tag{23}$$

with the Cholesky decomposition of C

$$LL^T = C \tag{24}$$

consequently derives the vector x from a set of (asymptotically) uncorrelated variables u with identical standard deviations. Now it is assumed that the joint density function of the components of u is a function of the distance to the origin only, i.e.

$$p_U(u) = f(u^T u) \tag{25}$$

This implies rotational symmetry in the n-dimensional space which is equivalent to the normalized vector $u/\sqrt{u^T u}$ being uniform on the unit sphere. This assumption is true in the deterministic limit as the noise intensity D_0 tends to zero. In this case the differential equation (1) simply becomes

$$\dot{x} = Ax \tag{26}$$

and the p-th norm $\|x\|^p$ as defined by eq(8) is (asymptotically) governed by the relation

$$\|x\|^p = \|x_0\|^p \, e^{p\mu/2\, t} \tag{27}$$

where μ is defined by the eigenvalue problem eq(17). Since this relation holds for arbitrary p the invariant measure on the unit sphere can be completely described by second moments only, i.e. in terms of C as given in eq(22). The shape of f(.) in the above equation is irrelevant for the last expectation in eq(18) since any radial dependence appears in both numerator and denominator and thus will cancel. This is clearly seen by rewriting E_2 in terms of u:

$$E_2 = E\left[\left(\frac{u^T L^T (B^T M + MB) L u}{u^T L^T M L u}\right)^2\right] \tag{28}$$

This expectation is evaluated by Monte Carlo simulation, i.e. by generating samples of u according to the density as specified by eq.(25) and averaging the results from eq(28). It can be interpreted as an average over an n-dimensional elliptical surface defined by the quadratic form $u^T L^T M L u$. Hence the procedure can be viewed as a generalization to the stochastic averaging concept.

For computational convenience, a multinormal density $p_U(u)$ is used in this simulation. However, any sampling density which is rotationally symmetric in the u-space, will give the same answer.

3 NUMERICAL EXAMPLES

3.1 Comparison to available exact results

In the following the sample stability boundaries for the simple oscillator as obtained by Mitchell and Kozin 1974 are compared to approximate results from the suggested approach.
The matrices A and B are given by

$$A = \begin{bmatrix} 0 & 1 \\ -1 & -2\zeta \end{bmatrix}, \quad B = \begin{bmatrix} 0 & 0 \\ 1 & 0 \end{bmatrix} \tag{29}$$

in which unit circular frequency is assumed and ζ denotes damping ratio. The stability boundaries for this case are given in Fig.1.

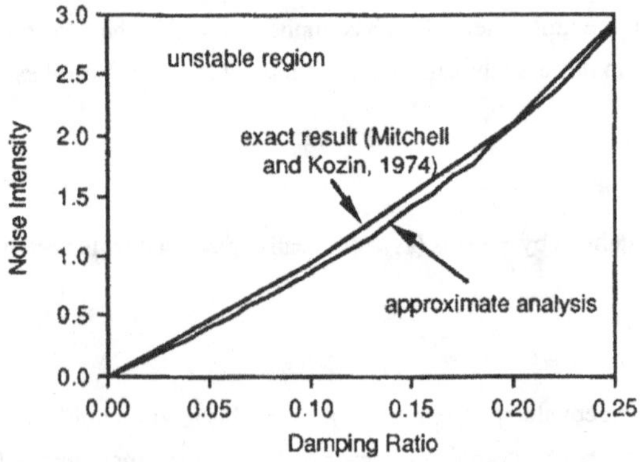

Fig. 1 Stability Boundaries for Simple Oscillator with Random Stiffness

It is found that for low levels of noise, i.e. for small D_0, the results agree very well with the exact ones. In fact, for the region as shown in Fig.1 they are almost indistinguishable.

3.2 Bridge Stability Analysis

The motion stability of a 2-degree-of-freedom bridge model with one torsional DOF α and one heaving DOF h as sketched in Fig.2 is investigated.
Assuming linear behavior in both structural as well as fluid-structure-interaction effects the equation of motion can be written in the form of eq(1) where the state vector contains 12 variables (Bucher and Lin 1988a). The first four state variables are $\alpha, \dot{\alpha}$, h and $\dot{h}$, i.e. they represent the structural motion whereas the

remaining 8 state variables model the fluid-structure-interaction. The oncoming wind velocity U(t) is assumed to be stationary with mean value $\bar{U}$:

$$U(t) = \bar{U}[1 + \xi(t)] \qquad (30)$$

The wind velocity enters the equation of motion as random parametric excitation. The elements of the 12 x 12 matrices A and B are given in Bucher and Lin 1988a.

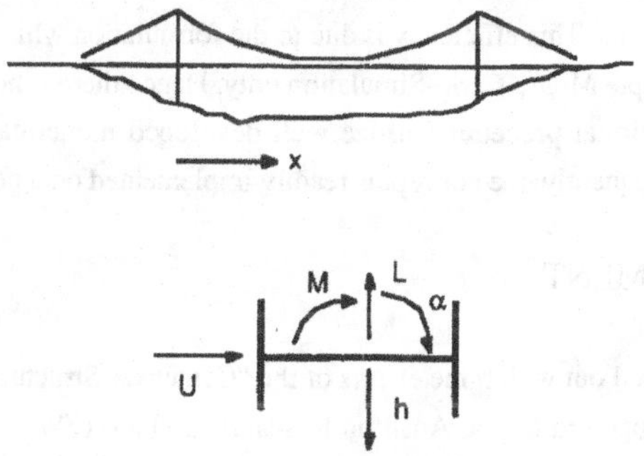

Fig.2 Bridge model with two mechanical degrees of freedom

The resulting sample stability boundary is shown in Fig.3 along with the narrower second moment stability boundary. For comparison, the exact sample stability boundary as obtained from Monte Carlo simulation using 1000 time steps ($\Delta t = 0.1$ sec) is plotted along with the approximate result. The ratio of computation time between direct Monte Carlo solution and the approximate result is 100 for this case.

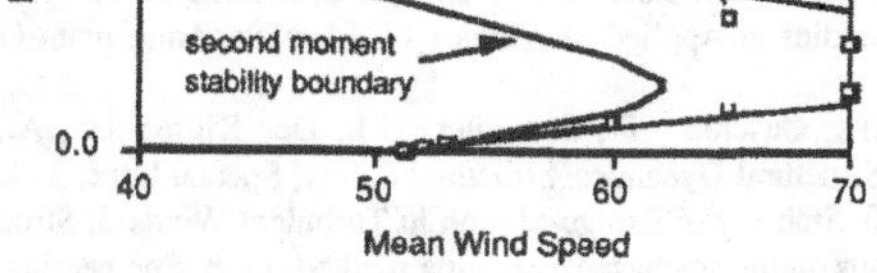

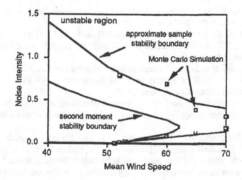

Fig.3 Stability boundaries of 2-DOF bridge model

The results clearly indicate a pronounced stabilizing effect of turbulence, especially at a noise level of D_0 = 0.2. In that region the onset of instability (flutter) is postponed to considerably higher wind speed as compared to the deterministic case $D_0 = 0$. This effect has also been found in experimental studies, e.g. Irwin and Schuyler 1978.

The difference between the sample stability criterion and the second moment stability is quite large in this region indicating that the second moment criterion yields an overly conservative stability boundary.

4 CONCLUSIONS

An efficient computational procedure to determine the top Lyapunov exponent and, consequently, the sample stability of multi-degree-of-freedom system was suggested. The application to several numerical examples show both accuracy when compared to available exact results and numerical efficiency when applied to higher order systems. This efficiency is due to the formulation which requires the solution of an eigenvalue problem and simple Monte-Carlo-Simulation only. Hence there is no need for dynamic analysis. Both parts of the computational procedure utilize well developed numerical algorithms thus reducing computational efforts. Consequently, the concept is readily implemented on a personal computer.

5 ACKNOWLEDGEMENT

This research has been carried out within the efforts of the "Center for Structural Dynamics" (Director G.I. Schuëller) and has been supported by the Austrian Research Council (FWF) under Contract No. S30-08 which is gratefully acknowledged by the author.

REFERENCES

Ariaratnam, S.T., Xie, W.-C. 1988. Dynamic snap-buckling of structures under stochastic loads. S.T. Ariaratnam, G.I. Schuëller, I. Elishakoff (eds) Stochastic Structural Dynamics, Elsevier Applied Science Publ, Barking, UK, pp 1 - 20.

Arnold, L., Oeljeklaus, E., Pardoux, E. 1986. Some papers on Lyapunov exponents. L. Arnold, V. Wihstutz (eds.). Almost sure and moment stability for linear Ito equations. Springer Lecture Notes in Mathematics No.1186.

Bucher, C.G., Lin, Y.K. 1988a. Stochastic Stability of Bridges Considering Coupled Modes. J. Eng. Mech. Proc. ASCE.14: 2035-2054.

Bucher, C.G., Lin, Y.K. 1988b. Effect of spanwise correlation of turbulence field on the motion stability of long-span bridges. J. of Fluids and Structures 2: 437-451.

Bucher, C.G. 1990. Reliability of Bridges in Turbulent Wind. A.H-S. Ang, M. Shinozuka , G.I. Schuëller (eds.).Structural Safety and Reliability, Proc. ICOSSAR'89, 5th Intern. Conf. on Structural Safety and Reliability, San Francisco, August 7-11. 1989. Vol.1: 103-106. New York:ASCE.

Infante, E.R. 1968. On the stability of some linear nonautonomous random systems. ASME. J. Appl. Mech. 35: 7-12.

Irwin, A.P.A.H., Schuyler, G.D. 1978. Wind Effects on a Full Aerolastic Bridge Model. ASCE Spring Convention and Exhibit. Preprint 3268 Pittsburg, Pennsylvania, April 24-28.

Kozin, F. 1986. Some results on stability of stochastic systems, I. Elishakoff, R.H. Lyon (eds.) Random Vibration - Status and Recent Developments, Studies in Applied Mechanics 14, Elsevier, Amsterdam-Oxford-New York-Tokyo, pp 163 - 191.

Lin, Y.K., Kozin, F., Wen, Y.K., Casciati, F., Schuëller, G.I., Der Kiureghian, A.,Ditlevsen, O., Vanmarcke, E.H. 1986.Methods of Stochastic Structural Dynamics. Structural Safety, Special Issue. 3+4: 231-267.

Lin, Y.K., Ariaratnam, S.T. 1980. Stability of Bridge Motion in Turbulent Winds. J. Struct. Mech. 881: 1-15.

Lin, Y.K. 1986. Some observations on the stochastic averaging method. Prob. Engineering Mechanics 1: 23-27.

Mitchell, R.R., Kozin, F.1974. Sample Stability of Second Order Linear Differential Equations With Wide Band Noise Coefficients. SIAM. J. Appl. Math. 27 (No.4):571-605.

Shih, T-S., Lin, Y.K. 1982.Vertical Seismic Load Effect on Hysteretic Columns. J. Engr. Mech. Div.Proc. ASCE. 108, No. EM": 242-254.

Sri Namachchivaya, N., Lin, Y.K. 1988. Application of stochastic averaging for nonlinear dynamical systems with high damping. Probabilistic Engineering Mechanics 3: 159-167.

Wedig, W.V. 1991. Dynamic stability of beams under axial forces - Lyapunov exponents for general fluctuating loads. Structural Dynamics, W.B. Krätzig et al. (eds.), Balkema, Rotterdam, 141-148.

Lyapunov Exponents of Control Flows[1]

FRITZ COLONIUS, UNIVERSITÄT AUGSBURG
WOLFGANG KLIEMANN, IOWA STATE UNIVERSITY

Abstract. The use of Lyapunov exponents in the theory of dynamical systems or stochastic systems is often based on Oseledeč's Multiplicative Ergodic Theorem. For control systems this is not possible, because each (sufficiently rich) control system contains dynamics that are not Lyapunov regular. In this paper we present an approach to study the Lyapunov spectrum of a nonlinear control system via ergodic theory of the associated control flow and its linearization. In particular, it turns out that all Lyapunov exponents are attained over so called chain control sets, and they are integrals of Lyapunov exponents on control sets with respect to flow invariant measures, whose support is contained in the lifts of control sets to $\mathcal{U} \times M$, where $\mathcal{U}$ is the space of admissible control functions and M is the state space of the system. For the linearization of control systems about rest points the extremal Lyapunov exponents are analyzed, which leads to precise criteria for the stabilization and destabilization of bilinear control systems, and to robustness results for linear systems. The last section is devoted to a nonlinear example, where we combine the analysis of the global controllability structure with local linearization results and Lyapunov exponents to obtain a complete picture of control, stabilization and robustness of the system.

Keywords. nonlinear control systems, dynamical systems, control sets, chain control sets, linearization, ergodic theory, stabilization, robustness

1980 *Mathematics subject classifications:* 93C10, 58F10, 58F11, 93B05, 93D20

1. Introduction

The local study of ordinary differential equations and smooth dynamical systems via linearization techniques and Lyapunov exponents goes back to Lyapunov (1892) and is exemplified e.g. in the books by Nemytskii and Stepanov (1960), Cesari (1971), Hahn (1967), Mañé (1987) and many others. In the time dependent case, Oseledeč's multiplicative ergodic theorem (1968) shows, how to obtain results with probability one about Lyapunov regular points, i.e. about Lyapunov exponents, invariant manifolds, exponential stability, and behavior under small perturbations. Likewise, entropy theory, bifurcation theory, strange attractors etc. can be closely related to Lyapunov exponents of dynamical systems. While Oseledeč's theorem is also a convenient starting point for the local analysis of nonlinear stochastic systems in the form of stochastic flows (compare e.g. many contributions in this volume), this is not the case for control systems.

Consider a control system of the form

$$\dot{x}(t) = X_0(x(t)) + \sum_{i=1}^{m} u_i(t) X_i(x(t)), \quad x(0) = x_0 \in M, \quad t \in \mathbb{R} \tag{1.1}$$

[1]Research supported in part by NSF grant no. DMS 8813976 and DFG grants no. Co124/6-1 and Co124/8-1.

on a finite dimensional, smooth manifold M with smooth vectorfields $X_0, \ldots, X_m$. The $u_i(\cdot)$ are the admissible control functions in $\mathcal{U} = \{u\colon \mathbb{R} \to U \subset \mathbb{R}^m$, locally integrable$\}$.

Typical questions in control theory include: Given two points $y, z \in M$, does there exist a control function $u \in \mathcal{U}$ such that $y = x(t, z, u)$? Does there exist a control $u \in \mathcal{U}$ such that the system can be stabilized at a given trajectory, e.g. a rest point? Is this stabilization robust with respect to small (or large) perturbations of the vector fields? If the trajectory of $\dot{x}(t) = X_0(x(t))$ has a complicated behavior, does there exist a $u \in \mathcal{U}$ such that the corresponding solution of (1.1) has a simple (desired) behavior? Can these goals be achieved via feedback controls, i.e. $u = f(x)$? Compare e.g. Wonham (1979), Isidori (1989), or Nijmeijer and van der Schaft (1990) for a range of control theoretic topics.

Many techniques in control theory are global in nature, compare the literature mentioned above. If linearization techniques are used (e.g. around rest points for all $X_0 \ldots X_m$), then the system is usually linearized with respect to x and u, yielding a linear control system of the form

$$\dot{x}(t) = Ax(t) + Bu(t). \tag{1.2}$$

But the typical problems listed above show that linearization with respect to u is often not appropriate, because one looks for the existence of some $u \in \mathcal{U}$ with a desired property, which is a global question in u. On the other hand, linearization with respect to x around a rest point leads to a bilinear control system

$$\dot{x}(t) = A_0 x(t) + \sum_{i=1}^{m} u_i(t) A_i x(t). \tag{1.3}$$

Since control systems contain matrix functions that are not Lyapunov regular, one cannot use Oseledeč's theorem or similar approaches to describe the Lyapunov spectrum of (1.3) and its implications for the local behavior of the nonlinear system. Other spectral concepts, like the dynamical spectrum (compare e.g. Sacker and Sell (1978) and Johnson et al. (1987)) or the Morse spectrum based on Morse decompositions of projected, linearized flows (compare Sections 3. and 4.) are too crude to describe the local behavior of (1.1) appropriately.

In this paper we present an approach to the study of global and local open loop properties of control systems, which is based on the concept of control flows on $\mathcal{U} \times M$ for the global ideas, and on the induced flows on $\mathcal{U} \times TM$ and $\mathcal{U} \times \mathbb{P}M$ for local analysis. (Here TM and $\mathbb{P}M$ are the tangent and the projective bundle over M.) It turns out that a careful study of the control properties of the systems on M and on $\mathbb{P}M$, using control and chain control sets, and the approximation of spectral values via Lyapunov exponents of regular elements (using ergodic theory or approximation by piecewise constant, periodic solutions) allows the characterization of Lyapunov exponents for (topologically) "thick" sets in $\mathcal{U} \times M$. Our techniques are a combination of methods from the theory of dynamical systems and from control theory, with many of them being inspired by ideas from the theory of stochastic flows.

Our goal is to develop techniques to obtain exact criteria for nonlinear control systems with respect to the problems mentioned above. For linear systems of the form (1.2) with $U = \mathbb{R}^m$ methods from linear algebra together with Lyapunov functions, Riccati equations etc. can be used to solve this problem, and the explicit constructions yield feedback controls via gain matrices automatically. But if $U \subset \mathbb{R}^m$ is bounded, then the problem becomes 'nonlinear' (compare Section 5.) and many techniques break down, i.e. yield only sufficient conditions. For nonlinear systems, Lyapunov function techniques still work for some problems (compare e.g. Sontag (1983)). Other authors have developed approaches to stabilization using center manifolds of reference systems (see e.g. Byrnes and Isidori (1989), Knobloch (1988)), and the feedback problem is treated e.g. in Sussmann (1979) and Sontag (1989). The approach presented here uses control flows and Lyapunov exponents as a unifying tool.

In Section 2. we describe the set up and the notations used throughout this paper, in particular control flows on $\mathcal{U} \times M$, linearized flows on $\mathcal{U} \times TM$ and their projections to $\mathcal{U} \times \mathbb{P}M$ are introduced. Since our results rely on some previous work of the authors, we review (with minor extensions) in Section 3. control sets, chain control sets and their characterizations for the flows presented before. Section 4 introduces Lyapunov exponents for (1.1) and characterizes the Lyapunov spectrum over control and chain control sets in terms of ergodic theory for the associated flows. In Section 5 we demonstrate for the problems of stabilization of bilinear systems and for robustness of linear systems that the Lyapunov exponent approach does yield exact criteria for control systems (and also for classes of stochastically perturbed systems). Finally in Section 6. the simple nonlinear, controlled Verhulst equation is treated with respect to controllability, stabilization and robustness, using the results from Sections 3.–5. The approach developed there can be used for general nonlinear control systems with one-dimensional state space.

2. Control Systems, Control Flows, and their Linearizations

In this section we set up the basic concepts and notations, which will be used throughout the paper. We will consider the following class of nonlinear control systems on a para-compact connected d-dimensional Riemannian C^∞ manifold M

$$\dot{x}(t) = X_0(x(t)) + \sum_{i=1}^{m} u_i(t) X_i(x(t)), \quad t \in \mathbb{R}, \quad x(0) = x_0 \in M, \tag{2.1}$$

where the $X_0, \ldots, X_m$ are C^∞ vectorfields and the admissible controls are $u := (u_i) \in \mathcal{U} := \{u \colon \mathbb{R} \to U, \text{ locally integrable}\}$ with $U \subset \mathbb{R}^m$ compact and convex. We assume that for all $u \in \mathcal{U}$ and $x_0 \in M$ the equation (2.1) has a unique solution $\varphi(\cdot, x_0, u)$, defined on $\mathbb{R}$ with $\varphi(0, x_0, u) = x_0$. Although most of our results remain true (with appropriate modifications) for more general systems of the type $\dot{x}(t) = X(x(t), u(t))$, (2.1) is a particularly nice and popular class of nonlinear control systems, because the controls appear linearly, compare e.g. Isidori (1989) and Nijmeijer and van der Schaft (1990) for basic concepts and results in nonlinear systems theory.

Because the admissible control functions in (2.1) are time dependent, for each $u \in \mathcal{U}$ the vector field on the right hand side is time dependent, and thus a flow for these systems has to be constructed over the space $\mathcal{U} \times M$. This can be done as follows: On $\mathcal{U}$ consider the natural shift defined by

$$(2.2) \qquad \theta: \mathbf{R} \times \mathcal{U} \to \mathcal{U}, \quad \theta_t u(\cdot) = u(t + \cdot).$$

Then θ defines a continuous dynamical system on $\mathcal{U}$, if we equip $\mathcal{U}$ with the weak* topology of $L^\infty(\mathbf{R}, \mathbf{R}^m) = \left(L^1(\mathbf{R}, \mathbf{R}^m)\right)^*$. $\mathcal{U}$ with this topology is compact and metrizable. Note that the weak* topology on $\mathcal{U}$ is well suited for the study of (2.1), because convergence of $u_n \to u$ in $\mathcal{U}$ implies uniform convergence of $\varphi(\cdot, x_0, u_n) \to \varphi(\cdot, x_0, u)$ in M on compact time intervals, compare Colonius and Kliemann (1990^a) for these facts and a more detailed study of the shift space $(\mathcal{U}, \theta)$.

To the control system (2.1) we associate a control flow in the following manner:

$$(2.3) \qquad \phi: \mathbf{R} \times \mathcal{U} \times M \to \mathcal{U} \times M, \quad \phi_t(u, x) = (\theta_t u, \varphi(t, x, u)),$$

where θ_t is the shift on $\mathcal{U}$, and $\varphi(\cdot, x, u)$ denotes the solution of equation (2.1) as above. ϕ is a skew product flow on $\mathcal{U} \times M$, because $\phi_{t+s} = \phi_t \circ \phi_s$, and $\varphi(t + s, x, u) = \varphi(t, \varphi(s, x, u), \theta_s u)$. In fact, ϕ defines a continuous dynamical system on $\mathcal{U} \times M$, compare Lemma 3.4 in Colonius and Kliemann (1990^a).

If $\mathcal{U}$ carries a θ-invariant probability measure, then (2.3) can be interpreted as a stochastic flow, see Colonius and Kliemann (1990^c) for the set up and some results on ergodic theory of control and stochastic flows, which can be proved in this unifying framework.

In this paper we are interested in linearization techniques for nonlinear control systems. Linearizing the system (2.1) with respect to the state variable x, we obtain a system defined on the tangent bundle TM:

$$(2.4) \qquad \begin{aligned} (\dot{T}x)(t) &= TX_0(Tx) + \sum_{i=1}^m u_i(t)TX_i(Tx), \quad t \in \mathbf{R}, \\ (Tx)(0) &= (x_0, v_0) \in T_{x_0}M, \text{ the tangent space at } x_0 \in M, \end{aligned}$$

where for a smooth vectorfield X on M its linearization is denoted by $TX = (X, DX)$. Locally this means: If $X_j = \sum_{k=1}^d \alpha_{kj}(x)\frac{\partial}{\partial x_k}$, denote the Jacobian of the coefficient functions by $A_j(x) = \left(\frac{\partial \alpha_{kj}(x)}{\partial x_e}\right)$. Then $TX_j(x, v) = (\alpha_j(x), A_j(x)v)$, and (2.4) is a pair of coupled differential equations, given locally by

$$(2.5) \qquad \begin{aligned} \dot{x}(t) &= \alpha_0(x) + \sum_{i=1}^m u_i(t)\alpha_i(x), \quad x(0) = x_0, \\ \dot{v}(t) &= A_0(x)v + \sum_{i=1}^m u_i(t)A_i(x)v, \quad v(0) = v_0. \end{aligned}$$

Note that we have linearized (2.1) only with respect to the state $x \in M$, and not with respect to the control $u \in U$. (Linearization with respect to x and u is common practice in the control theory literature, but we are interested in results that are global in $u \in \mathcal{U}$.)

In particular, if $x(t) \equiv x_0 \in M$ is a rest point for each vectorfield X_i, $i = 0, \dots, m$, then the linearized equation is a bilinear control system. (Linearization with respect to x and u leads to linear control systems, which for unbounded control values $u \in \mathbb{R}^m$ can be treated by methods from linear algebra.)

The system (2.4) induces a control system on the projective bundle $\mathbb{P}M$, given by

$$(2.6)\qquad \begin{aligned} (\mathbb{P}\dot{x})(t) &= \mathbb{P}X_0(\mathbb{P}x) + \sum_{i=1}^{m} u_i(t)\mathbb{P}X_i(\mathbb{P}x), \quad t \in \mathbb{R}, \\ (\mathbb{P}x)(0) &= (x_0, s_0) \in \mathbb{P}_{x_0}M, \text{ the projective space at } x_0 \in M, \end{aligned}$$

where $\mathbb{P}X$ is the projection of a vector field TX on TM onto $\mathbb{P}M$, i.e. the $\mathbb{P}X_j$ read locally

$$(2.7)\quad \mathbb{P}X_j(x,s) = (\alpha_j(x), h(A_j(x),s)), \quad h(A_j(x),s) = \left[A_j(x) - s^T A_j(x)s \cdot id\right] s.$$

Here ^{T} denote transposition, and id is the $(d \times d)$ identity matrix. The trajectories of the control system (2.4) will be denoted by $T\varphi(\cdot, Tx, u)$, and those of (2.6) by $\mathbb{P}\varphi(\cdot, \mathbb{P}x, u)$.

Note that using the same construction, it is also possible to lift the system (2.1) to $\mathbb{F}M$, the flag manifold over M, $G\ell(M)$, the frame bundle over M, or $\mathcal{O}(M)$, the orthonormal frame bundle over M (see e.g. San Martin and Arnold (1986) or San Martin (1986)), but we will not make explicit use of these systems in this paper.

Just as the underlying control system (2.1), the linearized system (2.4) on TM and the projected system (2.6) on $\mathbb{P}M$ give rise to associated flows

$$(2.8)\qquad T\phi\colon \mathbb{R} \times \mathcal{U} \times TM \to \mathcal{U} \times TM, \quad T\phi_t(u, Tx) = (\theta_t u, T\varphi(t, Tx, u)),$$

$$(2.9)\qquad \mathbb{P}\phi\colon \mathbb{R} \times \mathcal{U} \times \mathbb{P}M \to \mathcal{U} \times \mathbb{P}M, \quad \mathbb{P}\phi_t(u, \mathbb{P}x) = (\theta_t u, \mathbb{P}\varphi(t, \mathbb{P}x, u)),$$

which are also skew product flows, defining continuous dynamical systems on $\mathcal{U} \times TM$ (and $\mathcal{U} \times \mathbb{P}M$, respectively). In this paper we will be concerned mainly with the Lyapunov exponents of (2.8) and their characterization via (2.9). This analysis will require some characterizations of the flows $T\phi$ and $\mathbb{P}\phi$ in terms of concepts from the theory of dynamical systems. These results are presented in the next section.

3. Control Sets, Chain Control Sets, and Subbundle Decompositions

Consider again the underlying control system on the d-dimensional manifold M

$$(2.1)\qquad \dot{x}(t) = X_0(x(t)) + \sum_{i=1}^{m} u_i(t)X_i(x(t)), \quad t \in \mathbb{R}, \quad x(0) = x_0 \in M,$$

with $u \in \mathcal{U}$. In order to avoid degenerate situations, we will assume throughout the rest of this paper the following integrability condition

$$(\mathrm{H})\qquad \dim \mathcal{LA}\{X_0 + \sum u_i X_i;\ (u_i) \in U\}(x) = d \quad \text{for all } x \in M,$$

where for a set $\mathcal{X}$ of vectorfields on M, $\mathcal{LA}\{\mathcal{X}\}$ denotes the Lie algebra generated by $\mathcal{X}$. Associated with (2.1) are the positive semigroup $\mathcal{S}^+$, the negative semigroup $\mathcal{S}^-$, and the group $\mathcal{G}$ of solution diffeomorphisms generated by piecewise constant controls: Denote by $N := \{X_0 + \sum u_i X_i;\ (u_i) \in U\}$ the admissible (time independent) right hand sides of (2.1), and define

$$\begin{aligned} \mathcal{S}^+ &= \{\exp(t_n Y_n) \circ \cdots \circ \exp(t_1 Y_1);\ Y_i \in N,\ t_i \geq 0,\ i = 1, \ldots, n \in \mathbb{N}\}, \\ \mathcal{S}^- &= \{\exp(t_n Y_n) \circ \cdots \circ \exp(t_1 Y_1);\ Y_i \in N,\ t_i \leq 0,\ i = 1, \ldots, n \in \mathbb{N}\}, \\ \mathcal{G} &= \{\exp(t_n Y_n) \circ \cdots \circ \exp(t_1 Y_1);\ Y_i \in N,\ t_i \in \mathbb{R},\ i = 1, \ldots, n \in \mathbb{N}\}. \end{aligned} \tag{3.1}$$

The subsets of $\mathcal{S}^+$, $\mathcal{S}^-$, $\mathcal{G}$ with $\sum_{i=1}^{n} |t_i| \leq t$ are denoted by $\mathcal{S}^+_{\leq t}$, $\mathcal{S}^-_{\leq t}$, $\mathcal{G}_{\leq t}$, and similarly for $\sum |t_i| = t$ by $\mathcal{S}^+_t$, $\mathcal{S}^-_t$, $\mathcal{G}_t$. Under Hypothesis (H) we have for the system (2.1):

3.1. LEMMA.

(i) *$\mathcal{G}$ acts transitively on M, i.e. $\mathcal{G}x = M$ for all $x \in M$.*

(ii) *int $\mathcal{S}^+_{\leq t}x \neq \phi$, and int$\mathcal{S}^-_{\leq t}x \neq \phi$ for all $t > 0$, i.e. the reachable set up to time $t > 0$ (and the controllable set up to time $-t < 0$) has nonvoid interior on M for all $x \in M$. This property is often called 'local accessibility' in the control theory literature.*

(iii) *The (positive) orbits up to time $t > 0$ on M, defined by*

$$\mathcal{O}^+_{\leq t}(x) = \{y \in M;\ \text{there exist } u \in \mathcal{U} \text{ and } 0 \leq s \leq t \text{ with } y = \varphi(s, x, u)\}$$

have nonvoid interior in M, and $\overline{\mathcal{S}^+_{\leq t}x} = \overline{\mathcal{O}^+_{\leq t}(x)}$ for all $t > 0$, all $x \in M$. Similarly for the negative orbits $\mathcal{O}^-_{\leq t}(x)$.

For a proof see e.g. Isidori (1989) and Nijmeijer and van der Schaft (1990).

Lemma 3.1 shows that under integrability of N the manifold M is the state space of appropriate dimension for the system (2.1), if (H) holds. This does not imply, however, that (2.1) is completely controllable on M. We define the subsets, in which (2.1) is controllable, as follows:

3.2. DEFINITION: A set $D \subset M$ is called a control set of (2.1), if

(i) for every $x \in D$ there exists $u \in \mathcal{U}$ such that $\varphi(t, x, u) \subset D$ for all $t \geq 0$,

(ii) for all $x \in D$ we have $\overline{\mathcal{O}^+(x)} \supset D$,

(iii) D is maximal with respect to the properties (i) and (ii).

If, furthermore, $\overline{D} = \overline{\mathcal{O}^+(x)}$ for all $x \in D$, then D is called an invariant control set.

We are, in particular, interested in control sets with nonvoid interior, and for these sets the following results hold:

3.3. LEMMA.

(i) *Let $D \subset M$ be a control set with* int $D \neq \phi$. *Then*

(a) *for all $x \in D$, $y \in$* int *D there exist $t \geq 0$ and $g \in S^+_{\leq t}$ such that $y = gx$,*

(b) $\overline{\text{int } D} = \overline{D}$.

(ii) (a) *Control sets are pairwise disjoint, and in general neither open nor closed. A control set with nonvoid interior is closed iff it is invariant.*

(b) *If M is compact, then there exist at least one invariant and one open control set in M, the invariant control sets have nonvoid interior, i.e. are closed.*

A proof of these facts can be found e.g. in Colonius and Kliemann (1989).

The control sets of (2.1) are ordered in the following way: Let D_1 and D_2 be control sets, then we define

(3.2) $$D_1 \prec D_2 \text{ if there exist } x \in D_1 \text{ and } y \in D_2 \text{ with } y \in \overline{O^+(x)}.$$

The relation (3.2) defines an order on the control sets, where the closed (i.e. invariant) control sets are maximal elements of $\prec$, the open control sets are minimal elements. If M is compact, then the maximal (minimal) elements are exactly the closed (open) control sets, compare Colonius and Kliemann (1990[c]), Lemma 3.11.

For this paper, we are interested in control sets (with nonvoid interior) mainly for three reasons:

(i) Control sets $D \subset M$ can be lifted to invariant sets $\mathcal{D} \subset \mathcal{U} \times M$, and the flow ϕ, defined in (2.3), restricted to $\mathcal{D}$ is a dynamical system with several interesting properties.

(ii) Over the control sets $D \subset M$, the control structure of the system on the projective bundle $\mathbb{P}M$ (compare (2.6)) can be described, which allows us to give a characterization of the Lyapunov exponents.

(iii) The control sets $_{\mathbb{P}}D \subset \mathbb{P}M$ of (2.6) can be lifted to $\mathcal{U} \times \mathbb{P}M$, where an ergodic theory for Lyapunov exponents can be developed.

As a next step, we will therefore study the lifts of control sets and the control structure of (2.6).

For a control set $D \subset M$, with int $D \neq \phi$, we define its lift to $\mathcal{U} \times M$ by

(3.3) $$\mathcal{D} = c\ell\{(u,x) \in \mathcal{U} \times M;\ \varphi(t,x,u) \in \text{int } D \text{ for all } t \in \mathbb{R}\},$$

where the closure is taken with respect to the weak* topology in $\mathcal{U}$ and the given topology on M. Since $\mathcal{D}$ is a closed, ϕ-invariant set, we can consider the dynamical system $(\mathcal{D}, \phi|_{\mathcal{D}})$. Using this flow, control properties of (2.1) can be described in terms of concepts from the theory of dynamical systems, such as topological mixing, topological transitivity, chaos, compare Colonius and Kliemann (1990[a]), Section 3.

In order to analyze the linearized systems (2.4) and (2.6), we will need a condition similar to (H) for the system on $\mathbb{P}M$:

(PH) $$\dim \mathcal{LA}\{\mathbb{P}X_0 + \sum u_i \mathbb{P}X_i;\ (u_i) \in U\}(x,v) = 2d - 1 \text{ for all } (x,v) \in \mathbb{P}M.$$

Compare San Martin and Arnold (1986) for a detailed discussion of Assumption (PH) and its consequences.

3.4. REMARK: If $C \subset M$ is an invariant control set of (2.1), then, under (H), int $C \neq \phi$. Hence, under (PH), all the consequences of the Lemmas 3.1 and 3.3 hold for the control sets of (2.6) on the state space $\mathrm{Pint}C$, the projective bundle over the manifold $\mathrm{int}C \subset M$. In particular, if the system (2.1) is completely controllable on M, then $C = \mathrm{int}C = M$, and we can consider the system on $\mathbb{P}M$. This is the case studied by San Martin and Arnold (1986) and by San Martin (1986) in a stochastic context.

The following theorem is a slight generalization of Theorem 3.10 in Colonius and Kliemann (1990[b]), using the ideas in San Martin and Arnold (1986) and San Martin (1986). It shows that over the invariant control sets of (2.1) the control structure of (2.6) is uniform:

3.5. THEOREM. *Assume that (PH) holds, and let C be a compact invariant control set of (2.1). Then it holds for the projected system (2.6) on* $\mathrm{Pint}C$*:*

(i) *There are k control sets ${}_{\mathbb{P}}D_i$ with nonvoid interior in* $\mathrm{Pint}C$ *and $1 \leq k \leq d$; we call these the main control sets of the system.*
(ii) *The main control sets are linearly ordered, where the order is defined as in (3.2), where $\mathcal{O}^+$ now denotes the positive orbit of (2.6). We enumerate the sets such that ${}_{\mathbb{P}}D_1 \prec_{\mathbb{P}} {}_{\mathbb{P}}D_2 \prec_{\mathbb{P}} \cdots \prec_{\mathbb{P}} {}_{\mathbb{P}}D_k$. In particular, ${}_{\mathbb{P}}D_k$ is the unique invariant control set over C.*

A more explicit description of the control sets ${}_{\mathbb{P}}D_i$ is given in Section 5. for systems linearized around a rest point.

Now consider the linearized system (2.4) on the tangent bundle TM. We can extend, at least over invariant control sets C of (2.1), the ${}_{\mathbb{P}}D_i$ to ${}_{T}D_i \subset TM$ by taking in each T_xM the corresponding subspaces over $\mathbb{P}_x D_i := {}_{\mathbb{P}}D_i|_{T_xM}$. For a control set ${}_{\mathbb{P}}D$ in $\mathbb{P}M$, and its extension ${}_{T}D$ in TM define its lift to $\mathcal{U} \times \mathbb{P}M$ (and $\mathcal{U} \times TM$ respectively) to be

$$(3.4) \qquad {}_{\mathbb{P}}\mathcal{D} = cl\{(u, \mathbb{P}x);\ \mathbb{P}\varphi(t, Px, u) \in \text{int } {}_{\mathbb{P}}D \text{ for all } t \in \mathbb{R}\},$$

and similarly for ${}_{T}\mathcal{D}$. The question arises, whether the ${}_{T}\mathcal{D}$'s define an invariant subbundle decomposition of $(\mathcal{U} \times TM, T\phi)$. Unfortunately, this is in general not true, as the following simple example shows:

3.6. EXAMPLE:

Consider the following 2-dimensional system, linearized around a rest point (hence we give only the second component of (2.5))

$$(3.5) \qquad \dot{v}(t) = \begin{pmatrix} 1 & 0 \\ 0 & 0 \end{pmatrix} v(t) + u_1(t) \begin{pmatrix} 0 & 1 \\ 1 & 0 \end{pmatrix} v(t) + u_2(t) \begin{pmatrix} 0 & 0 \\ 0 & 1 \end{pmatrix} v(t)$$

with $U = \left[0, \frac{1}{2}\right] \times [1, 2]$. To describe the control sets of the projected system on $\mathbb{P}$, parametrize $\mathbb{P}$ via the angle as $\mathbb{P} = \{\eta;\ \frac{-\pi}{2} < \eta \leq \frac{\pi}{2}\}$. Then the set ${}_{\mathbb{P}}D_1 = \left(-\frac{\pi}{4}, 0\right)$ is the open main control set, and ${}_{\mathbb{P}}D_2 = \left[\frac{\pi}{4}, \frac{\pi}{2}\right]$ is the closed main control set. These sets are connected by a continuum of control sets (with empty interior), which are rest points on $\mathbb{P}$ of the equation corresponding to the controls $u_1(t) \equiv 0$, $u_2(t) \equiv 1$, compare

Example 5.8 in Colonius and Kliemann (1990[b]). Now note that for constant controls $u_1(t) \equiv \alpha$, $u_2(t) \equiv \beta$ the eigenvalues of the systems matrix $\begin{pmatrix} 1 & \alpha \\ \alpha & \beta \end{pmatrix}$ are given by

$$\lambda_{1,2} = \frac{1+\beta}{2} \pm \sqrt{\alpha^2 - \beta + \frac{1}{4}(1+\beta)^2},$$

and we have one dimensional (real) eigenspaces for all $(\alpha, \beta) \in U$, if $\alpha \neq 0$ and $\beta \neq 1$. For $\alpha = 0$, $\beta = 1$ the (generalized) eigenspace is $\mathbb{R}^2$. Therefore the sets ${}_T\mathcal{D}_1$ and ${}_T\mathcal{D}_2$ do not define a subbundle decomposition of $(\mathcal{U} \times TM, T\phi)$, because subbundles are necessarily constant dimensional.

In order to obtain the finest subbundle decomposition of $(\mathcal{U} \times TM, T\phi)$, which respects the control structure of the system (2.6) on $\mathbb{P}M$, we have to introduce generalizations of control sets, called chain control sets:

3.7. DEFINITION: A set $E \subset M$ is called a chain control set of the system (2.1) if

(i) for every $x \in E$ there exists $u \in \mathcal{U}$ such that $\varphi(t, x, u) \in E$ for all $t \in \mathbb{R}$,

(ii) for all $x, y \in E$, all $\varepsilon > 0$, and all $T > 0$ there are $k \in \mathbb{N}$, $x_0, \ldots, x_k \in M$, $u_0, \ldots, u_{k-1} \in \mathcal{U}$ and $t_0, \ldots, t_{k-1} \geq T$ with $x_0 = x$, $x_k = y$ and $d(\varphi(t_i, x_i, u_i), x_{i+1}) < \varepsilon$ for $i = 1, \ldots, k-1$. Here d denotes the Riemannian metric on M.

(iii) E is maximal with respect to these properties.

3.8. LEMMA.

(i) *Chain control sets are closed, connected, and pairwise disjoint.*

(ii) *For every control set $D \subset M$ there exists a unique chain control set E with $D \subset E$.*

(iii) *Chain control sets are ordered by the relation*

$$E_i \prec E_j \text{ if there exists } (u, x) \in \mathcal{U} \times M \text{ with } \omega^*(u, x) \subset E_i \text{ and } \omega(u, x) \subset E_j, \tag{3.6}$$

where $\omega^*(u, x)$ denotes the α-limit set of the trajectory $\varphi(\cdot, x, u)$, and $\omega(u, x)$ its ω-limit set.

For a proof compare Colonius and Kliemann (1990[a]).

3.9. REMARK: While each control set is contained in some chain control set, it is possible that one chain control set may contain several control sets (even with nonvoid interior), compare e.g. Example 3.6, where $E = \mathbb{P}^1$ is the chain control set. Also points that are in no control set, may be contained in a chain control set. Control sets are chain control sets, if they satisfy a certain isolation property, see the discussion in Colonius and Kliemann (1990[a]), Section 4.

Define the lift of a chain control set $E \subset M$ of (2.1) to $\mathcal{U} \times M$ by

$$\mathcal{E} = \{(u, x) \in \mathcal{U} \times M;\ \varphi(t, x, u) \in E \text{ for all } t \in \mathbb{R}\}. \tag{3.7}$$

Similarly for the projected linearized system (2.6) on $\mathbb{P}M$ we can define its chain control sets, which we denote by ${}_{\mathbb{P}}E \subset \mathbb{P}M$. ${}_{\mathbb{P}}E$ can be lifted to $\mathcal{U} \times \mathbb{P}M$ as above, and we define the extensions to $\mathcal{U} \times TM$ by

$$(3.8) \quad {}_{T}\mathcal{E} = \{(u, Tx) \in \mathcal{U} \times TM;\ (u, Tx) \notin Z \implies \mathbb{P}\varphi(t, Tx, u) \in {}_{\mathbb{P}}E \text{ for all } t \in \mathbb{R}\},$$

where Z is the zero section of the bundle TM. The following theorem gives the desired decomposition result:

3.10. THEOREM. *Assume that M is compact, and that M is the chain control set of (2.1).*

(i) *The control system (2.6) on $\mathbb{P}M$ has ℓ chain control sets ${}_{\mathbb{P}}E_1, \ldots, {}_{\mathbb{P}}E_\ell$ with $1 \leq \ell \leq d = \dim M$. The order defined in (3.6) is a linear order, and we enumerate the sets such that ${}_{\mathbb{P}}E_1 \prec_{\mathbb{P}} \cdots \prec_{\mathbb{P}} {}_{\mathbb{P}}E_\ell$.*

(ii) *The lifts ${}_{T}\mathcal{E}_i$, $i = 1, \ldots, \ell$ are invariant subbundles of $\mathcal{U} \times TM$ with $\mathcal{U} \times TM = {}_{T}\mathcal{E}_1 \oplus \cdots \oplus {}_{T}\mathcal{E}_\ell$.*

(iii) *$\{{}_{\mathbb{P}}\mathcal{E}_1, \ldots, {}_{\mathbb{P}}\mathcal{E}_\ell\}$ is the (unique) finest Morse decomposition of the flow $(\mathcal{U} \times \mathbb{P}M, \mathbb{P}\phi)$.*

A proof of these results can be found in Colonius and Kliemann (1990^b), Theorem 4.9.

Combining Theorems 3.5 and 3.10 we obtain the following

3.11. COROLLARY. *Assume that* (PH) *holds, and let C be a compact, invariant control set of (2.1). Then we have for the projected system (2.6) on* $\mathbb{P}\text{int}C$*: Every chain control set ${}_{\mathbb{P}}E_j$ contains a main control set ${}_{\mathbb{P}}D_i$. In particular, $1 \leq \ell \leq k \leq d$ and* int ${}_{\mathbb{P}}E_j \neq \phi$ *for all $j = 1, \ldots, \ell$.*

The proof is an obvious generalization of Theorem 5.5 in Colonius and Kliemann (1990^b). A more precise characterization of the ${}_{\mathbb{P}}D_i$ and ${}_{\mathbb{P}}E_j$ in terms of eigenspaces of matrices in the systems group of the linearized system will be given in Section 5.

Using the preparatory material in this section, we now turn to the discussion of the Lyapunov exponents of (2.1) and their characterization via ergodic theory.

4. Ergodic Theory for Lyapunov Exponents.

Consider again the nonlinear control system

$$(2.1) \qquad \dot{x}(t) = X_0(x(t)) + \sum_{i=1}^{m} u_i(t) X_i(x(t)), \quad t \in \mathbb{R}, \quad x(0) = x_0 \in M$$

and the induced systems (2.4) on TM, and (2.6) on $\mathbb{P}M$. Again in order to avoid degenerate situations, we will assume that the Lie algebra condition (PH) on $\mathbb{P}M$ holds. Recall from Section 2. that the linearized flow $T\phi$ on $\mathcal{U} \times TM$ is of the form $T\phi = (\phi, D\varphi)$, where $D\varphi$ is a cocycle associated with ϕ.

The (forward) Lyapunov exponent of the control flow ϕ on $\mathcal{U} \times M$ at $(u, x) \in \mathcal{U} \times M$ in the direction of $v \in T_x M$, $v \neq 0$, is then defined by

$$\lambda(u, x, v) = \limsup_{t \to \infty} \frac{1}{t} \log \|D\varphi(t, x, u)v\|. \tag{4.1}$$

Similarly, backward Lyapunov exponents $\lambda^-(u, x, v)$ for $t \to -\infty$ are defined. Note that, in general, the following facts hold for Lyapunov exponents (even of linear systems): $\lambda(u, x, v) \neq \lambda^-(u, x, v)$, the lim sup is not a lim, and (exponential) stability of the system does not necessarily imply stability of small perturbations of the vectorfield, see e.g. Cesari (1971) or Hahn (1967) for examples in the context of ordinary differential equations.

In order that these three properties do hold, Lyapunov has introduced the concept of regularity. This concept is best expressed in Oseledeč multiplicative ergodic theorem (1968), where stationarity of the underlying flow ϕ (i.e. the existence of a ϕ-invariant probability measure P) is assumed, together with a certain integrability condition on the cocycle. In this case, there is a ϕ- invariant set of full P-measure, such that all points in this set are Lyapunov regular, compare e.g. Ruelle (1979) for a presentation of this theory, which is one cornerstone in modern entropy theory (cp. Mañé (1987)), ergodic and stability theory of stochastic systems (cp. Carverhill (1985)), Arnold and Wihstutz (1986)), stochastic bifurcation theory (cp. Arnold and Boxler (1989)), and other areas, as also demonstrated by this proceedings volume. In the stochastic case, where the flow ϕ is induced by a nondegenerate diffusion process (over the Wiener space $(\Omega, \mathcal{F}, P)$) with invariant probability measure μ on M, it can actually be shown, for all v fixed, that the Lyapunov exponents are $P \otimes \mu$-almost surely constant, i.e. with probability one only one Lyapunov exponent will be realized, compare Arnold et al. (1986) for the linear case, and San Martin and Arnold (1986) for the nonlinear situation.

For control systems, the situation is completely different, because one always has to deal with nonregular points: Let the vector fields $X_0, \ldots, X_m$ in (2.1) be linear, i.e. the system is of the form

$$\dot{x}(t) = A_0 x(t) + \sum_{i=1}^{m} u_i(t) A_i x(t) \tag{4.2}$$

with non zero constant $d \times d$ matrices $A_0 \ldots A_m$. Let the space of admissible control values $U \subset \mathrm{R}^m$ be the product of intervals I_i, i.e. $U = \mathop{\mathrm{X}}_{i=1}^{m} I_i$, such that at least one I_i has nonvoid interior. Then there exists an admissible control function $u(t) = (u_i(t)) \in \mathcal{U}$, such that the matrix function $A_0 + \sum_{i=1}^{m} u_i(t) A_i$ is not Lyapunov regular, compare e.g. Cesari (1971) or Hahn (1967). Since we are interested in the entire Lyapunov spectrum for all $(u, x, v) \in \mathcal{U} \times TM$, $v \neq 0$, we cannot use versions of Oseledeč's theorem. In the following, we will relate the Lyapunov spectrum of (2.1)

$$\Sigma = \{\lambda(u, x, v);\ (u, x, v) \in \mathcal{U} \times TM,\ v \neq 0\} \tag{4.3}$$

to the control sets of the projected, linearized system (2.6) on $\mathbf{P}M$ via ergodic theory. More detailed information about the linear case (4.2) is presented in Section 5.

Our starting point is the following formula for the Lyapunov exponents

$$\lambda(u,x,v) = \limsup_{t\to\infty} \frac{1}{t} \int_0^t q\left(u(\tau), \mathrm{P}\varphi(\tau, \mathrm{P}x, u)\right) d\tau, \tag{4.4}$$

where again $\mathrm{P}\varphi(\cdot, \mathrm{P}x, u)$ denotes the solution of the projected system (2.6) on $\mathrm{P}M$, with initial value $\mathrm{P}\varphi(0, \mathrm{P}x, u) = (x,v) \in \mathrm{P}M$, i.e. $v \in \mathrm{P}_x M$. Note that $\lambda(u,x,\alpha v) = \lambda(u,x,v)$ for all $\alpha \in \mathrm{R}$, $\alpha \neq 0$, and hence it suffices to consider (4.4) on $\mathcal{U} \times \mathrm{P}M$. The function q is of the form

$$q(u,x,v) = \langle v, \nabla X_0(x)\rangle + \sum_{i=1}^{m} u_i \langle v, \nabla X_i(x)\rangle, \tag{4.5}$$

where $\langle \cdot,\cdot\rangle$ denotes the (Riemannian) inner product on TM, and ∇ is the Riemannian covariant derivative (compare e.g. Baxendale (1986) for details).

From Formula (4.4) it is clear that the Lyapunov exponents are defined on the tail of the trajectory $(u(\cdot), \mathrm{P}\varphi(\cdot, \mathrm{P}x, u))$, therefore we first examine the ω-limit sets of the flow $\mathrm{P}\phi$ on $\mathcal{U} \times \mathrm{P}M$. In the following, we will assume that the manifold M is compact, but our results generalize immediately to the case, where $\cup \{\overline{D}$, D is a control set of (2.1)$\}$ is a bounded set in the manifold metric on M, compare Colonius and Kliemann (1990^c^).

4.1. Definition: Let (S, Ψ) be a continuous dynamical system on a compact metric space S. The ω-limit set of $x \in S$ is defined as

$$\omega(x) := \{y \in S;\ \text{there exists } t_k \to \infty \text{ in } \mathrm{R} \text{ with } \Psi(t_k, x) \to y\}.$$

For a control set ${}_{\mathrm{P}}D$ of (2.6), not necessarily with nonvoid interior, we define its positive lift to $\mathcal{U} \times \mathrm{P}M$ as

$$ {}_{\mathrm{P}}\mathcal{D}^+ := cl\{(u, \mathrm{P}x) \in \mathcal{U} \times \mathrm{P}M;\ \mathrm{P}\varphi(t, \mathrm{P}x, u) \in {}_{\mathrm{P}}D \text{ for all } t \geq 0\}. \tag{4.6}$$

Under our assumption that M is compact, we obtain the following properties of the limit sets of $\mathrm{P}\phi$:

4.2. Lemma.

(i) *For all $(u, \mathrm{P}x) \in \mathcal{U} \times \mathrm{P}M$ the limit sets $\omega(u, \mathrm{P}x)$ are connected, compact, and $\mathrm{P}\phi$-invariant, hence they contain minimal $\mathrm{P}\phi$-invariant sets.*

(ii) *For all $(u, \mathrm{P}x) \in \mathcal{U} \times \mathrm{P}M$ there exists a chain control set ${}_{\mathrm{P}}E$ of (2.6) such that $\omega(u, \mathrm{P}x) \subset {}_{\mathrm{P}}\mathcal{E}$, in particular $\pi_{\mathrm{P}M}\omega(u, \mathrm{P}x) \subset {}_{\mathrm{P}}E$.*

(iii) *For all $(u, \mathrm{P}x) \in \mathcal{U} \times \mathrm{P}M$ there exists a control set ${}_{\mathrm{P}}D$ of (2.6) such that $\omega(u, \mathrm{P}x) \cap {}_{\mathrm{P}}\mathcal{D}^+ \neq \phi$, in particular $\pi_{\mathrm{P}M}\omega(u, \mathrm{P}x) \cap {}_{\mathrm{P}}D \neq \phi$.*

(iv) *Let $\mathcal{W} \subset \mathcal{U} \times \mathrm{P}M$ be a minimal $\mathrm{P}\phi$-invariant set, then there exists a control set ${}_{\mathrm{P}}D$ of (2.6) such that $\mathcal{W} \subset {}_{\mathrm{P}}\mathcal{D}^+$.*

(v) *Let ${}_{\mathrm{P}}D \subset \mathrm{P}M$ be a control set of (2.6), then for any $\mathrm{P}y \in \overline{{}_{\mathrm{P}}D}$ there exists $(u, \mathrm{P}x) \in \mathcal{U} \times \mathrm{P}M$ such that $\mathrm{P}y \in \pi_{\mathrm{P}M}\omega(u, \mathrm{P}x)$.*

(vi) *The set $\{(u,x) \in \mathcal{U} \times M;\ \pi_M\omega(u,x) \subset \mathrm{int}C$ for some invariant control set $C\}$ is open and dense in $\mathcal{U} \times M$.*

The proof of (i)–(v) can be found in Colonius and Kliemann (1990[a], Lemma 5.3), and (vi) is proved in Colonius and Kliemann (1991[a]).

Next we construct and characterize $\mathsf{P}\phi$-invariant probability measures via the Krylov-Bogolyubov construction: Let $C(\mathcal{U} \times \mathsf{P}M, \mathsf{R})$ denote the continuous functions from $\mathcal{U} \times \mathsf{P}M$ into R. For $(u, \mathsf{P}x) \in \mathcal{U} \times \mathsf{P}M$ consider the Cesaro limits for time sequences $t_k \to \infty$

$$\lim_{t_k \to \infty} \frac{1}{t_k} \int_0^{t_k} F(\theta_\tau u, \mathsf{P}\varphi(\tau, \mathsf{P}x, u))\, d\tau = \int_{\mathcal{U} \times \mathsf{P}M} F(v, y) d\mu_{u,\mathsf{P}x} \tag{4.7}$$

for all $F \in C(\mathcal{U} \times \mathsf{P}M, \mathsf{R})$. We will use the following notations for a continuous dynamical system (S, Ψ):

$$\begin{aligned} &\mathcal{M}_\Psi^+\text{: set of } \Psi\text{-invariant probability measures for } t \geq 0, \\ &\Sigma_\Psi^e = \{s \in S;\ \text{the Krylov-Bogolyubov measure } \mu_s \text{ is independent of the} \\ &\qquad\qquad \text{sequence } t_k \text{ and ergodic}\}, \\ &\Sigma_\Psi^s = \{s \in \Sigma_\Psi^e;\ s \in \operatorname{supp} \mu_s\}. \end{aligned} \tag{4.8}$$

4.3. Remark: The following properties of Ψ-invariant measures are well known for compact spaces S, compare e.g. Mañé (1987), Chapter II.6:

(i) $\Sigma_\Psi^s \neq \phi$, and Σ_Ψ^s has total measure with respect to $\mathcal{M}_\Psi^+$, i.e. $\mu(\Sigma_\Psi^s)^c = 0$ for all $\mu \in \mathcal{M}_\Psi^+$. (Here A^c denotes the complement of a set A.)

(ii) Each $\mu \in \mathcal{M}_\Psi^+$ has an ergodic decomposition: Each $F \in L^1(S, \mu)$ is μ_s-integrable for μ-almost all $s \in \Sigma_\Psi^s$ and $\int \left(\int_S F\, d\mu_s \right) d\mu = \int_S F\, d\mu$.

In Crauel (1986) a characterization of (stochastic) Lyapunov exponents in terms of invariant measures was given for linear stochastic systems. We will now generalize these results to nonlinear control systems and, with the help of Lemma 4.2, provide additional insight into the support of invariant measures, and thus into the structure of realizable Lyapunov exponents. In order to use the Krylov- Bogolyubov measures of the control flow $\mathsf{P}\phi$, we first have to lift the function $q\colon U \times \mathsf{P}M \to \mathsf{R}$, defined in (4.5) to a function on $\mathcal{U} \times \mathsf{P}M$: Define for $(u, \mathsf{P}x) \in \mathcal{U} \times \mathsf{P}M$

$$Q(u, \mathsf{P}x) = \liminf_{h \downarrow 0} \frac{1}{h} \int_0^h q(u(\tau), \mathsf{P}\varphi(\tau, \mathsf{P}x, u))\, d\tau. \tag{4.9}$$

4.4. Lemma.

(i) *The function $Q\colon \mathcal{U} \times \mathsf{P}M \to \mathsf{R}$ is measurable and bounded.*

(ii) *For all $(u, \mathsf{P}x) \in \mathcal{U} \times \mathsf{P}M$*

$$\limsup_{t \to \infty} \frac{1}{t} \int_0^t q(u(\tau), \mathsf{P}\varphi(\tau, \mathsf{P}x, u))\, d\tau = \limsup_{t \to \infty} \frac{1}{t} \int_0^t Q(\mathsf{P}\phi_\tau(u, \mathsf{P}x))\, d\tau.$$

PROOF:

(i) Boundedness of Q follows from the boundedness of q on the compact space $U \times \mathbb{P}M$. Furthermore, Q is the "lim inf" of continuous functions $(u, \mathbb{P}x) \mapsto \frac{1}{h}\int_0^h q(u(\tau), \mathbb{P}\varphi(\tau, \mathbb{P}x, u))\, d\tau$ and therefore is measurable.

(ii) It suffices to show that for all $T > 0$ we have $\int_0^T q\, d\tau = \int_0^T Q\, d\tau$: By definition it holds that

$$\int_0^T Q(\mathbb{P}\phi_\tau(u, \mathbb{P}x))\, d\tau = \int_0^T \liminf_{h \downarrow 0} \frac{1}{h} \int_0^h q(u(\tau+\sigma), \mathbb{P}\varphi(\tau+\sigma, \mathbb{P}x, u))\, d\sigma d\tau,$$

and the function $t \mapsto q(u(t), \mathbb{P}\varphi(t, \mathbb{P}x, u))$ is Lebesgue-integrable on $[0, T]$. Therefore, for almost all $\tau \in [0, T]$ the limit $\lim_{h\downarrow 0} \frac{1}{h}\int_0^h q(u(\tau+\sigma), \mathbb{P}\varphi(\tau+\sigma, \mathbb{P}x, u))\, d\sigma$ exists and equals $q(u(\tau), \mathbb{P}\varphi(\tau, \mathbb{P}x, u))$, which proves the lemma.

∎

4.5. THEOREM. *For each $(u, \mathbb{P}x) \in \mathcal{U} \times \mathbb{P}M$ there exists an invariant probability measure $\mu_{u,\mathbb{P}x}$ with supp $\mu_{u,\mathbb{P}x} \subset \omega(u, \mathbb{P}x)$ such that*

(i) $\lambda(u, x, v) = \int_{\omega(u,\mathbb{P}x)} Q(w, \mathbb{P}y) d\mu_{u,\mathbb{P}x} = \int_{\omega(u,\mathbb{P}x)} \lambda(w, \mathbb{P}y) d\mu_{u,\mathbb{P}x}$,

(ii) *there exists a chain control set ${}_{\mathbb{P}}E$ of (2.6) such that for all $(w, \mathbb{P}y) \in$ supp $\mu_{u,\mathbb{P}x}$ we have $\mathbb{P}\varphi(t, \mathbb{P}y, w) \in {}_{\mathbb{P}}E \cap \pi_{\mathbb{P}M}\omega(u, \mathbb{P}x)$ for all $t \geq 0$,*

(iii) *for $\mu_{u,\mathbb{P}x}$-almost all $(w, \mathbb{P}y)$ there exists a control set ${}_{\mathbb{P}}D$ of (2.6) with $\mathbb{P}\varphi(t, \mathbb{P}y, w)$ ${}_{\mathbb{P}}D \cap \pi_{\mathbb{P}M}\omega(u, \mathbb{P}x)$ for all $t \geq 0$,*

(iv) *if $\mu_{u,\mathbb{P}x}$ is ergodic, then the control set ${}_{\mathbb{P}}D$ in (iii) is unique, and $\lambda(w, \mathbb{P}y)$ is constant for $\mu_{u,\mathbb{P}x}$- almost all $(w, \mathbb{P}y) \in \mathcal{U} \times \mathbb{P}M$.*

PROOF:

(i) This follows directly from Lemma 4.4 and Birkhoff's ergodic theorem.

(ii) is an immediate consequence of Lemma 4.2(ii).

(iii) follows from Lemma 4.2(iii) and the fact that for all $(u, \mathbb{P}x) \in \mathcal{U} \times \mathbb{P}M$ there exists a set $\Gamma \subset$ supp $\mu_{u,\mathbb{P}x}$ with $\mu_{u,\mathbb{P}x}\Gamma = 1$, such that for all $(w, \mathbb{P}y) \in \Gamma$ there is a control set ${}_{\mathbb{P}}D \subset \mathbb{P}M$ with $\mathbb{P}\varphi(t, \mathbb{P}y, w) \subset {}_{\mathbb{P}}D \cap \pi_{\mathbb{P}M}\omega(u, \mathbb{P}x)$ for all $t \geq 0$, compare Theorem 5.5(ii) in Colonius and Kliemann (1990^a).

(iv) This is implied by Lemma 4.2(iv), and again Birkhoff's ergodic theorem.

∎

The following result is now a direct consequence of Theorem 4.5 and Remark 4.3. In particular, it shows that all Lyapunov exponents can be obtained as integrals over Lyapunov exponents of regular elements.

4.6. COROLLARY.

(i) *For all* $(u, \mathrm{P}x) \in \sum^{e}_{\mathrm{P}\phi}$ *we have* $\lambda(u,\mathrm{P}x) = \lim\limits_{t\to\infty} \frac{1}{t}\int\limits_0^t Q(\tau)d\tau = \lim\limits_{t\to\infty} \frac{1}{t}\int\limits_0^t q(\tau)d\tau =$ $\lambda(w,\mathrm{P}y)$ *for* $\mu_{u,\mathrm{P}x}$*-almost all* $(w,\mathrm{P}y)$.

(ii) *For all* $(u,\mathrm{P}x) \in \mathcal{U}\times \mathrm{P}M$ *it holds that*

$$\lambda(u,\mathrm{P}x) = \int\limits_{\omega(u,\mathrm{P}x)} \lambda(w,\mathrm{P}y)d\mu_{u,\mathrm{P}x},$$

where

$$\lambda(w,\mathrm{P}y) = \lim_{t\to\infty}\frac{1}{t}\int\limits_0^t q(\tau)d\tau.$$

Next we define the maximal and minimal Lyapunov exponent of the control system (2.1), which are realizable over a given set ${}_{\mathrm{P}}A \subset \mathrm{P}M$:

$$\begin{aligned} \kappa({}_{\mathrm{P}}A) &= \sup\{\lambda(u,\mathrm{P}x);\ \mathrm{P}\varphi(t,\mathrm{P}x,u) \in {}_{\mathrm{P}}A \text{ for all } t \geq 0\} \\ \kappa^*({}_{\mathrm{P}}A) &= \inf\{\lambda(u,\mathrm{P}x);\ \mathrm{P}\varphi(t,\mathrm{P}x,u) \in {}_{\mathrm{P}}A \text{ for all } t \geq 0\}. \end{aligned} \tag{4.10}$$

4.7. THEOREM. *Let* ${}_{\mathrm{P}}\mathcal{A}^+ = c\ell\{(u,\mathrm{P}x) \in \mathcal{U}\times \mathrm{P}M;\ \mathrm{P}\varphi(t,\mathrm{P}x,u) \in {}_{\mathrm{P}}A$ *for all* $t \geq 0\}$ *denote the positive lift of* ${}_{\mathrm{P}}A \subset \mathrm{P}M$ *to* $\mathcal{U}\times\mathrm{P}M$, *and assume* ${}_{\mathrm{P}}\mathcal{A}^+ \neq \phi$. *Then there exist ergodic* $\mathrm{P}\phi$*-invariant measures* $\mu_{\mathrm{P}A}$ *and* $\mu^*_{\mathrm{P}A}$ *with supp* $\mu_{\mathrm{P}A} \subset {}_{\mathrm{P}}\mathcal{A}^+$ *and supp* $\mu^*_{\mathrm{P}A} \subset {}_{\mathrm{P}}\mathcal{A}^+$ *such that*

$$\kappa({}_{\mathrm{P}}A) = \int\limits_{\mathcal{U}\times\mathrm{P}M} Q(w,\mathrm{P}y)d\mu_{\mathrm{P}A},$$

$$\kappa^*({}_{\mathrm{P}}A) = \int\limits_{\mathcal{U}\times\mathrm{P}M} Q(w,\mathrm{P}y)d\mu^*_{\mathrm{P}A}.$$

Furthermore, there are $(u,\mathrm{P}x) \in {}_{\mathrm{P}}\mathcal{A}^+$, $(u^*,\mathrm{P}x^*) \in {}_{\mathrm{P}}\mathcal{A}^+$ *such that* $\mu_{\mathrm{P}A} = \mu_{u,\mathrm{P}x}$ *and* $\mu^*_{\mathrm{P}A} = \mu_{u^*,\mathrm{P}x^*}$ *where* $\mu_{u,\mathrm{P}x}$ *and* $\mu_{u^*,\mathrm{P}x^*}$ *denote again Krylov-Bogolyubov measures from* $(u,\mathrm{P}x)$ *and from* $(u^*,\mathrm{P}x^*)$.

PROOF: Note first of all that, because ${}_{\mathrm{P}}\mathcal{A}^+$ is closed and positively $\mathrm{P}\phi$-invariant, we have $\omega(u,\mathrm{P}x) \subset {}_{\mathrm{P}}\mathcal{A}^+$ for all $(u,\mathrm{P}x) \in {}_{\mathrm{P}}\mathcal{A}^+$. Hence, by Corollary 4.6.

$$\kappa({}_{\mathrm{P}}A) = \sup\left\{\int\limits_{{}_{\mathrm{P}}\mathcal{A}^+} Qd\mu_{u,\mathrm{P}x};\ (u,\mathrm{P}x) \in \Sigma^s_{\mathrm{P}\phi} \cap {}_{\mathrm{P}}\mathcal{A}^+\right\}.$$

Now the Krylov-Bogolyubov measures from points in $\Sigma^s_{\mathrm{P}\phi}$ are extremal points of the convex set $\mathcal{M}^+_{\mathrm{P}\phi}$ (compare Mañé (1987), Section II.2), hence the measures from points in $\Sigma^s_{\mathrm{P}\phi} \cap {}_{\mathrm{P}}\mathcal{A}^+$ are extremal in $\{\mu \in \mathcal{M}^+_{\mathrm{P}\phi};\ \mathrm{supp}\ \mu \subset {}_{\mathrm{P}}\mathcal{A}^+\}$, which is a closed, convex set, hence there exist $\mu_{\mathbf{P}A}$ and $\mu^*_{\mathbf{P}A}$ as required. This proves the formula for $\kappa({}_{\mathbf{P}}A)$, and similarly for $\kappa^*({}_{\mathbf{P}}A)$.

By Remark 4.3 we have $\int_{{}_{\mathbf{P}}A^+} Q\, d\mu_{\mathbf{P}A} = \int_{{}_{\mathbf{P}}A^+} (\int Q\, d\mu_{u,\mathbf{P}x})\, d\mu_{\mathbf{P}A}$ with $(u, \mathbf{P}x) \in \Sigma^s_{\mathbf{P}\phi}$ almost surely with respect to $\mu_{\mathbf{P}A}$. Because $\mu_{\mathbf{P}A}$ is ergodic, $\int Q\, d\mu_{u,\mathbf{P}x}$ is $\mu_{\mathbf{P}A}$-almost everywhere constant, i.e. $\int Q\, d\mu_{\mathbf{P}A} = \int Q\, d\mu_{u,\mathbf{P}x}$ for $\mu_{\mathbf{P}A}$-almost all $(u, \mathbf{P}x)$, which proves the other claim of the theorem.

4.8. REMARK: (On time reversal and backward Lyapunov exponents.)

Consider now the control system (2.1) on the entire time axis $\mathbf{R}$. Its associated time reversed system reads with $x^*(t) = x(-t)$:

$$\dot{x}^*(t) = -X_0(x^*(t)) - \sum_{i=1}^{m} u_i(t) X_i(x^*(t)),\ t \in \mathbf{R},\ x(0) = x_0 \in M. \tag{4.11}$$

Define the backward Lyapunov exponents of (2.1) by

$$\lambda^-(u, x, v) = \limsup_{t \to -\infty} \frac{1}{t} \log \|D\varphi(t, x, u)v\|. \tag{4.12}$$

Note that

$$\begin{aligned}
\lambda^*(u, \mathbf{P}x) &:= \liminf_{t \to \infty} \frac{1}{t} \int_0^t q^*(u(\tau), \mathbf{P}\varphi^*(\tau, \mathbf{P}x, u))\, d\tau \\
&= \liminf_{t \to \infty} \frac{1}{t} \int_0^t -q(u(\tau), \mathbf{P}\varphi^*(\tau, \mathbf{P}x, u))\, d\tau \\
&= -\limsup_{t \to -\infty} \frac{1}{t} \int_0^t q(u(\tau), \mathbf{P}\varphi(\tau, \mathbf{P}x, u))\, d\tau,
\end{aligned}$$

where q^* and $\mathbf{P}\varphi^*$ correspond to the time reversed system (4.11). In particular, if the lim inf in the definition of $\lambda^*(u, \mathbf{P}x)$ is a limit, then $-\lambda^*(u, \mathbf{P}x)$ is a backward Lyapunov exponent of (2.1). Now for the time reversed system (4.11) all results developed above hold with the obvious changes, and the assumption $(\mathbf{P}H)$ holds for (4.11) iff it holds for (2.1).

We obtain from an extension of Birkhoff's ergodic theorem (cp. Mañé (1987), Corollary II.1.4): If the Krylov-Bogolyubov measure $\mu_{u,\mathbf{P}x}$ is ergodic, then

$$\lambda(w, \mathbf{P}y) = \lim_{t \to \infty} \frac{1}{t} \int_0^t Q(\tau) d\tau = \lim_{t \to \infty} \frac{1}{t} \int_0^t Q(-\tau) d\tau = \lambda^-(w, \mathbf{P}y)$$

for $\mu_{u,\mathbf{P}x}$-almost all $(w, \mathbf{P}y)$. Therefore, if $(u, \mathbf{P}x) \in \operatorname{supp} \mu_{u,\mathbf{P}x}$, then $\lambda(u, \mathbf{P}x) = \lambda^-(u, \mathbf{P}x)$, which extends the results of Corollary 4.6 and Theorem 4.7. Note, however, that if $\lambda(u, x, v) = \lambda^-(u, x, v)$ for some $v \in T_x M$, then a basis $\{v_1, \ldots, v_d\}$ in

T_xM need not exist with $\lambda(u,x,v_i) = \lambda^-(u,x,v_i)$ for all $i = 1,\dots,d$, i.e. the point $(u,x) \in \mathcal{U} \times M$ need not be Lyapunov regular for the flow ϕ.

4.9. REMARK: The connection of our results obtained so far with the multiplicative ergodic theorem of Oseledeč is as follows: Consider the point $(u, \mathbb{P}x) \in \mathcal{U} \times \mathbb{P}M$ and the Krylov-Bogolyubov measure $\mu_{u,\mathbb{P}x}$. The projection of $\mu_{u,\mathbb{P}x}$ onto $\mathcal{U} \times M$ is a ϕ-invariant measure, denoted by $\mu_{u,x}$. According to Oseledeč's theorem, there exists a set $\Gamma \subset \operatorname{supp} \mu_{u,x}$ with $\mu_{u,x}\Gamma = 1$, such that all points in Γ are Lyapunov regular. (Note that the set Γ may be very thin from a topological point of view, compare e.g. Mañé (1987), p. 264.) Theorem 4.5 and Corollary 4.6 describe the way, in which the Lyapunov exponents $\lambda(u,x,v)$ can be obtained from the finite Lyapunov spectrum of regular elements of the flow $(\mathcal{U} \times M, \phi, \mu_{u,x})$. In particular, if $\mu_{u,\mathbb{P}x}$ is ergodic, then $\mu_{u,x}$ is ergodic, and hence the set $\{\lambda(w,y,v);\ (w,y) \in \Gamma\}$ consists of at most d numbers, which are also the Lyapunov spectrum of (u,x).

If the measure $\mu_{u,x}$ is not ergodic, then the (finitely many) numbers $\{\lambda(u,x,v);\ v \neq 0\}$ are $\mu_{u,\mathbb{P}x}$-averages over the Lyapunov spectrum of Γ, which now may depend on $(w,y) \in \Gamma$.

We are now ready to apply our results to the analysis of the Lyapunov spectrum

$$\Sigma = \{\lambda(u,x,v);\ (u,x,v) \in \mathcal{U} \times TM,\ v \neq 0\} \tag{4.3}$$

of the control system (2.1). In control theory, one considers the dynamics, i.e. the vector fields $X_0,\dots,X_m$, and the set of control values, i.e. $U \subset \mathbb{R}^m$, as given. The problem then is to find an admissible control function $u \in \mathcal{U}$, such that a certain goal (like controllability, stabilization, etc.) is achieved from an initial point $x \in M$. The goal in our context is to describe the Lyapunov exponents that can be realized from a point $x \in M$ using all $u \in \mathcal{U}$. The standing assumption is still that $(\mathbb{P}H)$ holds, and that M is compact. We start analyzing the Lyapunov spectrum of (2.1) by considering it from the point of view of chain control sets.

Let $E \subset M$ be a chain control set of (2.1) and denote

$$\mathcal{P}E = \{\mathbb{P}E;\ {}_{\mathbb{P}}E \text{ is a chain control set of (2.6) and } \pi_{M\mathbb{P}}E \subset E\}. \tag{4.13}$$

Define the corresponding spectrum over E as

$$\Sigma(E) = \{\lambda(u,x,v);\ (u,x) \in \mathcal{E} \subset \mathcal{U} \times M,\ v \in T_xM\}. \tag{4.14}$$

For a chain control set ${}_{\mathbb{P}}E \subset \mathbb{P}M$ of (2.6) denote its spectral interval by

$$I({}_{\mathbb{P}}E) = [\kappa^*({}_{\mathbb{P}}E), \kappa({}_{\mathbb{P}}E)], \tag{4.15}$$

where κ and κ^* are defined in (4.10). We have the following result:

4.10. THEOREM.

(i) $\Sigma(E) \subset \cup\{I(_{\mathbf{P}}E);\ _{\mathbf{P}}E \in \mathcal{P}E\} =: I(E)$.

(ii) $\Sigma \subset \cup\{I(E);\ E$ is a chain control set of (2.1)$\}$.

(iii) *For a chain control set $E \subset M$ of (2.1), let "$\prec_{\mathbf{P}}$" be the order in $\mathcal{P}E$ as defined in (3.6). Then for two sets $_{\mathbf{P}}E_1$, $_{\mathbf{P}}E_2$ in $\mathcal{P}E$ we have: $_{\mathbf{P}}E_1 \prec_{\mathbf{P}}\ _{\mathbf{P}}E_2$ implies $I(_{\mathbf{P}}E_1) \le I(_{\mathbf{P}}E_2)$, in the sense that $\kappa^*(_{\mathbf{P}}E_1) \le \kappa^*(_{\mathbf{P}}E_2)$ and $\kappa(_{\mathbf{P}}E_1) \le \kappa(_{\mathbf{P}}E_2)$.*

(iv) *In particular, if M is the chain control set of (2.1), then $\Sigma \subset \bigcup_{j=1}^{\ell} I(_{\mathbf{P}}E_j)$, where the $_{\mathbf{P}}E_j$ are as in Theorem 3.10.*

PROOF:

(i) Denote by $\pi\colon \mathbf{P}M \to M$ the projective bundle over M, and note that $(u,x) \in \mathcal{E}$ implies $\mathbf{P}\varphi(t,(x,v),u) \in \pi^{-1}[E]$ for all $t \in \mathbf{R}$, all $v \in \mathbf{P}_x M$. By Lemma 4.2(ii) we therefore have that $\omega(u,x,v) \subset {_{\mathbf{P}}\mathcal{E}}$ for some chain control set $_{\mathbf{P}}E$ in $\mathcal{P}E$. Theorem 4.5(i) now says that $\lambda(u,x,v) \in I(_{\mathbf{P}}E)$.

(ii) For $(u,x) \in \mathcal{U} \times M$ the ω-limit set $\omega(u,x)$ is contained in some lifted chain control set $\mathcal{E}$ and $\pi_M \omega(u,x) \subset E$ by Lemma 4.2(ii). Hence the proof goes through as in (i).

(iii), (iv) This follows from Theorem 3.10 and its proof in Colonius and Kliemann (1990[b]).

∎

The spectral intervals $I(_{\mathbf{P}}E)$ over different chain control sets need not be disjoint, as the following simple example shows:

4.11. EXAMPLE:

Consider the following 2-dimensional linearized system

$$\dot{v}(t) = u_1(t)\begin{pmatrix}1 & 0\\ 0 & 1\end{pmatrix} v(t) + u_2(t)\begin{pmatrix}0 & 1\\ 0 & 0\end{pmatrix} v(t) + u_3(t)\begin{pmatrix}0 & 0\\ 1 & 0\end{pmatrix} v(t) \tag{4.16}$$

with $U = [0,2] \times [\frac{1}{2},1] \times [\frac{1}{2},1]$. The control sets of the projected system on the projective space $\mathbf{P}$ in $\mathbf{R}^2$ are given by

$$_{\mathbf{P}}D_1 = \pi_{\mathbf{P}}\left\{\begin{pmatrix}v_1\\ v_2\end{pmatrix} \in \mathbf{R}^2;\ v_2 = \alpha v_1,\ \alpha \in \left(-\sqrt{2}, -\sqrt{\frac{1}{2}}\right)\right\},$$

$$_{\mathbf{P}}D_2 = \pi_{\mathbf{P}}\left\{\begin{pmatrix}v_1\\ v_2\end{pmatrix} \in \mathbf{R}^2;\ v_2 = \alpha v_1,\ \alpha \in \left[\sqrt{\frac{1}{2}}, \sqrt{2}\right]\right\},$$

where $\pi_{\mathbf{P}}$ denotes the natural identification of points in $\mathbf{R}^2$ as subspaces, i.e. as elements in $\mathbf{P}$. The chain control sets are

$$_{\mathbf{P}}E_1 = {_{\mathbf{P}}\overline{D}_1}, \quad _{\mathbf{P}}E_2 = {_{\mathbf{P}}D_2},$$

compare Colonius and Kliemann (1990[c], Theorem 3.16) for a general technique to compute control sets and chain control sets for systems with 1-dimensional state space, and Colonius and Kliemann (1991[b]) for the details on this example.

For constant $u \in U$ denote by $A(u) = \begin{pmatrix} u_1 & u_2 \\ u_3 & u_1 \end{pmatrix}$ the possible right hand sides of (4.16). Let $\lambda_1(u)$ and $\lambda_2(u)$ be the (real parts of the) eigenvalues of $A(u)$ with $\lambda_1(u) \leq \lambda_2(u)$. Then $\{\lambda_1(u),\ u \in U\} = [-1, \frac{3}{2}]$ and $\{\lambda_2(u);\ u \in U\} = [\frac{1}{2}, 3]$. Since these sets are contained in the spectral intervals over ${}_{\mathbf{P}}E_1$ and ${}_{\mathbf{P}}E_2$, we obtain $I({}_{\mathbf{P}}E_1) \supset [-1, \frac{3}{2}]$ and $I({}_{\mathbf{P}}E_2) \supset [\frac{1}{2}, 3]$, in particular $I({}_{\mathbf{P}}E_1)$ and $I({}_{\mathbf{P}}E_2)$ overlap in an entire interval.

Theorem 4.10 says that, in order to find the entire Lyapunov spectrum Σ of a nonlinear control system (2.1), one only has to consider the spectrum over the chain control sets E of (2.1). In particular, take any $x \in M$ and any chain control set $E \subset M$ with $\overline{\mathcal{O}^+(x)} \cap E = \phi$. Then $\{\lambda(u, x, v);\ u \in \mathcal{U},\ v \in T_xM\} \cap I(E) \neq \phi$, i.e. some Lyapunov exponent in $I(E)$ can be realized from $x \in M$, if E can approximately be reached from x. A similar statement holds for $\{\lambda(u, x, v);\ u \in \mathcal{U}\}$ with given $(x, v) \in \mathbf{P}_xM$, with respect to reachability in the projected system (2.6). This, of course, does not mean that all Lyapunov exponents in $I(E)$ can be realized from $x \in M$, if $\overline{\mathcal{O}^+(x)} \cap E \neq \phi$, compare e.g. Colonius and Kliemann (1991[b]) for a counter example. However, the control structure of (2.1) and (2.6) gives more information on the Lyapunov exponents that can be realized from $x \in M$ or $\mathbf{P}x \in \mathbf{P}_xM$. This is the topic we will discuss throughout the rest of this section.

Let $D \subset M$ be a control set of (2.1) with int $D \neq \phi$, and denote

(4.17) $$\mathcal{P}D = \{{}_{\mathbf{P}}D;\ {}_{\mathbf{P}}D \text{ is a control set of (2.6) with } \text{int}\, {}_{\mathbf{P}}D \neq \phi \text{ and } \pi_M {}_{\mathbf{P}}D \subset D\}.$$

Define the spectrum over ${}_{\mathbf{P}}D$ and D by

(4.18) $$\begin{aligned} \Sigma({}_{\mathbf{P}}D) &= \{\lambda(u, x, v);\ (x, v) \in {}_{\mathbf{P}}D \text{ and } \pi_{\mathbf{P}M}\omega(u, x, v) \subset \text{int} D\}, \\ \Sigma(D) &= \cup\{\Sigma({}_{\mathbf{P}}D);\ {}_{\mathbf{P}}D \in \mathcal{P}D\}. \end{aligned}$$

4.12. THEOREM.

(i) *The closure of the $\Sigma({}_{\mathbf{P}}D)$ are intervals. Furthermore, for all $(x, v) \in {}_{\mathbf{P}}D$ we have $\{\lambda(u, x, v);\ u \in \mathcal{U}\} \supset \Sigma({}_{\mathbf{P}}D)$, i.e. the Lyapunov exponents in $\Sigma({}_{\mathbf{P}}D)$ can be realized from all $(x, v) \in {}_{\mathbf{P}}D$.*

(ii) *For all $(x, v) \in \mathcal{O}^-({}_{\mathbf{P}}D)$ we have $\{\lambda(u, x, v);\ u \in \mathcal{U}\} \supset \Sigma({}_{\mathbf{P}}D)$, where $\mathcal{O}^-({}_{\mathbf{P}}D)$ denotes the (open) negative orbit of the set ${}_{\mathbf{P}}D$ for the control system (2.6).*

(iii) *Let "$\prec_{\mathbf{P}}$" be the order on the control sets of (2.6). If for some $(x, v) \in \mathbf{P}M$ we have $\mathcal{O}^+(x, v) \cap {}_{\mathbf{P}}D$ then $\{\lambda(u, x, v);\ u \in \mathcal{U}\} \supset \cup\{\Sigma({}_{\mathbf{P}}D');\ \text{int}{}_{\mathbf{P}}D' \neq \phi$ and ${}_{\mathbf{P}}D \prec_{\mathbf{P}} {}_{\mathbf{P}}D'\}$. In particular for each $(x, v) \in \mathbf{P}M$ there exists an invariant control set ${}_{\mathbf{P}}C \subset \mathbf{P}M$ such that $\{\lambda(u, x, v);\ u \in \mathcal{U}\} \supset \Sigma({}_{\mathbf{P}}C)$.*

(iv) *For $x \in M$ let $D \subset M$ be a control set with $\mathcal{O}^+(x) \supset D$, and int $D \neq \phi$. Then $\{\lambda(u, x, v);\ u \in \mathcal{U},\ v \in \mathbf{P}_xM\} \supset \Sigma(D)$, and similarly for the results in (iii).*

(v) *The set $\{(u, x, v) \in \mathcal{U} \times TM;\ \lambda(u, x, v) \in \Sigma({}_{\mathbf{P}}C)$ for some invariant control set ${}_{\mathbf{P}}C$ in $\mathbf{P}M\}$ contains an open and dense subset of $\mathcal{U} \times TM$.*

PROOF:

(i) For the first claim see Colonius and Kliemann (1991[b]). Let $\lambda = \lambda(u', x', v') \in \Sigma(_{\mathbf{P}}D)$, i.e. $(x', v') \in \operatorname{int} {_{\mathbf{P}}D}$. Then, by Lemma 3.3(i), for each $(x, v) \in {_{\mathbf{P}}D}$ there exists a control $w \in \mathcal{U}$ such that $\mathrm{P}\varphi(t', (x, v), w) = (x', v')$. Define the composite control w' as

$$w'(t) = \begin{cases} w(t) & \text{for } t \in [0, t'] \\ u'(t - t') & \text{for } t \in (t', \infty). \end{cases}$$

Then $\lambda(w', x, v) = \lambda(u', x', v')$.

(ii) This is proved in the same manner as (i).

(iii) The first statement follows from the definition of the order $\prec_{\mathbf{P}}$, compare Definition (3.2). The statement about invariant control sets follows from the discussion after (3.2) and the fact that under (PH) invariant control sets of (2.6) always have nonvoid interior, compare Lemma 3.3.

(iv) follows directly from (i)–(iii).

(v) By Lemma 4.2(vi) and Theorem 4.5(v) the set $\{(u, x, v) \in \mathcal{U} \times TM;$ $\pi_{\mathbf{P}M}\omega(u, x, v) \subset \operatorname{int}_{\mathbf{P}}C$, ${_{\mathbf{P}}C}$ some invariant control set in $\mathbf{P}M\}$ is open and dense. For all these (u, x, v) one has $\lambda(u, x, v) \in \Sigma(_{\mathbf{P}}C)$.

∎

4.13. COROLLARY. *Let C be a (compact) invariant control set of* (2.1). *Then for each* $(x, v) \in \mathrm{P}\operatorname{int}C$, *the projective bundle over* int C, *there exists a unique index i_0 such that* $i_0 = \min\{i \in \{1 \dots k\};\ \mathcal{O}^+(x, v) \cap {_{\mathbf{P}}D_i} \neq \phi\}$, *where the* ${_{\mathbf{P}}D_i}$ *are defined in* Theorem 3.5. *Furthermore,* $\{\lambda(u, x, v);\ u \in \mathcal{U}\} \supset \Sigma(_{\mathbf{P}}D_i)$ *for all* $i \geq i_0$, *and* $\Sigma(_{\mathbf{P}}D_i) \leq \Sigma(_{\mathbf{P}}D_j)$ *if* $i \leq j$, *with the order defined as in* Theorem 4.10.

The proof follows directly from Theorem 3.5 and Theorem 4.12(iii).

Because all control sets are contained in some (unique) chain control set (compare Lemma 3.8), each set $\Sigma(_{\mathbf{P}}D)$ (and $\Sigma(D)$) is contained in some $I(_{\mathbf{P}}E)$ (and $\Sigma(E)$, respectively). In particular, over an invariant control set C of (2.1), we have by Corollary 3.11 that each $I(_{\mathbf{P}}E_j)$, $j = 1, \dots \ell$ contains some $\Sigma(_{\mathbf{P}}D_i)$ for $i = 1 \dots k$.

As the result of our discussion we see that the entire Lyapunov spectrum of a nonlinear control system (2.1) under Assumption (PH) can be found in spectral intervals over the chain control sets of (2.6). By Theorem 4.7, the boundary points of these intervals are themselves Lyapunov exponents, corresponding to some ergodic Krylov-Bogolyubov measures of $(\mathcal{U} \times \mathbf{P}M, \mathbf{P}\phi)$. On the other hand, the set of $(u, x, v) \in \mathcal{U} \times TM$ such that $\lambda(u, x, v)$ is attained over some invariant control set ${_{\mathbf{P}}C}$ in $\mathbf{P}M$ contains an open and dense subset of $\mathcal{U} \times TM$.

Linearizations of nonlinear control systems around rest points lead to bilinear control systems, as described in Section 2. For these systems, the picture is much more complete, compare Colonius and Kliemann (1990[d], 1991[b]). In the next section, we will apply some of our results to the study of typical problems in the systems theory of bilinear systems (and linear feedback systems). A nonlinear example is considered in Section 6.

5. Some Applications to Linear and Bilinear Systems.

Bilinear systems appear as linearization (with respect to the state variable) around a rest point of the nonlinear control system (2.1). Also, as we will see below, they are the feedback models of linear control systems. Lyapunov exponents of bilinear systems indicate the stability, stabilizability, and robustness properties of linear systems.

Consider the bilinear control system with bounded controls

$$\dot{x}(t) = A_0 x(t) + \sum_{i=1}^{m} u_i(t) A_i x(t), \quad t \in \mathbb{R}, \quad x(0) = x_0 \in \mathbb{R}^d \tag{5.1}$$

where $(u_i) = u \in \mathcal{U}$ as defined in Section 2. The $A_0, \ldots, A_i$ are given constant $d \times d$-matrices. (In Colonius and Kliemann (1990^d), a more general system with bounded and unbounded controls is treated.)

In linear systems theory for equations of the type

$$\dot{x}(t) = Ax(t) + Bu(t), \quad y(t) = Cx(t), \quad u(t) \in U \subset \mathbb{R}^m$$

(with A $d \times d$, B $d \times m$, C $k \times d$ constant matrices), one is usually interested in a feedback control $u(t) = F(t)Cx(t)$ to achieve a certain goal, like stabilization. The systems equation then reads

$$\dot{x}(t) = Ax(t) + BF(t)Cx(t)$$

which is a system of the form (5.1). If the feedback matrix $F(t)$ can take any value in the $m \times k$-dimensional space, methods from linear algebra can be used to solve the typical control problems, see e.g. Wonham (1979). If, however, the set of admissible feedback values is bounded, nonlinear techniques, like e.g. the ones developed here, are appropriate.

As in Section 2. we consider the projected system on the projective space $\mathbb{P}^{d-1}$ in $\mathbb{R}^d$, and equation (2.7) reads now

$$\dot{s}(t) = h(u(t), s(t)) = h_0(s) + \sum_{i=1}^{m} u_i(t) h_i(s) \tag{5.2}$$

with $h_j(s) = (A_j - s^T A_j s \cdot \mathrm{Id})s, \quad j = 0, \ldots, m.$

The Lyapunov exponents of (5.1) are for $(u, x) \in \mathcal{U} \times \mathbb{R}^d$, $x \neq 0$

$$\begin{aligned} \lambda(u, x) &= \limsup_{t \to \infty} \frac{1}{t} \log |\varphi(t, x, u)| \\ &= \limsup_{t \to \infty} \frac{1}{t} \int_0^t q(u(\tau), s(\tau, s, u))\, d\tau, \quad s = \frac{x}{\|x\|_2} \in \mathbb{P}^{d-1}, \end{aligned} \tag{5.3}$$

where $|\cdot|$ is any norm in $\mathbb{R}^d$, and $q(u, s) = q_0(s) + \sum_{i=1}^{m} u_i(t) q_i(s)$, with $q_j(s) = s^T A_j s$ for $j = 0, \ldots, m$.

NOTATION: In this section we will use the variable x for the state of the bilinear system (this was denoted by v in Section 2.), $\varphi(\cdot,\cdot,\cdot)$ for the solution of (5.1) (which corresponds to $T\varphi$ in Section 2.), s and $s(\cdot,\cdot,\cdot)$ for the state on the projective space, and for the solution of (5.2) respectively (which correspond to $v \in \mathbb{P}_x M$ and $\mathbb{P}\varphi$ in Section 2.). Furthermore, the projective space $\mathbb{P}^{d-1}$ in $\mathbb{R}^d$ will be denoted by $\mathbb{P}$. These changes are made so that we can use the usual notation in the literature for linear and bilinear systems. Similarly, for control sets (and chain control sets) on $\mathbb{P}$ we will simply use D instead of $_{\mathbb{P}}D$ (and E instead of $_{\mathbb{P}}E$, respectively).

First of all, we will characterize the control sets and chain control sets of (5.2) more precisely via generalized eigenspaces of elements in the systems semigroup. Denote again by $\mathcal{S}^+$ the positive semigroup of (5.1) (compare (3.1)), and note that $g \in \mathcal{S}^+$ acts on $\mathbb{P}$ in a natural way via $s \mapsto \frac{1}{|gs|} gs$. For $g \in \mathcal{S}^+$ denote by $u(g) \in \mathcal{U}$ a (not necessarily unique) periodic control function, associated with g. For a main control set D of (5.2) (i.e. a control set with nonvoid interior, compare Theorem 3.5) define

$$D(u(g)) = \{x \in \mathbb{P};\ s(t,x,u(g)) \in \overline{D} \text{ for all } t \in \mathbb{R}\}. \tag{5.4}$$

Then we have the following supplement to Theorems 3.5 and 3.10:

5.1. THEOREM. *Assume that* Assumption (H) *holds for the system (5.2).*

(i) *For every $g \in \operatorname{int}_{\mathcal{G}}\mathcal{S}$ and every $\lambda \in$ spec g there exists a main control set $D \subset \mathbb{P}$ such that the (generalized) eigenspace $E(g,\lambda)$ satisfies $\mathbb{P}(E(g,\lambda)) \subset \operatorname{int} D$; where for a subspace $X \subset \mathbb{R}^d$ its projection onto $\mathbb{P}$ is denoted by $\mathbb{P}(X)$. Furthermore, the interior of the main control sets consists exactly of those elements $x \in \mathbb{P}$, which are eigenvectors for a (real) eigenvalue of some $g \in \operatorname{int} \mathcal{S}^+$.*

(ii) *For every $g \in \mathcal{S}^+$ and every $\lambda \in$ spec g there exists a main control set D with $\mathbb{P}(E(g,\lambda)) \cap \overline{D} \neq \phi$, and for every main control set D and every $g \in \mathcal{S}^+$ there is a $\lambda \in$ spec g with $\mathbb{P}(E(g,\lambda)) \cap \overline{D} \neq \phi$.*

(iii) *For every $g \in \operatorname{int}\mathcal{S}^+$ and every main control set we have $D(u(g)) = \mathbb{P}(\oplus_\lambda E(g,\lambda))$, where the sum is taken over all $\lambda \in$ spec g with $\mathbb{P}(E(g,\lambda)) \subset \operatorname{int} D$.*

(iv) *Let E be a chain control set of (5.2) and denote by $\mathcal{E}$ its lift to $\mathcal{U} \times \mathbb{P}$. Then $\mathcal{E} = cl\{(u(g),x) \in \mathcal{U} \times \mathbb{P};\ g \in \operatorname{int} \mathcal{S}^+ \text{ and } x \in \bigoplus D_i(u(g)) \text{ for } i \text{ with } D_i \subset E\}$.*

The proof of this theorem can be found in Colonius and Kliemann (1990[b]), Theorem 3.10, Theorem 3.13, and Theorem 5.6. (Note that under Assumption (H) $\operatorname{int}_{\mathcal{G}}\mathcal{S} \neq \phi$.) This theorem allows us to reduce the analysis of (main) control sets and chain control sets to eigenspaces of elements in the systems semigroup, and it also implies that the Lyapunov spectrum over the main control sets can be approximated by (real parts of) the eigenvalues of the $g \in \operatorname{int} \mathcal{S}^+$, i.e. by the Lyapunov exponents corresponding to piecewise constant, periodic control functions. But the right hand side of (5.1) for this class of controls is Lyapunov regular. We continue by stating some implications of these facts for the stabilization of (5.1). We will be concerned here only with the extremal Lyapunov exponents of (5.1), the complete picture is described in Colonius and Kliemann (1991[b]), where also the existence of stabilizing feedback controls for bilinear systems is considered.

Denote, as in (4.18), by $\Sigma(D)$ the Lyapunov spectrum over a main control set D of (5.2). By Theorem 3.5 the main control sets of (5.2) are linearly ordered $D_1 \prec \cdots \prec D_k$, and by Lemma 3.3 we have: $D_1 =: C^-$ is the open control set, $D_k =: C$ is the closed, invariant control set (with int $C \neq \phi$). It can actually be shown (compare Colonius and Kliemann (1991[b]) that the spectral sets $\Sigma(C)$ and $\Sigma(C^-)$ are intervals, and by Theorem 4.7, their boundary points are Lyapunov exponents corresponding to ergodic, invariant measures of the flow $(\theta_t, s(t,\cdot,u))$ on $\mathcal{U} \in \mathrm{P}$. Define the following extremal Lyapunov exponents

(5.5) $$\kappa = \sup \Sigma(C), \quad \tilde{\kappa} = \inf \Sigma(C), \quad \kappa^* = \inf \Sigma(C^-).$$

As the following result shows, these three exponents can be characterized as global quantities of the control system (5.1), while $\sup \Sigma(C^-)$ has to be described through an invariant set of the flow on $\mathcal{U} \times \mathrm{P}$, compare again Colonius and Kliemann (1991[b]).

5.2. THEOREM. *Assume* Assumption (H) *for the system (5.2).*

(i) $\kappa = \sup_{x \in \text{int } C} \sup_{u \in \mathcal{U}} \lambda(u,x) = \sup_{x \neq 0} \sup_{u \in \mathcal{U}} \lambda(u,x) = \sup_{u \in \mathcal{U}} \lambda(u,x)$ *for all* $x \neq 0$,

(ii) $\tilde{\kappa} = \sup_{x \in \text{int } C} \inf_{u \in \mathcal{U}} \lambda(u,x) = \sup_{x \neq 0} \inf_{u \in \mathcal{U}} \lambda(u,x)$,

(iii) $\kappa^* = \inf_{x \in C^-} \inf_{u \in \mathcal{U}} \lambda(u,x) = \inf_{x \neq 0} \inf_{u \in \mathcal{U}} \lambda(u,x)$.

(iv) *For each* $\varepsilon > 0$ *and all* $x \in \mathrm{P}$, $\tilde{x} \in \text{int } C$, $x^* \in C^-$ *there exist* $g, \tilde{g}, g^* \in \text{int } \mathcal{S}^+$ *such that*

$$\lambda(u(g),x) > \kappa - \varepsilon, \quad \lambda(u(\tilde{g}),\tilde{x}) < \tilde{\kappa} + \varepsilon, \quad \lambda(u(g^*),x^*) < \kappa^* + \varepsilon.$$

The proof of this theorem is contained in Sections 4. and 5. of Colonius and Kliemann (1990[d]).

5.3. REMARK:

(i) It follows from Theorem 5.2 that $\kappa^* \leq \tilde{\kappa} \leq \kappa$. But the intervals $\Sigma(C)$ and $\Sigma(C^-)$ can overlap, i.e. $\tilde{\kappa} = \inf \Sigma(C) < \sup \Sigma(C^-)$ is possible, compare Example 4.11.

(ii) The growth rates κ and $\tilde{\kappa}$ can be realized from each $x \neq 0$: For κ this is Theorem 5.2(i), and for $\tilde{\kappa}$ the statement follows from Theorem 4.12 and Theorem 5.2(ii). However, in general κ^* cannot be realized from $x \notin C^-$, compare Example 3.6. Note that in this example we have only one chain control set $E = \mathrm{P}$, and thus the only spectral interval over a chain control set is in this case $I(\mathrm{P}) = [\kappa^*, \kappa]$, $I(\mathrm{P})$ being defined in (4.15). Therefore we see that the intervals of Lyapunov exponents over chain control sets can be too big to characterize precisely the stabilization behavior of (5.1). (Section 6. in Colonius and Kliemann (1990[d]) contains an example, where $C \cap \overline{C^-} \neq \phi$, $E = \mathrm{P}$, and κ^* can be realized only from $x \in C^-$.)

(iii) Theorem 5.2(iv) says, in particular, that there exist Lyapunov regular matrix functions, whose Lyapunov exponents are arbitrarily close to the extremal exponents κ (or $\tilde{\kappa}$, κ^*). This means that small perturbations of these functions will have the same stability behavior, compare Hahn (1967), Theorem 65.3.

As a consequence of Theorem 5.2 we obtain the following result on stabilization and destabilization:

5.4. COROLLARY.

(i) *The system (5.1) is (exponentially) destabilizable via an open loop control in $\mathcal{U}$ from some $x \neq 0$ (and hence from all $x \neq 0$) iff $\kappa > 0$.*

(ii) *The system (5.1) is (exponentially) stabilizable via an open loop control in $\mathcal{U}$ from all $x \neq 0$ iff $\tilde{\kappa} < 0$, and from some $x \neq 0$ (and hence from all $x \in C^-$) iff $\kappa^* < 0$.*

(iii) *The system (5.1) is (exponentially) stabilizable (or destabilizable) iff there exists $u \in \mathcal{U}$ with $A_0 + \Sigma u_i(t)A_i$ Lyapunov regular such that $\dot{x} = (A_0 + \Sigma u_i(t)A_i)x$ is exponentially stable (or unstable, respectively).*

Note that this corollary applies immediately to the problem of (de-)stabilization of linear systems via bounded, time varying output feedback.

It remains to compute the quantities κ, $\tilde{\kappa}$, κ^* from (5.5). In specific cases, this can be done explicitly (compare e.g. the linear oscillator with controlled restoring force in Section 6 of Colonius and Kliemann (1990^d)), but in general one needs numerical procedures to compute these numbers. The problem can actually be formulated as an optimal control problem with infinite time, average cost criterion (compare Colonius and Kliemann (1989)), for which algorithms are available.

5.5. REMARK: The problem of stabilization of bilinear (or, in general, of nonlinear) control systems can also be approached via high gain techniques, which require fast, unbounded controls. Results in this direction have been obtained e.g. by Meerkov (1980), Bellman et al. (1985, 1986), Knobloch (1988), Colonius and Kliemann (1990^d, Section 7), or by Arnold et al. (1983) and Arnold (1989) in a stochastic context.

The remainder of this section is devoted to another important topic in systems theory, namely robustness, which can be approached e.g. via Lyapunov exponents. Assume we are given a linear control system

$$(5.6) \qquad \dot{x}(t) = (A_0 + A(t))x + Bu, \quad t \in \mathbb{R}, \quad x(0) = x_0 \in \mathbb{R}^d,$$

where $A(t)$ represents the uncertainty about some (or all) parameters of the given systems matrix A_0. The problem is to find criteria, under which for all uncertainties the system is stable (for $B = 0$) or stabilizable (e.g. via output feedback). We will restrict ourselves here to the stability question (for stabilization criteria using Lyapunov exponents see Colonius and Kliemann (1990^f)).

More precisely, define the uncertainty range $U_\rho \subset \mathbb{R}^m$ for $\rho \geq 0$ by

$$U_\rho = \{u \in \mathbb{R}^m \, ; \, |u| \leq \rho\},$$

where $|\cdot|$ denotes any norm in $\mathbb{R}^m$ (often the Euclidean norm or the interval norm $u_i \in [\rho\alpha_i, \rho\beta_i]$, $\alpha_i < 0 < \beta_i$ for $i = 1 \dots m$, are used.) The uncertain system (without input) is then modeled as

$$(5.7_\rho) \qquad \dot{x}(t) = \left(A_0 + \sum_{i=1}^{m} u_i(t)A_i\right)x =: A(u)x, \quad (u_i(t)) \in U_\rho.$$

If A_0 is a stable matrix (i.e. all real parts of its eigenvalues are negative), then the system (5.7) is stable for all uncertainties, if ρ is small enough. The problem is to find the smallest ρ, such that for some uncertainty with values in U_ρ (5.7) becomes unstable, which leads to the following definition:

Denote by $\mathcal{U}_\rho = \{u\colon \mathbf{R} \to U_\rho;$ locally integrable$\}$ the set of uncertainties of size ρ, and again by $\lambda(u,x) = \limsup_{t\to\infty} \frac{1}{t} \log |\varphi(t,x,u)|$ the Lyapunov exponents of the solutions of (5.7_ρ).

5.6. DEFINITION: Let $A_0 \in g\ell(d,\mathbf{R})$ be a stable matrix. Then the (Lyapunov) stability radius of A_0 with respect to the uncertainty structure in (5.7_ρ) is defined by

$$r_L(A_0) = \inf\{\rho \geq 0;\ \sup_{x\neq 0}\ \sup_{u\in\mathcal{U}_\rho} \lambda(x,u) \geq 0\}$$
$$= \inf\{\rho \geq 0;\ \kappa_\rho \geq 0\},$$

where κ_ρ is defined as in (5.5) with $\mathcal{U}_\rho$ as the set of admissible controls.

Similarly, instability radii can be defined using $\tilde{\kappa}$ and κ^*. In the literature one finds a wide variety of stability radii, e.g. for only constant uncertainties $u(t) \equiv u \in U_\rho$, or for $U_\rho \subset \mathbf{C}^m$, or for $\{U_\rho;\ \rho \geq 0\}$ just an increasing family of subsets in $\mathbf{R}^m$, etc. (compare e.g. Hinrichsen and Pritchard (1990), or Colonius and Kliemann (1990e)). In this brief exposé we will only be concerned with stability radii, and only with those that are defined through all time-varying uncertainties with values in U_ρ given as above.

We will again assume a nondegeneracy condition on the vector fields of the projected system on $\mathbf{P}$, which in this context reads (with $h(u,s)$ defined as in (5.2)):

(H_ρ) $\qquad \dim \mathcal{LA}\{h(u,\cdot);\ u \in U_\rho\}(s) = d-1$ for all $s \in \mathbf{P}$ and some (and hence all) $\rho > 0$.

We first note some uniformity and smoothness properties of the stability radius r_L:

For a function $u \in \mathcal{U}_\rho$ denote by $\Phi_u(t,s)$ the fundamental matrix of (5.7_ρ), and define the Bohl exponent for this equation by

$$k_B(u) = \limsup_{s,t-s\to\infty} \frac{1}{t-s} \log \|\phi_u(t,s)\|.$$

This exponent indicates uniform asymptotic stability of a linear, time varying differential equation, while the largest Lyapunov exponent $k_L(u)$ indicates asymptotic stability, hence, in general, $k_B(u) > k_L(u)$. This is not true for the corresponding stability radii, compare Colonius and Kliemann (1990^e), Theorem 5:

5.7. PROPOSITION. $r_L(A) = r_B(A) := \inf\{\rho \geq 0;\ \sup_{u\in\mathcal{U}_\rho} k_B(u) \geq 0\}$.

Furthermore, the maximal Lyapunov exponent $k_L(u)$ need not be continuous, nor even semi continuous in $u \in U$ or $u \in \mathcal{U}$. However, under Assumption (H_ρ), this effect is smoothed out in the extremal exponents κ_ρ for $\{U_\rho,\ \rho \geq 0\}$ as above:

5.8. PROPOSITION. *The function $\rho \mapsto \kappa_\rho$ is continuous and increasing. In particular, the set $\{\rho \geq 0;\, \kappa_\rho = 0\}$ is a closed, connected subset of $[0,\infty)$ and $r_L(A) = \inf\{\rho \geq 0;\, \kappa_\rho = 0\}$.*

The result is proved in Colonius and Kliemann (1990^f). Note that $\rho \mapsto \kappa_\rho$ can be constant on intervals in $[0,\infty)$, i.e. the function need not be strictly increasing.

The following result on the uniformity of $r_L(A)$ with respect to the initial value $x \neq 0$ is an immediate consequence of Theorem 5.2: Denote by $\kappa_\rho(x) := \sup_{u \in \mathcal{U}_\rho} \lambda(u,x)$ the maximal Lyapunov exponent of (5.7_ρ), which can be realized from a point $x \neq 0$.

5.9. PROPOSITION. *For all $x \in \mathrm{R}^d$, $x \neq 0$ we have $r_L(A) = \inf\{\rho \geq 0;\, \kappa_\rho(x) = 0\}$.*

5.10. REMARK: Willems and Willems (1983) proved several results about the robustness of linear systems with respect to stochastic uncertainties. In our context, we have the following results, compare Colonius and Kliemann (1990^e) for the precise set up. Let M be a compact smooth manifold and $D_\rho \colon M \to g\ell(d,\mathrm{R})$ a family of smooth maps with $D_\rho[M] = U_\rho$. Denote by Stat the set of all stationary processes with values in M, and by Diff the stationary, ergodic, nondegenerate diffusion processes on M. For a process ξ_t in Stat let $\varphi(t,x,\xi_t)$ be the solution of $\dot{x} = A(D_\rho(\xi_t))x$ (compare (5.7_ρ) for the definition of $A(.)$), and define $\lambda_\rho(\xi_t,x) = \limsup_{t\to\infty} \frac{1}{t}\log|\varphi(t,x,\xi_t)|$,

$$\kappa_\rho(\xi_t) = \text{ess sup} \sup_{x\neq 0} \lambda_\rho(\xi_t,x),$$

where the essential supremum is taken over the measure induced by ξ_t in $\mathcal{U}$,

$$g_\rho(\xi_t,p) = \sup_{x\neq 0} \limsup_{t\to\infty} \frac{1}{t} \log E\left(|\varphi(t,x,\xi_t)|^p\right).$$

Then we have for stationary processes in Stat

$$\begin{aligned} r_L(A) &= \inf\{\rho \geq 0;\; \sup_{\xi_t \in \mathrm{Stat}} \kappa_\rho(\xi_t) \geq 0\} \\ &= \sup\{\rho \geq 0;\; \sup_{\xi_t \in \mathrm{Stat}} g_\rho(\xi_t,p) \leq 0 \text{ for all } 0 < p < \infty\}, \end{aligned}$$

and for processes ξ_t in Diff:

$$\begin{aligned} r_L(A) &= \sup\{\rho \geq 0;\, g_\rho(\xi_t,p) \leq 0 \text{ for all } 0 < p < \infty\} \\ &\leq \inf\{\rho \geq 0;\; \sup_{\xi_t \in \mathrm{Diff}} \kappa_\rho(\xi_t) \geq 0\}, \end{aligned}$$

where we conjecture that the last inequality is actually an equality. (The connection of these results with large deviation theory is explained in Colonius and Kliemann (1990^e).) These findings show that the Lyapunov stability radius of (5.7) (and therefore also the Bohl stability radius by Proposition 5.7) is also the stability radius for stochastic uncertainties, both in the pathwise and in the pth-moment sense. This is just one way, in which Lyapunov exponents for control systems and for stochastic systems lead to a unifying approach.

6. A Nonlinear Example: The Controlled Verhulst Equation.

In this section we will analyze a simple one-dimensional nonlinear example under the aspects discussed in Section 5. for linear and bilinear systems, where we will make use of the results in Sections 3. and 4.

Consider the controlled Verhulst equation in $\mathbb{R}^1$

$$\dot{x}(t) = X_0(x(t)) + u(t)X_1(t) = \alpha x(t) - x(t)^2 + u(t)x(t), \tag{6.1}$$

with $\alpha \in \mathbb{R}$ and $u(t) \in [A, B] \subset \mathbb{R}$. For $u \equiv 0$, this equation undergoes a transcritical bifurcation at $\alpha = 0$, where the rest point $x^0 = 0$ changes from stable to unstable as α increases. We are here concerned with stabilization and robustness properties of (6.1), in particular in the vicinity of the bifurcation point.

6.1. REMARK: Equation (6.1) can be solved explicitly: In the half spaces $(-\infty, 0)$ and $(0, \infty)$ set $y = \frac{1}{x}$, which leads to the linear differential equation $\dot{y} = -(\alpha + u(t))y + 1$. Retransformation of the solution of this equation leads to

$$\varphi(t, x, u) = \frac{\exp\left\{\alpha t + \int_0^t u(\tau)d\tau\right\}}{\frac{1}{x} + \int_0^t \exp\left\{\alpha s + \int_0^s u(\tau)d\tau\right\} ds}$$

for all $u \in \mathcal{U}$ and all $x \neq 0$. From this expression, the asymptotic behavior of the solutions, including their Lyapunov exponents and possible finite explosion times can be computed. We will not use this solution (except for one statement in Remark 6.2.), but rather present a method that relies only on the zeros of the right hand side of (6.1) for constant $u \in U$, and which, therefore, can be used for any one-dimensional control system.

For one-dimensional systems it is convenient to picture the systems dynamics with rest points and signs of the right hand sides in the $U \times \mathbb{R}$ plane, which yields in the case of equation (6.1) the following 'bifurcation diagram':

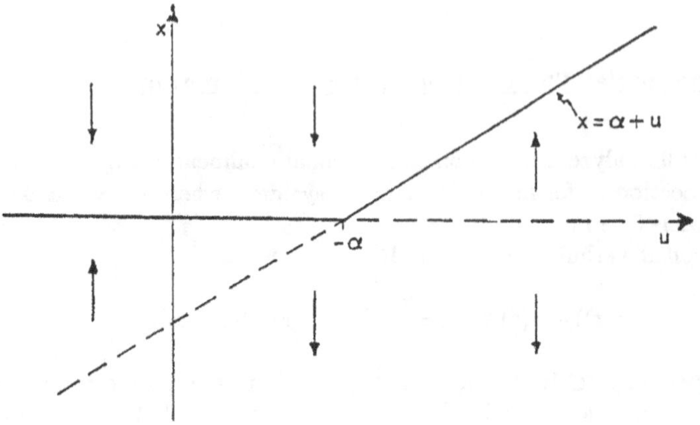

Here, for constant $u \in U$, the arrows indicate the direction of the vectorfields, the solid line corresponds to stable rest points, and the broken line to unstable ones. Define the corresponding two functions of zeros by

$$(6.2)\qquad \begin{aligned} z_1(u) &= \begin{cases} \alpha + u & \text{for } u \leq -\alpha \\ 0 & \text{for } u \geq -\alpha \end{cases} \\ z_2(u) &= \begin{cases} 0 & \text{for } u \leq -\alpha \\ \alpha + u & \text{for } u \geq -\alpha. \end{cases} \end{aligned}$$

We start by analyzing the control sets, chain control sets, and their respective regions of attraction for the system (6.1). In Kliemann (1980) a general approach for finding the control sets of one-dimensional control systems was described, and extended in Colonius and Kliemann (1990^c) to chain control sets. (For this approach, Assumption (H) is not needed, and, in fact, it is violated here at the rest point $x^0 = 0$.) Using the results in Colonius and Kliemann (1990^c), we obtain:

$$(6.3)\qquad \begin{aligned} D_1 &= (z_1(A), z_1(B)) \quad \text{if } A < -\alpha, \\ D_2 &= \{0\}, \\ D_3 &= \begin{cases} (z_2(A), z_2(B)] & \text{if } A \leq -\alpha,\ B > -\alpha, \\ [z_2(A), z_2(B)] & \text{if } A > -\alpha, \end{cases} \end{aligned}$$

and

$$(6.4)\qquad \begin{aligned} E_1 &= \overline{D_1} \quad \text{if } B < -\alpha \\ E_2 &= \begin{cases} D_2 & \text{if } B < -\alpha \text{ or } A > -\alpha \\ [z_1(A), z_2(B)] & \text{otherwise} \end{cases} \\ E_3 &= D_3 \quad \text{if } A > -\alpha. \end{aligned}$$

The sets D_i (and E_i) for $i = 1, \ldots, 3$ are the possible control sets (and chain control sets, respectively) of (6.1).

The regions of attraction $\mathcal{A}(D_i)$ of the control sets (and similar for the chain control sets) are defined by

$$\mathcal{A}(D_i) = \{(u, x) \in \mathcal{U} \times \mathsf{R};\ \pi_{\mathsf{R}}\omega(u, x) \subset D_i\}. \tag{6.5}$$

Note that, in general, there may exist $(u, x) \in \mathcal{U} \times \mathsf{R}$, whose ω-limit set intersects several control sets. These points are in no region of attraction, but they are a 'thin' set according to Lemma 4.2(vi). For chain control sets this cannot happen, compare Lemma 4.2(ii).

We have defined the regions of attraction of (chain) control sets as subsets of $\mathcal{U} \times \mathsf{R}$, and not of the state space R for reasons that will become clear from the following results for the system (6.1);

(i) If a control set is also a chain control set E, then there exists a compact neighborhood N of E such that for all $(u, x) \in \mathcal{U} \times N$ we have either $\omega(u, x) \subset \mathcal{E}$ or $\omega(u, x) \cap \mathcal{E} = \phi$, i.e. $\mathcal{E}$ is either an attractor or a repeller of the flow $(\mathcal{U} \times \mathsf{R}, \phi)$. (For the definition of $\mathcal{E}$ compare (3.7).)

(ii) If a control set D is open and $E = \overline{D}$ is a chain control set, then there is a (compact) neighborhood N of E such that for all $(u, x) \in \mathcal{U} \times (N \setminus E)$ we have $\omega(u, x) \cap \mathcal{E} = \phi$, i.e. $\mathcal{E}$ is a repeller. In this case, we have the following control behavior in D:

(a) For all $u \in \mathcal{U}$ there exist $x, y, z \in \overline{D}$ with
- $\phi(t, x, u) \in \overline{D}$ for all $t \geq 0$,
- $\lim_{t\to\infty} \varphi(t, y, u) = 0 \notin \overline{D}$,
- $\lim_{t\to\infty} \varphi(t, z, u) = -\infty \notin \overline{D}$,

and these are the only three possibilities for points in $\mathcal{U} \times \overline{D}$.

(b) For all $x \in D$ there exist $u, v, w \in \mathcal{U}$ such that
- $\pi_{\mathsf{R}}\omega(u, x) \subset D$,
- $\pi_{\mathsf{R}}\omega(v, x) = \{0\}$,
- $\pi_{\mathsf{R}}\omega(w, x) = \{-\infty\}$.

Note that, because of the continuity of the flow ϕ, if $\pi_{\mathsf{R}}\omega(v, x) = \{0\}$ or $\pi_{\mathsf{R}}\omega(w, x) = \{-\infty\}$ then the same holds for open neighborhoods of x, and of v or w.

In particular, (a) and (b) show that regions of attraction have to be defined on $\mathcal{U} \times \mathsf{R}$ and that these sets can have a very complicated structure.

(iii) It remains to discuss the invariant control sets C, which are not closed, and the variant control sets D with $0 \in \overline{D}$. The closures of these sets are not chain control sets, and hence their lifts to $\mathcal{U} \times \mathsf{R}$ are neither attractors nor repellers. (Note that the corresponding chain control sets $E_2 = [z_1(A), z_2(B)]$ are neither attractors nor repellers, too.) For the invariant control sets C we have in this case:

(a) For all $x > 0$ there exist $u, v, w \in \mathcal{U}$ such that
- $\pi_{\mathsf{R}}\omega(u, x) \subset \operatorname{int} C$,

- $\pi_{\mathbf{R}}\omega(v,x) = \{0\}$,
- $\pi_{\mathbf{R}}\omega(w,x) = C \cup \{0\}$,

and there are $u \in \mathcal{U}$ such that for all $x > 0$ one has $\omega(u,x) = \{0\}$.

Similarly, we obtain for the variant control sets D: The possible limit behavior from $x \in D$ is the same as in (ii)(b), and there are $u \in \mathcal{U}$ such that for all $x > 0$ one has $\omega(u,x) = \{-\infty\}$.

The proof of (i)–(iii) follows from the construction of (chain) control sets in Colonius and Kliemann (1990^c), and Proposition 4.8. therein.

6.2. REMARK: Let D be a control set of (6.1), then for all θ-invariant (probability) measures ρ on $\mathcal{U}$ there exists a ϕ-invariant measure $\mu(du,dx) = \mu_u(dx)\rho(du)$ on $\mathcal{U} \times \mathbf{R}$ with supp $\mu \subset \mathcal{D}^+$, and all ϕ-invariant probability measures have support in $\cup\{\mathcal{D}^+ ; D$ is a control set$\}$, compare Colonius and Kliemann (1990^c), Theorem 4.3 and Corollary 4.9. In fact, it can be shown using the explicit solution in Remark 6.1, that for each $u \in \mathcal{U}$ there are exactly two ergodic invariant probability measures of the type μ_u, compare Arnold and Boxler (1990). These findings are a starting point for a bifurcation theory of stochastic and control systems.

We now turn to the question of stabilization at the rest point $x^0 = 0$. This will be done via linearization around x^0, which together with the global control analysis above will give a complete picture in Proposition 6.3.

We have $DX_0|_{x=0} = \alpha$, $DX_1|_{x=0} = 1$, and hence the linearized system at x^0 reads

$$\dot{v}(t) = (\alpha + u(t))\, v(t), \quad u(t) \in [A,B]. \tag{6.6}$$

Since the system (6.1) is one-dimensional, we do not need the projected system (2.6), but we can analyze (6.6) directly:

For constant $u \in [A,B]$ the linearized system is (exponentially) stable iff $u < -\alpha$, (then the Lyapunov exponent is $\lambda(u) = \alpha + u$), and (exponentially) unstable, iff $u > -\alpha$. For $u = \alpha$ we have $\lambda(u) = 0$, a bifurcation takes place in (6.1) at this point, and $x^0 = 0$ is attractive in (6.1) for $x > 0$. Because of the monotonicity of the vector field $X_0 + uX_1$ in u, it suffices for the analysis of the stabilization behavior of the nonlinear system to consider constant controls. We obtain the following results:

6.3 PROPOSITION.

(i) *The system (6.1) is (locally exponentially) stabilizable at $x^0 = 0$ iff $A < -\alpha$, the maximal rate of convergence is $\lambda(A) = \alpha + A$.*

(ii) *If (6.1) is stabilizable at $x^0 = 0$, then the maximal stabilization manifold (for $U = [A,B]$) is $M^s = (z_1(A), +\infty)$. For $x \notin M^s$ and all $u \in \mathcal{U}$, $u \neq A$ on a set of positive Lebesgue measure, we have $\varphi(t,x,u) \to -\infty$.*

(iii) *For $u = A$ Lebesgue almost everywhere, the system is stabilizable at $x^0 = 0$ from $x \in (0,\infty)$*

(iv) *The system is not exponentially stabilizable at $x^0 = 0$, iff $A \geq -\alpha$ (for all $x \neq 0$).*

The proof is a simple application of linearization theory with stable and unstable manifolds (for constant $u \in U$), together with the global results (i)–(iii) above. It is interesting to note that for $B > -\alpha$ there exists a second invariant control set, denoted by D_3 in (6.3), which also has a nonvoid region of attraction with points outside of D_3 by (iii) above. This rises the question, whether the system (6.1) can be stabilized e.g. at rest points or periodic solutions in D_3.

Finally, we discuss briefly some robustness properties of the nonlinear system (6.1). Consider the system

$$(6.7\rho) \qquad \dot{x}(t) = \alpha x(t) - x(t)^2 + u(t)x(t), \quad u(t) \in [\rho A, \rho B], \quad A < 0 < B, \quad \rho \geq 0,$$

where now $u \in \mathcal{U}_\rho$ is interpreted as an uncertainty in the parameters of the vector field $X_0 = \alpha x - x^2$, compare (5.7ρ) for the set up in the linear case. For the stable rest point $x^0 = 0$ of X_0 (i.e. $\alpha < 0$) a local and a global stability radius can be defined: Denote by $\lambda(u)$, $u \in \mathcal{U}$ again the Lyapunov exponents of the linearized system (6.6), and define the local stability radius by

$$(6.8) \qquad r_L(x^0) = \inf\{\rho \geq 0;\ \sup_{u \in \mathcal{U}_\rho} \lambda(u) \geq 0\}.$$

This radius corresponds to the existence of a stable manifold M^s with $x^0 \in \operatorname{int} M^s$, but it does not indicate, how large this stable manifold is, i.e. for which $x \in \mathbb{R}$ the long term behavior of $\varphi(t, x, u)$ is the same as that of $\varphi(t, x, u \equiv 0)$ with respect to the stability of x^0. We therefore define global stability radii for $x \in \mathbb{R}$ by

$$(6.9) \qquad r(x^0; x) = \inf\{\rho \geq 0;\ \text{there exists } u \in \mathcal{U} \text{ with } \pi_{\mathbb{R}}\omega(u, x) \neq \{x^0\}\}.$$

Note that for linear systems the stability radius is always independent of $x \neq 0$ by Theorem 5.2(i), whereas the instability radii, based on $\tilde{\kappa}$ and κ^*, may depend on x, compare Colonius and Kliemann (1990^c), Corollary 4 and Example 6.2, as well as Corollary 5.4. above.

From Proposition 6.3 we obtain immediately:

6.4. PROPOSITION. *Consider the nonlinear, uncertain system (6.7ρ) with $\alpha < 0$.*

(i) $r_L(x^0) = -\frac{\alpha}{B}$

(ii) $r(x^0; x) = \begin{cases} -\frac{\alpha}{B} & \text{for } x > 0 \\ \frac{-\alpha + x}{B} & \text{for } x \in (\alpha, 0) \\ 0 & \text{for } x \leq \alpha. \end{cases}$

Similarly, local and global instability radii, i.e. the smallest ρ such that (6.7ρ) is stabilizable for $\alpha > 0$, can be analyzed.

7. References

Arnold, L. (1989). *Stabilization by noise revisited.* preprint, University of Bremen, FRG.

Arnold, L., P. Boxler (1989). *Eigenvalues, bifurcation and center manifolds in the presence of noise.* In: C. Dafermos (ed.) EQUADIFF '87, M. Dekker.

Arnold, L., P. Boxler (1990). *Stochastic bifurcation: Instructive examples in dimension one.* in: Wihstutz, V. (ed). Proceedings of the Conference on Stochastic Flows, Charlotte, NC, March 1990. Birkhäuser.

Arnold, L., H. Crauel, V. Wihstutz (1983). *Stabilization of linear systems by noise.* SIAM J. Control Optim. **21**, 451–461.

Arnold, L., W. Kliemann, E. Oeljeklaus (1986). *Lyapunov exponents of linear stochastic systems.* In: Arnold, Wihstutz (1986), 85–125.

Arnold, L., V. Wihstutz (eds.) (1986). Lyapunov Exponents. Lecture Notes in Mathematics No. 1186, Springer.

Baxendale, P. (1986). *The Lyapunov spectrum of a stochastic flow of diffeomorphisms.* In: Arnold, Wihstutz (1986), 322–337.

Bellman, R., J. Bentsman, S. M. Meerkov (1985). *On the vibrational stabilizability of nonlinear systems.* J. Optim. Theory Appl. **46**, 421–430.

Bellman, R., J. Bentsman, S. M. Meerkov (1986). *Vibrational control of nonlinear systems: Vibrational stabilizability.* IEEE Trans. Aut. Control **AC-31**, 710–716.

Byrnes, C., A. Isidori (1989). *New results and examples in nonlinear feedback stabilization,* Systems and Control Leters **12**, 437–442.

Carverhill, A. (1985). *Flows of stochastic dynamical systems: ergodic theory.* Stochastics **14**, 273–317.

Cesari, L. (1971). Asymptotic Behavior and Stability Problems in Ordinary Differential Equations. Springer, 3rd ed.

Colonius, F., W. Kliemann (1989). *Infinite time optimal control and periodicity.* Appl. Math. Opt. **20**, 113–130.

Colonius, F., W. Kliemann (1990^a). *Some aspects of control systems as dynamical systems.* Submitted.

Colonius, F., W. Kliemann (1990[b]). *Linear control semigroups acting on projective space.* Submitted.

Colonius, F., W. Kliemann (1990[c]). *Remarks on ergodic theory for stochastic and control flows.* in: Wihstutz, V. (ed). Proceedings of the Conference on Stochastic Flows, Charlotte, NC, March 1990. Birkhäuser.

Colonius, F., W. Kliemann (1990[d]). *Maximal and minimal Lyapunov exponents of bilinear control systems.* To appear in J. Diff. Equations.

Colonius, F., W. Kliemann (1990[e]). *Stability radii and Lyapunov exponents.* In: Control of Uncertain Systems (D. Hinrichsen, B. Martensson, eds.), Birkhäuser, 19–55.

Colonius, F., W. Kliemann (1990[f]). *Stabilization of uncertain linear systems.* To appear in: Modeling and Control of Uncertain Systems (G. DiMasi, A. Gombani, A. Kurzhanski, eds.), Birkhäuser.

Colonius, F., W. Kliemann (1991[a]). *Limit behavior and control sets of nonlinear control systems.* In preparation.

Colonius, F., W. Kliemann (1991[b]). *The Lyapunov spectrum of bilinear control systems.* In preparation.

Crauel, H. (1986). *Lyapunov exponents and invariant measures of stochastic systems on manifolds.* In: L. Arnold, V. Wihstutz (eds.).

Hahn, W. (1967). Stability of Motion. Springer.

Hinrichsen, D., A. J. Pritchard (1990). *Destabilization by output feedback.* preprint, University of Bremen, FRG.

Isidori, A. (1989). Nonlinear Control Theory. Springer. 2nd ed.

Johnson, R. A., K. J. Palmer, G. R. Sell (1987). *Ergodic properties of linear dynamical systems.* SIAM J. Math. Anal. **18**, 1–33.

Kliemann, W. (1980). *Qualitative Theory of Nonlinear Stochastic Systems* (in German). Ph. D. dissertation. Bremen, FRG.

Knobloch, H. W. (1988). *Stabilization of control systems by means of high gain feedback.* In: Feichtinger, G. (ed). Optimal Control Theory and Economic Analysis 3. North Holland, 153–173.

Lyapunov, M. A. (1893). *General problems in the stability of motion* (in Russian). Comm. Soc. Math. Kharkow. (reprinted in Ann. Math. Studies **17**, Princeton (1949)).

Mañé, R. (1987). Ergodic Theory and Differentiable Dynamics. Springer.

Meerkov, S. M. (1980). *Principle of vibrational control: Theory and applications.* IEEE Trans. Aut. Control **AC-25**, 755–762.

Nemytskii, V. V., V. V. Stepanov (1960). Qualitative Theory of Differential Equations. Princeton University Press. (Russian edition 1949).

Nijmeijer, H., A. J. van der Schaft (1990). Nonlinear Dynamical Control Systems. Springer.

Oseledeč, V. I. (1968). *A multiplicative ergodic theorem. Lyapunov characteristic numbers for dynamical systems.* Trans. Moscow Math. Soc. **19**, 197–231.

Ruelle, D. (1979). *Ergodic theory of differentiable dynamical systems.* Publ. I.H.E.S. **50**, 275–320.

Sacker, R. J., G. R. Sell (1978). *A spectral theory for linear differential systems.* J. Differential Equations **37**, 320–358.

San Martin, L. A. B. (1986). *Invariant Control Sets on Fibre Bundles.* Ph.D. dissertation, Warwick, England.

San Martin, L. A. B., L. Arnold (1986). *A control problem related to the Lyapunov spectrum of stochastic flows.* Matematica Aplicada e Computacional **5**, 31–64.

Sontag, E. D. (1983). *A Lyapunov-like characterization of asymptotic controllability.* SIAM J. Control Optim. **21**, 462–471.

Sontag, E. D. (1989). *Feedback stabilization of nonlinear systems.* In: Kaashoek, M.A., J.H. van Schuppen, A.C.M. Ran (eds.), Robust Control of Linear Systems and Nonlinear Control, Proceedings of the International Symposium MTNS–89, Vol. II, Birkhäuser.

Sussmann, H. J. (1979) *Subanalytic sets and feedback control.* J. Differential Equations **31**, 31–52.

Willems, J. L., J. C. Willems (1983). *Robust stabilization of uncertain systems.* SIAM J. Control Optim. **21**, 352–374.

Wonham, W. M. (1979). Linear Multivariable Control: A Geometric Approach. Springer.